Methods in Mammalian Embryology

Methods in
Mammalian Embryology

Edited by

Joseph C. Daniel, Jr.

University of Colorado

W. H. Freeman and Company

San Francisco

Printed in the United States of America

Library of Congress Catalog Card Number: 76–116894
International Standard Book Number: 0–7167–0819–1

1 2 3 4 5 6 7 8 9 10

Dedicated to an expert in mammalian embryology—
the mother of my children.

Contents

viii   *Contents*

Preface

The field of experimental mammalian embryology is expanding so rapidly that it is difficult to find details of the highly specialized techniques which typically are either incompletely described or found scattered throughout the literature. It is hoped that this book will satisfy the need which has been expressed by many potential workers in this field for details of these techniques.

The methods are presented so as to be useful to both students and experienced investigators in those branches of zoology, medicine, veterinary medicine, and animal science that are concerned with mammalian development. The book is designed so that the chapters are related according to the developmental sequence.

Authors of individual chapters were encouraged to emphasize specific methods and to evade description of those biochemical, physiological, histological and endocrinological techniques that are already in common use, except in those cases in which they are specifically modified for applicability in the mammalian embryo. Among the chapters, one will find some minor duplication of techniques where they are presented for different purposes and express the individual author's preference for the method he is describing. The chapters present truly some of the best current techniques used in research in mammalian embryology.

To organize effectively the preparation of a single volume which includes contributions from over forty separate authors requires the peak of cooperation of each individual. As editor, I wish to thank each of the contributors who so generously provided that cooperation.

November, 1970   JOSEPH C. DANIEL, JR.

Introduction

BY BENT G. BÖVING

Methods link nature and understanding. In one direction we use methods to gain an understanding of nature, in the other we use our understanding to develop new methods. The first we call science, the other technology. But since the methods of science are technological and technology is the legitimate offspring of science, the distinction is less vital than the combination, which has, perhaps, the greatest potential of any rational enterprise of man. Cultivating that potential in the field of embryology is the primary concern of the writers—and readers—of this volume. But no potential is unlimited, and the limitations of methods are pertinent to setting goals as well as to reaching them. To introduce a collection of methods it is appropriate to consider their limitations.

The growth of scientific understanding *is* limited by the methods available to enhance it, *can* be limited by failure to use the best method to promote it, and *may* not only be limited but inhibited by the improper application of even the most appropriate method. These limitations are being narrowed by the contributors to this volume: not content merely to create new methods, they are here making them accessible to others and offering the benefit of their experience to ensure that the proper methods will be chosen and, once chosen, will be applied correctly.

Mammalian embryologists have a special problem: the embryo is not readily seen or manipulated in its natural environment—the dark interior of a gestation sac and uterus. There are three ways to get at the embryo directly: the uterus may be breached and the embryo removed for study, the uterus may be breached and the embryo studied *in situ,* or the embryo and the uterus may be removed and studied together, usually after preservation.

Anatomical methods can preserve, describe, and analyze the structural aspects of the natural situation. Specimens exhibiting structural differences can be arranged in a sequence of stages of increasing complexity. Such an array conveys a sense of development, and, if the times of both fertilization and death are known for at least some of the specimens, a time scale for development can be estimated. Such a description of major anatomy by stage and age provides points of reference for descriptions or analyses of the development of special structures as studied macroscopically, histologically, electronmicrographically, histochemically, or chemically. These extensions of embryological description are important because they keep the frontier of knowledge moving by providing detail, discrimination, or quantitation. But just as histology has been defined wryly as "the science of reproducible artifacts," so all the methods that interpose lethal manipulations between the living object and the observation yield understanding with restricted or even uncertain relevance to the natural situation. They offer no reply to the taunt of Mephistopheles: "You men who study life begin by killing it!"

In vitro methods do provide a reply—they permit studying living structures without first killing them—but some questions about artifacts remain. Does the specimen survive by altering its normal functioning to compensate for the abnormal situation? One may legitimately dodge that question by declaring that potentialities rather than normal manifestations are being studied. Is the surviving specimen just dying more slowly than it would without the culture method? This question is irrelevant if study is restricted to the identification of factors that are essential for survival. And either question—or any similar ones—can be diverted by saying that if the cultivated specimen behaves normally when returned to its normal environment, then it was possibly normal all the time. Alternatively, one can compare the structure of a cultured specimen with that of a specimen that developed in the normal situation. But even close agreement is suspect because lethal methods must be used to prepare the specimens for comparison, and it seems likely that some of the cultured specimen's deviation from natural development will be too subtle for detection by lethal methods. For example, *in vitro* methods cannot be expected to imitate the natural motions of blood circulation or the molecular motions that depend on them, and it is not certain that anatomical or chemical methods would reveal all the consequences. Thus, although *in vitro* methods escape the taunt of Mephistopheles, they are not without shortcomings that might be the object of further derision: "You men who study nature begin by denaturing it!" Accordingly, if the biologist's objective is to understand nature and not just collect data, he will wish to study his subject not only alive but in its natural context.

All methods of observation affect the living embryological system that is under scrutiny, but *in vivo* methods do so to a markedly less degree than others. *In vivo* methods approach the philosophical ideal of keeping all the unknown variables natural while one aspect of nature is observed. But the goal of absolute nondisturbance of the living system is not attained; the *in vivo* methods may avoid overt disruptions of the organism but they initiate a series of covert disruptions.

Although there is no single technique or no single category of methods that provide ideal access to the living embryo, methods may be combined to balance shortcomings or artifacts. If the combination is used frequently, the complementary combination tends to be thought of as a single method with a single result. A fundamentally different reason for applying more than a single method is to compare the results of one method with the results of another. If the difference between methods is small and random, as in the simple replication of one method, then similar results confirm the

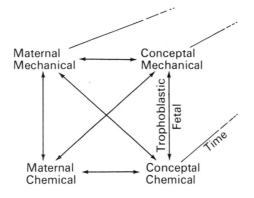

FIGURE I-1

This diagram suggests that mammalian embryology encompasses mechanical and chemical aspects of both mother and conceptus, their interactions, and the changes with time. Fitting traditional subdisciplines on the diagram points out the limitations of their outlook. Thus, endocrinology is essentially the left vertical interaction and chemical embryology the right one, whereas classical descriptive embryology is only the right upper and outer corner.

reproducibility and proper execution of the methods. If a difference between methods is recognized but thought to be insignificant, then similar results not only confirm the reproducibility and proper execution as before but they also confirm the assumption that the difference between the methods is trivial. If the difference between methods is fundamental, then similar "results" verify only reproducibility and execution, they indicate nothing about differences being trivial. Fundamentally different methods can not, in fact, give comparable observations, and what are compared are not observations but interpretations. By comparing interpretations, one goes beyond checking methods and their execution; one checks the most likely source of error —the reasoning.

As the increasing diversity of methods is applied to an increasing diversity of subjects, appreciation grows for the interrelationships as well as for the details constituting the developmental process. Once, sweeping pronouncements about *the* mammalian embryo were based on studies of one or two species. Then it became a mark of sophistication to recognize species differences and the need for comparative study. Now the science of mammalian embryology, like the conceptus itself, is not only cleaving, differentiating, and growing, but metamorphosing from an anatomical subspecialty to a comprehensive, holistic science.

The purview of the embryologist broadens, because new methods provide new viewpoints as well as new data. The architecture of science may thereby be fitted ever more closely to the architecture of nature, provided that the major constituents of nature—so far as we may see them—are reflected in the planning of our research.

Methods in Mammalian Embryology

1

Observing Ovulation and Egg Transport

RICHARD J. BLANDAU
Department of Biological Structure
School of Medicine
University of Washington
Seattle, Washington 98105

Although important advances have been made in the past two decades toward an understanding of the mechanism of ovulation and egg transport in mammals, the physiological and endocrinological details of these processes remain largely unknown.

One reason for this hiatus in our knowledge stems from the technical difficulty of observing ovulation and egg transport simultaneously. An investigator who wishes to observe these phenomena directly in the living animal is confronted with the difficult problem of invading the peritoneal cavity while maintaining its environmental conditions in as normal a physiological state as possible. Many investigators have emphasized that the ovulatory process is exceedingly sensitive to environmental conditions and may be inhibited by anesthetics and a great variety of other insults to which animal tissues may be exposed.

The literature that deals with attempts to unravel the endocrine basis for ovulation and egg transport is contradictory and confusing. A basic problem that confronts the investigator is that physiological levels have not been established for the various hormones that are used. Furthermore, there is

The techniques described in this chapter were developed under research grants from the National Institutes of Health, United States Public Health Service. The author wishes to express his appreciation to Mrs. Elizabeth Phinney and Mr. Roy Hayashi for the many helpful suggestions in developing the techniques.

significant animal variation in the sensitivity of the target organs to hormone action. The injection of unphysiological doses of hormones results in pharmacological reactions that are quite different from the normal.

This chapter deals with a method for observing ovulation and egg transport in living rats, rabbits, and monkeys. The basic technique has been developed in our laboratories and refined by experience and by the suggestions of various individuals over many years. The procedures are described in sufficient detail so that a beginning graduate student can become familiar with the technique quickly and be able to make any modifications necessary to suit his own particular problems.

The method for preparing the animal for observation varies with the species. Thus, the procedure followed for several different animals will be described separately.

PREPARATION OF THE RAT

INDUCTION OF OVULATION

Ovulation may be induced in the rat by injecting gonadotropin preparations into immature females by using the technique of Rowlands (1944). Rats of the Wistar or Sprague-Dawley strains, weighing 35 to 40 g, are injected subcutaneously in the nape of the neck with 20 I.U. of pregnant mare's serum (PMS), gonadotropin preparation (Equinex, Ayerst). Fifty-four hours later the animals are injected subcutaneously with 20 I.U. of a chorionic gonadotropin preparation (A.P.L., Ayerst). The animals are not fed after the last injection but receive water *ad libitum*. Rats injected according to this schedule ovulate approximately 10 hours after the last injection. In our experience long-acting anesthetics tend to inhibit ovulation in the rat, requiring the spinal-cord section to be performed as follows.

SPINAL-CORD SECTION

The animal is anesthetized with ether, placed on its abdomen on a cork board, and the vertebral column exposed between the fifth and seventh cervical vertebrae. After incising the shaved skin at the nape of the neck, the vertebral column can be exposed, with minimal bleeding, by blunt dissection. The various layers of tissue are held aside with retractors made from small paper clips whose ends have been bent to an acute angle. These are attached to slightly stretched rubber bands that are held in place with thumb tacks driven into the cork board. Approach to the vertebral column is made under a low-power binocular dissecting microscope. A sharp iridectomy knife is inserted between the fifth and sixth cervical vertebrae, sectioning the cord. The operative area is packed immediately with Gelfoam (Upjohn) to control bleeding; the muscles, connective tissue, and skin are sutured to hold the Gelfoam in place. With practice the cord can be sectioned in less than five minutes. To be certain adequate respiration is

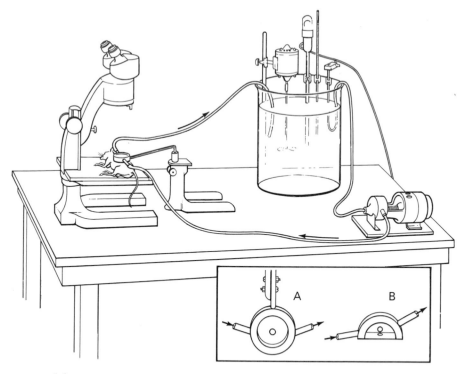

FIGURE 1-1

A diagrammatic representation of the method for observing ovulation and egg transport in the rat. Two observation chambers, A and B, are shown in the box below. Chamber A is shown in detail in Fig. 1-2; the general photographic setup is shown in Fig. 1-3.

maintained, an intratracheal tube is inserted directly into the trachea below the thyroid. The tube is held in place with sutures around the trachea and is then attached to a respiration pump. Oxygen may be added to the air supply by attaching an 18-gauge hypodermic needle to the end of some gum-rubber tubing attached to an oxygen tank. The needle is then inserted through the tubing of the respiration pump.

EXPOSING THE OVARY AND OVIDUCT

The ovaries and oviducts are exposed by making an oblique incision, 1 cm long, in the lateral body wall in the angle between the last rib and the muscles of the back. A generous area of skin surrounding the site of the incision is clipped free of hair, shaved and then covered with a thick layer of Celvacene[1] vacuum grease.

The animal is placed on its side upon an expanded stage of a dissecting microscope (see Fig. 1-1). An observation chamber (see Figs. 1-1, A, B and 1-2) is fastened to a mechanical ratchet and brought over the animal so that the opening in the bottom is in direct line with the incision. As the container is lowered the periovarial adipose tissue is grasped with a fine

---

[1] Consolidated Vacuum Corporation, Rochester, New York.

4

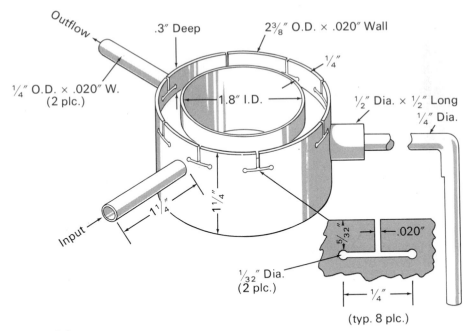

FIGURE 1-2

One type of double-walled chamber that is sewn into the abdominal wall. The inner wall is made of very thin copper sheeting that transmits heat readily.

forceps and gently drawn into the chamber, carrying the ovary and oviduct with it. Fortunately, in most rats the mesovaria and mesosalpinges are pendulous enough to be readily mobilized without stretching the blood vessels within them. Great care must be taken not to pull excessively on the mesenteries or blood vessels since they are injured easily. The chamber is brought into contact with the greased skin, and a waterproof seal is made by pressing the skin against the paraffin covering the bottom of the dish with a tongue depressor. A fine needle (see page 13 for preparation of anchoring needles) is passed through the mesometrium in the region of the uterotubal junction, thereby anchoring it to the paraffin. The ovary of the rat is enclosed within a delicate membrane of the periovarial sac. To expose the ovary and the infundibulum of the oviduct the periovarial sac is cut near its junction with the oviduct with a finely pointed electrosurgical knife attached to a Bovie Electro-Surgical Unit. The blood vessels passing over the surface of the periovarial sac are cauterized by a minimal electric current that is necessary to occlude them from the Bovie Unit. After the periovarial sac has been incised it is pulled back over the surface of the ovary and the tissues are immediately covered with Hank's or Eagle's tissue-culture medium that has been warmed in a water bath to 38°C. The ovary can be anchored to the paraffin base by inserting fine needles through the edges of the periovarial sac and into the paraffin.

The tissue-culture medium bathing the ovary and oviduct is maintained at body temperature by forcing water from a constant-temperature bath through the outer jacket of the dish (see Fig. 1-2). Respiratory movement

FIGURE 1-3

General arrangement of equipment for observing and photographing ovulation and egg transport in mammals. The animal is placed on a movable stage. The camera is attached to a movable ratchet so that it can be focused easily. The low-power microscope, affixed to a long arm, can be swung quickly into position for observation.

may force fluids back and forth through the opening in the bottom of the chamber making observations difficult and photography impossible. To remedy this problem vaseline in a syringe is brought to body temperature and injected into the opening, thereby sealing it.

The ovaries and oviducts of rats prepared in this manner may be observed continuously for six to eight hours. If cinematographic records of ovulation or egg transport are desired an Arriflex camera may be used as shown in Figure 1-3. A viewfinder attached to the camera allows for full visualization of the field at all times during the photography.

PREPARATION OF THE RABBIT

INDUCTION OF OVULATION

Ovulation may be induced in the rabbit by several different methods. Normal ovulation may be evoked in mature females that have been caged separately for at least 21 days by mating them with a vigorous male: ovulation will usually occur approximately 10 hours after such a mating.

Ovulation may be induced more conveniently by the use of gonadotropins. In our laboratories sexually mature females weighing 8 to 10 pounds, which have been caged separately for at least 21 days, are injected intravenously with 50 I.U. of Follutein (Squibb). Such animals will ovulate 12 to 15 eggs approximately ten hours after the hormone is injected.

Forty to eighty eggs per ovulation may be obtained by subcutaneously injecting separately caged females for three successive days with 25 units of Pergonal (Cutter) per day. Eight to ten hours before the ovaries are to be examined the animal is injected with 50 to 100 I.U. of Follutein. (This technique was suggested by Dr. S. J. Behrman of the University of Michigan, Ann Arbor, Michigan.)

The reproductive tracts of female rabbits should be exposed for observation 1 or 2 hours before ovulation is expected. Rabbits that have been on a regular diet, *ad libitum,* until they reach the desired weight of 8 to 10 pounds are excessively fat, and their oviducts are buried in large masses of adipose tissue. To obviate this, rabbits that have reached a weight of 8 to 10 pounds should be placed on a complete but restricted diet for several weeks before the observations are to be made. This reduces the amount of fat deposited in the mesenteries of the reproductive tract and makes the ovaries and oviducts much more visible and accessible but does not interfere with normal or induced ovulation and egg transport.

ANESTHESIA

The animal is not fed the night before it is to be used. The next morning it is restrained in a rabbit box, the hair is clipped from the right ear, and the area of skin over the marginal vein is shaved. To control anesthesia an indwelling polyethylene tube is inserted into the marginal ear vein in the following manner:

1. Two drops of 0.5% pontocaine hydrochloride (Winthrop) are dropped into each eye to avoid possible irritation from Xylene, which is applied to the ear in order to dilate the blood vessels.

2. A 20-gauge thin-wall lumbar-puncture needle on a 5-cc syringe is inserted into the marginal ear vein, and approximately 1.4 cc of sodium pentobarbital (Diabutal: 60 mg/ml) is injected rapidly. The normal beveled edge of the lumbar-puncture needle is not sufficiently acute or sharp to penetrate the vein easily but can be made so by increasing the angle of the bevel with a power hand tool to which is attached a cuttle bone disc. The beveling and sharpening is done under a binocular dissecting microscope.

With the needle held in place the anesthetized rabbit is removed from the box and laid on her side on a table. The syringe is removed from the needle and an intramedic polyethylene tube (0.024 inch O.D.)[2] is threaded through the needle and into the vein for 3 or 4 inches (see Fig. 1-4). The

---

[2] Intramedic Polyethylene tubing (PE 10): I.D. .011 inch—O.D. .024 inch (Clay Adams).

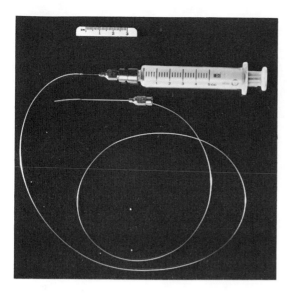

FIGURE 1-4

The lumbar puncture needle which is inserted into the marginal ear vein and the polyethylene tubing which is threaded into the ear vein. A 27-gauge hypodermic needle is inserted into the end of the tubing and is used to introduce saline and anesthetics.

tubing is held in place and the needle is completely withdrawn from the vein and from the tubing. Bleeding around the tubing is controlled by placing a paper clip on the ear for a few minutes. A 27-gauge hypodermic needle is inserted into the end of the polyethylene tubing and a small amount of saline contained in a syringe is injected through the tubing to ascertain patency. The animal is put on its back and the hair on the abdominal wall and the tracheal region is clipped with animal hair clippers. All loose hair is picked up by a vacuum cleaner. The animal is now removed to the operating table and a 5-ml Tomac Star Interfit Lock Control Syringe containing 30 mg/ml Diabutal (half the strength used for the original anesthesia) is attached to the ear tubing to control the depth of the anesthesia. The use of dilute Diabutal gives much better control of the depth of anesthesia.

The next step is to insert an intratracheal tube. This is done by incising the skin and subcutaneous connective tissue over the trachea and separating the strap muscles by blunt dissection. When the trachea is exposed, the cartilaginous rings just below the thyroid are cut with fine-pointed scissors. The trachea is then grasped with a mouse-toothed forceps, and a beveled stiff-wall tube (5 mm OD) is inserted into it and tied in place. The tube, which is anchored securely by stitching it to the skin of the cheek, is attached immediately to the respirator pump, which supplies a mixture of air and oxygen adequate to maintain the normal pink appearance of the tissues. The subcutaneous tissues are closed with a few interrupted sutures and the skin is closed with skin clips.

INSERTING THE OBSERVATION CHAMBER

An observation chamber must now be inserted in the abdominal wall (the general dimensions and configuration of the chamber are shown in

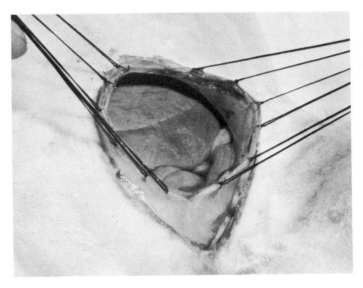

FIGURE 1-5

A 2.5 inch incision in the abdominal wall just above the right ovary and oviduct. The 12 ties are in place and the opening is ready to receive the chamber.

Figure 1-2). A 2.5 inch diagonal incision is made in the abdominal wall just over the left oviduct and ovary. Twelve threads, 12 to 14 inches long and approximately 8 mm apart, are inserted into the muscles of the wall with single-gauge Berbecker Eye Needles, half curved (see Fig. 1-5). Inexpensive Belding Corticelli pure silk buttonhole twist thread, size D, is excellent for this purpose. Each thread is tied securely in place; the remainder is used to anchor the muscle wall to the chamber. During the entire operative procedure the abdominal viscera are kept moist by frequent applications of Gey's solution.

If the bladder contains urine, it is emptied by inserting an 18-gauge hypodermic needle and withdrawing the fluid by negative pressure. The rectum is often filled with fecal pellets which may interfere with the observations or may press upon blood vessels as the chamber is lowered into position. These pellets may be removed by making a longitudinal slit in the antimesometrial border and taking them out one at a time with moistened blunt forceps. The incision is then closed with a curved eye needle and fine silk thread.

POSITIONING THE REPRODUCTIVE TRACT

When the chamber is in place, the abdominal viscera may obscure the view of the reproductive tract. To avoid such interference the viscera are withdrawn from the peritoneal cavity into an artificial pouch. A 2 inch incision is made high on the right side of the abdomen just to the right of the large mammary vein. A hole, 2 inches in diameter, is cut in the center of a 14 inch square plastic sheet. The sheet is so arranged that the hole in its center lies over the incision, and the edges of the cut muscle and those of the hole in the plastic sheet are approximated and joined by skin clips (see Fig. 1-6). If this is done correctly a leakproof seal will be made. The clips are covered with a thick layer of vaseline. A piece of vaseline-soaked gauze is placed over the plastic sheeting and a hole is cut in its center, directly above the hole in the plastic sheet are approximated and joined by skin clips (see cavity, care being taken to avoid interference with the blood supply. The edges of the plastic sheet are brought together and clamped, making a pouch with a waterproof seal. The animal is now moved from the operating table to the observation rack (see Fig. 1-3). The chamber is lowered over the lower abdominal incision. The long threads, shown in Figure 1-5, are used to pull and attach the muscle over the edge of the chamber. The threads are then inserted into the appropriate slots in the upper edge of the chamber and tied securely in place (see Figs. 1-2 and 1-7). The skin is brought over the edge of the chamber by inserting a few stitches in the appropriate places, anchoring them in the slots, and tying them securely. Melted vaseline is injected between the chamber and the muscle anchored to it, thereby making a waterproof seal.

The chamber may now be raised or lowered at will. The contoured opening in the bottom of the dish is positioned over the reproductive tract. Gey's balanced salt solution, maintained at body temperature, is used generously to flood the area, and is aspirated as often as necessary to remove any blood.

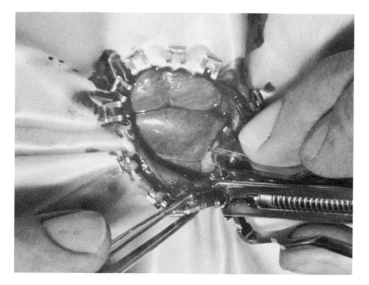

FIGURE 1-6

An opening in the abdominal wall high on the right side of the abdomen. The edges of the cut muscle and plastic sheet are approximated and sealed by skin clips as shown. After the skin clips are covered by a thick layer of vaseline the viscera are drawn from the peritoneal cavity and sealed in the plastic pouch.

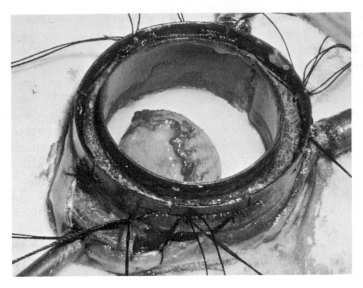

FIGURE 1-7

*The chamber in the anterior abdominal wall. The threads shown
in Fig. 1-5 have been tied in place and the skin is being brought
up to the edge of the chamber to make a watertight seal.*

PREPARATION OF MOISTENING FLUID

Ordinary physiologic saline is to be avoided as a moistening medium. Experience has shown that if normal saline is used for more than a short period of time it causes leakage of plasma from the blood vessels and deposition of fibrin on the surfaces of the tissues. We have found that Gey's balanced salt solution, as modified by Pomerat, is an excellent moistening solution. It is made up as a 10× concentrated stock solution, two separate solutions being prepared, since mixing of the various salts before autoclaving results in precipitation.

Solution A

$NaCl$ . . . . 80.0 g
KCl . . . . 3.8 g
$Na_2HPO_4 \cdot 7H_2O$ . . . . 3.0 g
KH_2PO_4 . . . . 0.25 g

These salts are dissolved in 1000 ml of glass-distilled water in the order given. A 2000-ml flask is convenient for continuous mixing.

Solution B

$MgCl_2 \cdot 6H_2O$ . . . . 2.1 g
$CaCl_2 \cdot 2H_2O$ . . . . . 1.3 g

These salts are dissolved in the order given in 1000 ml of glass-distilled water. Again, a 2000-ml flask is convenient for continuous mixing.

One hundred ml each of solutions A and B are dispersed separately into 125 ml serum bottles with "sleeve-type" stoppers and "one-piece" aluminum caps that are crimped with a hand crimper. The bottles are autoclaved at 15 lb pressure for 60 minutes (slow exhaust).

The glucose component for Gey's balanced salt solution is made up as a 10% solution and filtered to sterilize because autoclaving occasionally causes caramelization. Fifty grams of glucose are dissolved in 500 ml of glass-distilled water. This solution is sterilized by Millipore filtration and stored in the refrigerator.

The last ingredient needed is a 10% solution of sodium bicarbonate, which can be purchased already prepared and kept refrigerated. Before use the solution should be examined carefully for crystals; if present, they must be dissolved completely.

By preparing these concentrated stock solutions the technician can make 10 liters of Gey's balanced salt solution before having to reweigh the ingredients. Thus, much time and effort are saved.

Once all of the concentrated solutions have been prepared and sterilized, regular Gey's balanced salt solution can be made. Measure 100 ml each of solutions A and B into a 1000 ml sterile graduated cylinder, add 40 ml of the 10% glucose solution, fill the cylinder to the 1000 ml mark with glass-distilled water, and then add 2.5 ml of the 10% bicarbonate solution. If the bicarbonate solution is added before the other components of Gey's balanced salt solution are diluted, a precipitate will form, and the entire solution will have to be discarded. The completed solution is stored in sterile 1000 ml polyethylene bottles in the refrigerator until ready to be used, when it should be warmed to 37°C.

Observing Egg Transport

A diagrammatic representation of the various subdivisions of the reproductive tract of the rabbit as they may be positioned in the chamber is shown in Figure 1-8. A simple method for positioning the reproductive tract for observation is to make a longitudinal incision in the peritoneum as it ascends from the mesosalpinx onto the lateral body wall. Since there are relatively few blood vessels in this membrane it may be cut with little or no bleeding. Gently pulling the cut edge of the mesentery medially mobilizes the mesovarium and mesosalpinx. This extensive membrane may be used to position the oviduct and ovary within the chamber. The mesentery is anchored to the paraffin in the bottom of the chamber by very sharp, small tungsten needles (see page 13), and the chamber is then lowered into the peritoneal cavity, care being taken to insure that the blood vessels coursing through the mesenteries are not stretched or broken. Vaseline which has been brought to its melting point is drawn into a warmed 10-cc, three-fingered syringe, and injected gently below the tract and edges of the dish in order to seal the entire area from the peritoneal cavity. The Gey's solution may then be retained within the dish. Sealing off the dish also greatly reduces the disturbances caused by the respiratory movements. The fluid within the dish is maintained at body temperature (±0.05°F) by pumping warmed water through the jacket (see Fig. 1-2).

To observe ovulations directly the mesotubarium superius and the fimbria attached to it are gently drawn to one side, exposing the ovarian

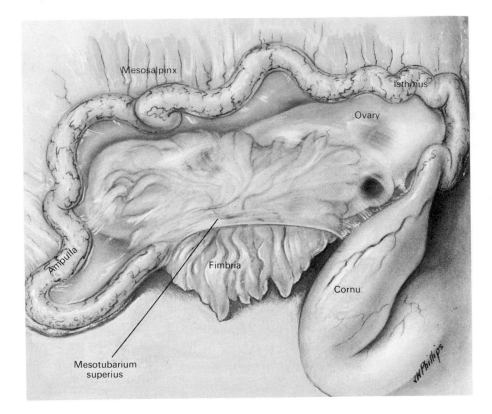

FIGURE 1-8
A diagrammatic representation of the ovary, the oviduct, and associated mesenteries after being put in the chamber.

surface. Since the mesotubarium superius is a thin, transparent membrane, ovulations may be seen by direct inspection.

Egg transport, particularly through the ampulla, may be studied both qualitatively and quantitatively by observing gonadotropin-induced ovulations directly or by recovering freshly ovulated eggs from donors and applying them to the edge of the fimbria. Donor eggs, embedded in their cumuli, are submerged in a dish of physiological saline to which a few drops of 0.2% methylene blue dissolved in 0.85% NaCl may be added. The movements of such supravitally stained eggs and cumuli are followed readily.

Rabbits may be anesthetized approximately an hour before the onset of ovulation without inducing the inhibition of ovulation that the anesthetic induces in the rat. Cord section is somewhat more difficult to perform in the rabbit, but its usefulness for long term observations should be evaluated. If the animal is to remain in good physiological condition it is important to pay strict attention to the level of anesthesia. With the indwelling ear catheter and the use of 30 mg/ml Diabutal a minimal but complete anesthesia can be maintained for many hours. During observations the animal should rest on a heating pad, which helps maintain body temperature (see Fig. 1-3).

Photographic records are made with a 16-mm Arriflex camera that is suspended above the preparation and can be brought into focused position quickly (see Fig. 1-3). A variety of Micro-Tessar and Luminar lenses are available and Eastman Ektachrome Commercial (ECO) film is used in our laboratory. The best light source has been found to be two Zeiss 12-volt, 60-watt, high-intensity lamps; these may be focused easily and give excellent color rendition.

If egg transport is to be studied in the ovariectomized rabbit it is essential that removal of the ovaries does not injure the infundibulum nor cause adhesions. Ovariectomies can be performed in rabbits without injury to the fimbria by exposing each ovary through a lateral incision. Under a binocular dissecting microscope the blood vessels entering the ovaries are identified, and each is occluded by applying a coagulating current from a Bovie Electro-Surgical Unit. When all blood vessels have been occluded the ovary is cut away with a fine electro-surgical knife. At no time is the oviduct handled with forceps. Sterile technique is used during this entire procedure, and the reproductive tract is kept moist by dripping Gey's solution onto it. After ovariectomy the rabbit is injected intravenously with 150,000 units Bicillin (Wyeth) and 2 mg Decadron (Dexamethasone 21-phosphate; Merk, Sharp and Dohme). These compounds control infection and reduce adhesions significantly.

PREPARATION OF THE MONKEY

Egg transport in several pigtail monkeys has been followed at midcycle by the same technique described for the rabbit. An observation chamber with a larger opening in its bottom is placed in the midline just above the pubic symphysis. Thus, both ovaries and oviducts may be viewed by slightly shifting the chamber to the right or left. The animal is pre-anesthetized with 1 mg/kg Sernylan (Phencyclidine hydrochloride; Parke Davis). Anesthesia is induced by injecting nembutal through an indwelling catheter in the femoral vein at the rate of 16 mg/kg, but only until the anesthetic takes effect. By controlling the anesthetic carefully the animal may be maintained in good condition for many hours.

PREPARATION OF ANCHORING NEEDLES

Needles fine enough for anchoring thin tissues and for delicate dissections are often difficult to obtain commercially, but extremely fine-pointed, reasonably rigid needles that do not corrode during submersion in various fluids or during sterilization can be prepared easily in any laboratory. (The original suggestion for the preparation of the needles comes from Dr. John Boling, professor of biology, Linfield College, McMinnville, Oregon.)

Tungsten wire[3] of either .05 mil or .01 mil diameter is clamped into a

---

[3] North American Philips Co., Inc., Lewiston, Maine.

hemostat, which is in turn attached to one of the electric wires plugged into a powerstat that is set to operate at ten volts. The second wire from the powerstat is attached to a copper container, about the size of a 200 ml beaker, which is filled with a concentrated solution of sodium hydroxide. The tip of the tungsten wire is etched electrolytically by dipping it into the sodium hydroxide solution. By removing the tungsten wire from the bath after about 30 seconds and examining its etched tip under a binocular dissecting microscope, needles of just the right fineness and sharpness can be prepared. The needles then may be cut to the desired length and inserted into a suitable needle holder.

BIBLIOGRAPHY

Asdell, S. A. (1962) The mechanism of ovulation. In: The Ovary. Chap. 8, p. 435. Ed. by Solly Zuckerman. Academic Press, New York.

Black, D. L., and S. A. Asdell (1958) Transport through the rabbit oviduct. Am. J. Physiol., 192:63.

Blandau, R. J. (1955) Ovulation in the living albino rat. Fertility Sterility, 6:391.

———— and R. E. Rumery (1963) Measurements of intrafollicular pressure in ovulatory and preovulatory follicles of the rat. Fertility Sterility, 14:330.

Christiansen, J. A., C. E. Jensen, and F. Zachariae (1958) Studies on the mechanism of ovulation: Some remarks on the effect of depolymerization of high-polymers on the preovulatory growth of follicles. Acta Endocrinol. (Kobenhavn), 29:115.

Claesson, L. (1947) Is there any smooth musculature in the wall of the graafian follicle? Acta Anat. (Basel), 3:295.

Clewe, T. H., and L. Mastroianni, Jr. (1958) Mechanisms of ovum pickup. I. Functional capacity of rabbit oviducts ligated near the fimbria. Fertility Sterility 9:13.

Espey, L. L., and H. Lipner (1963) Measurements of intrafollicular pressures in the rabbit ovary. Am. J. Physiol., 205:1067.

Greenwald, G. S. (1961) A study of the transport of ova through the rabbit oviduct. Fertility Sterility, 12:80.

Harper, M. J. K. (1961) The mechanisms involved in the movement of newly ovulated eggs through the ampulla of the rabbit Fallopian tube. J. Reprod. Fertility, 2:522.

Lipner, H. J., and B. A. Maxwell (1960) Hypothesis concerning the role of follicular contractions in ovulation. Science, 131:1737.

Parker, G. H. (1931) The passage of sperms and eggs through the oviducts in terrestrial vertebrates. Phil. Trans., 219:381.

Rowlands, I. W. (1944) The production of ovulation in the immature rat. J. Endocrinol., 3:384.

VanDemark, N. L., and A. N. Moeller (1951) Speed of spermatozoan transport in reproductive tract of the estrous cow. Am. J. Physiol., 165:674.

Westman, A. (1926) A contribution to the question of the transit of the ovum from ovary to uterus in rabbits. Acta Obstet. Gynecol. (Scandinavia), 5:1.

Zachariae, F., and C. E. Jensen (1958) Studies on the mechanism of ovulation. Histochemical and physico-chemical investigations on genuine follicular fluids. Acta Endocrinol. (Kobenhavn), 27:343.

2

Collection and Preservation of Spermatozoa

N. L. FIRST
University of Wisconsin
Madison, Wisconsin 53706

The major requisites for successful experimentation involving insemination and fertilization are the collection and preservation of spermatozoa and the accomplishment of either *in vivo* or *in vitro* fertilization. Only the most successful methods of semen collection and preservation for each of several species will be considered in the following discussion. As it is impossible to consider all mammalian species we will discuss only those most commonly used for embryological study: cattle, sheep, swine, monkeys, rabbits, rats, mice, guinea pigs, and dogs. The most appropriate method of semen collection varies from species to species. Some of the more commonly employed methods are: recovery of semen from the vagina after mating, use of an artificial vagina, electroejaculation, ejaculation by anesthesia, recovery of spermatozoa from the extirpated epididymis or testis, digital manipulation of the penis, and rectal massage of the ampulla and vas deferens.

CATTLE

The earliest method used to collect semen was recovery from the vagina after natural mating. This method has limited value because the ejaculate is contaminated with secretions from the cow, and because only partial recovery is possible.

Some semen can be obtained from the bull by rectal massage of the ampulla, seminal vesicles and vas deferens, but the concentration and yield

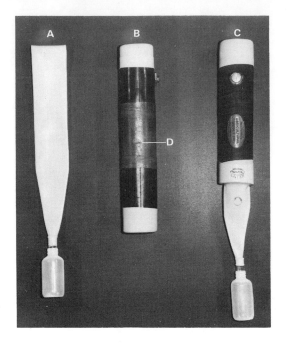

FIGURE 2-1

*Artificial vaginas for the bull.
(A) Inner liner for Cornell
model artificial vagina.
(B) Cornell model artificial
vagina, approximately 60 cm
long and 6.5 cm in diameter.
(C) Danish model artificial
vagina, approximately 40 cm
long and 6.5 cm in diameter.
(D) Pressure escape valve.*

of spermatozoa are low and many samples are contaminated with urine. This method should be used only with badly crippled bulls.

The two most successful methods for collecting bull semen are use of the artificial vagina and electroejaculation.

Artificial vagina

Use of artificial vaginas entails the following steps: (1) preparation of equipment and the bull, (2) teasing or sexual excitement, (3) mounting of a cow, steer, bull, or dummy, (4) insertion of the erect penis into a rubber artificial vagina as the bull mounts, and (5) the ejaculation of semen into a graduated test tube affixed to the artificial vagina.

Two models of artificial vagina are commonly used (see Fig. 2-1). The longer Cornell model protects the ejaculate from temperature shock by surrounding the collection tube with a warm liner. The Danish model is shorter and easier to use, but if ejaculation takes place in a cold environment the semen collection tube must be protected with some type of insulation. A protective water jacket can be easily prepared by surrounding the collection tube with a larger plastic tube sealed at the lower end and filled with water at 36°C: the plastic tube also reduces the danger of breaking the unprotected collection tube, and it is hence advisable to use it under all conditions. A separate artificial vagina for each bull will prevent bacterial contamination between bulls.

The parts of the artificial vagina must be clean, sterile, and dry before assembly. After much use the collection funnel may no longer grasp the collection tube. When this happens, it should be replaced with a new funnel or held in place with a plastic-coated wire. It should not be attached with a rubber band because as the bull thrusts his penis may separate the collection tube from the funnel, and the rubber band may become affixed around the penis; it may even amputate part of the penis.

After the artificial vagina is assembled, the space between the jacket and the liner is filled with water at about 55°C, to provide a temperature of 42° to 44°C in the vaginal lumen. Since the ejaculatory response by the bull is the result of both temperature and pressure stimuli, air is often pumped into the water jacket to increase pressure. Pressure should be carefully regulated, for if it is too great, the ejaculate may be deposited on the inner liner and not in the collection tube, or the inner liner may be forced from the jacket, allowing the water to escape. One way of regulating it is by providing an escape valve consisting of a piece of soft rubber tubing that fits over a small hole in the jacket (see Fig. 2-1B). If the artificial vagina is too long the ejaculate may be deposited in the inner liner; if it is too short the glans penis can hit the collection tube, thereby forcing it from the collection funnel, and perhaps causing injury to the penis. After the vagina is assembled, a sterile glass rod is used to apply sterile lubricant (K-Y jelly) to the inner liner. Some bulls prefer an inner liner of rough texture, and a trial may be necessary to determine a bull's preference.

Sexual preparation is essential before semen can be collected with the artificial vagina, and adequate preparation has been shown to result in a greater ejaculatory yield of spermatozoa (Hale and Almquist, 1960). The sexual excitement can be provided by exposing the bull to a teaser animal without allowing him to mount, or by allowing him to mount without ejaculation. Bulls can be sexually excited by cows, steers, or other bulls, and they will mount either sex (see Fig. 2-2). Some teaser animals are capable of eliciting greater sexual response from a male than others, and this does not necessarily depend on the sex of the teaser. Bulls that become apathetic about a frequently used teaser can usually be excited by a new teaser (Hale and Almquist, 1960). For best response from the bull, the artificial vagina and the technician making the collection should always be the same. The teaser and teasing environment can, and sometimes must, be varied to maintain sex drive or to excite slow bulls.

Semen is collected by holding the artificial vagina in line with the thrust of the bull's penis as he attempts intromission, intercepting his thrust, and immediately after he ejaculates tilting the artificial vagina downward to allow the semen to run into the collection tube (see Fig. 2-2). Best results are obtained when the technician does not touch the penis of the bull, although he may, with some bulls, have to guide the penis by grasping the sheath lightly. If the bull thrusts, but fails to ejaculate, it is likely that the temperature or pressure of the artificial vagina is not correct. The artificial vagina is the method of choice when healthy bulls can be handled and trained.

Electroejaculation

Electroejaculation is particularly useful with range bulls, which cannot easily be trained to mount a teaser animal with a man nearby. When this

FIGURE 2-2
Collection of semen from the bull by use of the artificial vagina.

method is to be used the first requisite is a strong squeeze chute or a stock, to restrain the bull. During electroejaculation, the shoulders of the bull thrust forward with considerable pressure as skeletal muscles of the hind limbs undergo contraction. Sometimes the hind limbs will not support the bull. If a squeeze chute is not used, the technician should, to protect his arm, collect ejaculated semen with a rubber funnel or short artificial vagina mounted on a 40 to 60 cm shaft.

Several models of electroejaculators are available. The most commonly used commercial electroejaculators in the United States are modifications of units developed by Dziuk *et al.* (1954 a,b) and Marden (1954). The Dziuk model has a bipolar rectal electrode and produces an alternating current of 60 cps. With the Marden model, the frequency can be varied between 15 and 90 cps, and the rectal probe contains alternately charged longitudinal brass electrodes. Rowson and Murdoch (1954) modified the rectal probe by placing finger electrodes over the gloved hand of a technician. This allowed more accurate placement of the stimuli, resulting in less muscle contraction in the hind limbs.

The stress, discomfort, and skeletal muscle contractions accompanying electroejaculation have been reduced by tranquilization before ejaculation with .04 to .09 mg of a tranquilizer (Pfizer PUD-5) per kg of body weight (Wells *et al.,* 1966).

Before electroejaculation is attempted the bull's prepuce should be washed and clipped free of hair, feces must be removed from the rectum, and the probe lubricated with a good conductor. Erection should take place before ejaculation, thereby avoiding contamination of the ejaculate by bacteria in the prepuce. Erection and accessory sex gland secretion are induced by low voltage stimulation, and ejaculation by increasing the voltage. Excessive initial stimulation can prevent erection, and may even fail to cause ejaculation. The most satisfactory procedure is to increase the voltage slowly, in two-volt increments at five-second intervals. When erection occurs, the voltage is increased to 10 to 15 volts and ejaculation results. The most successful operators are those who observe the response of the bull and increase stimulation accordingly. The position of the probe in the rectum can also influence erection and ejaculatory response: it should be firmly pressed against the ventral wall of the rectum. Experience will help determine the best depth of rectal insertion for each bull.

The ejaculate is collected with a short artificial vagina and collection funnel. The jacket of the artificial vagina need not be filled with water, but the collection tube affixed to the collection funnel should be insulated or protected by a plastic jacket containing warm water (36°C). Like the jacket used with the Danish artificial vagina, this protects the spermatozoa from temperature shock and prevents breakage of the collection vessel.

Semen samples obtained from the bull with an electroejaculator are usually of larger volume but of lower sperm concentration than those obtained by other methods. The electroejaculate has a lower fructose and citric acid concentration and a higher pH (Lutwak-Mann and Rowson, 1953; Austin, Hupp, and Murphree, 1962). Manton (1956) found that the pregnancy rate after insemination with bull semen collected by electroejaculation did not differ from that achieved after insemination with semen collected by an artificial vagina.

Procedures for collection of bull semen have been reviewed and described by Salisbury and VanDemark (1961), Herrick and Self (1962), Melrose (1962), and Foote and Trimberger (1968).

SHEEP

Semen is collected from rams either by use of an artificial vagina or by electrical stimulation. Rams are easily trained to serve an artificial vagina if the mount is a teaser ewe in estrus. If estrual ewes are not available, estrus can be induced in the ewe by injecting intramuscularly 10 mg of progesterone in oil daily for 10 to 14 days, and 36 hours after the last injection injecting 2 mg of estradiol benzoate in oil, or a longer acting estrogen. This treatment should keep the ewe in estrus for approximately one week.

The training of wild or timid rams can be facilitated by allowing the ram to mate in the same area and under the same environmental conditions in which semen collection will be attempted. The training of rams has been discussed extensively by Miller (1961) and Salamon and Lindsay (1961).

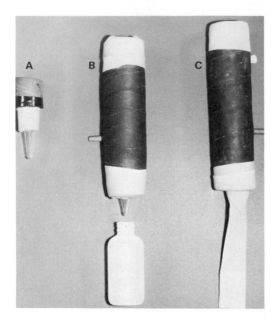

FIGURE 2-3

(A) Artificial vagina for rabbits, 6 cm long, inside diameter of larger casing is 2.5 cm and inside diameter of smaller casing is 1.8 cm. (B) Artificial vagina for rams, 20 cm long and 6 cm diameter. (C) Artificial vagina for boars, 22 cm long and 5 cm diameter.

Artificial vagina

A satisfactory artificial vagina for use with a ram (see Fig. 2-3b) is essentially the same as that for a bull and is used in the same way. Rams are very sensitive to the temperature and pressure of the vagina. Best results are obtained when the internal temperature of the artificial vagina is approximately 42°C, and the pressure approximately 50 mm Hg. If the temperature is cooler, the ram may not ejaculate. Because the small ejaculate characteristic of the ram is easily lost if smeared across the liner of the artificial vagina, the pressure of the artificial vagina must be low enough, and the vagina short enough, that the end of the penis can reach the collection funnel and vessel. Extreme precaution must be exercised to prevent temperature shock, since the small volume of the ram ejaculate will change temperature rapidly. The collection vessel can be protected within the warm artificial vagina or by a warm water jacket. The water jacket around the collection tube should be 36°C at the time of ejaculation. Before collection, the artificial vagina should be held at an angle, preventing the flow of lubricant into the collection vessel. After ejaculation, it must be tilted to allow all the ejaculate to flow to the base of the collection vessel, and the collection vessel immediately removed. The ejaculate may also be collected in a small sherry glass which is inserted into the artificial vagina and replaces the collection funnel and tube (Miller, 1961).

Electroejaculation

The ram responds well to electrical stimulation, and electroejaculation has been used for some time as a means of collecting ram semen (Gunn, 1936).

The early methods of Gunn involved stimulation between a lumbar needle electrode and a rectal electrode. This resulted in severe contraction of skeletal muscles and necessitated restraining the ram on its side for collection. The electroejaculators previously mentioned for bulls, when equipped with a small bipolar rectal probe or longitudinal electrode probe provide excellent ejaculatory response with less immobilization of the hind limbs. For semen collection rams can be restrained in a crate, or restrained on their side. Erection should occur prior to ejaculation. If erection does not occur, the ram's penis can be manually extended by straightening the sigmoid flexure and held by a loop of cotton gauze behind the glans penis. Electroejaculation usually occurs at less than 10 volts. The ejaculate is collected in a graduated 15 ml centrifuge tube surrounded with warm water (36°C at ejaculation) to prevent temperature shock. On occasion, the ram will urinate when electroejaculated. The semen collector should avoid covering the penis with the collection funnel during urination. Collection equipment contaminated with urine must be cleaned or replaced before semen is collected.

Although electroejaculation is used frequently for quick and convenient semen collection, some rams will not respond and the quality of the semen has been reported by some workers to be inferior to that collected by artificial vagina. This method is most useful when semen is collected infrequently, especially from rams of insufficient libido to serve an artificial vagina.

Salamon and Morrant (1963) report that collection by the artificial vagina resulted in greater sperm concentration and semen, which appeared to result in a greater lambing rate than semen obtained by electroejaculation. Lambing rate is more highly correlated with semen quality when ram semen is collected by artificial vagina than when collected by electroejaculation (Hulet *et al.*, 1964). The results of most investigators indicate that an individual collection with an artificial vagina is smaller in volume with a higher concentration of spermatozoa, and more successive ejaculates can be collected by artificial vagina before the ram fails to ejaculate. This subject was reviewed by Emmens and Robinson (1962).

Swine

Semen is easily collected from the boar by a modified artificial vagina or by grasping the end of the penis with the hand (see Fig. 2-4). Unlike the bull and ram, which ejaculate after temperature and pressure stimuli are applied to the penis, the sexually excited boar ejaculates only when the corkscrew shaped glans penis is firmly grasped. In natural mating, the penis is firmly locked in the cervix of the sow.

Although the boar learns easily, he is also a creature of habit. Unmated young boars older than seven months are easily trained to mount a dummy sow. Use of the dummy sow by other boars, sexual excitement by exposure to an estrual female, or the desire to fight after exposure to another male, all increase the young boars' desire to mount.

Older boars, sexually experienced with sows only, often express no desire

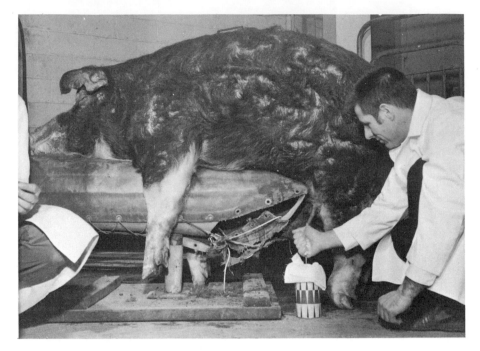

FIGURE 2-4
Collection of semen from a boar using the hand-grasp method.

for the dummy sow. With patience, they can be trained for collecting semen with the artificial vagina after having mounted an estrual sow. The estrual sow is placed adjacent to the dummy sow, and while the boar is ejaculating, the front legs are lifted to the dummy sow. The estrual sow is then removed. After several such experiences, the boar is usually willing to mount the dummy sow and ejaculate.

The prepuce of the boar often contains great quantities of urine and deteriorating cellular debris as well as a large microbial population which may contaminate the ejaculate (Aamdal and Hogset, 1957; Aamdal *et al.,* 1958). When the artificial vagina is used, it should not be applied until the boar has thrust several times and is ready to ejaculate. Neither should it be allowed to touch the prepuce. Holding the entrance downward will prevent preputial contaminants from entering. A suitable artificial vagina is shown in Figure 2-3c. It is prepared and filled with warm water as described for the bull. The pressure must be sufficiently low to allow the penis to extend through the artificial vagina.

After the boar has thrust several times against the dummy, the penis is directed into the artificial vagina. The penis should completely penetrate and extend at least two to five inches beyond the main body of the artificial vagina, which allows the glans penis to be grasped manually through the thin rubber collection funnel.

The hand-grasp method obviates the artificial vagina, and normally results in an ejaculate free of preputial contamination (see Fig. 2-4). The

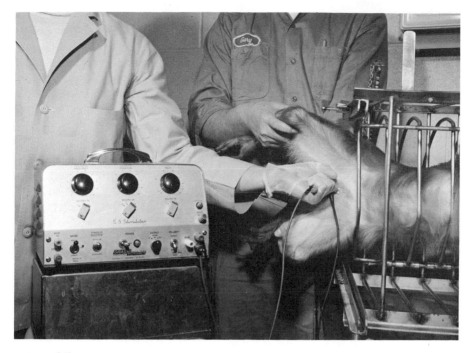

FIGURE 2-5

Collection of semen from the monkey by electroejaculation. Electrodes are placed at the glans and at the base of the penis. The ejaculate is collected in a graduated 15 ml tube.

boar must be very excited and thrust several times before the penis is grasped by the technician. The hand must be warm and may be covered by a rubber glove. Normally, six to seven minutes are required to complete the ejaculatory process (Ito *et al.*, 1948).

Once ejaculation is started, the boar will remain quiet until completion. Pulsation or relaxation of pressure on the end of the boar's penis will result in cessation of the flow of the sperm-rich fraction, and with the subsequent return to thrusting by the boar, increased seminal fluid emission and renewed secretion of the gelatinous plug will follow.

The ejaculate consists of three fractions. The first contains mostly seminal fluids and some of the gelatinous pellet-like material from the bulbo-urethral glands (Mann, 1964); this presperm fraction has a high bacterial count, very few or no spermatozoa, and may be discarded. The gel-plug can be eliminated from the ejaculate during collection by filtering through two to four layers of cotton mesh (see Fig. 2-4). The second fraction is the sperm-rich portion, which is of a creamy white color and contains approximately 80% of the spermatozoa (Self, 1959). The third, or postsperm fraction, is a clear fluid with some gelatinous pellets. Usually a well-trained boar emits only one sperm-rich fraction at a collection. Some boars, if interrupted during collection, will ejaculate a series of two or three sperm-rich fractions separated by seminal-fluid emission. It has been estimated that the presperm fraction comprises 5 to 20% of the total ejaculate volume, the sperm-rich 30 to 50%, and the postsperm fraction 40 to 60% (McKenzie *et al.*, 1938).

The boar ejaculate is large in volume and changes temperature more slowly than ejaculates of most species. Temperature shock can occur however, but may be prevented by collecting the ejaculate in a widemouth vacuum flask that has been prewarmed to 36°C. The semen must be removed from the flask soon after collection, since prolonged storage at this temperature may result in deterioration of the spermatozoa (Anand, 1967).

The composition of the boar ejaculate has been extensively reviewed by Mann (1964). Collection, storage and evaluation procedures have been reviewed by Rowson (1962), Foote and Trimberger (1968), Herrick and Self (1962), Aamdal (1964), Borton *et al.* (1965), and Polge (1956).

MONKEY

Very little success had been achieved collecting semen from the monkey until Mastroianni and Manson (1963) discovered that an ejaculate could be obtained by stimulation with electrodes placed on the penis. The monkey is restrained in a restraining rack as shown in Figure 2-5. Gentle traction is applied to the glans penis. A strip of aluminum foil 2 cm wide is applied around the base of the shaft and held with a clamp electrode. The penis is moistened with saline and a second electrode is held with a gloved hand against the ventral aspect of the glans near the frenulum. Intermittent charges of 20 volts are delivered at a frequency of 10 to 20 impulses per second, for durations of 25 to 50 milliseconds. Monophasic alternating current is supplied by a model S-5 stimulator, which is produced by the Grass Instrument Co., Quincy, Mass. If ejaculation does not occur within 1 or 2 minutes, the voltage is increased gradually to 40. These workers obtained 37 specimens in 40 attempts, and there was no indication that monkeys became refractory to repeated ejaculation at intervals of 2 to 3 days.

Monkey semen is liquid when ejaculated, but coagulates within seconds afterward. Reliquefaction is slow; only 30% of the ejaculate is liquid 30 minutes after emission (Mastroianni and Manson, 1963).

Kirton (1967) modified this method to enhance reliquefaction. The ejaculate is collected into 0.5 ml of warm, physiological saline which contains 1% trypsin. Liquefaction may also be accomplished by collecting the ejaculate into diluent media containing 2% α-chymotrypsin (Hoskins and Patterson, 1967). The collection vessel and saline solution should be at a temperature of 36°C at the time of ejaculation to avoid temperature shock to the spermatozoa.

Semen has been successfully collected from the squirrel monkey by the use of rectal electrodes, after the monkey has been lightly tranquilized (Bennett, 1967). Both monkeys and chimpanzees have been ejaculated by electroejaculation (Weisbroth and Young, 1965; Roussel and Austin, 1967 and Hoskins and Patterson, 1967).

Rabbit

The rabbit is commonly used for reproductive physiology studies and semen is easily obtained from the male. The most effective collection device for this species is the artificial vagina (see Fig. 2-3a) .

Bucks will mount a teaser doe readily, and some will mount a restrained buck. The arm of the technician holding the teaser should be covered with a long glove for protection. The teaser may be replaced with a dummy after the male is trained. The dummy can be an arm covered with rabbit skin or a rabbit-skin glove, which is worn fur outwards. The artificial vagina is held by the gloved or skin-covered hand (White, 1955) .

Several different models of artificial vagina have been developed for the rabbit (White, 1955; Walton, 1958; Bredderman et al., 1964) . Two major requirements for satisfactory collection are a warm artificial vagina and the deposition of semen directly into the collection tube. Ten percent or more of the ejaculated spermatozoa are lost when spermatozoa are deposited on the rubber liner (Bredderman et al., 1964) . The artificial vagina should not exceed 6 cm in length.

An artificial vagina that can be adjusted in length for different males was developed by Bredderman et al. (1964) . It is constructed from two pieces of reinforced rubber casing, each 3.5 cm long. The larger casing has an inside diameter of 2.5 cm; the smaller casing fits tightly within the larger (see Fig. 2-3a) After assembly the two casings are cemented together and a liner of thin surgical tubing, 1.5 to 2 cm in diameter and 11 cm long, is inserted through the heavy tubing, stretched over both ends, and held firmly with large rubber bands. The casing can be filled with water (50°C) before the inner liner is secured or a filling hole can be provided in the casing. Bredderman et al. (1964) devised a method to provide warm, assembled artificial vaginas which are instantly ready for use by filling the casing with diethylene glycol and warming the entire apparatus to 50°C in an incubator. Such artificial vaginas can be washed without disassembly or removal of the diethylene glycol, which is replaced approximately once per month.

The collection funnel and vessel are a 15 ml graduated centrifuge tube cut back to approximately 7 ml capacity. The cut-and-polished end is inserted into the artificial vagina to the depth desired for each male. The buck ejaculates directly into the collection tube. The artificial vagina should be lightly lubricated with a sterile inert jelly (K-Y jelly or a drop of glycerol) .

During semen collection, the artificial vagina is held between the hind legs of the doe. When the male mounts, his penis enters the open, lubricated end and ejaculation occurs. The number of spermatozoa in the ejaculate is greater when the buck is restrained or allowed a false mount before ejaculation (Macmillan and Hafs, 1967) . The second of two successive ejaculates contains more spermatozoa (Desjardins, et al., 1965) and is of greater volume (Amann, 1966) .

Approximately one to two minutes are required to collect the ejaculate. If there is delay in collecting the ejaculate, the artificial vagina must be rewarmed. Most bucks will not ejaculate if the temperature of the artificial vagina is below 50°C. To avoid temperature shock to the spermatozoa, the artificial vagina should be held so that a warm hand covers the collection tube. After the ejaculate is collected, the tube must be removed promptly, and slow cooling or processing of the sample for preservation initiated.

Rat, Mouse, and Guinea Pig

Spermatozoa have been obtained from rodents by recovery from the extirpated epididymis and vas deferens, recovery from the uterus after natural mating or by electroejaculation.

Epididymal and vas deferens recovery

This method necessitates killing the male or destroying its reproductive capability. It does yield, however, a concentrated sample of good quality that is free of coagulum. It was used by Dziuk and Runner (1960) to obtain spermatozoa from mice, and the same procedure may be used for other species. Males are killed and dissected in a warm room or chamber to prevent temperature shock of the spermatozoa during removal. Spermatozoa may be stripped from the excurrent ducts into a physiological solution of reconstituted skim milk (9.5% solution heated in a boiling water bath for 10 minutes and cooled to room temperature), or they may also be removed by macerating strips of the trimmed, cleaned epididymis in a warm physiological solution. When a storage period greater than two hours is desired, milk or egg yolk diluent-extended samples can be cooled slowly to 7°C. The extended-and-cooled sperm cells can thereby remain capable of fertilization for 24 hours or more.

Uterine recovery

Baker (1962) obtained spermatozoa from the excised uterus of female mice killed immediately after mating. With this method, the content of the uterus is flushed into a warm depression slide with 0.24 to 0.33 ml of a 9.5% solution of reconstituted dried skim milk. The skim milk is prepared as previously described and should be at 37°C at the time of use. The content of the depression slide is gently mixed. The sample collected in this manner can be used immediately, or stored for 24 hours by slowly cooling it to 7°C. This method may also be used for other species and it may be used with diluents other than milk.

Electroejaculation

Disagreement exists among research workers regarding the best method for electroejaculation of rodents and therefore, several successful methods will be described. The guinea pig was the first species subjected to electroejaculation. Battelli (1922) caused guinea pigs to ejaculate by applying 30 volts of alternating current (47 cycles/minute) at the base of the brain. This was a severe shock, and as the ejaculate coagulated after such procedure less severe methods were sought.

Dalziel and Phillips (1948) developed less severe procedures for electroejaculation of guinea pigs. These procedures were modified slightly by Freund (1958), who used them to collect 110 ejaculates from 18 mature albino guinea pigs. Freund's primary purpose was to test procedures for the liquefaction of guinea pig semen. These procedures have since been used by Freund in several experiments (Freund, 1962; Freund and Borrelli, 1964, 1965).

The guinea pig is clipped in the lumbar region and electrode jelly is applied. It is strapped on its back to an animal board through which is inserted a round copper lumbar electrode (1.75 cm disk). An anal electrode (a smooth brass tube of 0.4 cm diameter and 8 cm length) is lubricated and inserted 4 to 6 cm into the rectum. The stimulus is produced by an oscillator and amplifier which deliver a square wave of 25 volts rms at 1,000 cps. The stimulus is applied intermittently by an interval timer with periods of three seconds on and 12 seconds off. Three to 14 shocks are required for ejaculation. The semen is collected in a 5 ml vessel which contains a mg of chymotrypsin in a ml of phosphate buffer. Freund (1958) found chymotrypsin the most effective liquefaction agent of several enzymes tried.

In many laboratories, it would be convenient to use one electroejaculator for several species. Scott and Dziuk (1959) ejaculated guinea pigs, mice and rats with the electroejaculator previously used for cattle and sheep (Dziuk et al., 1954a, 1954b). The following are their procedures.

The male is restrained on a board without anesthesia, the stimulus is applied to rats and guinea pigs with a bipolar rectal electrode which is 1.1 cm in diameter with electrodes 2.5 cm apart. The rectal probe is lubricated and inserted into the rectum until the tip electrode is near the fourth lumbar vertebra. Sodium chloride is added to the lubricant to increase conductivity. The voltage is gradually changed from zero to 2 or 3 volts and back to zero over a 4 to 5 second period, followed by a 5 to 10 second interval of no stimulus. This is repeated, increasing the voltage each time, until for the rat and mouse, a peak of 10 to 15 volts is reached at the sixth stimulus. Stimuli are continued at this level until semen is obtained or until about twenty-five stimuli have been applied. For the guinea pig 3 or 4 stimuli, reaching a 5 to 10 volt peak, are applied. The penis is then manually extended and light digital manipulation is applied to its base. Ejaculation is spontaneous and accompanied by body tremor and a clucking sound. Semen can be washed from the penis and collected into a buffer or diluent containing chymotrypsin as described by Freund (1958) or collected in a small amount of warm suspending media in a depression slide. Scott and Dziuk (1959) avoided coagulation of the ejaculate and obtained satisfactory ejaculates from guinea pigs by collecting it directly in a 3% solution of sodium citrate. However, coagulum formed in the urethra of rats and mice during electroejaculation, caused uremia and resulted in the death of many of the males. Uremia can be prevented if the coagulating gland (Lawson and Sorensen,

1964) or coagulating gland and seminal vesicles (Scott and Dziuk, 1959) are surgically removed one week before electroejaculation.

The presence or absence of the coagulum in the ejaculate may depend on the characteristics of the current applied to the animal. Snyder (1966) found that coagulum was not present after electroejaculation of the mouse with a stimulator set to deliver intermittent charges of 50 to 80 volts at a frequency of 80 pulses/second and a duration of four milliseconds; but coagulum was present when stimuli of 100 pulses per second and 80 volts were applied. Coagulum-free semen has been obtained from rats with intact coagulating glands by Birnbaum and Hall (1961), who used a Wave Forms Audio Oscillator Model 401A with a rectal probe similar to that described by Scott and Dziuk (1959). The coagulum-free ejaculates were achieved when the rats received stimuli of 30 cps and a maximum current of 2 volts. However, if the maximum voltage was less than one, ejaculation did not occur and, when it was greater than three, coagulum was present. The methods described by these authors require precision in instrumentation and technique, and allow very little chance for biological variation in response threshold between different males. The removal of coagulating glands before electroejaculation of rats and mice is therefore preferred.

The most extensive experiment involving electroejaculation of rats is that of Lawson *et al.* (1967). They modified the methods that Scott and Dziuk (1959) used for rats, mice, and guinea pigs, and successfully collected 295 ejaculates from 25 male rats. The following are their procedures.

Coagulating glands are surgically removed from male rats approximately 30 days before semen collection. The rats are restrained in a jacket developed by Lawson *et al.* (1966) and placed on a collection platform which contains the semen collection vessel. This method eliminates need for anesthesia during collection and leaves the animal much quieter than a restraining board. All hair is then clipped from the pubic area.

The electroejaculator produces a 60-cycle alternating current which can be varied from 0 to 25 volts. Stimuli are delivered through a small bipolar rectal probe, 0.3 cm in diameter and 9 cm long. The two electrodes of the probe are separated by 1.9 cm.

The anal canal is voided of fecal material and the lubricated probe is inserted into the rectum so that the terminals are in proximity to the animal's prostate and seminal vesicles. The first electrical stimulus is low to avoid frightening the animal. With the thumb and fingers at the base of the tail, the operator can feel the vibrations of the animal resulting from the stimuli. These reflex vibrations should not be of a violent nature. The probe is moved in order to obtain the desired reaction. Stimuli of less than 0.5 volt are applied and this voltage is gradually increased until ejaculation occurs. After the initial stimulation, the prepuce is manually depressed to extrude the penis. The stimuli are then produced in rhythmic waves, with repeated periods of stimulation and nonstimulation. If the animal responds too strongly to a stimulus, the electrode is reseated and stimulations continued. Some animals respond and ejaculate from as little as 0.5 volt, while others require as much as 12 volts. In most instances, 3 to 4 volts is sufficient. The animals become conditioned to the procedure after several

TABLE 2-1

The ejaculate

Species	Volume (ml)	Spermatozoa/ml (million)	Spermatozoa/ Ejaculate (billion)	Adapted from
Cattle	5–8	1000–1800	5–15	Foote and Trimberger (1968)
Sheep	0.7–2.0	2000–5000	2–5	Mann (1964)
Swine	150–500	25–300	6–75	Mann (1964)
Rabbit	0.4–6.0	50–350	0.05–0.35	Mann (1964)
Dog	2–15	60–300	0.5–2.7	Mann (1964)
Monkey		90–800		Mastroianni and Manson (1963)
Guinea pig	0.4–0.8	5–17	0.003–0.010	Mann (1964)
	0.5		0.007	Freund and Borrelli (1964)
			0.068	Scott and Dziuk (1959)
Rat	0.023	340	0.008	Lawson *et al.* (1967)
			0.063	Scott and Dziuk (1959)
Mouse	0.003	593	0.002	Snyder (1966)

collections, and the clipping and handling of the animal seems to condition it for more satisfactory collection.

The ejaculate consists of one or two small drops of semen. This can be washed from the glans penis with one of the appropriate diluents shown in Tables 2-1 and 2-2. Approximately 3 ml of diluent are used in the flushing process. Although the coagulating gland is removed, a small soft plug sometimes forms and should be removed by stripping the urethra shortly after stimulation is begun. Low voltages result in fewer and smaller plugs of coagulum, higher voltages cause larger and firmer plugs. Three to four minutes are required to restrain and ejaculate each male.

Since these procedures and the electroejaculator model are similar to those used by Scott and Dziuk (1959) they should be applicable to mice and guinea pigs as well as rats. A 1.1 cm bipolar electrode should be used with the guinea pig. Similar electroejaculation methods have been applied to the chinchilla (Dalziel and Phillips, 1948; Hillemann, *et al.*, 1963 and Healey and Weir, 1967).

Dog

Semen can be obtained from dogs by the use of electroejaculation, an artificial vagina, or digital manipulation of the penis. When these methods were tried and compared, digital manipulation was preferred (Baucher *et al.*, 1958; Macpherson and Penner, 1967).

TABLE 2-2

Composition of buffers and diluents for semen storage at 5° to 7°C

Ingredient	Cattle	Sheep	Swine	Primates
Buffers				
Sodium citrate dihydrate (g) $(Na_3C_6H_5O_7 \cdot 2H_2O)$	14.5	28.0		
Sodium bicarbonate (g)	2.1		2.1	
Potassium chloride	0.4			
Sodium L–a–aminoglutarate (g)				30.0
Citric acid (g)	1.0			
Glucose (g)	3.0	8.0[1]	42.9	
Skim milk powder (g)				
Glycine (g)	9.4			
Sulfanilamide (g)	3.0			
Distilled water (ml Pinal/vol.)	1000	1000	1000	1000
Diluents				
Buffer (% by vol.)	80	80	70	78
Egg yolk (% by vol.)	20	20	30	20
Dihydrostreptomycin ($\mu g/ml$)	1000	1000	1000	1000
Diluents for freezing—replace part of buffer with glycerol (% by vol.)	7	7		14

SOURCE: For cattle: Foote *et al.* (1960); for sheep: Lunco *et al.* (1961); for swine: Dziuk (1958); and for primates: Roussel and Austin (1967).
[1] 8 grams of fructose or 5 grams of arabinose can be substituted for glucose.

Semen can be obtained by this method without the presence of a female teaser. However, Baucher *et al.* (1958) found considerably more spermatozoa in the ejaculate when dogs were teased by an estrual bitch before and during semen collection. With this method, the penis of the dog is manipulated through the sheath, using a brisk forward and backward motion until erection is achieved, after which digital pressure is maintained posterior to the bulbus glandis. Ejaculation is induced by additional light manipulation of the penis or by touching the glans while the bulb is firmly locked. Semen can be collected in a petri dish containing warm diluent (Macpherson and Penner, 1967) or in a funnel attached to a 15 ml centrifuge tube (Baucher *et al.*, 1958). The collection tube should be warmed and protected as described for the bull and ram to avoid temperature shock.

The ejaculate of the dog consists of three fractions. The first fraction, which is obtained in 30 to 50 seconds, is small in volume and contains very few spermatozoa. The second fraction, obtained in 50 to 80 seconds, is also small in volume but has a dense milky appearance and contains most of the spermatozoa. The third fraction, collected during a period of 3 to 30 minutes, is a clear fluid of large volume (Harrop, 1962). The sperm rich fraction is saved. For a more extensive discussion of collection and preservation of dog semen see Harrop (1962), Baucher *et al.* (1958), and Foote (1964a,b).

EXOTIC SPECIES

Electroejaculation has been attempted in species other than those already discussed and has been particularly useful with exotic and wild species. Healey and Sadleir (1966) have described the construction of a variety of rectal electrodes, which could be used with animals of differing sizes.

THE EJACULATE

The volume of the ejaculate, its concentration and the total number of spermatozoa are extremely variable within species. These characteristics depend in part on: (1) the method of semen collection, (2) the level of sexual excitement before ejaculation, (3) the breed or strain involved, (4) the age of the male and (5) the frequency of ejaculation. The frequency of ejaculation is not always known to the research worker. Males should be confined in individual cages to prevent pederasty. Even then males of some species—cattle, swine, and monkeys—may masturbate, and individually caged rats have been observed to ejaculate spontaneously in their cages almost daily (Orbach *et al.*, 1967). Additionally, when part of the ejaculate is retained within the artificial vagina or when spermatozoa are trapped in the coagulum, the result may be a reduced number of spermatozoa. For these reasons, there is great variation in the average sperm count reported in the literature. Table 2-1 contains approximate values for each species.

Sometimes semen samples that are more highly concentrated than the normal ejaculate are desired. These can be obtained by collecting only the sperm rich fraction in species such as the dog and swine, where a fractionated ejaculate occurs, by collecting epididymal spermatozoa, or, in swine, by blocking the secretions of the urethral and bulbo-urethral glands with atropine (Dziuk and Mann, 1963).

PRESERVATION OF SPERMATOZOA

Spermatozoa can be preserved for one to two hours in several physiological buffers such as Lockes or Ringers solution provided they are maintained near body temperature to avoid temperature shock. Many species can be stored for one to four days at 5 to 15°C, and others may be stored for several months at −80 to −195°C. Cold shock, which results from rapid cooling, causes irreversible damage to spermatozoa and must be avoided with all methods of preservation (Chang and Walton, 1940; Mann and Lutwak-Mann, 1955; Mann, 1964). Considerable protection against cold shock is provided by the inclusion of lipoproteins containing lecithin in the diluent media (Mayer and Lasley, 1945; Kampschmidt *et al.*, 1953; Blackshaw and Salisbury, 1957). Egg yolk is commonly used to provide the needed protection (Phillips and Lardy, 1940).

TABLE 2-3

Composition of buffers and diluents for semen storage at room temperature or above

| | Amount Contained in 100 ml of Diluent | | |
| | Norman (1964) | | Mann (1964) |
Ingredient	NJ–1[1]	NJ–2[1]	Modified K–H–R– Phosphate[2]
Glucose	500.0 mg	500.0 mg	
Calcium nitrate, tetrahydrate	482.0 mg	—	
Calcium chloride, anhydrous	—	226.0 mg	
Magnesium sulfate, heptahydrate	45.0 mg	—	30.0 mg
Magnesium chloride, hexahydrate	—	37.0 mg	
Sodium citrate, diydrate	2.2 gm	2.2 gm	
Sodium sulfate	34.0 mg	—	
Sodium phosphate, heptahydrate			550.0 mg
Sodium chloride	—	14.0 mg	70.0 mg
Sodium bicarbonate			20.0 mg
Potassium nitrate	14.0 mg	—	
Potassium chloride	11.0 mg	21.0 mg	36.0 mg
Potassium phosphate			16.0 mg
Monosodium phosphate, monohydrate	3.0 mg	3.0 mg	
Sulfanilamide	300.0 mg	300.0 mg	
Penicillin-G-sodium	60.0 mg	60.0 mg	
Dihydrostreptomycin sulfate	135.0 mg	135.0 mg	
Polymixin-B-sulfate	10.0 mg	10.0 mg	
Catalase (sterile aqueous solution)	15,000 units	15,000 units	
Mycostatin	1000 units	1000 units	
Egg yolk	5.0 ml	5.0 ml	

[1] pH of diluent adjusted to 7.4 with 10% solution of sodium hydroxide.
[2] Adjust pH of diluent to 7.4 with 1 N HCl before making to volume.

To reduce the risk of cold shock, diluents containing lipoproteins are preferred to physiological buffers for all species mentioned except the rabbit. With the addition of glycerol after the diluted sample has cooled, many of these same diluents, or a milk diluent, can be used as media in which spermatozoa are frozen. Several diluents, their composition, and the species in which they can be used are shown in Tables 2-2 and 2-3.

The NJ1 diluent has been successfully used for room temperature storage of the semen of cattle and primates, and the NJ2 has been used with rabbits, swine, cattle and primates. NJ diluents can also be used for storage at 5° to 10°C if they contain egg yolk (Norman, 1964).

Milk has been used as a diluent for the semen of cattle, sheep, mice, and monkeys, and is satisfactory for one day storage of swine semen, but it is not a good diluent for rabbit semen. Either homogenized milk, pasteurized skim milk without added solids, or reconstituted skim milk (9.5% solids) may be used. It should be heated to 92°C over boiling water and held at 92° to 95°C for 10 minutes, slowly cooled to room temperature, filtered and antibiotics added (Almquist *et al.,* 1954).

For extensive and specialized information on spermatozoa preservation, the reader is referred to the works of Salisbury and Van Demark (1961), Maule (1962), Parkes (1960), Mann (1964), and Aamdal (1964).

Biologists will not be surprised that the collection and preservation of semen is as much an art as a science. Spermatozoa are fragile creatures produced by temperamental beasts. Their collection requires great patience and a thorough understanding of the behavior patterns of the animal being studied. Their preservation in the laboratory, as in nature, is achieved only by tender and loving care.

BIBLIOGRAPHY

Aamdal, J. (1964) Artificial insemination in the pig. Proc. 5th Int. Congr. Animal Reprod. (Trento), 4:147.

———— and I. Hogset (1957) Artificial insemination in swine. J.A.V.M.A. 131:59.

————, I. Hogset, O. Sveberg, and N. Koppang (1958) A new type of artificial vagina and a new collection technique for boar semen. J.A.V.M.A., 132:101.

Almquist, J. O., R. J. Flipse, and D. L. Thacker (1954) Diluters for bovine semen. IV. Fertility of bovine spermatozoa in heated homogenized milk and skim milk. J. Dairy Sci., 37:1303.

Amann, R. P. (1966) Effect of ejaculation frequency and breed on semen characteristics and sperm output of rabbits. J. Reprod. Fertility, 11:291.

———— and J. R. Lambiase, Jr. (1967) The male rabbit. I. Changes in semen characteristics and sperm output between puberty and one year of age. J. Reprod. Fertility, 14:329.

Anand, A. S., W. G. Hoekstra, and N. L. First (1967) Effect of aging of boar spermatozoa on cellular loss of DNA. J. Animal Sci., 26:171.

Austin, J. W., E. W. Hupp, and R. L. Murphree (1962) Comparison of quality of bull semen collected in the artificial vagina and by electro-ejaculation. J. Dairy Sci., 44:2292.

Baker, R. D. (1962) Effect of sperm numbers on fertility in two inbred strains of mice and their F_1 hybrid. Progress report, Jackson Memorial Laboratory Bar Harbor, Maine.

Battelli, F. (1922) A method of obtaining complete emission of the liquid of the seminal vesicles of the guinea pig. (Trans. Title) Comp. Rend. Soc. Phys. d' Hist. Nat. (Geneva)., 39:73.

Bennett, J. P. (1967) Semen collection in the squirrel monkey. J. Reprod. Fertility, 13:353.

Birnbaum, D., and T. Hall (1961) An ejaculation technique for rats. Anat. Record, 140:49.

Blackshaw, A. W., and G. W. Salisbury (1957) Factors influencing metabolic activity of bull spermatozoa. II. Cold-shock and its prevention. J. Dairy Sci., 40:1099.

Borton, A., A. Jaworski, and J. E. Nellor (1965) Factors influencing the fertility of naturally and artificially mated swine. Mich. State Univ. Agr. Exp. Sta. Res. Bull., 8.

Baucher, J. H., R. H. Foote, and R. W. Kirk (1958) The evaluation of semen quality in the dog and the effects of frequency of ejaculation upon semen quality, libido and depletion of sperm reserves. Cornell Vet., 48:67.

Bredderman, P. J., R. H. Foote, and A. M. Yassen (1964) An improved artificial vagina for collecting rabbit semen. J. Reprod. Fertility, 7:401.

Chang, M. C., and H. Walton (1940) The effects of low temperature and acclimatization on the respiratory activity and survival of ram spermatozoa. Proc. roy. Soc. B., 129:517.

Dalziel, C. F., and C. L. Phillips (1948) Electric ejaculation, determination of optimum electric shock to produce ejaculation in chinchillas and guinea pigs. Am. J. Vet. Res., 9:225.

Desjardins, C., K. T. Kirton, and H. D. Hafs (1965) Rabbit sperm output after varying ejaculation frequencies. J. Animal Sci., 24:916 (abstr.) .

Dziuk, P. J. (1958) Dilution and storage of boar semen. J. Animal Sci., 17:548.

———, E. F. Graham, and W. E. Petersen (1954a) The technique of electro-ejaculation and its use in dairy bulls. J. Dairy Sci., 37:1035.

———, E. F. Graham, J. D. Donker, G. B. Marion, and W. E. Petersen (1954b) Some observations in collection of semen from bulls, goats, boars and rams by electrical stimulation. Vet Med., 69:455.

——— and T. Mann (1963) Effect of atropine on the composition of semen and secretory function of male accessory organs in the boar. J. Reprod. Fertility, 5:101.

——— and M. N. Runner (1960) Recovery of blastocysts and induction of implantation following artificial insemination of immature mice. J. Reprod. Fertility, 1:321.

Emmens, C. W., and T. J. Robinson (1962) The sheep. In: The Semen of Animals and Artificial Insemination. Ed. by J. P. Maule, Commonwealth Agricultural Bureau, Edinburgh.

Foote, R. H. (1964a) The effects of electrolytes, sugars, glycerol and catalase on survival of dog sperm stored in buffered-yolk mediums. Am. J. Vet. Res., 25:32.

——— (1964b) Extenders for freezing dog semen. Am. J. Vet. Res., 25:37.

———, L. C. Gray, and D. C. Young (1960) Fertility of bull semen stored up to four days at 5°C. in 20% egg yolk extenders. J. Dairy Sci., 43:1330.

——— and G. W. Trimberger (1968) Artificial Insemination. In: Reproduction in Farm Animals. Ed. by E. S. E. Hafez. Lea and Febiger, Philadelphia, 442p.

Freund, M. (1958) Collection and liquefaction of guinea pig semen. Proc. Soc. Exp. Biol. (New York) , 98:538.

——— (1962) Initiation and development of semen production in the guinea pig. Fertility Sterility, 13:190.

——— and F. J. Borrelli (1964) Semen characteristics in the guinea pig. Proc. 5th Inter. Cong. Animal Reprod. Artif. Insem. (Trento) , 3:452.

——— and F. J. Borrelli (1965) The effects of x-irradiation on male fertility in the guinea pig semen production after x-irradiation of the testis, of the body, or of the head. Radiation Res., 24:67.

Gunn, R. M. C. (1936) Fertility in sheep. Artificial production of seminal ejaculation and the characters of the spermatozoa contained therein. Bull. Coun. Sci. Industr. Res. (Australia) , 94:116.

Hale, E. B., and J. O. Amquist (1960) Relation of sexual behavior to germ cell output in farm animals. J. Daily Sci., 43:145 (Suppl.) .

Harrop, A. E. (1962) Artificial insemination in the dog. In: The Semen of Animals and Artificial Insemination. Ed. by J. P. Maule. Commonwealth Agricultural Bureau, Farnham Royal Bucks, England. 420p.

Healey, P., and R. M. F. S. Sadleir (1966) The construction of rectal electrodes for electro-ejaculation. J. Reprod. Fertility, 11:299.

Healey, P., and B. J. Weir (1967) A technique for electro-ejaculation in chinchillas. J. Reprod. Fertility, 13:585.

Herrick, J. B., and H. L. Self (1962) Evaluation of fertility in the bull and boar. Iowa State University Press, Ames, Iowa, 148p.

Hillemann, H. H., A. I. Gaynor, and A. Dorsch (1963) Artificial insemination in chinchillas. J. Small Anim. Pract., 3:77.

Hoskins, D. D., and D. L. Patterson (1967) Prevention of coagulum formation with recovery of motile spermatozoa from Rhesus monkey semen. J. Reprod. Fertility, 13:337.

Hulet, C. V., W. C. Foote, and R. L. Blackwell (1964) Effects of natural and electrical ejaculation on predicting fertility in the ram. J. Animal Sci., 23:418.

Ito, S., T. Niwa, and A. Kudo (1948) Studies on the artificial insemination in swine. I. On the method of collection of semen and the condition of ejaculation. Zootech. Exp. Sta. Res. Bull. no. 55. (Chiba).

Kampschmidt, R. F., D. T. Mayer, and H. A. Herman (1953) Lipid and lipoprotein constituents of egg yolk in the resistance and storage of bull spermatozoa. J. Dairy Sci., 36:733.

Kirton, K. T. (1967) Personal communication.

Lawson, R. L., and A. M. Sorensen, Jr. (1964) Ablation of the coagulating gland and subsequent breeding in the albino rat. J. Reprod. Fertility, 8:415.

———, S. Barranco, and A. M. Sorensen, Jr. (1966) A device to restrain the mouse, rat, hamster, and chinchilla to facilitate semen collection and other reproductive studies. Lab. Animal Care., 16:72.

———, G. M. Krise, and A. M. Sorensen, Jr. (1967) Electro-ejaculation and evaluation of semen from the albino rat. J. Appl. Physiol., 22:174.

Lunca, M., V. Otel, M. Paraschivescu, and O. Seserman (1961) Progresses realized by artificial insemination in sheep with diluted semen. Proc. 4th Int. Congr. Anim. Reprod. (The Hague)., 4:876.

Lutwak-Mann, C., and L. E. A. Rowson (1953) The chemical composition of the presperm fraction of bull ejaculate obtained by electrical stimulation. J. Agric. Sci., 43:135.

Macmillan K. L., and H. D. Hafs (1967) Semen output of rabbits ejaculated after varying sexual preparation. Proc. Soc. Exp. Biol. and Med., 125:1278.

Macpherson, J. W., and P. Penner (1967) Canine reproduction. I. Reaction of animals to collection of semen and insemination procedures. Canad. Journ. Comp. Med. and Vet. Sci., 31:62.

Mann, T. (1964) The Biochemistry of Semen and of the Male Reproductive Tract. John Wiley, New York. 493p.

——— and C. Lutwak-Mann (1955) Biochemical changes underlying the phenomenon of cold shock in spermatozoa. Arc. Sci. Biol., 39:578.

Manton, V. J. A. (1956) The fertility of bovine semen collected by electro-ejaculation. Vet. Rec., 68:1015.

Marden, W. G. R. (1954) New advances in the electro-ejaculation of the bull. J. Dairy Sci., 37:556.

Mastroianni, L., Jr., and W. A. Manson, Jr. (1963) Collection of monkey semen by electro-ejaculation. Proc. Soc. Exp. Biol. Med., 112:1025.

Maule, J. P. (1962) The semen of animals and artificial insemination. Commonwealth Agricultural Bureau, Farnham Royal, Buck., England. 420p.

Mayer, D. T., and J. F. Lasley (1945) The factor in egg yolk affecting the resistance, storage potentialities, and fertilizing capacity of mammalian spermatozoa. J. Animal Sci., 4:261.

McKenzie, F. F., J. C. Miller, and L. C. Bauguess (1938) The reproductive organs and semen of the boar. Mo. Agr. Exp. Sta. Res. Bull. no. 279.

Melrose, D. R. (1962) Artificial insemination in cattle. In: Semen of Animals and Artificial Insemination. Ed. by J. P. Maule. Commonwealth Agricultural Bureau, Edinburgh.

Miller, S. T. (1961) Ram management for artificial insemination. In: Artificial Breeding of Sheep in Australia. Ed. by E. M. Roberts, New South Wales University Press, Sydney, Australia, 223p.

Norman, C. (1964) Further studies on the preservation of mammalian sperm at variable temperatures. Proc. 5th Int. Cong. Animal Reprod. and Art. Insem. (Trento), 4:269.

Orboch, J., M. Miller, A. Billimoria, and N. Salkkhan (1967) Spontaneous seminal ejaculation and genital grooming in rats. Brain Res., 5:550.

Parkes, A. S. (1960) Marshal's Physiology of Reproduction. vol. I: part II. Longmans, Green and Co. London. 877p.

Phillips, P. H., and H. A. Lardy (1940) A yolk-buffer pabulum for the preservation of bull semen. J. Dairy Sci., 36:173.

Polge, C. (1956) Artificial insemination in pigs. Vet. Record, 68:62.

Rousell, J. D., and C. R. Austin (1967) Preservation of primate spermatozoa by freezing. J. Reprod. Fertility, 13:333.

Rowson, L. E. A. (1962) Artificial insemination in the pig. In: The Semen of Animals and Artificial Insemination. Ed. by J. P. Maule. Commonwealth Agricultural Bureau, Farnham Royal Bucks., England. 420p.

———— and M. I. Murdoch (1954) Electrical ejaculation in the bull. Vet. Record, 66:326.

Salamon, S., and A. J. Morrant (1963) A comparison of two methods of artificial breeding in sheep. Australian J. Exp. Agr. Animal Husbandry, 3:72.

Salamon, S., and D. R. Lindsay (1961) The training of rams for the artificial vagina with some observation on ram behavior. In: Artificial Breeding of Sheep in Australia. Ed. by E. M. Roberts. New South Wales University Press, Sydney, Australia. 223p.

Salisbury, G. W., and N. L. VanDemark (1961) Physiology of Reproduction and Artificial Insemination of Cattle. W. H. Freeman and Company, San Francisco. 639p.

Scott, J. V., and P. J. Dziuk (1959) Evaluation of the electro-ejaculation techniques and spermatozoa thus obtained from rats, mice and guinea pigs. Anat. Record, 133:655.

Self, H. L. (1959) The problem of storing and inseminating boar semen. Ann. De Zootech., Supp. 121.

Snyder, R. L. (1966) Collection of mouse semen by electro-ejaculation. Anat. Record, 155:11.

Walton, A. (1958) Improvement in the design of an artificial vagina for the rabbit. J. Physiol. 143:26.

Weisbroth, S., and F. A. Young (1965) The collection of primate semen by electro-ejaculation. Fertility Sterility, 16:229.

Wells, M. E., W. N. Philpot, S. D. Musgrave, E. W. Jones, and W. E. Brock (1966) Effect of method of semen collection and tranquilization on semen quality and bull behavior. J. Dairy Sci., 49:500.

White, I. G. (1955) The collection of rabbit semen. Australian J. Exp. Biol. Med. Sci. 33:367.

3

Techniques and Criteria Used in the Study of Fertilization

J. M. BEDFORD
Department of Anatomy and
The International Institute for
the Study of Human Reproduction,
Columbia University,
College of Physicians and Surgeons,
New York, New York 10032

The freshly ovulated ovum in most mammals is stimulated to develop beyond the metaphase of the second meiotic division by entry of the fertilizing sperm into the vitellus. Immediately after its entry, the posterior region of the sperm head begins to change form and undergoes nuclear swelling, presumably stimulated by substances in the ooplasm. After decondensation the sperm nucleus then transforms to become the definitive male pronucleus. Coincident with the onset of changes in the fertilizing sperm head, the second maturation division is completed and a second polar body is usually extruded from the vitelline surface into the perivitelline space. Chromosomes that are retained within the ovum undergo a period of transformation to become the female pronucleus. The male and the female pronucleus move into apposition in the center of the egg and, after several hours, the nuclear membranes undergo partial fusion allowing combination of their genetic material: a process termed *syngamy*. This development marks the completion of fertilization.

Heteroploidy, parthenogenetic activation, and *in vitro* fertilization have been produced by experimental manipulation of the mammalian egg. In recent years fertilization has been used as the primary indicator or means of assay in studies of sperm capacitation. Although early embryologists were able to analyze correctly the ordered sequence and significance of the events

Many of the methods and techniques mentioned here have been developed or improved in the laboratory of M. C. Chang, to whom the author is most grateful. Thanks are extended to Dr. R. W. McGaughey for the photographs in Figures 3-4 and 3-7, 8, 9, 10, as well as for details of the technique used for study of egg chromosomes, and to Dr. R. H. F. Hunter for the photograph in Figure 3-1, and for comments on the ovum of the pig. Partial support for this chapter has been available from Grants 2-T01-GM-00256 7-R01-CA-10749 and HD-03623 of the National Institutes of Health, United States Public Health Service.

which occur during fertilization, a survey of subsequent literature suggests that in some, more recent, experimental studies the state of the ovum may well have been misinterpreted: a fragment of the first polar body being considered as the second polar body, or degenerative fragmentation in an unfertilized egg being mistaken for the normal cleavage which follows fertilization. It should be made clear that accurate assessment of the state of certain ova, even by the experienced investigator, may still prove difficult and is sometimes not feasible. In the last two decades, however, techniques for handling, fixing, and staining of whole ova have been improved. Furthermore, the accumulation of knowledge of ovum behavior in a number of species, together with the introduction and improvement of the phase-contrast microscope, has allowed precise assessment of the state of most ova in either normal or experimental conditions. Recently, important information about the events involved in fertilization has been gained with the electron microscope, but this knowledge is still incomplete.

The various aspects of fertilization have been discussed in detail from a theoretical standpoint by Rothschild (1956), Austin and Walton (1960), Hancock (1962) Austin (1961a, 1965), Monroy (1966), and Metz and Monroy (1967, 1969). This chapter seeks to outline in detail the common procedures and criteria used in the handling and assessment of mammalian ova. Major emphasis is given to events which occur in the single-celled ovum prior to the first cleavage division, since it is at this stage that the various anomalies of fertilization can most easily be recognized. Many of the data or idiosyncrasies mentioned here relate to the rabbit and other common laboratory animals, and to a lesser degree to domestic species.

RECOVERY OF TUBAL OVA

The recovery of tubal ova has been discussed by Austin (1961a) and by Blandau (1961). Ova of the rabbit, ferret, domestic animals, and man, are recovered by flushing the fallopian tube from its uterine end. The tube is first removed from the body cavity with about one inch or less of uterine horn; the latter acts as a guiding funnel during insertion of the pipette through the uterotubal junction. The tip of the pipette should be fire-polished and about 0.5 mm in diameter. Before inserting the pipette, the fallopian tube must be straightened by dissection, and the mesosalpinx and fat close to the uterotubal junction should be removed because corrugation of the muscular isthmus impedes entry of the pipette. When, on occasion, the uterotubal junction offers too great a resistance, or when the tube has been cut at the junction leaving no guide funnel, the pipette can be inserted into the ampulla via the fimbria, and the tube is flushed in the direction of the uterus.

The tube is flushed best vertically with 1 to 4 ml of fluid medium (depending on the species) which is collected in a flat-bottomed watch glass. The ova collected are then identified with a stereo dissecting microscope. Gentle vibration of the watch glass will tend to move the eggs and tubal fragments to the center of the glass. The number of rabbit ova found in the flushings 11 to 12½ hours after an ovulatory stimulus is sometimes less than would be expected judging from the number of ruptured follicles. Missing ova may sometimes be found with cumulus cells adhering to the surface of the ruptured follicle; for this reason, the ovaries should be placed in a fluid medium immediately after their separation from the fimbria.

In the case of rodents, ovum recovery must be carried out entirely with the aid of the stereo dissecting microscope. The ova remain in the cumulus oophorus for several hours, and immediately after ovulation they may still be found in the periovarian sac: in this case the ova can be recovered simply by rupturing the sac in a few drops of medium. After passing into the ampulla of the oviduct, the egg mass can be seen for two or three hours through the distended oviducal wall, and will pass out into the surrounding medium if the tube is incised with a thin Graafe knife, scalpel, or a dissecting needle, over or near the point of distension. At a later stage, when the eggs have become dispersed along the tube, their collection may be accomplished by incising the tube at various points and squeezing out the tubal contents with needles, or by flushing the tube. This requires dissection, straightening of the initial segments of the tube, and the use of pipettes of very small dimensions.

REMOVAL OF FOLLICLE CELLS

The presence of cumulus or corona cells around the ovum tends to obscure features which ordinarily become apparent after fixing and staining of the egg. These cells should be removed in cases where details such as the swelling sperm head, the fertilizing sperm tail, or the distinctive chromatin of the polar bodies are to be identified. The granulosa cells which surround the follicular oocyte are not always removed by enzymes, and must sometimes be dislodged with dissecting needles. Both the cumulus oophorus and corona radiata are easily removed from rodent, pig, cow, and sheep eggs by hyaluronidase prepared from bovine or ovine testis (0.1% solution of the preparation containing 500 units/mg). Very high concentrations of this preparation will also tend to remove the zona pellucida, presumably through the action of proteolytic contaminants since chromatographically pure hyaluronidase solutions do not affect the zona. Hyaluronidase will bring about dispersion of the cumulus cells but not of the corona radiata of the rabbit ovum, which can be removed either by shaking the ova vigorously for 30 to 60 seconds in a small siliconed tube (with a rubber cap treated with silicone to prevent sticking), or by vigorously propelling the ova in-and-out of a fine pipette. Prior treatment of the ova with hyaluronidase facilitates removal of the surrounding corona cells. Recently, it has been shown that the corona radiata may be removed in two hours from hyaluronidase-treated ova, without mechanical agitation, by exposure to bicarbonate concentrations of 66 m-eq/l, at 37.5°C (Stambaugh, Noriega and Mastroianni, 1969). Nothing is known of the viability of ova after such treatment. Natural or spontaneous loss of follicle cells from ova is hastened by fertilization (Dickmann, 1963). There is considerable variation between species in the time at which the follicle cells usually disappear—cow and sheep ova apparently lose these cells very soon after ovulation; in the unfertilized rabbit ovum dispersion is not complete until 6 to 10 hours after ovulation, in the hamster until 12 to 18 hours (Yanagimachi and Chang, 1961), and in the ferret as long as 78 hours after ovulation (Chang and Yanagimachi, 1963).

HANDLING AND MOUNTING OVA

Before mounting, ova are collected with dissecting needles into a compact group. Glass mounting slides are prepared by washing in alcohol, followed by application of four wax spots (approximately nine parts of vaseline to one part of paraffin wax) in a square, which act to support the alcohol-cleaned cover glass at its four corners. Ova are transferred in a drop of fluid with a pipette whose internal diameter is two or three times that of the ova. After transfer to the slide, the ova are again grouped and a cover glass is placed upon the wax spots; the glass must not touch the droplet containing the eggs until it is pressed down onto the wax at each corner. Initial contact between the fluid and the cover glass should be controlled under a wide-field dissecting microscope, as the fluid tends to be drawn by surface tension to the edge of the slide. Pressure is then applied to the cover glass to anchor the eggs on the slide. Because of the danger of egg rupture at this stage the anchoring procedures should be carried out carefully at higher magnification after transfer of the slide to a phase microscope.

EXAMINATION OF LIVING OVA

Mounted ova may be rolled by a carefully controlled horizontal movement of the cover glass; this becomes impossible however should the egg stick to either glass surface. By use of the rolling technique, the polar bodies (see Fig. 3-1), and perivitelline sperm (see Fig. 3-2) can be brought into view; better assessment may also be made of the position of a particular sperm, i.e. whether it is lying on the surface of, or within, the zona pellucida. By rolling the egg, sperm in the perivitelline space or the zona pellucida can be brought into sharp focus with the vitelline circumference, and yet appear to be in the zona, rather than on the surface.

Living pronuclei, especially in rodent ova, are best seen using positive phase optics. Larger eggs, from rabbits and domestic animals, usually require a flattening of the ovum (by pressure on the cover slip) to obtain a clear image of the living pronucleus (see Fig. 3-3). Even this procedure, however, does not always render a clear image of the pronucleus—particularly in horse and pig ova, which have a dense lipid-filled ooplasm.

FIXATION OF OVA FOR
PHASE CONTRAST MICROSCOPY

Commonly used fixatives consist of various proportions of glacial acetic acid and 95% or absolute alcohol; a mixture of one part acetic acid to

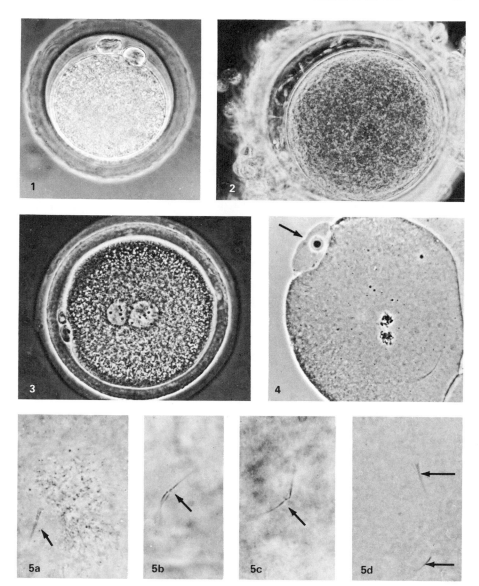

FIGURES 3-1—3-5

1. *Phase-contrast photomicrograph of a living fertilized ovum of guinea pig. Note the first and second polar bodies, and the distinct perivitelline space which becomes apparent after contraction of the vitellus at fertilization.* **2.** *Fertilized rabbit ovum recovered six hours after fertilization; many sperm are in the perivitelline space.* **3.** *Living fertilized rabbit ovum recovered about eight hours after ovulation. This shows the male and female pronucleus, and the first and second polar bodies in the perivitelline space.* **4.** *Fertilized ovum of the mouse after fixation and staining shows the fertilizing sperm tail, the central chromosomes of the first mitotic division, and the dense chromatin normally seen in the second polar body (arrowed). When present, the first polar body displays the chromatin in a more dispersed form.* **5.** (a–d). *Examples of the fertilizing sperm tail in rabbit ova, stained with lacmoid. In* (a) *the filaments of the tail have not split, and this lies in apposition to the pronuclei (seen here as a collection of stained granules). In* (b) *and* (c) *the main filaments have split and are joined only in the region of the centriole (arrowed). In* (d) *is shown an example of monospermic fertilization in which the single fertilizing tail has become separated into two filaments—this contrasts with Fig. 3-14, which shows two sperm tails within a polyspermic egg.*

three parts alcohol, has been found to be suitable. Fixation for 15 to 60 minutes is adequate for rabbits and rodents, but pig ova require periods of up to 24 hours to obtain adequate clearing. In such fixatives the contour of the zona pellucida is not preserved, although main features such as the metaphase spindle or pronuclei become clearly visible. The shape of the zona pellucida is better maintained if the ova are fixed for several hours in 10% neutral formalin, then immersed in 95% alcohol for 6 to 12 hours in the case of larger ova (rabbit, ferret, man, etc.), and for 15 to 60 minutes with the smaller ova of laboratory rodents.

The slides upon which the eggs are mounted can be stacked vertically in Coplin jars during fixation, but this method risks losing the cover glass. If rare or unusual eggs are involved, fixation can be achieved on the microscope stage by allowing the fixing fluid, which evaporates readily, to run under the cover glass.

STAINING OF OVA

Staining of fixed ova brings out clearly the meiotic chromosomes as well as those which form in the center of the egg prior to the first cleavage. After staining, the polar bodies can be characterized more definitely (see below), and the fertilizing sperm tail (see Figs. 3-4 and 3-5) and any sperm lying about the ovum (see Fig. 3-6) become visible. The stains commonly used include lacmoid, made up as a 0.25 to 1% solution in 45% acetic acid (weaker solution for rodent ova); 0.5% carmine or 1% orcein in 45% acetic acid have been used successfully for the study of pig and sheep ova—the latter stains tend to leach out more easily than does lacmoid, and become useful where overstaining may occur. One can observe the staining process under the phase microscope by allowing the stain to be drawn beneath the cover slip as the fixative evaporates—lacmoid is quickly taken up by the chromatin of the polar bodies, by chromosomes in the egg, and by sperm on the surface of the zona or in the perivitelline space. Staining for half an hour is recommended if permanent preparations are to be made; or, if the tail of the fertilizing sperm is to be identified. Excess stain is removed from eggs by washing with absolute alcohol for about thirty seconds, and then with 45% acetic acid for up to five minutes. Semipermanent preparations can be made by sealing the edges of the cover glass with nail varnish.

EMBEDDING AND SECTIONING OF OVA

Light Microscope

Present standards of phase microscopy, with available techniques for fixing and staining of ova, now make it generally unnecessary to section ova for routine light microscope examination: the state of the egg and relationships within the egg can usually be determined by rolling the whole egg under the phase microscope. If sections of ova are required, fixation is

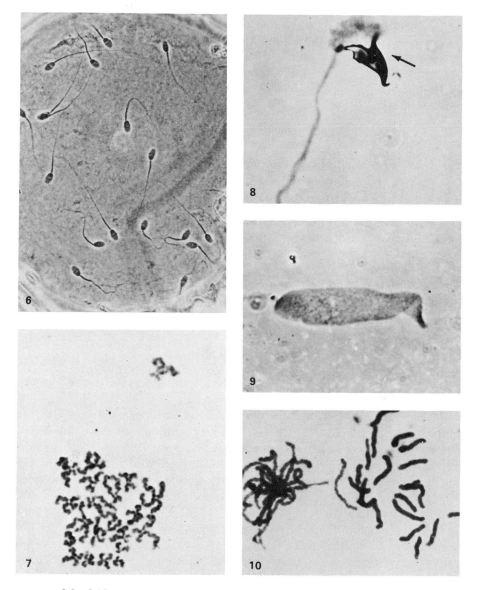

FIGURES 3-6—3-10

6. *Sperm on the surface of a fertilized rabbit ovum, stained with lacmoid.* **7.** *Chromosomes of the second maturation spindle of a rabbit ovum, prepared according to the smear technique described in the text.* **8.** *Fertilizing mouse sperm about two hours after entry into the egg. Note the persistence of the dense anterior part of the sperm head, and the smoky disintegration of the posterior region. Part of the dense material is contributed by the perforatorium.* **9.** *Swelling sperm nucleus in a hamster ovum, stained with lacmoid.* **10.** *Condensation of male and female chromosomes in late pronuclei of a fertilized mouse ovum, prepared by smear technique.*

carried out in Helly or Bouin for 5 or 6 hours. At this stage, light staining with dilute Delafield's haematoxylin in water allows the pronuclei to be seen; they can then be used for orientation before the ova are dehydrated and embedded in paraffin (Amoroso and Parkes, 1947). In the rabbit it is recommended that the mucoprotein coat around the ovum be removed with fine needles after fixation and before embedding the ovum in paraffin—this facilitates sectioning, which can be performed without difficulty at 5 to 10 microns. Sectioning of ova may also be carried out easily at 5 microns or less after fixation and embedding for electron microscopy in epoxy resins, such as Epon *812.*

ELECTRON MICROSCOPE

The study of the ultrastructure of fertilization and its early development is an important area of investigation, and requires excellent fixing and sectioning techniques. Several studies have been published of the events at fertilization as seen in the electron microscope (Hadek, 1963; Austin, 1968: Bedford, 1968; Barros and Franklin, 1968). Techniques for electron microscopy are discussed by Enders in Chapter 16. For ultrathin sectioning, particularly where efforts are being made to locate the fertilizing sperm at or soon after penetration, employ a diamond knife and automatic microtome since, ideally, this combination allows serial sectioning of whole ova without interruption.

SMEAR TECHNIQUE FOR THE STUDY OF EGG CHROMOSOMES

The chromosomes of the second meiotic and the first cleavage spindle can be seen clearly after fixation and staining, but their disposition in the intact egg does not allow counting and study of individual chromosomes. A procedure which permits the study of individual mouse egg chromosomes has been introduced by A. K. Tarkowski; it has also been developed for studies of eggs from the gerbil, hamster, mouse, ferret, and the rabbit by R. W. McGaughey.

Eggs are treated, if necessary, to remove follicle cells, either by physical means or with hyaluronidase. The eggs are then placed in 0.7% sodium citrate (hypotonic) for 3 to 10 minutes, depending upon the size of the ovum, and the swollen eggs are then transferred to a clean glass slide and arranged in a single line. Excess citrate solution must be removed from around the eggs with small strips of filter paper; a microdrop of 25% acetic alcohol (15% solution for mouse eggs) is placed beside the row and allowed to flow over the line of eggs; 2 to 5 microdrops are then added. As the acetic alcohol evaporates, the eggs gradually flatten; this results in a spreading of the nuclear elements. The dry preparation is hydrolysed in 0.1*N* HCl, stained with basic fuchsin, and mounted as a permanent preparation in the usual way. Such preparations remain in good condition for at least one year.

This technique has proved suitable for the study of the first maturation division chromosomes of follicular oocytes in the mouse, hamster and ferret, and of second maturation spindle chromosomes (see Fig. 3-7) and transformation of the fertilizing sperm head (see Figs. 3-8 and 3-9) in the mouse, hamster, Mongolian gerbil, and rabbit. It has also allowed inspection of the chromosomes which form immediately prior to the first cleavage division in mouse and hamster eggs (see Fig. 3-10). (Techniques for the study of chromosomes are considered also by Kinsey in Chapter 17.)

CRITERIA OF FERTILIZATION

The genetic apparatus of the ovum in most mammals emerges from its dictyate resting state only hours before ovulation. It is at this time that the first maturation division is completed, and the first polar body is extruded. The second maturation proceeds as far as metaphase; the spindle of this stage lies tangentially to the surface in nonactivated ova of most species. Exceptions to this include the dog, fox, and possibly the horse, in which the ovum is said to be ovulated and penetrated as a vesicular oocyte. The unfertilized ovum in mammals is thus generally characterized by a peripheral metaphase spindle with one polar body in its vicinity, lying in the perivitelline space. Aging of the unfertilized egg may be accompanied by changes in some organelles in certain species (see below). Rotation of the spindle from a tangential to a radial position (see Fig. 3-11) may be a sign of activation in the freshly ovulated ovum, but this should be treated with caution and cannot be used alone. In the unfertilized ovum, the perivitelline space is minimal, but becomes larger following contraction of the vitellus at fertilization. Again, this single criterion cannot be used alone with confidence since vitelline retraction is a well-known correlate of the beginning of degeneration in aging eggs.

The best criterion of fertilization is that of implantation and continued fetal development following transfer of experimental ova, after termination of their fertilizable life, back into a foster mother. Recourse to this type of procedure is not always possible, and is not necessary in many instances where a comprehensive cytological examination of ova can be performed. The most convincing cytological evidence of fertilization can be obtained several hours after sperm penetration, when the male and female pronuclei have had time to develop, but before onset of the first cleavage division.

The following criteria are important in establishing the occurrence of fertilization in single-celled ova, though not all are necessarily appropriate for every species:

1. Presence of a fertilizing sperm tail.
2. Swelling sperm head in ooplasm.
3. Male and female pronucleus.
4. Definitive first and second polar body.
5. Presence of sperm in the zona pellucida or the perivitelline space.
6. Appearance of the ooplasm.
7. Activation of the egg involving loss of cortical granules and rotation of the maturation spindle.

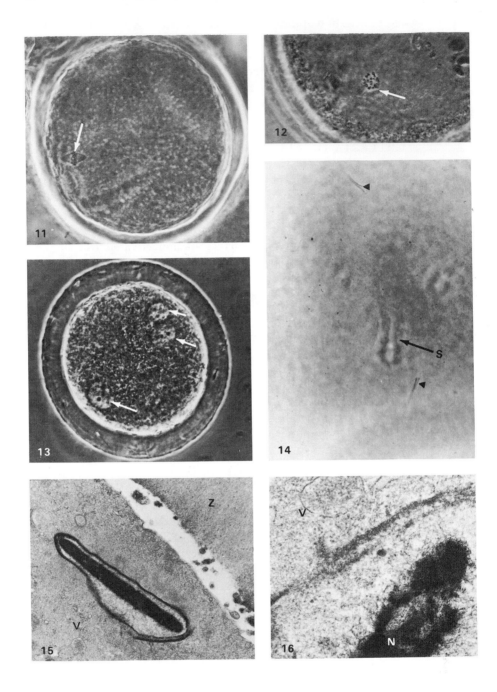

FERTILIZING SPERM TAIL

The tail of the fertilizing sperm normally enters the vitellus in many, though not all, mammals (see Austin and Walton, 1960). The fertilizing tail can often be identified in one-celled ova, and can sometimes be found in one blastomere after cleavage has occurred. Because of its

FIGURES 3-11—3-16

11. *Unfertilized rabbit ovum 10 hours after ovulation, lightly stained with lacmoid. The second maturation spindle (arrowed) has come to lie in a radial position. Adoption of the radial from the resting tangential position is the first sign of activation, but may occur spontaneously in the absence of sperm penetration.* **12.** *Polar view of a second maturation spindle (arrowed) in an unfertilized rabbit ovum. Phase contrast; stained with lacmoid after fixation.* **13.** *Rabbit egg recovered about eight hours after ovulation. The three distinct pronuclei (arrowed) probably resulted from penetration of the egg by two sperm (dispermy).* **14.** *Polyspermic rabbit ovum stained with lacmoid, showing the outline of two sperm tails in the ooplasm. Note that neither tail has split completely (see Fig. 3-5d). The shadow out of focus is that of a sperm lying on the surface of the egg.* **15.** *Electron photomicrograph of fertilizing rabbit sperm head in the vitellus (v). The head is beginning to swell and areas of reduced density appear as light spots in the dark nucleus.* **16.** *Higher magnification of the fertilizing sperm head in a rabbit ovum. This shows the fibrillar disintegration of the dense sperm nucleus which precedes the formation of the definitive male pronucleus.*

relatively large size, the sperm tail can be easily distinguished in fertilized rodent eggs after fixation and staining (see Fig. 3-4)—Yanagimachi and Chang (1964) show clearly outlined sperm tails within fresh hamster eggs which were merely compressed between cover glass and slide, and viewed with a phase-contrast microscope. Identification of the fertilizing tail in rabbit ova is more difficult, and must be performed under oil immersion with a phase-contrast microscope after staining and differentiation (see Fig. 3-5). Hours after entrance into the rabbit egg, the sperm tail often becomes separated from the male nucleus and begins to split. Occasionally the two halves become completely separated (see Fig. 3-5d); the slender fragments must not be interpreted as being two sperm tails, which are seen only in cases of polyspermy. It is not difficult to distinguish the fertilizing sperm tail from those lying on the surface of the zona or in the perivitelline space: the fertilizing tail comes into focus in the ooplasm between the upper and lower focal planes of the egg surface and has a fragile, lighter appearance probably caused in part by loss of the cell membrane during its entry into the vitellus.

In ova which have a dense ooplasm, as in the pig, detection of the fertilizing sperm tail presents a greater problem and often is not possible; Hancock (1961) was able to find the tail in only 10 of 175 ova penetrated.

SWELLING SPERM HEAD

This parameter may be used during the first hours after sperm entry. The sperm head begins to swell soon after penetration into the vitellus and this often occurs first in the posterior half of the sperm nucleus (see Fig. 3-8). In the early stages the fertilizing head becomes slightly larger than those lying outside the vitellus and its outline appears less distinct. Electron micrographs indicate that the reduction in density of the swelling nucleus is due to the decondensation of the densely packed nuclear material (see Figs. 3-15 and 3-16). The fertilizing sperm head can sometimes be seen in fresh rodent eggs, when compressed, but can be seen satisfactorily in rabbit and other large eggs only after staining. Examples in various laboratory species are shown in a light-microscope study by Chang and Hunt (1962).

MALE AND FEMALE PRONUCLEI

The time of appearance of definitive pronuclei varies with different species, being 4 or 5 hours after sperm penetration in the rabbit and laboratory rodents, and about eight hours in the sheep, cow, and pig. It is sometimes difficult to ascertain the gender of either pronucleus. In the early formative stages, the sperm tail lies close to the male pronucleus, but these often become separated at a later stage. In the mouse and rat, the male pronucleus is much larger than the female, and in the rabbit it tends to be larger but is not always so; this disparity does not hold for many other species. Differences have been observed in the staining reactions of the respective pronuclei, in the rabbit, pig and hamster; in the rabbit and pig, chromatin material in the female pronucleus tends to condense first on the side nearest to the male elements.

The presence of two pronuclei does not necessarily denote normal fertilization: it must be considered with other criteria such as the presence of polar bodies, tendency for spontaneous activation in a particular species, and time relationships, for example, of the stage of development in relation to the estimated time of ovulation. The unfertilized ovum of the hamster often undergoes spontaneous activation with the formation of two pronuclei—on occasion two polar bodies may also be seen in such ova. In these cases, however, activation does not occur until after the egg has lost its fertilizability at which time the normally fertilized egg would have entered the first cleavage stage (Yanagimachi and Chang, 1961).

Abnormal numbers of pronuclei are sometimes seen. There are various reasons for this: entry of more than one sperm into the vitellus (see Figs. 3-13 and 3-14); failure of extrusion of the second polar body; fragmentation of the female pronucleus prior to syngamy (see Fig. 3-19). These conditions apparently tend to follow fertilization of aging ova in some species (rat, rabbit, pig, sheep) but not in others (cow—Thibault, 1967). Ova displaying only one pronucleus may arise as a result of spontaneous activation with or without extrusion of the second polar body, or by failure of development of the male pronucleus after sperm penetration. Androgeny, with expulsion of all female chromosomes, has been brought about in the rat by administration of colchicine $2\frac{1}{2}$ hours after mating (Piko and Bomsel-Helmreich, 1960). On occasion, immature vesicular oocytes may be released by the ovary—these cells possess one central nucleus and may be comparable with, or rather smaller than, the mature ovum (see Fig. 3-17), but lack the first polar body.

The question of abnormalities of fertilization will not be considered further here—this topic is discussed at length by Austin and Walton in Marshall (1960, Chap. 10) and by Austin (1969).

POLAR BODIES

In most mammals the resting oocyte is activated several hours before ovulation, and the first reduction division and the extrusion of the first polar body occur before release of the egg from the follicle. In the rabbit, guinea pig and hamster the first polar body normally remains as a stable structure in the perivitelline space until after the first cleavage division. In some cases, however, the first polar body will divide spontaneously (see Fig. 3-18) ; these fragments can usually be distinguished by the absence of, or nontypical configuration of, their chromatin (see below). In the rat and often in the mouse the first polar body disappears soon after its extrusion, and in the deermouse (Marston and Chang, 1966) it appears only as a faint cytoplasmic globule.

The second polar body is extruded from the vitellus 1 or 2 hours after fertilization; this is approximately coincident with observable swelling of the fertilizing sperm head as seen in the light microscope (Odor and Blandau, 1951). In the rabbit the first and second polar bodies can usually be distinguished after staining—the chromatin of the first polar body appears typically as a number of scattered particulate fragments, whereas the second characteristically displays a somewhat dense chromatin mass, which often appears moon-shaped (see Fig. 3-20). In the mouse and hamster also, the first and second polar bodies may be distinguished from each other since the latter tends to form a compact nucleus (see Fig. 3-4). In species such as the ferret, the situation is more difficult since there is no consistent morphological difference between the first and second polar bodies; division of the first polar body occurs quite frequently in both fertilized and unfertilized ferret ova.

In the electron microscope the first and second polar body may be readily distinguished by the presence of cortical granules in the first polar body (see Fig. 3-22). These granules are not present in the second polar body, but remain in the first for at least several hours after fertilization.

SPERM ON, OR WITHIN, THE ZONA PELLUCIDA

Fertilized ova commonly show sperm present on the surface of, or within the substance of, the zona pellucida. The presence of sperm on the zona surface (see Fig. 3-6) should be noted and taken into account when assessing ova, but can be considered as no more than an indication of the probability of fertilization. In some species supplementary sperm often appear in the perivitelline space (see Fig. 3-2). Perivitelline sperm are a normal feature of fertilized ova in the rabbit, mole and pika, and a few are sometimes seen in the rat, ferret, guinea pig and mouse. The existence of sperm in the perivitelline space, though highly indicative, cannot be taken as unequivocal evidence of fertilization. After the sperm were treated with

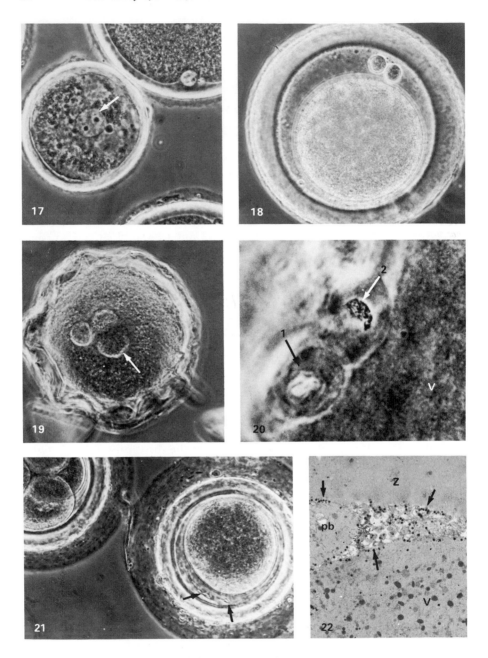

hyaluronidase inhibitors, ova were found where penetration of the zona had occurred without penetration or activation of the vitellus (Parkes, Rogers, and Spensley, 1954). The untreated rabbit ovum in Figure 3-21 had about 12 normal sperm in the perivitelline space, and yet none had penetrated the vitellus which contained the metaphase spindle of the second maturation division. Supplementary sperm are rarely seen in ova of the hamster and the domestic species when they are fertilized soon after ovulation. It appears, however, that the strong zona reaction of hamster egg begins to decline with

FIGURES 3-17—3-22

17. *Rabbit vesicular oocyte, ovulated together with normal fertile rabbit ova. This oocyte was smaller than the mature ova and possessed no polar body. Note the vesicular resting nucleus (arrowed) which resembles the pronuclei formed at fertilization, by having more than one nucleolus. More commonly the nucleus of the resting oocyte contains one large dense nucleolus.* **18.** *Unfertilized rabbit ovum recovered about 10 hours after ovulation. The first polar body has divided spontaneously, and the fragments could at this stage be mistakenly considered as the first and second polar bodies.* **19.** *Fertilized rabbit ovum fixed in acetic alcohol but not stained. The presence of three pronuclei is due to fragmentation of the female pronucleus (the male pronucleus is arrowed and is identified by the proximity of the sperm tail after staining).* **20.** *Phase-contrast photograph of the first and second polar bodies of the rabbit ovum. This shows the dense, often moon-shaped, chromatin mass of the second polar body which is seen after staining with lacmoid.* **21.** *Living rabbit ova seen in the phase-contrast microscope; the ovum on the left has been normally fertilized and has cleaved as far as the four-cell stage. The ovum on the right has several sperm in the perivitelline space (arrowed) but is not fertilized (the metaphase spindle was not activated and remained in the tangential position).* **22.** *Electron micrograph of unfertilized rabbit egg. This shows the dense rounded cortical granules (arrowed) at the periphery of the first polar body (pb), in the villi, and on the surface of the vitellus.*

aging (approximately 15 hours after ovulation), as several supplementary sperm were found in 24% of ova from hamsters inseminated 12 hours after the time of ovulation (Yanagimachi and Chang, 1961).

ACTIVATION OF THE EGG

Although definite evidence of fertilization must depend on other criteria outlined above, some indication of activation in single-celled denuded rabbit ova can be obtained with the low-power stereo microscope. If the egg remains unfertilized, the process of aging is very often accompanied by development of a coarse appearance of the ooplasm in which floccules appear—this contrasts markedly with the appearance of the activated (fertilized) egg, whose cytoplasm presents a bland uniform appearance. In the dissecting microscope, ova containing pronuclei typically display a lighter zone in the center of the egg. This type of assessment of ova in watch glasses under low power may be helpful where it becomes necessary to transfer single-celled fertilized ova to a foster mother or to different culture conditions, and avoids the disturbance of mounting onto a glass slide.

It is believed that the cortical granules at the periphery of the vitellus break down and disappear very soon after the egg becomes activated by entry of the sperm (Szollosi, 1967). The changes occurring in the vitelline surface at this time probably ensure the block to polyspermy. In most species it is extremely difficult to see cortical granules without an electron microscope (see Fig. 3-22); their presence or absence is an extremely useful indicator in electron microscope studies of early fertilization, since examination of a very few sections allows an assessment of whether or not the egg has been activated. The presence of intact granules indicates that penetration of the vitellus has not yet occurred, though sperm may be found within the zona of these eggs. Eggs which have been penetrated by sperm show no (or only very few) cortical granules other than those which remain in the first polar body.

The metaphase spindle lies tangential to the surface of the ovum in freshly ovulated ova of most mammals, and its rotation to a radial position (see Fig. 3-11) is one of the first signs of activation in some species. Such rotation may occur spontaneously, and in some species may be the normal orientation of the second maturation spindle. Thus rotation of the spindle cannot be considered as more than an adjunct to other, more definite, criteria of activation and fertilization.

ASSESSMENT OF OVA AFTER CLEAVAGE

Cleavage has been the most widely used criterion of fertilization; this is particularly so in species in which it has not been possible to estimate the exact time of ovulation. This criterion does not provide the unequivocal evidence supplied by the presence of a fertilizing sperm head or tail, but in many species a normal cleavage pattern with stained nuclei of uniform appearance (mitotic figures appear in a proportion of blastomeres in the late morula and early blastocyst) gives a reasonable indication that fertilization has occurred.

The onset of cleavage has not been determined accurately for more than a few species: cleavage occurs about 12 hours after penetration in the rabbit; 15 to 17 hours after ovulation in the rat, mouse and hamster; 22 to 24 hours after ovulation in the gerbil; 16 to 18 hours after sperm entry in the sheep; 20 to 24 hours after sperm entry in the cow; and 20 to 24 hours after mating in the sow. In the early stages of development that follow the first cleavage division, ova with three, five, or seven blastomeres should not be considered as abnormal without other evidence of abnormality since asynchronous cleavage has been observed in normally fertilized ova. In assessing ova it must be remembered that the cleavage rate sometimes varies with the species. According to Hancock (1961), fertilized pig ova usually do not reach the eight-cell stage until 72 to 96 hours after mating; rabbit ova, which develop at a quicker rate, reach the eight-cell stage about 20 hours after fertilization and develop into an early blastocyst after 96 hours (Gregory, 1930). Within the rodent family (see Austin, 1961) the mouse egg appears to cleave at a slightly faster rate than that of the rat or hamster; the eight-cell stage is reached about 47 hours after ovulation in the mouse, and about 60 hours afterward in the rat and hamster. A similar discrepancy is seen in the time at which the blastocyst first becomes recognizable; this is about 63 hours after ovulation in the mouse, and about 80 hours in the rat and hamster.

Cleavage at the two- and four-cell stages can be used with reasonable confidence as an indication of fertilization in some species, but careful assessment with particular reference to time/development relationships must be made in others which show a propensity for spontaneous activation. Activation of rabbit ova may be achieved by parthenogenetic agents but spontaneous cleavage *in vivo* is rare in this species, and occurs long after the expected time of the first cleavage. The spindle in unfertilized rabbit ova may sometimes remain for up to 38 hours before fragmenting, but persists for a short time only in the unfertilized hamster ovum which very often develops pronuclei or a cleavage spindle, though actual cleavage of these activated ova is rare. In the ferret, the second maturation spindle may persist for 64 hours or more, but spontaneous cleavage ultimately occurs in approximately 50% of the ova (Chang, 1957). A type of spontaneous cleavage has been reported in rat ova, particularly in the immature animal stimulated to ovulate as a result of gonadotropin injection; here, however, the nucleus is often absent or abnormal thus making it possible to distinguish fragmenting ova from those undergoing normal development after fertilization (Austin, 1949). Up to 80% of unfertilized pig ova recovered from the uterus more than 48 hours after ovulation show varying degrees of fragmentation—some of these may resemble normally cleaving ova (Dziuk, 1960).

EXPERIMENTAL MANIPULATION

Sperm Capacitation

It has been shown in several mammals that some physiological change (capacitation) occurs in sperm within the female reproductive tract that enables the sperm to fertilize the egg. This capacitation is apparent and is studied most easily in the rabbit. Samples of capacitated rabbit sperm are usually obtained by flushing the uterine horns 10 to 12 hours after natural mating, but recent data suggests that samples for use in *in vitro* fertilization experiments are better taken at a later time. A blunt needle is inserted through the cervix and 2 to 4 ml of a physiological fluid is injected into the horn and then withdrawn. It is not necessary to clamp the uterotubal junction of the estrous animal during this procedure. If serum is used in the flushing medium it must be pre-heated at 60°C for 15 minutes, or it can be stored for several days to destroy the spermicidal factor. Capacitated sperm may be concentrated by centriguation at 1000 to 3000 g for 2 or 3 minutes without apparently harming the fertility of the viable sperm.

There is at present no way of obtaining consistently good samples of capacitated rabbit sperm; samples obtained after mating are variable in concentration, but direct instillation of sperm into the uterine lumen usually provokes increased leucocytosis and phagocytosis, and many are lost also via the cervix. Recent experimental evidence (Soupart and Orgebin-Crist, 1966) indicates that interference involving ligation tends to inhibit the capacitating activity of the uterine environment. As yet, there is no method for capacitation of rabbit sperm *in vitro*. In my experience the best uterine sperm samples are obtained usually, though not consistently, from females showing an immediate lordosis reaction to the male; stud males should not be used for mating more than once every three days.

The capacitation requirement of hamster sperm seems to be less stringent than that of the rabbit. A low percentage of eggs were fertilized *in vitro* by hamster sperm removed from the epididymis (Yanagimachi and Chang, 1964) ; it must be presumed that in this case capacitation was achieved by factors associated with the cumulus mass or the eggs themselves. A higher fertilization rate was obtained in these experiments with samples collected from the hamster uterus about 4 or 5 hours after mating. More recently, an equally high fertilization percentage has been achieved with epididymal sperm alone when incubated with mature follicular oocytes (Barros and Austin, 1967) . It thus appears that hamster sperm may be capacitated completely by the agency of factors present in the follicle (Yanagimachi, 1969a, b) .

Assessment of Capacitation

As yet, there is no specific test for capacitation, and fertilization is the only parameter by which capacitation may be judged. Assessment of the state of capacitation of sperm is best made by *in vitro* fertilization, where no contribution can be made by the ovum donor. This is now a reasonable proposition in the hamster, and in the rabbit (see below) . Capacitation has been judged in some experiments in the rabbit by placing fresh unfertilized ova in the capacitation site, i.e. in the uterus or fallopian tube (Bedford, 1969) . The state of capacitation is usually judged, however, by the ability of sperm to fertilize after instillation into the oviduct of a recipient doe, late in the fertilizable life of the egg. A significant percentage of tubal eggs can still be fertilized when good capacitated sperm samples are inseminated into the ampulla of the rabbit fallopian tube as late as 15 hours after an ovulation injection; insemination at this time minimizes the possibility of some phase of capacitation being able to occur in the tube of the recipient, before the egg loses its fertilizability.

This assay procedure is carried out as follows: the flank is shaved and the fallopian tube exposed through a flank incision about 1½ inches below the iliac crest. Not more than 0.025 ml of sperm suspension is put into the ampulla via the fimbriated opening; a greater volume tends to disturb the tubal environment and sometimes leads to a loss of ova. The tube is replaced in the abdomen and left for at least 6 to 10 hours, by which time the male and female pronuclei will have developed in fertilized eggs. It is important, in assessing the eggs, that particular note be made of sperm associated with the egg surface. If unfertilized eggs have sperm adhering to the surface, it is reasonable to assume that viable sperm present at the site were not capable of fertilization. The absence of sperm on the surface of unfertilized eggs indicates that inadequate numbers of viable sperm were present at the site of fertilization, and that the failure of fertilization cannot then be ascribed merely to lack of capacitation. It is advisable in this type of experiment that one oviduct be used to test a sample of sperm known to be fully capacitated, and the other tube used for the experimental sample. This arrangement serves as a check on the operator's technique and as an indicator of the fertilizability of the recipient's ova at the time of tubal insemination.

Another approach to the study of capacitation in the rabbit involves the transfer of eggs to the site of sperm incubation; for example, eggs transferred to the uterine cavity are first penetrated 10 or 11 hours after mating, thereby showing that capacitation of most sperm requires at least this length of time in the uterus alone (Bedford, 1969). Using this technique, it has been reported that capacitation is not completed for about 19 or 20 hours in the uterus of the ovariectomised female rabbit (Bedford, 1970).

In Vitro FERTILIZATION

The use of *in vitro* fertilization techniques allows objective study of the parameters governing mammalian fertilization, and can be use as an assay system in studies of sperm capacitation.

Rabbit

The development of repeatable techniques for fertilization of rabbit ova *in vitro* rests largely upon the discovery of the significance of sperm capacitation in mammalian fertilization. The history of earlier experiments on this topic in the rabbit and other species has been discussed at length by Austin (1961b), and more recently by Chang (1968), Brackett (1970), and Thibault (1969).

In vitro fertilization of rabbit ova has now definitely been accomplished in several different laboratories (Dauzier and Thibault, 1956; Thibault and Dauzier, 1961; Chang, 1959; Bedford and Chang, 1962; Suzuki and Mastroianni, 1965; Brackett and Williams, 1965).

The methods that these authors used have now been improved (see below); they involve the incubation of freshly ovulated ova with sperm samples recovered from the uterus. In most cases uterine sperm were recovered 10 to 12 hours after mating, the assumption being that sperm are completely capacitated in 5 or 6 hours in the rabbit uterus. Recent results indicate, however, that many completely capacitated sperm do not begin to appear in the rabbit uterus until 10 or 11 hours after mating (Bedford, 1969). It is recommended, therefore, that uterine sperm be collected as late as 14 to 16 hours postcoitum. This has already been shown to be advantageous in the experiments of Seitz, Rocha, Brackett and Mastroianni (1970), who obtained the highest fertilization rate (90%) with sperm recovered from the uterus 16 to 18 hours after mating. Fresh ova in cumulus have been collected usually by flushing the oviducts soon after ovulation, i.e. 11 to 12 hours after mating or HCG injection. Thibault and Dauzier obtained their best results when the ova were washed with Locke's solution for 30 minutes or longer before semination, but other investigators have not found this to be an advantage. Fertilization has been achieved in Carrel flasks, on depression slides, and in glass tubes having an internal diameter of 1.5 to 3.0 mm which were sealed at both ends, after instillation of the gametes. Acidic saline with 5% heated rabbit serum, Locke's, Waymouth's medium, and oviduct fluids recovered 4 to 5 days after ligation of the fimbriated ostium, have all been used successfully to support fertilization *in vitro*.

In my opinion, the most consistently successful technique appears to be that devised recently by Seitz, Brackett, and Mastroianni (1970), in which

TABLE 3-1

Medium for in vitro *fertilization of rabbit ova*

Component	g/l	mM
NaCl	6.550	112.0
KCl	0.300	4.02
$CaCl_2 \cdot 2H_2O$	0.330	2.25
$NaH_2PO_4 \cdot H_2O$	0.113	0.83
$MgCl_2 \cdot 6H_2O$	0.106	0.52
$NaHCO_3$	3.104	37.0
Glucose	2.500	13.90
Crystalline bovine serum albumin	3.000	
Penicillin-G-Sodium	0.031	
Distilled H_2O to 1000 ml		

SOURCE: Brackett (1970).

ova recovered from the surface of the ovary are incubated under oil equili-brated with 5% CO_2 in air. The medium used (Brackett and Williams, 1968), is shown in Table 3-1. The pH of this medium at 37°C under an atmosphere of 5% CO_2 in air is 7.8. Ovum donors are superovulated with PMS and HCG (150 I.U. of PMS is given 72 to 96 hours before 75 I.U. of HCG). The tubal fimbriae are removed at laparotomy 9½ hours after HCG injection. Two hours later (about one hour after ovulation) the ovaries are taken, immersed in a bath at 37°C and the ova are dissected away from the surface of the ruptured follicles. These ova are placed in synthetic medium +20% rabbit serum until semination, when they are transferred to sperm samples flushed from the uterus with synthetic medium +20% serum. The gametes are brought together under oil that is equilibrated with 5% CO_2 in air, at 37°C, in a self-sealing culture dish. Sperm penetration normally occurs within 3 hours; the ova are transferred at about 5 hours to a culture medium containing 10% serum, or to Brinster's (1970) medium for rabbit ova which contains lactate or pyruvate, but no serum. The ova can then be transferred in the 2 to 4 cell stage to a synchronized foster mother; otherwise, cytological examination of the ova at about 10 hours after semination will allow the most accurate assessment of the normality of fertilization. At this time the male and female pronucleus and first and second polar bodies should be clearly visible, and the fertilizing sperm tail can often be located.

Hamster

The hamster has been found to be highly amenable to the study of *in vitro* fertilization. Fertilization of hamster ova *in vitro* was achieved first by Yanagimachi and Chang (1964), the best results (60% fertilization) being obtained with sperm samples recovered from the uterus 4 or 5 hours after mating. Ova are released 1 to 5 hours after ovulation (at 5 to 8 A.M. in hamsters maintained on a normal light regime) by dissection of the swollen ampulla of the fallopian tube on the warm stage of a stereo dissecting microscope. A small drop of sperm suspension is then added to the ova in a watch glass beneath warm mineral oil, and the gametes are incubated at 36° to 38°C. The sperm suspension is best made by mixing the dense sperm recovered from the uterus or cauda epididymidis with warm Tyrode solution which acts to disperse the sperm. Comparable fertilization results have been obtained by Barros and Austin (1967) using the same technique, but they obtained ova directly from mature follicles, and the sperm suspension was composed of epididymal sperm.

Subsequently, Yanagimachi (1969a) reported that hamster sperm can be completely capacitated in about 3 hours in hamster follicular fluid and rather less efficiently in follicular fluids from the mouse and rat. Bovine follicular fluid (Yanagimachi, 1969b), once detoxified by heating to 56°C for 30 minutes, was also able to capacitate sperm and support fertilization (95.5%) of hamster ova. Since Yanagimachi achieved the highest fertilization rates in the presence of high concentrations of follicular fluid, he concluded that this is an important condition for successful fertilization. By contrast, Bavister (1969) obtained a 96% fertilization rate using only a modified Tyrode medium containing 3 mg/ml sodium bicarbonate, 2.5 mg/ml bovine serum albumin and 10^{-4} M-sodium pyruvate. In this case the highest rate of fertilization was achieved when the medium was equilibrated with 5% CO_2 in air, at a pH of 7.6, and the eggs and sperm were incubated in 2 to 4 volumes of medium to one of cumulus. These latter results may imply that it is the physical conditions rather than specific factors which are most important for successful fertilization in the hamster.

Polyspermy and the entry of supplementary sperm into the perivitelline space occur frequently during *in vitro* fertilization in the hamster. These phenomena are seen very rarely *in vivo,* and they may be due to the unusually large numbers of sperm available in the vicinity of the hamster egg. So far, the conditions imposed during *in vitro* fertilization experiments in the hamster have been found to be inimical to subsequent development of the ova, and these must be transferred to fresh medium soon after fertilization to sustain their survival. Although some ova will then divide into two cells if suspended in a medium composed of Dulbecco's solution (ten parts), TC *199* (ten parts), and bovine serum (one part), none ever develop further than the two-cell stage. It has not been possible to transfer hamster ova fertilized *in vitro* successfully into foster mothers to establish whether or not they will develop normally.

Mouse

Fertilization of mouse eggs *in vitro* has been reported by Whittingham (1968). Ova were expressed into .5 ml of medium under paraffin oil from the oviducal ampulla, in superovulated randomly bred Swiss mice, 14 to 14½ hours after injection of HCG. The mice were superovulated with i/p

TABLE 3-2
Medium for in vitro fertilization of mouse ova

NaCl	68.49 mM
KCl	4.78 mM
CaCl$_2$	1.71 mM
MgSO$_4$	1.19 mM
KH$_2$PO$_4$	1.19 mM
NaHCO$_3$	25.07 mM
Na lactate	25.00 mM
Na pyruvate	0.50 mM
Glucose	5.56 mM
Bovine serum albumin	4.0 mg/ml
Streptomycin sulphate	50.0 μg/ml
Penicillin-G	100.0 I.U./ml

SOURCE: Whittingham (1968).

injection of 5 I.U. of PMS (Primantron: Schering) and 5 I.U. of HCG (Pregnyl: Organon), given 48 hours apart. The medium used in these experiments, a modification of that employed by Whitten and Biggers (1968) to culture mouse zygotes, is shown in Table 3-2. Sperm were collected from the uterus of mice mated one or two hours previously, the contents of the uterine horns being expressed into .5 ml of the medium in a watch glass, under 1 ml of paraffin oil. After agitation, approximately 50 μl of the sperm suspension was transferred to the ova under oil. The watch glass was incubated 4 to 5 hours at 37°C in a flowing atmosphere of humidified 5% CO$_2$ in air. The ova were cultured for 24 hours, as described below in Chapter 6 by Biggers, Whitten, and Whittingham. The occurrence of fertilization was confirmed in some instances by transfer of two-cell ova to synchronized genetically dissimilar foster mothers. Although the fertilization percentage was not high (10 to 40%) in this initial study, the technique can probably be improved without difficulty. It is not yet clear whether epididymal sperm can be used for fertilization under these conditions.

Rat

The fertilization of normal rat ova *in vitro* has not yet been achieved. However, Toyoda and Chang (1968), using epididymal sperm, have obtained consistent penetration (with some polyspermy), of rat ova in which the zona pellucida was first removed with chymotrypsin. Treatment with chymotrypsin did not appear to effect any significant degree of capacitation. This indicates that, in the rat at least, capacitation is concerned with the approach to and penetration of the zona pellucida, and is not necessary for union of the sperm with the vitelline surface.

INDUCTION OF HETEROPLOIDY

Heteroploidy in mammals is considered as a potentially lethal state for the embryo; it has been discussed at length by Beatty (1957), and by Bomsel-Helmreich (1965). Variation away from the euploid state is known to arise occasionally as a result of one of several different anomalous forms of fertilization (Austin and Walton, 1960), but there are ways in which these can be induced experimentally in a high percentage of mammalian ova. This is considered briefly below with reference to experimental suppression of polar-body formation, and induction of polyspermy.

SUPPRESSION OF POLAR-BODY FORMATION

The antimitotic action of colchicine has been used in a series of experimental studies to prevent extrusion of the second polar body during fertilization. In the rat, 70% polar body suppression was observed by Austin and Braden (1954) after intraperitoneal (i/p) injection of 0.05 to 0.1 mg colchicine/100 gm body weight. According to Piko and Bomsel-Helmreich (1960), after i/p injection of doses of 0.025 mg colchicine/100 gm body weight into female rats two hours after mating, 38% of ova had retained the second polar body. Timing is critical, however, for when this procedure was performed 2½ hours after mating, only 11% of the eggs showed this abnormality, whereas 28% were found to be androgenetic as a result of expulsion of all female chromatin from the egg. These androgenetic embryos presumably do not implant as no haploid embryos were found in these animals at any time after the beginning of nidation.

In the rabbit, 95% triploid embryos have been obtained after experimental administration of colcemid (Ciba) during *in vitro* fertilization. In this method an *in vitro* fertilization system is prepared according to the technique of Thibault and Dauzier (1961), and one hour after introduction of sperm, colcemid is introduced into the fertilization chamber, at 0.02 mg/ml; this apparently prevents the formation of the second polar body, the chromosomes of which are retained within the egg. Such triploid zygotes will implant when transplanted to a foster doe; some develop for as long as 16 days before dying *in utero*. When a relatively larger number of these triploid zygotes were transplanted, a few were able to survive up to 22 days (two-thirds of gestation) after fertilization (Bomsel-Helmreich, 1967). The cause of death of these normally formed embryos has not been established.

INDUCTION OF POLYSPERMY

Polyspermy is a potentially pathological state in mammalian eggs (see Piko, 1961). The incidence of polyspermy is usually very low but may be increased during fertilization of aging ova (Austin and Braden, 1953; Odor and Blandau, 1956). In such studies it is helpful to identify the fertilizing sperm tails at the pronuclear stage (see Fig. 3-14) since the incidence of digyny is believed to increase markedly during fertilization of aging ova in some species. Thibault (1967) reported both dispermy (21%) and digyny (52%) when aged rabbit eggs were exposed to aged capacitated sperm. In the studies of Piko and Bomsel-Helmreich (1960), who obtained about 10% dispermy in rats maintained at 40° to 41°C and mated at the end of estrus, triploid embroys were found in some of these females as late as the tenth day of gestation.

The incidence of polyspermy in the pig may be influenced by the systemic progesterone level. Polyspermy occurs commonly in pig ova artificially ovulated during the luteal phase of the cycle (Hunter, 1967). Day and Polge (1968) have found a high incidence of polyspermy in pigs injected subcutaneously with 100 mg progesterone in oil at various times before ovulation; in sows inseminated 12 to 24 hours before HCG-induced ovulation, injection of progesterone 12, 18, 24, and 36 hours before ovulation resulted in 18, 13, 40, and 36% polyspermy, respectively. In the above experiments ten or more sperm were sometimes observed in the ooplasm; the consequences of this state for syngamy have not yet been studied. Systemic progesterone does not appear to have a comparable effect in the rabbit or in the hamster.

In recent years attempts have been made to induce polyspermy in sea urchin eggs by direct microinjection of sperm into the ooplasm (Hiramoto, 1962). Although no activation of the injected sperm takes place in unfertilized eggs, such sperm are reported to develop to the pronuclear stage and to take part in syngamy in those eggs normally fertilized by other sperm. Polyspermy has also been induced in sea urchin eggs following treatment with various mercurial compounds (Runnstrom and Manelli, 1964), but neither of these experimental approaches has yet been reported using mammalian ova.

CONCLUDING COMMENTS

The foregoing discussion concerns itself primarily with methods for the study of fertilization, and of mammalian ova in normal or experimental conditions. Included in the bibliography are key references to a comprehensive account of the theoretical aspects of fertilization and other associated topics.

No attempt has been made to give an exhaustive account of the idiosyncrasies of ova in different species, but the specific examples given here point out that ova from different mammals do not necessarily behave in the same way, and that criteria or methods which serve in one species may be inappropriate for another. Current methods for *in vitro* fertilization of rabbit and hamster ova have now been refined to the point where they can be used for analysis of the parameters which govern mammalian fertilization. As an assay system for sperm capacitation, such techniques avoid possible contribution to the capacitation of the experimental sperm sample in the recipient test female; this has undoubtedly occurred sometimes, when such assays have been performed *in vivo*.

ADDENDUM

In recent studies aimed at characterizing the egg surface we have had to remove the zona pellucida. This can be performed with chymotrypsin (0.02 to 0.04 mg/ml) or pronase (0.5% solution), but these enzymes may also modify the vitelline surface. The zona can be removed physically from fresh eggs with sharp needles but because of its resilience this procedure often results in rupture of the vitellus. We have found that the granulosa cells and zona can very easily be removed with needles after fixation in gluteraldehyde for 10 to 20 minutes. Such treatment kills but does not alter the negative surface charge of cells and allows easy preparation of denuded intact ova.

Successful *in vitro* capacitation of mouse spermatozoa, with *in vitro* fertilization, has been reported by Iwamatsu and Chang (1969), *Nature*, 224:919. Capacitation was achieved using media containing human or bovine follicular fluids.

Studies on fertilization of denuded hamster eggs *in vitro* have shown that sperm will stick to and fuse with the vitelline surface (by the post-acrosomal region) only when capacitated. This indicates that capacitation of hamster sperm involves changes in the posterior as well as the rostral portion of the sperm head (Yanagimachi and Noda: 2nd meeting of Society for Study of Reproduction, Davis, Calif., Sept. 1969).

BIBLIOGRAPHY

Amoroso, E. C., and A. S. Parkes (1947) Effect on embryonic development of X-irradiation of rabbit spermatozoa *in vitro*. Proc. Roy. Soc. B., 134:57–78.

Austin, C. R. (1949) The fragmentation of eggs following induced ovulation in immature rats. J. Endocrinol., 6:104–110.

———— (1961a) The Mammalian Egg. Blackwell, Oxford.

———— (1961b) Fertilization of mammalian eggs, *in vitro*. Intern. Rev. Cytol., 12:337–356.

———— (1965) Fertilization. Prentice Hall, Englewood Cliffs, N.J.

———— (1968) Ultrastructure of fertilization. Holt, Rhinehart and Winston, New York.

———— (1969) Variations and anomalies of fertilization. In: Fertilization, vol. II, chap. 10. Ed. by C. B. Metz and A. Monroy. Academic Press, New York, pp. 437–465.

————and A. W. H. Braden (1954) Induction and inhibition of the second polar division in the rat egg, and subsequent fertilization. Australian J. Biol. Sci., 7:195–210.

———— and A. Walton (1960) Fertilization. In: Marshall's Physiology of Reproduction. Third ed., vol. I, part II, chap. 10. Longmans, London.

Barros, C., and Austin, C. R. (1967) *In vitro* fertilization and the sperm acrosome reaction in the hamster. J. Exp. Zool., 317–323.

Barros, C., and L. E. Franklin (1968) Behaviour of the gamete membranes during sperm entry into the mammalian egg. J. Cell. Biol., 37:C13-C18.

Bavister, B. D. (1969) Environmental factors important for *in vitro* fertilization in the hamster. J. Reprod. Fertility, 544–545.

Beatty, R. A. (1957) Parthenogenesis and polyploidy in mammalian development. Cambridge Univ. Press.

Bedford, J. M. (1968) Ultrastructural changes in the sperm head during fertilization in the rabbit. Am. J. Anat., 123:329–357.

——— (1969) Limitations of the uterus in the development of the fertilizing ability (capacitation) of spermatozoa, J. Reprod. Fertility, 8:19–26. (Suppl.)

——— (1970) The influence of oestrogen and progesterone on sperm capacitation in the reproductive tract of the female rabbit. J. Endocrinol., 46:191–200.

——— and Chang, M. C. (1962) Fertilization of rabbit ova *in vitro*. Nature, 193:898–899.

Blandau, R. J. (1961) Biology of eggs and implantation. In: Sex and Internal Secretions. Vol. II, chap. 14. Ed. by W. C. Young. Williams and Wilkins, Baltimore.

Bomsel-Helmreich, O. (1965) Heteroploidy and embryonic death. In: Preimplantation Stages of Pregnancy. Ciba Foundation Symp., Little, Brown, Boston.

——— (1967) Thesis submitted for Doctor of Science, Univ. of Paris.

Brackett, B. G. (1970) Recent progress in investigations of fertilization *in vitro*. In: Blastocyst Biology. Ed. by R. J. Blandau. University of Chicago Press, Chicago.

——— and W. L. Williams (1965) *In vitro* fertilization of rabbit ova. J. Exp. Zool., 160:271–282.

——— and W. L. Williams (1968) Fertilization of rabbit ova in a defined medium. Fertility Sterility, 19:144–155.

Brinster, R. L. (1970) Culture of two-cell rabbit embryos to morulae. J. Reprod. Fertility, 21:17–22.

Chang, M. C. (1957) Natural occurrence and artificial induction of parthenogenetic cleavage of ferret ova. Anat. Record, 128:187–200.

——— (1959) Fertilization of rabbit ova, *in vitro*. Nature, 184:466–467.

——— (1968) *In vitro* fertilization of mammalian eggs. J. Animal Sci., 27:15–22. Suppl. 1.

——— and D. M. Hunt (1962) Morphological changes of sperm head in the ooplasm of mouse, rat, hamster and rabbit. Anat. Record, 142:417–426.

——— and R. Yanagimachi (1963) Fertilization of ferret ova by deposition of epididymal sperm into the ovarian capsule with special reference to fertilizable life of ova and the capacitation of sperm. J. Exp. Zool., 154:175–187.

Dauzier, L., and C. Thibault (1956) Récherche experimentale sur la maturation des gamètes mâles chez les mammifères par l'étude de la fécondation *in vitro* de l'oeuf de lapine. Proc. III Int. Congr. Animal Reprod., Cambridge.

Day, B. N., and C. Polge (1968) Effects of progesterone on fertilization and egg transport in the pig. J. Reprod. Fertility, 17:227–230.

Dickmann, Z. (1963) Denudation of the rabbit egg: time sequence and mechanism. Am. J. Anat., 113:303–316.

Dziuk, P. (1960) Frequency of spontaneous fragmentation of ova in unbred gilts. Proc. Soc. Exp. Biol. Med., 103:91–92.

Gregory, P. W. (1930) The early embryology of the rabbit. Contr. to Embryology, 125. Carnegie Inst. of Washington, Vol. 121.

Hadek, R. (1963) Submicroscopic changes in the penetrating spermatozoon of the rabbit. J. Ultrastruct. Res., 8:161–169.

Hancock, J. L. (1961) Fertilization in the pig. J. Reprod. Fertility, 2:307–331.

——— (1962) Fertilization in farm animals. Animal Breed. Abstr., 30:285–310.

Hiramoto, Y. (1962) Micro-injection of live spermatozoa into sea-urchin eggs. Exp. Cell Res., 27:416–426.

Hunter, R. H. F. (1967) Polyspermic fertilization in the pig. J. Exp. Zool., 165:451–459.

Marston, J. H., and M. C. Chang (1966) The morphology and timing of fertilization, and early cleavage in the Mongolian gerbil and deermouse. J. Embryol. Exp. Morphol., 15:169–191.

Metz, C. B., and A. Monroy (1967) Fertilization: Comparative morphology, Biochemistry and Immunology. Vol. I. Academic Press, N.Y.

———— (1969) Fertilization: Comparative morphology, Biochemistry and Immunology. Vol. II. Academic Press, New York.

Monroy, A. (1965) Chemistry and Physiology of Fertilization. Holt, Rinehart and Winston, New York.

Odor, L., and R. J. Blandau (1951) Observations on the formation of the second polar body in the rat ovum. Anat. Record, 110:329–340.

———— (1956) Incidence of polyspermy in normal and delayed matings in rats of the Wistar strain. Fertility Sterility, 7:456–467.

Parkes, A. S., H. J. Rogers, and P. C. Spensely (1954) Biological and biochemical aspects of the prevention of fertilization by enzyme inhibitors. Proc. Soc. Study Fertility, 6:65–80.

Piko, L. (1961) Polyspermie chez les animaux. Ann. Biol. Animal Bioch. Biophys., 1:324–384.

———— and O. Bomsel-Helmreich (1960) Triploid rat embryos and other chromosomal deviants after colchicine treatment and polyspermy. Nature, 186:737–739.

Rothschild, Lord (1956) Fertilization. Methuen, London.

Runnstrom, J., and H. Manelli (1964) Induction of polyspermy by treatment of sea urchin eggs with mercurials. Exp. Cell Res., 35:157–193.

Seitz, H. M., B. G. Brackett, and L. Mastroianni (1970) *In vitro* fertilization of ovulated rabbit ova recovered from the ovary. Biol. Reprod., 2:262–267

Seitz, H. M., G. Rocha, R. G. Brackett, and L. Mastroianni (1970) Influence of the oviduct on sperm capacitation in the rabbit. Fertility Sterility, 21:325–328.

Soupart, P., and M-C. Orgebin-Crist (1966) Capacitation of rabbit spermatozoa delayed *in vivo* by double ligation of uterine horns. J. Exp. Zool., 163:311–318.

Stambaugh, R., C. Noriega, and L. Mastroianni (1969) Bicarbonate ion: the corona cell dispersing factor of rabbit tubal fluid. J. Reprod. Fertility, 18:51–58.

Suzuki, S., and L. Mastroianni (1965) *In vitro* fertilization of rabbit ova in tubal fluid. Am. J. Obstet. Gynecol., 93:465–471.

Szollosi, D. (1967) Development of cortical granules and the cortical reaction in rat and hamster eggs. Anat. Record, 159:431–446.

Thibault, C. (1967) Analyse comparée de la fécondation et des ses anomalies chez la brébis, la vâche, et la lapine. Ann. Biol. Anim. Bioch. Biophys., 7:5–23.

———— (1969) *In vitro* fertilization of the mammalian egg. In: Fertilization. Vol. II. Ed. by C. B. Metz and A. Monroy. Academic Press, New York, pp. 405–435.

———— and Dauzier, L. (1961) Analyse des conditions de la fecondation *in vitro* de l'oeuf de la lapine. Ann. Biol. Anim. Bioch. Biophys., 1:277–294.

Toyoda, Y., and M. C. Chang (1968) Sperm penetration of rat eggs *in vitro* after dissolution of zona pellucida by chymotrypsin. Nature, 220:589–591.

Whittingham, D. G. (1968) Fertilization of mouse eggs *in vitro*. Nature, 220:592–593.

Yanagimachi, R. (1969a) *In vitro* capacitation of hamster spermatozoa by follicular fluid. J. Reprod. Fertility, 18:275–286.

———— (1969b) *In vitro* acrosome reaction and capacitation of golden hamster spermatozoa by bovine follicular fluid and its fractions. J. Exp. Zool., 170:269–280.

———— and M. C. Chang (1961) Fertilizable life of golden hamster ova and their morphological changes at the time of losing fertilizability. J. Exp. Zool., 14:185–204.

———— and M. C. Chang (1964) *In vitro* fertilization of golden hamster ova. J. Exp. Zool., 156:361–376.

4

Maximizing Yield and Developmental Uniformity of Eggs

ALLEN H. GATES
Department of Gynecology and Obstetrics
Stanford University School of Medicine
Stanford, California 94305

Although many investigators use the technique of inducing ovulation with gonadotropin injections to obtain mammalian ova for experimental studies, the types of animals reportedly used and the ages at which they are injected generally do not produce markedly greater numbers of eggs than are available from spontaneous ovulations. For example, the frequent reports of the use of randombred mice at about eight weeks of age indicate that the average number of eggs induced to ovulate is rarely more than three times that of uninjected adult mice. This chapter gives a detailed account of techniques which consistently yield an average of approximately 80 eggs (fertilized, if desired) from precisely timed ovulations.

The laboratory mouse (on which the techniques were developed) offers the advantage of surpassing all mammals in the wealth of information known about its mutant genes (now over 400 known) and their chromosomal linkages (more than 250 known). Already a fair number of mouse genes have been identified which influence known biochemical pathways, some of them similar to metabolic defects occurring in the human (Green, 1966). Furthermore, genetic uniformity, an important consideration in any embryological study, can best be provided by the many, readily available, inbred strains of mice and their F_1 hybrids.

Part of the work on which this chapter is based was supported by U.S. Public Health Service Research Grant HD 00611.

The procedure to be described for inducing superovulation may find its greatest usefulness in the expanding field of investigation into the biochemistry of the developing mammalian egg. Available techniques for microchemical analysis, employed to quantify and identify specific types of protein and RNA with a high degree of resolution, could require several thousand eggs per assay to give valid results. The superovulated prepuberal mouse can be an efficient and economical source of such quantities of eggs. Furthermore, fertilized mouse eggs lend themselves well to developmental studies, since a high degree of success has been achieved with present methods for their *in vitro* culture (Brinster, 1963) and their subsequent transplantation to the reproductive tracts of adult hosts (Gates, 1956).

The use of mice in good physical condition is of great importance to the success of superovulation and subsequent mating. To have adequate control over both the physical quality and the genetic uniformity of his animals, the investigator may well benefit by raising his own mice. However, the decision to establish an inbred colony should not be made hastily: inbred mice require high standards of nutrition and environmental control; these strains are quite susceptible to disease; and diligence must be exercised to assure the proper maintenance of pedigree records. For those who depend upon commercial suppliers of mice, the strains referred to in this chapter are obtainable from The Jackson Laboratory, Bar Harbor, Maine, unless other sources are specified.

INDUCTION OF SUPEROVULATION

CHOICE OF STRAIN

Strains differ in the maximal number of ova that can be superovulated (McLaren, 1967). Of the inbred strains tested, the BALB/c (and related F_1 hybrids) have consistently exhibited a superb response to superovulation (Gates and Runner, 1957). Two sublines (BALB/cGa and BALB/cGRr) and three hybrids derived from BALB/c (CAF$_1$, C129F$_1$ and CLGF$_1$) have all given high yields of eggs. Since there are sublines of BALB/c available which are outstanding in reproductive performance, this is an advantageous strain in terms of ease of raising large numbers of animals. (The author's subline originated from a BALB/c colony maintained by the Cancer Research Genetics Laboratory, University of California, Berkeley.) In experiments where maximum uniformity of superovulation response (both in number of eggs ovulated and in time of ovulation) is sought, it is recommended that F_1 hybrids derived from a cross between BALB/c and a second inbred strain be used.

INFLUENCE OF AGE AND WEIGHT

A major factor influencing the number of eggs superovulated (second in importance to the amount of follicle stimulating hormone [FSH] in circulation) is the developmental stage of the animal at the time of injection (Gates and Runner, 1957; Zarrow and Wilson, 1961). In strain BALB/c a wave of follicle maturation occurs between two and three weeks of age. At three weeks of age, the number of ovarian follicles with antra reaches a maximum of about 150 per female (Gates, unpublished). It is this vesicular stage in the development of the follicle which is capable of responding to stimulation by FSH. By inducing superovulation at three weeks of age, one may take advantage of the large numbers of follicles capable of stimulation. Between three and four weeks of age, a wave of follicular atresia occurs, and the number of follicles capable of responding to FSH decreases rapidly. After four weeks of age, high superovulation yields in response to gonadotropin treatment have not been consistently obtained. Effects of age similar to those for BALB/c females have been observed for the inbred strains 129 and LG and in the F_1 hybrids C129F_1 and CLGF$_1$, but not in the one outbred stock tested (J, which is maintained at the Institute of Animal Genetics in Edinburgh). Slight variations between strains occur in the age at which maximal superovulation can be obtained. For example, the age at initial treatment resulting in the greatest egg yield was found to occur in strain BALB/cGa at 21 days, in strain 129/RrGa at 24 days, and in the F_1 hybrid between these strains (C129F_1) at 20 days.

Animals that are retarded, as evidenced by lowered body weight, at the start of injections for superovulation tend to yield reduced numbers of eggs (Lang and Lamond, 1966). For injection at 21 days of age, we use BALB/c females averaging 13 gm and C129F_1 females averaging 14 gm (a standard that most commercial breeders would have difficulty meeting). A high quality of females used for superovulation is maintained through rigid standards of nutrition, disease control and culling at the time of weaning. At birth, litters are reduced to five or six, to assure that each offspring receives an ample supply of milk at nursing. The mothers are fed a high fat-content diet (Purina Mouse Chow). Infantile diarrhea, when detected, is controlled with tetracycline hydrochloride. The mice used for superovulation are generally weaned no more than one day before their initial hormone treatment.

FOLLICLE PRIMING HORMONE

The number of ova resulting from induced ovulations of prepuberal mice is most readily controlled by the quantity and quality of the *priming* hormone which is injected (Wilson and Zarrow, 1962). This preparation, containing FSH, brings about the preovulatory maturation of a new set of follicles. A variety of hormones can be used to prime the follicles for superovulation (Gates and Dronkert, 1965). The purest FSH preparations, such as NIH-FSH or Pergonal brand of human menopausal gonadotropin, have the disadvantage of a short half-life, thereby requiring from two to four injections if the maturation of a maximum number of follicles is desired. NIH-FSH also necessitates the addition of at least 5% luteinizing hormone (LH) for optimal stimulatory effect. Pregnant mares' serum (PMS) is by far the most efficacious of FSH-type hormones, requir-

TABLE 4-1

Influence of PMS dose on average number of eggs (±SE)[1]
ovulated by mice injected at 21, 25, or 29 days of age[2]

PMS Dose (I.U.)	Age at PMS Injection (days)		
	21	25	29
0	1.4 ± 1.0	4.2 ± 1.8	3.2 ± 1.3
1	15.2 ± 1.7	11.9 ± 1.0	10.6 ± 0.8
2	38.3 ± 3.8	24.8 ± 2.5	20.4 ± 1.8
4	67.4 ± 3.3	38.1 ± 2.8	28.7 ± 2.4
6	35.7 ± 7.5	34.5 ± 4.7	10.2 ± 2.6
8	1.4 ± 1.0	6.7 ± 2.2	6.2 ± 3.4

[1] SE: standard error.
[2] Assay animals: (CLGF$_1$ × 129) trihybrids. Injection procedure: PMS followed 42 hours later by 2 I.U. HCG. Experimental design: 3 replications of all 18 groups, with 11 females in each group.

ing only a single injection to maintain a proper stimulation of follicles up to the time of ovulation.

The ease with which the number of ova ovulated can be controlled by adjustment of the dose of PMS is demonstrated by the data presented in Table 4-1. Numbers of eggs comparable to those following spontaneous ovulation result from a priming injection of 1 I.U. PMS. At all prepuberal ages, and in all strains of mice tested by the author, the maximum numbers of eggs have been obtained with a dose of 5 I.U. PMS.[1]

The method of preparation of gonadotropins can have an appreciable effect on their activity. Like other proteins, the activity of these hormones can be reduced by adsorption to glass surfaces, by excessive mixing after dilution, and by an increased ratio of surface area to volume.

Virtually no difference can be detected between the effectiveness of intraperitoneal and subcutaneous routes of PMS administration; however, the former is used for convenience. In order to prevent leakage after injection, gonadotropins are injected in only 0.1 ml of physiological saline using 27- to 30-gauge needles.

OVULATORY HORMONE

The administration of a second gonadotropin, hereafter called the *ovulatory* hormone, is required for the rupture of the matured Graafian follicles. The ovulatory hormone should get into circulation quickly, and is therefore injected intraperitoneally. Any one of the various preparations that are high in luteinizing hormone (LH) content may be utilized. The one that is most convenient because of its commercial availability is human

[1] The author utilizes "Equinex," a brand of PMS manufactured by Ayerst Laboratories, Inc., New York, N.Y.

chorionic gonadotropin (HCG). After priming injections of up to 2.5 I.U. PMS, a 1 I.U. dose of HCG is sufficient to stimulate ovulation. At the optimal priming injection, 5 I.U. PMS, an ovulatory dose of 2.5 I.U. HCG should be given to assure ovulation.

The interval of time between the injection of the priming and the ovulatory hormones (PMS and HCG) which is best (in terms of number of eggs superovulated, and the degree of control over the time of ovulation and receptivity to a male) is between 42 and 52 hours. With shorter intervals the number of eggs superovulated is reduced. When intervals exceed 52 hours (especially in females which received PMS after noon), control over the timing of ovulation may be less precise because of the release of endogenous LH.

Conditions for Maximal Ovulation

In summary, the factors which have been described as influencing the number of ova superovulated are all readily controllable. High yields of eggs can be obtained repeatedly when optimum conditions for superovulation are met, as demonstrated by the following results. In a series of experiments, strain BALB/c females that were 21 days old and averaged 13 gm were injected with 5 I.U. PMS at 4 P.M., followed 45 hours later by 2.5 I.U. HCG; among a total of 35 females so treated, an average of 89.5 ($\pm$4) eggs were superovulated.

FACTORS INFLUENCING OVULATION TIME

The ability to accurately control the time of ovulation is of particular value in experiments in which one desires maximum uniformity of development of the eggs recovered from different females. A further advantage of being able to regulate the time of ovulation is that the collection of eggs of specific stages (e.g., late preovulation ovarian eggs or early postovulation oviducal eggs) can be made to conform to a time schedule that provides the greatest convenience for further observation or treatment of the eggs.

Release of Endogenous LH

LH release and ovulation can occur in response to a single injection of PMS in prepuberal female rats (Wagner and Brown-Grant, 1965) and mice (Gates, 1969). In C129F_1 hybrid mice, 98% of the females are able to release their own LH in response to a single injection of 2.5 I.U. PMS administered between 22 and 26 days after birth. However, at the dosage of PMS and at the age at injection recommended above for maximal ovulation, less than 50% of the females ovulate in the absence of HCG treatment. About 80% of C129F_1 females can be stimulated to release their own LH with 2.5 I.U. PMS as early as 17 days; however, the earliest age at which a mouse is able to release LH in response to PMS varies according to its strain. It is important to establish whether the particular strain being used is capable of responding to PMS stimulation by releasing sufficient endogenous LH to evoke ovulation in the absence of exogenous ovulatory hormone. If precise control over the time of ovulation is to be achieved, it is essential that HCG not be injected after the release of endogenous LH has occurred.

The correct timing for injection of the ovulatory hormone therefore requires knowledge of the time of LH release in the stock of animals used; this is regulated closely by the daily cycles of light and dark to which the animals have become accustomed. The users of mice from commercial breeders would be well advised to make sure that their animal suppliers maintain regular periods of light and dark in their breeding rooms. For at least two weeks prior to injections to induce superovulation, mice should be kept in rooms where the intervals of light and dark are carefully controlled. A daily period of darkness lasting 10 hours is recommended; in our colony, this extends from 7 P.M. to 5 A.M., with the midpoint at midnight. Under this lighting regime, prepuberal C129F$_1$ hybrid mice are neurally stimulated to release their own LH between 15 and 20 hours after the midpoint of the second period of darkness following injection of PMS (Gates, 1969). Thus, in females given PMS late on one afternoon, LH release occurs two days later between 3 and 8 P.M. This was demonstrated by the fact that in C129F$_1$ females, sedation with barbiturates during that interval resulted in blockage of the neural stimulus for LH release. (In inbred strain 129 and BALB/c females, the range in time of LH release appears to be somewhat greater.)

Timing of Ovulatory Hormone

In C129F$_1$ females, the range in times of onset of ovulation after endogenous LH release is from 1 to 4 A.M. It has been estimated that the average interval between neural activation of LH release and ovulation is 9 hours in these hybrid mice (Gates, 1969). However, the interval between the injection of ovulatory hormone and the induced ovulation is approximately 12 hours (see the following section). Consequently, in order to control the time of ovulation by exogenous ovulatory hormone, it is necessary to administer the hormone at least three hours before endogenous LH release (or at least 12 hours before the time of endogenous LH–induced ovulation).

A convenient schedule for the injection of gonadotropins is to administer PMS at 4 P.M., and HCG at noon two days later; this schedule satisfies two basic requirements: injection of the ovulatory hormone at least three hours prior to the release of endogenous LH; and provision for an optimal PMS-to-HCG interval.

For special experiments, it may be desirable to vary the time at which induced ovulation occurs. HCG can be given at any hour if the time of LH release is controlled either by being blocked with a barbiturate, or by altering the periods of light and dark. Using the first of these methods, endogenous LH may be blocked by injection of sodium phenobarbital (0.12 mg per gm of body weight) 15 hours after the midpoint of the second period of darkness following PMS injection; ovulatory hormone can then be injected at any time prior to reactivation of the LH release mechanism on the following afternoon. An example of the second method of controlling the time of LH release is to set the lights in the mouse room to go off automatically at 4 A.M. and go on at 2 P.M. For C129F₁ females maintained at least two weeks under such lighting, an injection of PMS late in the afternoon results in ovulation three days later, at about 11 A.M. Under the second system, HCG can be given at any time between noon and 11 P.M., two days after PMS.

DETERMINATION OF SUPEROVULATION TIME

The following experiments, indicating the timing of superovulation relative to the time of injection of ovulatory hormone, are based on recoveries of eggs from the ampulla of the fallopian tube. Although procedures were not employed specifically to determine the time of rupture of ova from the ovaries, there is apparently little delay in the interval between rupture of a follicle and passage of the ovum through the tubal infundibulum. In any case, it is of considerable practical importance to know the time of entry of ova into the ampulla, the site of fertilization in the mouse, especially in experiments in which one wishes to control accurately the time of fertilization and subsequent ovular development.

TECHNIQUE OF OVUM RECOVERY

The recovery of recently ovulated eggs from the tubal ampulla is facilitated by the fact that the eggs are held together in one clump, due to the stickiness of the cells that form their corona radiata. Thus, it is easy to see the clutch of eggs in the excised ampulla when viewed through a dissection microscope with a ground-glass plate on a transilluminated base. Recently ovulated eggs are readily removed from the ampulla, using two sharp dissection needles. A nick is made at one end of the ampulla, after which the eggs either flow out by themselves or are easily teased out with one of the needles into a drop of medium on a microscope slide. By drawing the clutch of ova past the edge of the drop it flattens sufficiently to permit observation of the ova with good resolution and clarity.

TIME OF ONSET OF OVULATION

The inbred and outbred stocks of mice investigated have not permitted the same degree of control over the time of ovulation as have F₁ hybrids. In experimental series (see Table 4-2) in which hybrid female mice had been treated at three weeks of age with 2 to 5 I.U. PMS followed two days later with 2.5 I.U. HCG (prior to time of endogenous LH release), no female had ovulated by 10 hours after injection of the ovulatory hormone. By 11 hours afterward, ovulation had begun in only one of 18 females. Twelve hours after HCG injection a majority of the females

TABLE 4-2

Time and rate at which ova reach the site of fertilization following superovulation

Time of Autopsy (hours after HCG)	Prepuberal Mice: Age at PMS					
	C129F$_1$ Hybrids[1] (19–22 days)			P, Q, R Hybrids[2] (18–23 days)		
	Number Autopsied	% with Ova	Average Ova Recovered (±SE)[3]	Number Autopsied	% with Ova	Average Ova Recovered (±SE)
10	6	0	0.0	—	—	—
11	10	10	1.3	8	0	0.0
12	10	100	27.4 ± 4.0	14	86	17.9 ± 3.1
13	10	100	54.1 ± 4.2	7	100	35.0 ± 3.5
14	10	100	62.4 ± 7.3	8	100	47.1 ± 7.7
>16	20	100	62.2 ± 3.2	24	100	56.8 ± 4.8

[1] Priming dose: 5 I.U. PMS; ovulatory dose: 2.5 I.U. HCG.
[2] Priming dose: 2 I.U. PMS; ovulatory dose: 2 I.U. HCG. Derivation of stocks: P = PCX × CBA, Q = PCX × KL, R = PCT × C3101F$_1$. P, Q, and R from Institute of Animal Genetics, Edinburgh.
[3] SE: standard error.

autopsied had begun ovulating, and ovulation was underway in all females that were examined 13 hours after treatment.

TIME OF COMPLETION OF OVULATION

The rate of progression of superovulation for a given group of gonadotropin-treated females is reflected in the average number of ova recovered at various times after HCG injection. For example, in C129F$_1$ hybrids (see Table 4-2) superovulation appeared to be virtually complete in females that were autopsied at 13 hours after HCG; at that time the average number of ova recovered was not significantly different from that among females that were autopsied later. Summarizing the data on the time of superovulation induced in immature C129F$_1$ mice, it appears that most of the ova arrive at the site of fertilization during the interval from 11 to 13 hours after injection of HCG.

PREMATURE OVULATION

In experiments in which mating is allowed to occur ten or more hours before the time of superovulation, the investigator should be aware that the spermatozoa may occasionally encounter ova from a premature ovulation. It has been found that in prepuberal females that are primed with 5 I.U. of PMS, approximately half of the injected mice may have from one to eight ova present in the ampulla at approximately 26 hours after PMS injection (Gates, unpublished). These eggs are normal in so far as they have the capacity of fertilization and early cleavage if mating takes place near the time when they are ovulated. However, under normal conditions of superovulation, when HCG is given 45 hours after the priming injection and mating takes place several hours later, the occasional eggs from these premature ovulations are too aged to be fertilized. Furthermore, these prematurely ovulated eggs are far along the fallopian tube by the time the HCG-induced ovulation occurs. There is some evidence that this premature ovulation may result from stimulation of the graafian follicles by the high content of LH present in the injected PMS.

FERTILIZATION AND DEVELOPMENT FOLLOWING SUPEROVULATION

REGULATION OF MATING TIME

When superovulation is induced in animals that are maintained under normal periods of light and dark (and HCG is injected in the early afternoon), it is a common practice to mate females late in the afternoon and to check them for the presence of vaginal plugs the following morning. (Since ovulation can be expected to occur shortly after midnight in these animals, mating them at a time close to the occurrence of ovulation obviously inconveniences the investigator.) However, an appreciable degree of developmental variability occurs among eggs from different females that are mated overnight; this is apparently the result of differences between females in the times of both coitus and ovulation (Gates, 1965).

By placing females with males within three hours of the time of ovulation, one maximizes both the uniformity in times of fertilization and the percentage of eggs fertilized. In experiments where maximal developmental uniformity of eggs is desired, one might consider blocking the release of endogenous LH with an injection of sodium phenobarbital (as mentioned earlier) and injecting ovulatory hormone at 10 P.M.; the ovulation then occurs the following morning at approximately 10. Females so treated could be put with males for copulation during the morning, thus minimizing the interval between ovulation and mating.

An alternative schedule, one which allows injections and matings to take place during normal working hours, is based upon the use of an animal room in which the periods of light and dark are reversed (e.g., lights off at 4 A.M., on at 2 P.M.). If PMS is administered at 5 P.M. to animals maintained in such a dark/light reversed room, HCG may be given two days later at 5 P.M., with ovulation being expected to occur at 5 A.M. the following morning. These females may be placed with males at 8 A.M. and checked for the presence of vaginal plugs before noon. Under this schedule, the eggs could be expected to be present in the ampulla of the fallopian tube at the time the spermatozoa arrive, thus maximizing the uniformity in time of fertilization. There is, furthermore, merit to deliberately having mating occur after the time of ovulation, for the cumulus cells surrounding the large mass of superovulated ova may consequently be sufficiently loosened to facilitate penetration by spermatozoa.

Frequency of Matings

The percentage of prepuberal females which will mate following superovulation is directly proportional to their physiological development. If one wishes to obtain matings during the age at which the maximal number of eggs can be superovulated (at three weeks of age), it is necessary to use females that are large and healthy. As mentioned previously, our strain BALB/c females weigh an average of 13 gm at three weeks of age. Approximately 60% of these females mate following superovulation at this age. Although the percentage of females which will mate increases with increasing age (approximately 75% at 3.5 weeks and 90% at 4 weeks) at time of superovulation, the lower numbers of eggs recoverable from older females may make their use for superovulation less desirable.

Recovery of Fertilized Eggs

By twelve hours after fertilization the eggs in the fallopian tubes of mated females are essentially devoid of cumulus cells. These denuded eggs can most readily be removed from the oviduct by flushing. This is accomplished by the use of a 30-gauge needle which has been blunted on a fine carborundum stone and attached to the end of a short 2 cc syringe. With the aid of a dissection microscope set up for transillumination and a magnification of ×25, the blunted needle is inserted into the oviduct's fimbriated end, which is clamped with very fine watchmaker forceps while the eggs are gently flushed. It is important that the oviduct should be removed carefully, without pinching or tearing it, so that an effective flush along the entire length of the tube may be accomplished.

Rate of Fertilization and Cleavage

In immature females that are allowed to mate after being induced to ovulate large numbers of eggs, fertilization appears to proceed normally. For example, among three-week-old BALB/c females induced to superovulate, approximately 95% of the eggs are fertilized (determined by the presence of pronuclei 12 hours after mating). Development proceeds to the stage of cleavage, i.e., the first mitotic division, in approximately 80% of the eggs. Thus, an average of 60 cleaved eggs can be recovered 36 hours after superovulation and copulation in three-week-old BALB/c females (Gates, 1965).

Later Ovular Development

Eggs resulting from superovulation can develop quite normally past the first cleavage, but the majority of them fail to do so if they are left in the reproductive tract of the sexually immature female. In these females, eggs are transported too rapidly to the uterus; as such, the premorula stages do not normally adapt to a uterine environment. One should not use superovulation of the immature mouse in experiments in which it is desirable to have development proceed *in vivo* to the morula or blastocyst stage. Superovulated eggs develop well, however, to these later stages if the eggs are removed from the immature donor female and subjected to *in vitro* culture. An average of 33 blastocysts are obtained from the superovulated immature BALB/c female when the eggs are removed at the late two-celled stage and subsequently cultured (Gates, 1965). Blastocysts resulting from induction of large numbers of eggs are of normal viability and developmental capacity. Thus, for example, among three-week-old strain BALB/c females that superovulate an average of more than 80 ova, the percentage of eggs which survive both *in vitro* culture and subsequent uterine transplantation is comparable to that of similarly treated eggs from spontaneously ovulated adult mice (Gates, unpublished).

SUMMARY

Large numbers of eggs may be obtained by hormonal induction of ovulation in the prepuberal mouse; the success with which this superovulation can be achieved is dependent upon the dosage and the timing of gonadotropin administration and upon the strain and the physiological development of the mouse used. The influence of each of these factors has been discussed and it has been shown that, under optimum conditions, average yields of 80 eggs per injected mouse may be obtained regularly.

It is possible to control accurately the times of superovulation and mating, thus being able to obtain large numbers of zygotes which vary little in time of fertilization. Under the conditions described, superovulation can be controlled to begin and to approach completion in all females between 11 and 13 hours after the injection of ovulatory hormone (which must be given three or more hours prior to the release of endogenous LH). The time of day when ovulation is induced to occur can be conveniently regulated, provided the time of release of endogenous LH is controlled either by barbiturate blockage or by altering the schedule of lighting in the animal room. The percentage of eggs fertilized is highest when mating occurs within three hours of ovulation. Approximately 95% of superovulated eggs become fertilized and 80% undergo cleavage when the procedures suggested are followed.

In conclusion, superovulation of the prepuberal mouse is highly recommended as a source of mammalian eggs in experiments requiring large numbers of eggs in early postovulation stages of development (from secondary oocyte to the four-celled stage). Later stages do not develop well in the reproductive tract of the prepuberal female. However, for experiments in which zygotes (at any given preimplantation stage of development) are required in large numbers for treatment *in vitro,* superovulation is an excellent source of the two-celled eggs which can be used for subsequent culturing to the blastocyst stage.

BIBLIOGRAPHY

Brinster, R. L. (1963) A method for *in vitro* cultivation of mouse ova from two-cell to blastocyst. Exp. Cell Res., 32:205–208.

Gates, A. H. (1956) Viability and developmental capacity of eggs from immature mice treated with gonadotropins. Nature, 177:754–755.

——— (1965) Rate of ovular development as a factor in embryonic survival. In: Preimplantation Stages of Pregnancy. Ed. by G. E. W. Wolstenholme. Churchill, London, pp. 270–293.

——— (1969) Time relationship between neural stimulus for LH-releasing factor and ovulation in the PMS-treated mouse. Am. Zoologist, 9:1080.

——— and A. Dronkert (1965) Gonadotropin preparations and superovulation in the mouse. Am. Zoologist, 5:148.

——— and M. N. Runner (1957) Influence of prepuberal age on number of ova that can be superovulated in the mouse. Anat. Record, 128–554.

Green, M. C. (1966) Genes and development. In: Biology of the Laboratory Mouse. Ed. by E. H. Green. McGraw-Hill, New York, pp. 329–336.

Lang, D. R., and D. R. Lamond (1966) Some factors affecting the response of the immature mouse to PMS and HCG. J. Endocrinol., 34:41–50.

McLaren, A. (1967) Factors affecting the variation in response of mice to gonadotrophic hormones. J. Endocrinol., 37:147–154.

Wagner, J. W., and K. Brown-Grant (1965) Studies on the time of luteinizing hormone release in gonadotrophin-treated immature rats. Endocrinology, 76:958–965.

Wilson, E. D., and M. X. Zarrow (1962) Comparison of superovulation in the immature mouse and rat. J. Reprod. Fertility, 3:148–158.

Zarrow, M. X., and E. D. Wilson (1961) The influence of age on superovulation in the immature rat and mouse. Endocrinology, 69:851.

5

Obtaining Eggs and Embryos from Sheep and Pigs

PHILIP J. DZIUK
Department of Zoology
University of Illinois
Urbana, Illinois 61803

Many different means have been used to recover sheep and pig eggs, with varying degrees of success. The methods presented in this paper are those that I have found to be consistently effective.

To obtain meaningful results on studies of eggs and embryos one must recover the eggs or embryos at an exact time relative to a fixed point in the egg's life. Such a point could be: initiation of the second meiotic division, ovulation, or fertilization. The second meiotic division of an oocyte in a mature follicle of a proestrous pig can be initiated by injecting 500 I.U. of human chorionic gonadotrophin (HCG) intramuscularly (Hunter and Polge, 1966). In the proestrus ewe, 250 I.U. of HCG is effective (Dziuk *et al.*, 1964). Ovulation of an egg containing a first polar body with the chromatin in the second meiotic metaphase occurs 40 hours after injection of HCG in pigs (Dziuk and Baker, 1962), and 25 hours after a similar injection in sheep (Dziuk, 1965).

Egg recovery is most efficiently performed when the age of the egg, and consequently its location in the reproductive tract, is known. The pig egg stays in the oviduct about 48 hours after ovulation; the sheep egg stays about 72 hours. Cleavage of the single-celled fertilized egg into two blastomeres occurs 18 hours after ovulation in the pig and 22 hours after ovulation in the sheep. The eggs are four-celled between 36 and 48 hours after ovulation (Baker, Dziuk, and Norton, 1967).

EGG RECOVERY *IN VITRO*

When the reproductive tract has been removed at slaughter or by laparotomy, the oviduct and uterus should first be dissected from any attached ligaments. Oviductal eggs can be recovered from the isolated oviduct by recovering and examining the fluid flushed through the oviduct from the infundibular end toward the isthmus with a 20 ml syringe and blunted needle. An alternative approach, if the eggs happen to be in the uterus, would be to flush the oviduct and the upper quarter of the attached uterine horn. The proportion of eggs recovered is highest when about 20 ml of fluid is flushed with enough force to distend the oviduct and uterus slightly. When the eggs are already in the uterus, 50 to 100 ml of fluid, vigorously infused, may be needed to distend the uterus sufficiently to recover the eggs. The distension is produced by temporarily blocking the exit of the fluid during the infusion; a vigorous flushing action results when the block is released. Ninety percent of the eggs can be recovered by this method.

Although many eggs are recovered in the first fraction of fluid, restriction of the volume of fluid will lead to a reduction in the proportion of eggs recovered. Eggs settle quickly in most physiological solutions and are usually found in a single plane at the bottom of the fluid, but occasionally an egg or blastocyst may become attached to a small air bubble and be carried to the surface. If the recovered fluid is cloudy or bloody, after allowing a minute or two for the eggs to settle, the upper layers of liquid can be drawn off and discarded. Additional clear fluid may then be used to dilute the remainder and again drawn off until the field is clear.

If the eggs are not to be kept alive, any physiological salt solution will suffice as a flushing medium. Eggs that are to be transferred to a recipient or that are to be cultured should be recovered in one of the more complex media containing protein (Dziuk *et al.*, 1964; Hancock and Hovell, 1962; see suggested equipment, p. 84). Eggs that are to be transferred should be recovered in fluid at or slightly below body temperature. Sudden changes in temperature should be avoided even though sheep and pig eggs tolerate storage between 22° to 24°C for at least an hour.

EGG RECOVERY *IN VIVO*

When the eggs are in the oviduct and it is important that the recovered eggs not be exposed to uterine sperm or fluids, the oviduct may be severed at the lower isthmus (see Figs. 5-1a, b). The isthmus is then freed of attached ligaments by blunt dissection for approximately 3 or 4 cm and the freed end placed in a 20 ml test tube (see Fig. 5-1c). Fluid is flushed from the ovarian end towards the isthmus (see Fig. 5-1d; Baker *et al.*, 1968). Eighty-five percent of the ovarian eggs should be recovered by this method (Hunter and Dziuk, 1968).

If it is important that the oviduct remain intact, the eggs may be flushed from the isthmus toward the infundibulum, which is suitably cannulated.

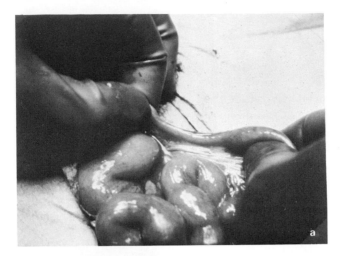

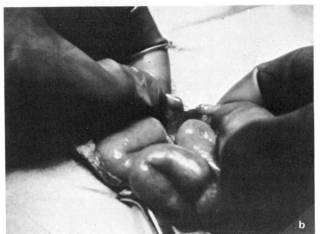

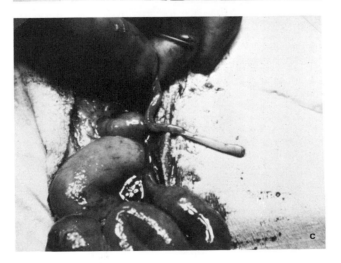

FIGURE 5-1
*Surgical techniques
for recovery of embryos
of sheep and pigs.*

Details of (a) through (c) are described in the text, p. 77.

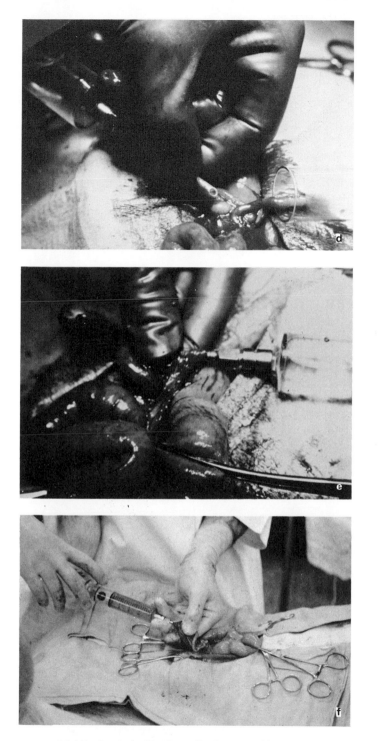

FIGURE 5-1 (d) through (f). Described on pp. 77, 82.

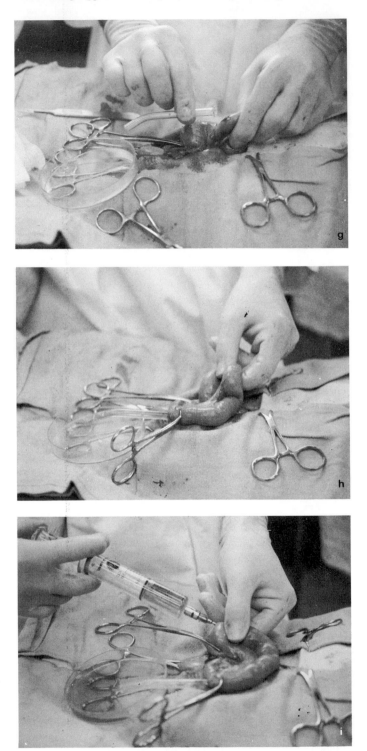

FIGURE 5-1 (g) through (i). Described on p. 82.

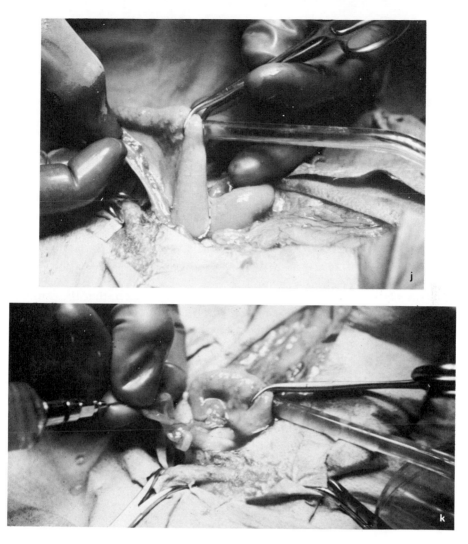

FIGURE 5-1 (j) and (k). Described on p. 82.

This can be done easily in sheep, but the oviduct of the pig must be penetrated via the uterotubal junction because it is ordinarily very difficult to force fluid from the uterus into the oviduct. This approach is most effective in both sheep (Moore and Rowson, 1960) and pigs (Vincent *et al.,* 1964) when the eggs comprise 2 or 4 cells; we have found the proportion of eggs recovered to be quite low when recovery is attempted within 12 hours of ovulation.

When the oviductal eggs can be flushed into the uterus without upsetting the experiment, and if eggs may be present in both the oviduct and upper uterus, a slightly different approach can be used (Dziuk *et al.,* 1964; Smidt *et al.,* 1965; Hancock and Hovell, 1962). An intestinal forceps is applied to the pig's uterine horn 15 to 20 cm from the junction with the oviduct (see

Fig. 5-1e). Twenty ml of fluid are then infused through the ovarian end of the oviduct into the uterus (see Fig. 5-1f), and the fluid forced toward the intestinal clamp by stripping the upper 5 cm of uterus towards the clamp. The fluid is held between the clamp and the point where the fingers have constricted the uterus. A puncture wound, 1 cm long, is made in the upper 5 cm of the uterus near the fingers (see Fig. 5-1g). The puncture is made with the handle of a scalpel or some other blunt instrument; this lessens the possibility of bleeding. A flared cannula about 1 cm in diameter, made of either glass or plastic, is inserted into the wound with the flared end in the uterus toward the trapped fluid and eggs. A towel clamp applied around the uterus and cannula keeps the cannula in place and prevents loss of fluid (see Fig. 5-1h). The fluid is then collected and an additional 20 ml of fluid is vigorously flushed toward the cannula by puncturing the uterus near the intestinal clamp with the blunt needle and syringe (see Fig. 5-1i). The puncture wound for the cannula can be sutured and the animal may thus be reused several times. Ninety percent of eggs are recovered by the method described.

In the ewe, essentially the same procedure is used, except that the cannula is put into the uterus 7 or 8 cm from the uterotubal junction before the infusion of any fluid (see Fig. 5-1j). The cannula is also open toward the oviduct and the fluid is flushed directly from the infundibulum, through the oviduct, into the uterus, and through the cannula (see Fig. 5-1k). The wound in the uterus can be sutured and animals may be reused several times.

Both pig and sheep embryos up to 20 days old can be recovered by the method just described, but to do so the entire length of uterus must be flushed. Movement of fluid and embryos through the uterus can be expedited by stripping the uterus gently between the fingers from the point of the fluid infusion toward the cannula.

FINDING AND HANDLING THE EGGS

When eggs are recovered *in vitro* from the oviducts the volume of flushing fluid may be small enough to be contained in a watch glass. Gently swirling the watch glass with the fluid and eggs will cause the eggs to congregate in the central part of the bottom. For larger volumes the most consistently effective dish for egg-search is a marked, plastic Petri dish that holds about 40 ml. The dishes can be purchased sterile or autoclavable; bottoms can be easily marked in squares or parallel lines. The distance between lines should approximate the diameter of the field of your binocular microscope at $\times 15$ magnification. The eggs are all on a single focal plane and the lines facilitate a quick, systematic and thorough search in a large volume. Debris and contaminating blood can be diluted or dispersed so as not to interfere with identification of eggs; there is no need to manipulate eggs or debris to separate them as is necessary with a watch glass (see Fig. 5-2).

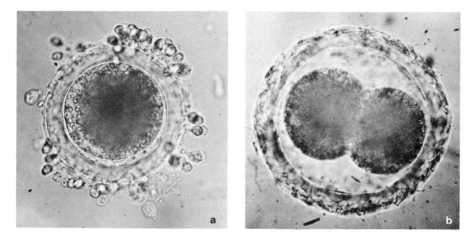

FIGURE 5-2

Typical pig eggs at one- and two-cell stages. Magnification a: ×115, b: ×165.

TRANSFER OF EGGS TO HOST

Recovered eggs are transferred to their hosts in a fine-bore, glass pipette that is attached to a 1-ml syringe. The pipette can be used to puncture the uterine wall or can be threaded part way down the oviduct, whichever method seems appropriate. Special care should be taken to insure that the pipette is in the uterine lumen and not in the uterine wall when the fluid and eggs are expelled. There seems to be no advantage in using less than 0.2 ml of fluid for transfer nor any disadvantage in using as much as 2 ml. Fifty percent of the eggs transferred should develop into fetuses.

MICROSCOPIC EXAMINATION

Eggs that are to be mounted should be picked up in a minimal amount of fluid. A microsyringe attached to a blunted 22-gauge needle, 2 cm long, is used. Eggs are deposited on a clean slide in a drop of fluid just large enough to cover the eggs. A strip of a paraffin-vaseline mixture, or stopcock grease, is put on each side of the eggs, parallel to the long axis of the slide. A drop of rubber cement is applied to two diagonally opposing corners of a coverslip which is then placed over the eggs and gently pressed over the paraffin strips until the eggs are just touching both the slide and coverslip. The amount of compression of the eggs is determined by the nature of the examination. Of course, rupture of the zona pellucida must be avoided if the cytoplasm is to be examined.

Fresh sheep and pig eggs are different from eggs of mice and rabbits; they contain opaque materials which obscure cytoplasmic structures, such as penetrated sperm or pronuclei, from observation by either bright field or phase contrast microscopy. Eggs must be fixed, cleared, and stained before examination. The slides are immersed in fixative, then removed, and small amounts of stain are pipetted onto the slide at the edge of the coverslip betwen the parallel strips of paraffin. The stain is drawn under the coverslip by touching a small piece of paper toweling to the opposite side of the coverslip. The coverslip can be sealed with melted paraffin or clear fingernail polish if the examination is to be so lengthy that the stain may evaporate.

The slide is then placed in a special supporting frame with the coverslip facing downward. The stained eggs are located under the low-power microscope and lines are drawn to them from the edge of the slide, or the eggs are circled with a diamond-point marking pencil. The lines can then be seen easily in the narrow beam of light from the condenser of the high-power microscope; the eggs are found without recording the position of the stage.

SUGGESTIONS FOR EQUIPMENT

In Vitro EGG RECOVERY

1. 20 ml syringe
2. Blunt 16-gauge needle, 4 cm long
3. Plastic petri dish 12 mm deep and 9 cm in diameter
4. Stereoscopic binocular microscope, 10X to 30X
5. Physiological saline solution

In Vivo EGG RECOVERY

1. 20-ml syringe with blunted 16-gauge needle, 4 cm long
2. Plastic (Nalgene, autoclavable) Petri dish, 12 mm deep, 10 cm in diameter
3. Glass cannula with flared end, 1 cm diameter, 8 cm long
4. Intestinal forceps
5. Backhaus towel clamp
6. TC medium 199 with sodium bicarbonate, Tyrodes with bovine serum albumin Fraction V, Brinster's medium, or sheep serum

EGG TRANSFER

1. 1-ml non-luer-lock tuberculin syringe with glycerin-coated barrel
2. Connecting Silastic, or plastic, tubing
3. Pasteur pipettes

MOUNTING AND FIXING

1. 1-ml micrometer syringe
2. Blunted 22-gauge needle, 2 cm long
3. Glass slides and coverslips 2 cm × 2 cm

4. Vaseline-paraffin mixture 50:50 put in a 3-ml syringe with a 17-gauge needle, 2 cm long
5. Frame to hold slide while marking the side opposite the coverslip
6. Pipette for staining
7. Paper toweling 1 cm × 2 cm
8. Rubber cement

BIBLIOGRAPHY

Baker, R. D., P. J. Dziuk, and H. W. Norton (1967) Polar body and pronucleus formation in the pig egg. J. Exp. Zool., 164:491–498.

——— (1968) Effect of volume of semen, number of sperm and drugs on transport of sperm in artificially inseminated gilts. J. Animal Sci., 27:88–93.

Dziuk, P. J. (1965) Timing of maturation and fertilization of the sheep egg. Anat. Record, 153:211–223.

——— and R. D. Baker (1962) Induction and control of ovulation in swine. J. Animal Sci., 21:697–699.

———, F. C. Hinds, M. E. Mansfield, and R. D. Baker (1964) Follicle growth and control of ovulation in the ewe following treatment with 6-methyl-17-acetoxyprogesterone. J. Animal Sci., 23:787–790.

———, C. Polge, and L. E. Rowson (1964) Intrauterine migration and mixing of embryos in swine following egg transfer. J. Animal Sci., 23:37–42.

Hancock, J. L., and G. J. R. Hovell (1962) Egg transfer in the sow. J. Reprod. Fertility, 4:195–201.

Hunter, R. H. F., and P. J. Dziuk (1968) Sperm penetration of pig eggs in relation to the timing of ovulation and insemination. J. Reprod. Fertility, 15:199–208.

Hunter, R. H. F., and C. Polge (1966) Maturation of follicular oocytes in the pig after injection of human chorionic gonadotrophin. J. Reprod. Fertility, 12:525–531.

Moore, N. W., and L. E. A. Rowson (1960) Egg transfer in sheep. Factors affecting the survival and development of transferred eggs. J. Reprod. Fertility, 1:332–349.

Smidt, D., J. Steinbach, and B. Scheven (1965) Modified method for the *in vivo* recovery of fertilized ova in swine. J. Reprod. Fertility, 10:153–156.

Vincent, C. K., O. W. Robison, and L. C. Ulberg (1964) A technique for reciprocal embryo transfer in swine. J. Animal Sci., 23:1084–1088.

6

The Culture of
Mouse Embryos *in Vitro*

J. D. BIGGERS
Division of Population Dynamics
The Johns Hopkins School of Hygiene
and Public Health
Baltimore, Maryland 21205

W. K. WHITTEN
The Jackson Laboratory
Bar Harbor, Maine 04609

D. G. WHITTINGHAM
Department of Anatomy
University of Cambridge
Cambridge, England

The viviparous mode of reproduction in the Marsupialia and the Eutheria places severe restrictions upon experimental studies on the control of early mammalian development. To analyze the many problems involved it is necessary to devise methods for the study of mammalian embryonic development outside of the maternal reproductive tract. The earliest attempts to do this were those of Schenk (1880) and Onanoff (1893) who both claimed to have observed limited cleavage of the early rabbit and guinea pig *in vitro*. Unfortunately both workers claimed their results occurred following fertilization of the ova *in vitro;* as this experiment has never been repeated, their findings are necessarily suspect. The earliest successful studies on the culture of cells and mammalian embryos *in vitro* occurred a few decades later; the beginning of tissue culture began with Ross Harrison in 1907, and mammalian embryos were cultured by Brachet in 1912.

The work of Dr. J. D. Biggers and his colleagues, summarized in this chapter, has been generously supported for several years by the Population Council, Inc. Dr. D. G. Whittingham was formerly a Pennsylvania Plan Fellow, University of Pennsylvania, and was recently a post-doctoral Fellow in Population Dynamics, at Johns Hopkins University, where he was supported by the National Institute of Child Health and Human Development (Grant #5 T01-HD-00109). Dr. W. K. Whitten's studies are supported by an allocation from Public Health Service General Research Support Grant #FR-05545 to the Jackson Laboratory, and Research Grant #HD-00473 from the National Institute of Child Health and Human Development. We wish to thank Dr. Charity Waymouth for helpful suggestions.

Despite these beginnings more than a half-century ago, progress in mammalian embryo culture has been slow. Only rabbit cleavage stages with the mucin coat intact had been cultured successfully prior to 1949, the year that Hammond demonstrated that certain stages of the mouse could also be cultured. Not until 1968 were two more species cultivated successfully—the domestic pig (Frederick and Polge, personal communication) and the prairie deermouse (*Peromyscus maniculatus bairdii*) (Mintz and Whitten, unpublished). To investigators who hoped for quick development of a general technique applicable to all species progress is disappointing. On the other hand, the evidence that has been gathered by experimenters suggests that different species have different nutritive requirements, and the discovery of these requirements and their developmental significance, presents an exciting challenge.

The two species whose embryos have been extensively cultured *in vitro* are the rabbit and mouse. Yet because the rabbit studies were largely done prior to 1949, and the mouse studies since that time, our knowledge of the two species is still far from comparable. Most of the basic work on the culture of rabbit ova was done in the 1930's and 1940's, largely by Pincus and his colleagues. Although biological media were used in this work [see Brinster (1964) for a thorough review of the literature] the possibility of using *in vitro* methods for the determination of the nutritional requirements of cleavage was recognized (Pincus, 1937). The demonstration that mouse ova would cleave in a chemically defined medium (Whitten, 1956a) paved the way for a much more rigorous approach to the problem, and since then the mouse has been studied almost exclusively using similar media. Only recently has some attempt been made to analyze similarly the requirements for cleavage in the rabbit (Daniel, 1965, 1967).

The purpose of this chapter is to describe the methods which have been used recently in the study of mouse cleavage stages *in vitro*. The procedures fall into two major categories—the culture of mammalian embryos completely *in vitro* and the culture of mammalian embryos in organ cultures of the fallopian tube. Our purpose is not to present a set of standardized recipes since the field of study is in a state of continual flux. It seems more useful to stress the concepts on which present methods are based, and their relationship to earlier work, in the hope that those who use these methods can adapt them with understanding to their own particular problems.

A recent summary of knowledge on reproduction in the mouse is given by Bronson, Dagg and Snell (1966), and only aspects pertinent to ovum culture will be mentioned in this article.

PRODUCTION OF MOUSE EMBRYOS AND COLLECTION OF FALLOPIAN TUBES

Choice of Donors

The mouse has a continuously polyestrous cycle, although the duration between consecutive estrous periods may be variable. Ova may be

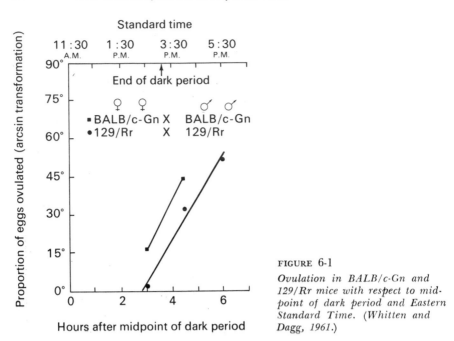

FIGURE 6-1

Ovulation in BALB/c-Gn and 129/Rr mice with respect to midpoint of dark period and Eastern Standard Time. (Whitten and Dagg, 1961.)

collected after natural mating or after superovulation with gonadotrophins.

Random bred animals have been used by most investigators to provide eggs for *in vitro* studies, but no attempt has been made to determine if some strains of animals provide eggs which are in any way superior for cultivation. The vigor and uniformity characteristic of F_1 hybrids of two inbred strains may not be expressed fully until after implantation which is too late for the experiments in question. However, these hybrids may be used to convey such advantages to the cytoplasm of the oocyte. In addition, F_1 females should ovulate more uniformly and perhaps earlier, and F_1 sperm effect earlier fertilization (Krzanowski, 1964). Experience with eggs from these animals indicates that a real advantage exists with fewer nondevelopers.

Some strain differences may influence the choice of donors; for example, the fallopian tubes of SJL/J mice are small and tightly coiled which makes it difficult to collect eggs, eggs of the 129/J strain have clear cytoplasm which facilitates observation of the nucleus, and dense granules develop in BALB/cDg eggs after fertilization.

NATURAL OVULATION

Mice should be kept under constant environmental conditions of light and temperature—in a typical scheme they are exposed to 14 hours of light (6 A.M. to 8 P.M.) and 10 hours of darkness, and are maintained at 70 to 75°C. Using this system, ovulation can be expected between 3 and 7 A.M. (see Fig. 6-1) (Whitten and Dagg, 1961). However the exact light/dark cycle can be adjusted for special purposes. If one wishes to observe the zygote just prior to first cleavage on the morning of a normal working day, a light cycle in which the middle of the dark period was set at 11 A.M. would be suitable. Large numbers of eggs may be obtained by pairing only those females judged to be in estrus from an examination of the vulva or of a vaginal smear. Alternately, the synchronization of the estrous cycles which occur on the third night after pairing, particularly of females caged in groups beforehand, may be utilized to increase the yield of eggs (Whitten, 1966). Some strains of mice are markedly different however, for example, the majority of BALB/cDg females will mate shortly after pairing with SJL/J males but not until later with males of their own strain.

SUPEROVULATION

To obtain fallopian tubes containing fertilized ova, young adult mice are superovulated by injections of pregnant mare serum gonadotropin (PMS) and human chorionic gonadotropin (HCG) (Fowler and Edwards, 1957). (In our work the gonadotrophins have been obtained from Organon, Inc., and Ayerst Laboratories.) This system produces an average yield of 20 to 30 ova per mouse, but may be affected by genetic factors. However, even within the same strain the number of ova obtained per mouse may vary considerably, and the doses of hormones have to be adjusted from time to time. The doses currently used are 2 to 10 International Units (I.U.) of PMS and HCG given intraperitoneally 44 to 48 hours apart in 0.2 ml of 0.9% sodium chloride. Edwards and Gates (1959) estimated the time of ovulation to be 12 ± 3 hours after the injection of HCG. The average times for the onset and completion of ovulation in a randombred strain of Swiss mice has been found to be 10.5 and 12.5 hours respectively after the injection of HCG (Whittingham, unpublished). Thus, to coordinate the ovulation of superovulated animals with the natural time of ovulation, the HCG injection is given at noon or early afternoon. Although some contend that many superovulated ova are abnormal, little evidence exists to support this belief (Gates, 1965).

As soon as the HCG is given, the females are placed with males; they are checked the following morning and those with vaginal plugs are assumed to have mated successfully. Because the time of fertilization cannot be known precisely due to the variation between the times of ovulation and sperm penetration, the injection of HCG is used as the zero reference time.

Fertilized ova may also be obtained by the superovulation of prepubertal mice (see Gates, p. 65 of this book).

SYNCHRONIZATION OF THE ESTROUS CYCLE

Low doses of gonadotrophins can be used to obtain mice at different stages of the estrous cycle at any given time. One to two I.U. of PMS intraperitoneally followed 48 hours later by 1 or 2 I.U. of HCG intraperitoneally are sufficient to synchronize the cycles, yet do not cause superovulation. The mice are in proestrus on the day of the second injection; metestrus-I, 22 hours later; metestrus-II, 46 hours later; diestrus-I, 70 hours later;

and diestrus-II, 94 hours later. Vaginal smears are used to check the stage of the cycle in each mouse using the criteria shown in Table 6-1.

Early Development of the Mouse

The classical work on the early development of the naturally ovulated mouse was done by Lewis and Wright (1935). The cleavage times were confirmed by Whitten and Dagg (1961) who used the midpoint of the dark period and not copulation as the reference time. The results (see Table 6-2) show that after the first cleavage division the blastomeres do not

TABLE 6-1
Injection schedule for synchronization of the Estrous cycle

Time (hours)	Injection of Gonadotrophin	Stages of Estrous Cycle	Vaginal Smears
−48	2 I.U. PMS		
−2	—	Proestrus I	Squamous epithelial cells— some cornified
0[1]	2 I.U. HCG		
+2	—	Proestrus II	
22	—	Metestrus I	Cornified, clumped epithelial cells
46	—	Metestrus II	Cornified epithelial cells + leucocytes + epithelial cells
70	—	Diestrus I	Few epithelial cells + leucocytes + mucus
94	—	Diestrus II	Few epithelial cells + many leucocytes + mucus

[1] Injection of HCG taken as zero reference time throughout this study.

TABLE 6-2
The early development of the mouse embryo under artificial illumination[1]

Stage	Hours
1–cell	0–29
2–cell	23–52
3–4 cells	42–59
5–8 cells	49–60
Morula	68–77[2]
Blastocyst	74–82

[1] The time measured from the midpoint of the dark period.
[2] Passage through the uterotubal junction.

TABLE 6-3

Volume and protein content of cleaving mouse embryos

Stage	Volume (μ^3)	Protein (ng/embryo)
1–cell	192,000	27.8 ± 0.51
2–cell	158,000	26.1 ± 0.41
4–cell	162,000	—
8–cell	138,000	23.4 ± 0.32
Morula	219,000	20.6 ± 0.74
Early blastocyst	—	23.9 ± 0.63
Late blastocyst	—	20.1 ± 0.74

SOURCE: For column one: Lewis and Wright (1935); for column two: Brinster (1967a).

necessarily divide synchronously, so that stages with any number of blastomeres are found. The asynchrony in development becomes more pronounced as cleavage proceeds. The embryo passes through the uterotubal junction at the morula or early blastocyst stage.

Table 6-3 shows the volume and total protein content of mouse embryos at different stages of development. In general, the values of both properties fall as cleavage proceeds. The volume is minimal at the eight-cell stage whereas the protein content is minimal at the morula stage. The lack of agreement in the two studies may denote a strain difference and this needs further investigation on the same animals. Nevertheless, both studies indicate a reduction in the mass of the embryo during its initial phases of development.

GENERAL CULTURE METHODS

In the culture of mammalian embryos completely *in vitro* or in organ cultures of fallopian tubes, standard tissue culture procedures are used. These procedures, and the principles upon which they are based, are fully described in the several well-known manuals on tissue culture (Parker, 1961; Paul, 1965). A monograph by New (1966) deals in general terms with the culture of embryos of vertebrates. The culture of mammalian embryos has also been reviewed recently by Mintz (1967) and Brinster (1968).

In this section only a few special topics will be discussed. These concern methods applicable to the culture of embryos both completely *in vitro* and in organ cultures of fallopian tubes.

CULTURE ROOMS AND THEIR ALTERNATIVES

For large-scale work it is useful to use a small tissue-culture room in which air pressure is slightly higher than that in the outside laboratory. Such rooms are normally equipped with ultraviolet sterilization lamps.

Small-scale work can be done readily under a bench-type hood (made of lucite) that is constructed to accommodate the dissecting microscope. Alternatively, the work can be done on an open bench top beneath a slowly moving curtain of filtered air. It is also possible to work on an open bench in a room in which air movement is minimal. This is especially easy if all solutions are protected by mineral oil.

STERILIZATION

Most instruments and glassware can be sterilized with dry heat for one hour at 150°C. The most convenient way to sterilize solutions is to pass them through sterile millipore filters (pore size 0.45μ). Aliquots of the solutions are stored in sterilized test tubes; they may be either deep frozen or kept in a refrigerator.

INCUBATORS

The importance of being able to regulate the composition of the gas phase in the culture of mammalian embryos and organ cultures has been amply confirmed. For mouse embryos an atmosphere consisting of 5% carbon dioxide in air or some artificial mixture of gases is essential. Two main methods may be used to supply the required atmospheric conditions: (1) place the culture dishes in a desiccator or an anaerobic jar and displace the air with the gas mixture each time the vessel is opened; (2) use a continuous flow system in which the mixture is continuously passed through the incubation chamber. In both methods the atmosphere must be kept saturated with water, especially if the medium is not covered by mineral oil (p. 96). In closed systems a tray of water must be placed in the chamber, whereas in continuous flow systems the gas mixture must be bubbled through water before it reaches the incubation chamber.

It is not widely appreciated how long it takes a chamber to reach the composition of a gas mixture fed into it and, if chambers are opened frequently, a high flow of gas is required to restore the specified atmosphere rapidly. Clearly the chamber should be as small as possible and preferably should be set aside for individual experiments or for the use of an individual investigator. Modern CO_2 incubators are ill-adapted for this purpose as well as being expensive. A more useful and often cheaper system is shown in Figure 6-2. Compressed air, either from a central supply or from a cylinder, is mixed with carbon dioxide using two rotameter flowmeters to give the desired mixture (usually 5% carbon dioxide in air). This mixture is led into the incubator, saturated with water vapor, then led into a drying cabinet that contains the culture chambers. These cabinets may be operated separately with their own flow-meters or they may be connected in series. Whatever the particular arrangement, the gas mixture flowing through the system is checked by bubbling the effluent gas through water. The cultures are incubated at 37°C.

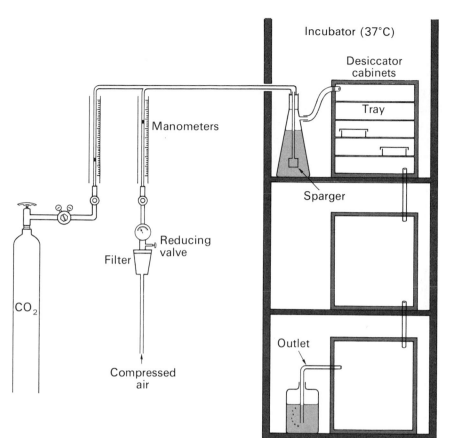

FIGURE 6-2

Equipment for incubating organ cultures in an atmosphere with a controlled variable content of carbon dioxide. Air from a central compressor is fed through a filter (Model F-300, Autofly Corp., Detroit) and a reducing valve to a manometer (Universal Flow Meter, #204, Saphire float, Matheson Company, East Rutherford, N.J.). Carbon dioxide is fed into a similar manometer (#203 Saphire float). These meters are adjusted, from charts supplied by the manufacturer, to provide 5% CO_2 in air, at a flow rate of 1 liter/min. The gas mixture is humidified and warmed by dispersing it through a flask of water kept inside the incubator. Dispersion is obtained with a fritted glass dispersion tube (Corning #39533-12 EC). The warm, humidified 5% CO_2 is passed through single or serially connected desiccator cabinets (Boekel type #4434-K, A. H. Thomas Co., Philadelphia). The outflow is checked by bubbling it through water. (From Biggers, 1965.)

SPECIFIC METHODS

Two major techniques have been used for the study of early mammalian development *in vitro*: (1) the culture of mammalian embryos independently of maternal tissue, and (2) the culture of embryos within organ cultures of fallopian tubes. The two techniques will be described separately.

Embryo Culture *in Vitro*

Collection of embryos

FERTILIZED SINGLE-CELL OVA (ZYGOTES). Fertilized single-cell ova are obtained from either naturally ovulated or superovulated mice on the morning a vaginal plug is found. Thus the embryos, athough several hours old when collected, are considerably younger than the age at which they are expected to undergo their first cleavage division. The mice are killed by cervical dislocation; their fallopian tubes are removed aseptically and placed in a drop of medium in a sterile glass Petri dish (60 mm in diameter). [A 4-cm-square cavity slide with a 3 cm diameter cavity may also be used (Adams, #A-1478).] The single-cell ova, surrounded by the cumulus, are released by incising the wall at the distal end of the ampullary region of the fallopian tube with a fine scalpel or needle. Since the cumulus mass is under pressure, it flows rapidly from the fallopian tube. The ova and cumulus are transferred with a finely drawn Pasteur pipette to a 13 x 100 mm test tube in one ml of medium, and stored under 5% carbon dioxide in air at 37°C in a water bath. Alternatively the ova and cumulus may be transferred to one ml of medium in an embryological watch glass. The ova from the mice in each experiment are pooled and divided among the experimental treatments. To do this the cumulus cells are removed with hyaluronidase.

A stock solution of hyaluronidase from bovine testis containing 300 U.S.P. units per ml is prepared. Twenty mg of hyaluronidase (Sigma Chemical Co.) and 200 mg of polyvinyl-pyrrolidone (PVP) (Plastone C, General Aniline and Film Corp.) are dissolved in 20 ml of Dulbecco's phosphate-buffered saline (PBS) at pH 7.2 (Dulbecco and Vogt, 1954). The solution is sterilized by passing it through a Millipore filter. One ml aliquots are pipetted into small sterile test tubes and stored before use at −20°C. If PBS is not available in the laboratory, the hyaluronidase can be dissolved in an aliquot of the culture medium. This solution should not be frozen. If the ova and cumulus are collected in a small test tube they should be transferred to an embryological watch glass by agitating the tube and pouring out its contents. One ml of the stock hyaluronidase solution is added to the embryological watch glass and the mixture is left at room temperature. If the enzyme is dissolved in the medium the ova should be incubated in mineral oil. Within several minutes the cumulus cells separate from the ova. By gently rotating the watch glass the ova are brought together and are then transferred to 1 ml of medium under 2 ml of mineral oil in a second watch glass. The washing procedure is repeated, two washes being sufficient to remove almost all the cellular debris from the ova.

The finely drawn Pasteur pipette is made by drawing out a commercially available sterile Pasteur pipette, the stem being heated by the small pilot flame of a bunsen burner. Only pipettes with a diameter slightly greater than that of the embryos or cumulus mass are used. The pipette is connected to a rubber tube by a sterile glass connector plugged with absorbent cotton, and is controlled either by mouth or by an Agla micrometer syringe. An alternative to the finely drawn Pasteur pipette is the braking pipette (see Glick and Holter, 1961, for a description of various types).

TWO-CELL EMBRYOS. Late two-cell embryos are flushed from the fallopian tubes of mated mice about 33 hours after the midpoint of the dark period, or from superovulated mice 44 to 48 hours after the injection of HCG. The tubes are removed as before, then flushed with medium from either the ampullary or uterine end. Flushing is done with a small syringe fitted with a 30-gauge needle. When flushing from the ampullary end, a blunt needle is inserted directly into the lumen; when flushing from the uterine end, a sharp needle with a short bevel is inserted through the wall of the isthmus. The embryos from the mice are collected in 2 ml of medium in either a small test tube or embryological watch glass. The test tube should be gassed frequently with 5% CO_2 in air, and the medium in the watch glass should be covered with one ml of mineral oil. The embryos are washed in at least two changes of medium before being allotted to different treatments.

EIGHT-CELL EMBRYOS, MORULAE, AND BLASTOCYSTS. These stages are collected in essentially the same manner as the two-cell embryos. The eight-cell embryos are obtained from superovulated mice about 70 hours after the injection of HCG or 56 to 60 hours after the midpoint of the dark period during which mating occurred. At this time a distribution of stages is obtained with the eight-cell stage being the mode. Morulae and early blastocysts are obtained about 98 hours after the injection of HCG. In collecting all these stages about 0.5 cm of uterus is left attached to the fallopian tube; the combined structure is then flushed. The uterine horn is flushed in the opposite direction. These procedures increase the yield of morulae and early blastocysts significantly since both exist in the fallopian tube and uterus at this period.

Removal of the zona pellucida

Removal of the zona pellucida, which is necessary in many experiments, can be done either mechanically (Tarkowski, 1963), or, more conveniently, with the enzyme pronase (Mintz, 1962; Gwatkin, 1964). A 0.25% solution of pronase is prepared by dissolving 50 mg pronase (Calbiochem) and 200 mg PVP in 20 ml PBS. The solution is allowed to stand for 30 minutes at room temperature before it is sterilized by passing it through a Millipore filter. One ml aliquots are stored at $-20°C$ in sterile test tubes. The solution should be filtered again immediately before use because a precipitate forms upon freezing. Ova are transferred to 1 ml of pronase in an embryological watch glass at room temperature. The zona pellucida disappears within 3 to 7 minutes and the ova are then washed twice in 2 ml of medium. Embryos without their zonas are very sticky and should be transferred with siliconized pipettes (Siliclad, 1% v/v solution, Clay-Adams Inc.).

Culture dishes

A variety of vessels have been used for the culture of mammalian embryos (Brinster, 1968). Of these only three have been used extensively for the culture of mouse embryos. These are: (1) the test-tube method, (2) the microdroplet method, and (3) the watch-glass method.

TEST-TUBE METHOD. One ml of freshly gassed medium is placed in glass or plastic 5 ml test tubes. The air space above the medium is flushed with 5% CO_2 in air and the tubes promptly sealed. Equal numbers of embryos are placed in the tubes which are again flushed with the gas mixture and placed in the incubator. After the appropriate culture period each tube is sharply agitated to dislodge the embryos from the bottom and the contents are poured into a cavity slide for examination. The embryos may be recovered and placed in another tube for further incubation. This method was described by Whitten (1956a).

MICRODROPLET METHOD. In this method, as originally described by Brinster (1963), droplets of medium are placed beneath a layer of mineral oil in a 15 x 60 mm plastic Petri dish. Mineral oil presents no significant barrier to the exchange of gases between the medium and atmosphere in the incubator. Several days before being used the mineral oil is equilibriated with the medium. To do this 10 to 15 ml of complete medium, or of only the salts of the medium, are mixed with 400 ml of light weight paraffin oil (Saybolt viscosity 125 to 135, Fisher, Inc.). The oil is sterilized by dry heat, 5% carbon dioxide in air is bubbled through it for 15 minutes, and it is then stored at 37°C in 5% CO_2 in air. Two days are allowed for separation of the components of the medium. Ten ml of sterilized equilibrated oil is placed in each dish, and the medium is pipetted as a microdrop (25 to 100μl) under the oil, using a finely drawn Pasteur pipette or syringe. Usually four drops of oil are placed in each Petri dish; equal numbers of embryos are then put into each of the drops (p. 106). The function of the mineral oil appears to be the limitation of evaporation and consequent concentration of the medium.

The advantages of this method are that the embryos can be observed easily under the microscope at any stage of the experiment and that small volumes of the media can be used. Nevertheless there are problems in the use of the method. From a purely practical point of view it is essential to obtain the correct grade of plastic Petri dish (Tissue Culture Dish, Falcon Plastics, #3002) and the correct batch of mineral oil. Batches of light oil vary in their ability to stabilize aqueous drops; the reason for this is unknown. It is thus necessary to test several batches of oil for this property before use.

A serious limitation of this method is the ready solubility of components of the medium in mineral oil. Many substances, such as steroids, are freely soluble in mineral oil, so if they are incorporated within the medium, they will equilibrate with the oil according to their partition coefficients between water and mineral oil. Recently Donahue and Stern (1968) have shown that pyruvate is very soluble in mineral oil and this may seriously interfere with quantitative studies of early cleavage. Unfortunately, the use of liquid silicones instead of mineral oil does not overcome this problem.

WATCH-GLASS METHOD. In one version of this method mouse embryos are cultured in one ml of medium under one ml of mineral oil in an embryological watch glass enclosed by a cover glass; this allows the cultivation of more embroys in one dish, and is therefore suitable for bulk production of blastocysts from earlier stages. Attempts to culture mouse embryos without the mineral oil have not been uniformly successful, even if a cover glass is held firmly on the watch glass. In another version, Mintz (1964) uses small watch glasses suspended in French-square bottles containing a bicarbonate-buffered salt solution (see p. 194).

ORGAN CULTURE OF FALLOPIAN TUBES

There are several methods available for the culture of organs *in vitro* (Moscona, Trowell, and Willmer, 1965). However, only two, both of which use liquid media, have been extensively used to study the development of cleavage stages. These are the "grid" method used by Biggers, Gwatkin, and Brinster (1962), and Pavlok (1967) to study entire explants of the mouse oviduct, and the "raft" method used by Whittingham (1967) to study isolated ampullary and isthmal regions *in vitro*.

Grid method

The method of Trowell (1954, 1959), modified by Jensen and Castellano (1960) and Jensen, Gwatkin, and Biggers (1964) consists of a stainless steel mesh grid (J. E. Franklin Company, Philadelphia) 20 x 20 x 3 mm high with arches cut away from two of the vertical walls to allow entrapped air to escape. The grids are cleaned, and then sterilized in a hot air oven (one hour at 150°C). A single grid is placed in a sterile Petri dish (60 mm diameter), and 10 ml of medium is pipetted over the platform. The height of the grid should be adjusted to allow the surface of the medium to reach the underside of the grid top. A piece of sterile tea-bag paper (#10-V-7-1/4 #1, C. H. Dexter and Sons, Inc., Windsor Locks, Connecticut) is placed on the grid, and capillarity draws the medium into the paper, holding it in place. A smaller commercial version of this chamber is available (Falcon Plastics, dish #3010; grid #3014).

The oviducts are removed from mice by cutting through the ovarian bursi as close as possible to the ovaries and then through the uterotubal junctions. The oviducts are then placed in a drop of medium in a sterile Petri dish and carefully uncoiled by clipping the mesentery. Each oviduct is stretched out gently across the tea-bag paper. If cultures of only the ampullary or isthmal regions are required the organ is cut at the ampullary-isthmal junction, and the portions are placed on the tea-bag paper. The cultures are incubated and, at the end of the culture period, the tubes are flushed with medium using a 30-gauge needle and the embryos collected.

Raft method

A modification of Chen's technique (1954) may be used to culture the complete fallopian tube or its ampullary and isthmal regions separately (Biggers and Gwatkin, 1961). The culture chamber consists of a Petri dish (60 mm in diameter), an annulus of absorbent synthetic sponge which allows observation of the explant under the microscope with transmitted light, and a small watch glass (40 mm in diameter) supported by the sponge. This chamber as a unit is sterilized with dry heat. A sufficient amount of sterile distilled water (2 ml) is placed in the bottom of the Petri dish to maintain the humidity of the chamber. One-half of medium is pipetted into the watch glass, and a piece of sterile, washed tea-bag paper is floated on top of the medium. The explants are placed on the paper and incubated. The medium is renewed every 48 hours by transferring the tea-bag paper plus the attached explant to a fresh culture chamber.

Comparison of the grid and raft methods

The raft method has certain advantages over the grid method. First, the volume of medium used for the culture of each explant is reduced tenfold. Second, the explant can be easily observed under the microscope. These advantages are particularly important in experiments in which ova are introduced into the explanted ampullary region.

MEDIA

Two main types of media are used in tissue culture work—biological and chemically defined. The design of chemically defined media has been regarded as a central problem of tissue culture since its conception (Lewis and Lewis, 1912). Some of the advantages of using chemically defined media are: (a) they are reproducible at different times and in different laboratories, (b) they can be varied in a controlled manner, and (c) they are free of enzyme activities which may interfere with the responses being studied. A chemically defined medium means that only highly purified reagents are used (this includes the composition of the gas phase), and that ideally the interactions between the constituents are understood. The medium is not chemically defined if it includes biological fluids or partially purified products, such as dialyzed serum. The characteristics and principles of the design of such media have been discussed from several points of view (Waymouth, 1954, 1965; Biggers, Rinaldini, and Webb, 1957; Biggers, 1963). The work on the culture of mouse embroys in organ cultures of adult fallopian tubes has used chemically defined media such as BGJ (Biggers, Gwatkin, and Heyner, 1961) and F10 (Ham, 1963) which do not require further discussion. It is necessary, however, to consider the problems of culturing mammalian embryos in chemically defined media in more detail.

MEDIA FOR THE CULTURE OF
MOUSE EMBRYOS *in Vitro*

Biological media

Only two investigators appear to have used biological media. Hammond (1949) was able to obtain blastocysts from the eight-cell stage using a mixture of thin egg-white, egg yolk and inorganic salts. The atmosphere was not enriched with carbon dioxide. Recently, Mintz (1964) has cultured late two-cell stages to blastocysts in a medium consisting of 50% fetal calf serum (Microbiological Associates) and 50% Earle's balanced salt solution, containing 0.002% phenol red, and one mg/ml L (+)-lactic acid. The pH was adjusted to 7.0 with 7.5% sodium bicarbonate under 5% carbon dioxide in air.

Chemically defined media

The first chemically defined medium for the culture of mammalian cleavage stages was described by Whitten (1956a), who was able to obtain mouse blastocysts from eight-cell stages. Its composition is shown in Table 6-4. It is based on Krebs-Ringer bicarbonate, supplemented with glucose, bovine plasma albumin and antibiotics. McLaren and Biggers (1958) produced blastocysts by this technique, transferred them to uterine foster mothers and produced normal offspring. Exactly the same method was used by Tarkowski (1961) to produce from eight-cell stages the first pre- and postnatal

TABLE 6-4
Media for the culture of early mouse embryos in mM

Component	Whitten (1956a)	Whitten (1957)	Brinster (1963)	Brinster (1965b)
Sodium chloride	118.46	118.46	109.23	119.32
Potassium chloride	4.74	4.74	4.78	4.78
Potassium dihydrogen phosphate	1.18	1.18	1.19	1.19
Calcium chloride	2.54	—	1.71	1.71
Magnesium sulphate	1.18	1.18	1.19	1.19
Calcium lactate (L+)	—	2.54	—	—
Sodium lactate (DL)	—	—	10.15	25.0
Sodium pyruvate	—	—	—	0.25
Sodium bicarbonate	24.88	24.88	25.07	25.07
Glucose	5.55	5.55	—	—
Crystalline bovine albumin	1 mg/ml	1 mg/ml	1 mg/ml	1 mg/ml
Penicillin	10 μg/ml	10 μg/ml	100 I.U./ml	100 I.U./ml
Streptomycin	10 μg/ml	10 μg/ml	50 μg/ml	50 μg/ml

mouse chimeras, now known as allophenic mice (see Mintz, p. 186). Since the description of the original medium several modifications have been made, and the more important of these are also shown in Table 6-4.

Whitten (1957a) demonstrated that the addition of calcium lactate allowed late two-cell embryos to develop into blastocysts. Brinster (1963) modified this medium, and clearly showed that lactate alone could supply the embryo during cleavage and that an energy source was essential. Brinster (1965a) subsequently demonstrated that pyruvate, oxaloacetate, and phosphoenolpyruvate could also substitute for lactate in the medium. The application of appropriate statistical procedures for the exploration of concentration response surfaces (Biggers and Brinster, 1965) finally resulted in the demonstration that the incorporation of pyruvate and lactate in the medium results in better yields of blastocysts from late two-cell embryos than is obtained with the optimum concentration of lactate alone (Brinster, 1965b). Biggers, Moore, and Whittingham (1965) transferred mouse blastocysts produced from two-cell stages in lactate or pyruvate into pseudopregnant uterine foster mothers and showed that normal fetuses would develop. Whittingham and Biggers (1967) subsequently found that mouse zygotes would readily cleave to the two-cell stage in the pyruvate/lactate medium but they would not in the medium containing lactate. These two-cell stages developed to blastocysts in organ cultures of fallopian tubes and into normal embryos on transfer to uterine foster mothers. However, it has not been possible to culture mouse embryos successfully between the first and second cleavage division. Later Biggers, Whittingham, and Donahue (1967) showed that only pyruvate and oxaloacetate could act as single energy sources for the first cleavage division, and therefore it is unnecessary to include lactate in the medium. It was also shown that isolated oocytes would undergo meiotic maturation to the metaphase II stage in the same medium.

The development of the media shown in Table 6-4 was the result of new discoveries on the special requirements of different stages of development, and the search for optimal concentrations by factorial experimentation. The composition of media must be expected to vary, and the choice of any particular one will depend on particular needs. It is also highly desirable to test the normality of cleavage stages produced in different media by seeing if normal individuals develop after transfer to uterine foster mothers.

Relatively few studies have been made on other components of the media. Brinster (1965d) has been able to culture two-cell embryos to blastocysts in a medium in which the bovine serum albumin is replaced with an equivalent mixture of free amino acids, if a nonprotein polymer was included. The best polymers were polyvinylpyrrolidone (Mol. wt. = 150,000) gum acacia, dextran, and Ficoll.

A medium useful for the production of numbers of well-developed blastocysts from the two-cell stage is shown in Table 6-5. This medium contains three energy sources. The calcium is added to the medium in the form of calcium lactate which has an advantage over the calcium chloride commonly used because the lactate is nonhygroscopic. If calcium chloride is used, a given concentration of calcium ions can be obtained by titrating a

TABLE 6-5

Standard egg culture medium

Component	Mol. Wt.	gm/l	mM	Milliosmols
NaCl	58.4	5.540	94.59	189.19
KCl	74.6	0.356	4.78	9.56
Ca-lactate · 5H$_2$O	308.3	0.527	1.71	5.13
KH$_2$PO$_4$	136.1	0.162	1.19	2.38
MgSO$_4$ · 7H$_2$O	246.5	0.294	1.19	2.38
NaHCO$_3$	84.0	2.106	25.07	50.14
Na-pyruvate	110.0	0.028	0.25	0.50
Na-lactate	112.1	2.416 (3.68 ml/1 of syrup)	21.58	43.10
Glucose	180.2	1.0	5.56	5.56
Crystalline bovine albumin		1.0		
Antibiotic stock solution[1]		1.0 ml		
Distilled water		1000 ml		
Total				308

[1] 100,000 I.U./ml penicillin and 50 mg/ml streptomycin. This stock solution is kept frozen in 1– ml lots.

stock solution of calcium chloride with silver nitrate. Figures 6-4 and 6-5 show the development of two-cell embryos to blastocysts in this medium.

OSMOLARITY. Table 6-5 shows the molecular weight and molar concentration of each component used to calculate the osmolarity of the medium. (The bovine plasma albumin and antibiotics are ignored in this calculation.) The osmolarity of a single component in dilute solution equals its molar concentration multiplied by the number of particles produced in solution. Complete ionization is assumed for all the salts and the corresponding osmolarities shown in Table 6-5. Thus the osmolarity of glucose equals its molar concentration since it does not ionize, and the osmolarity of calcium lactate is three times its molar concentration since it ionizes into three particles. The sum of the osmolarities of the individual components gives the total osmolarity of the medium. Further discussion of the osmotic properties of solutions from a physiological point of view is given by Brown (1965) . The total osmolarity of the medium in Table 6-5 is 308 milliosmols. If the concentrations of components are changed, an osmolarity of 308 milliosmols is maintained by adjusting the concentration of sodium chloride.

These calculations are based on the assumptions of complete ionization and no interaction between the different ions or between ions and other particles. There is no guarantee that these assumptions are valid for defined media and empirical measurements should be made. Nevertheless, Biggers

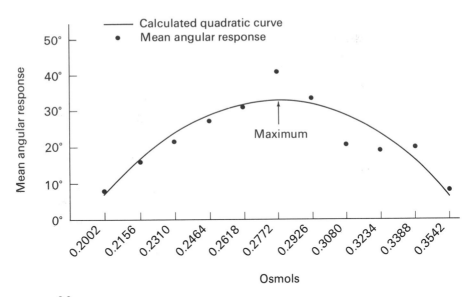

FIGURE 6-3

Effect of osmolarity on the development of two-cell mouse embryos. (From Biggers and Brinster, 1965.)

and Brinster (1965) derived a quadratic curve which adequately represented the responses of two-cell eggs cultured in solutions of ideal calculated osmolarity. The data and the curve are presented in Figure 6-3. The maximum response was given by an osmolarity of 277 milliosmols. More extensive data gave an estimate of the maximum response as 276 with fiducial limits of error $(P = 0.05)$ 272 to 280 (Brinster, 1965).

PREPARATION OF CHEMICALLY DEFINED MEDIA. Two methods may be used for the preparation of media: (1) bulk production with storage in small volumes, and (2) stock solutions from which the complete medium is prepared. The first method is ideally suited for a standard laboratory medium used for routine purposes by several investigators. The second method is preferable in investigative work where several variants of a medium need to be compared simultaneously.

To prepare one liter of the medium shown in Table 6-5, the solid components, except the crystalline bovine albumin, are weighed out and dissolved in 800 ml of distilled water. The lactate syrup is added and the volume made up to one liter. The crystalline bovine albumin is dissolved last in order to avoid frothing. One ml of the antibiotic stock solution is added to the medium, which is then made up to one liter with distilled water. The medium is sterilized by passage through a Millipore filter under positive pressure to minimize the loss of bicarbonate (Parker, 1961). The medium is then stored in small volumes for individual use. It is important to leave 30% of the volume of the bottle for gassing with 5% CO_2 in air; it is then stored in a refrigerator. The medium should not be shaken. The pH of the medium must be maintained between 7.2 and 7.4 by gassing with 5%

CO_2 in air each time the bottle is opened. The pH sould be checked with a pH meter.

To prepare small quantities of the pyruvate/lactate medium shown in Table 6-4, the *isoosmotic* stock solutions (308 milliosmols) shown in Table 6-6 are prepared. The solutions are made with distilled water containing one gm/liter bovine serum albumin, 10^5 U/liter penicillin, and 50 mg/liter streptomycin. The stock solutions are sterilized through Millipore filters and stored in the refrigerator. To prepare 13 ml of medium the aliquots shown in Table 6-6 are mixed and equilibrated with 5% carbon dioxide in air to a pH of 7.38. If necessary the mixture can be sterilized with a Millipore filter in a Swinney adaptor. If variations in the composition of the medium are required or if the volumes shown in Table 6-6 are altered, the osmolarity should be kept constant by adjusting the amount of sodium chloride added.

TABLE 6-6

Preparation of pyruvate/lactate medium

Solution	Component	Concentration gm/l	ml to make 13 ml
1	NaCl	5.546	5.90
2	KCl	0.356	0.40
3	$CaCl_2$	0.189	0.20
4	KH_2PO_4	0.162	0.10
5	$MgSO_4 \cdot 7H_2O$	0.294	0.10
6	$NaHCO_3$	2.106	2.10
7	Na-lactate	2.253	2.10
8	Na-pyruvate	0.028	2.10[1]

[1] Na-pyruvate is $0.00154M$ pyruvate in $0.154M$ NaCl.

THEORETICAL ASPECTS OF EMBRYO CULTURE

An embryo-culture system consists basically of two phases: the embryos and the surrounding medium. The two phases are not independent; they interact with each other from the moment the embryos are placed *in vitro*. If the system uses mineral oil a third phase is present, and the relation between the medium and the oil overlay must also be considered. Most experiments are designed to correlate the response of the embryos with the composition of the medium.

CRITERIA OF EMBRYO RESPONSE

Many types of observations can be made of the response of embryos in culture. In general they may be divided into three broad categories: (1) quantal (all or none) properties, (2) quantitative measurements of

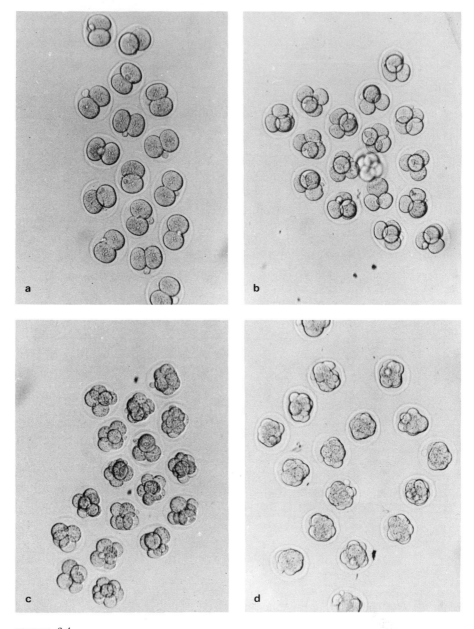

FIGURE 6-4

*Serial photographs of eggs collected at the two-cell stage, at 10 AM on the morning
following that on which vaginal plugs were observed, in hybrid* F_1 ♀ SJL/J ♂ ×
C57BL/10J, *females mated naturally with* F_1 ♀ BALB/cDg × ♂ 129/J *males. The eggs
were cultured in the medium given in Table 6-5, under paraffin at 37° C and photographed at 12 hourly intervals.* ×101.

 (a) *Two-cell eggs at beginning of* in vitro *culture.* (b) *Three- and four-cell eggs after
12 hours in culture.* (c) *Eggs at the four- to eight-cell stages after 24 hours in culture.*
 (d) *Morulae after 36 hours in culture.*

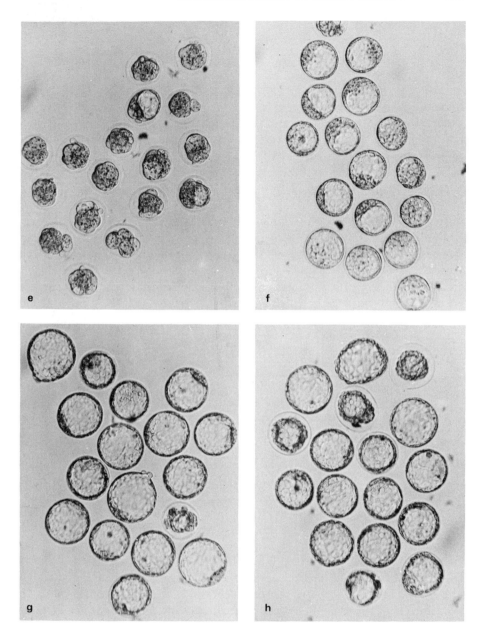

(e) *Morulae and early blastocysts after 48 hours.* (f) *Sixteen blastocysts and one doubtful developer after 60 hours.* (g) *Hatching blastocysts (top and center). Small cavity in doubtful developer at 72 hours.* (h) *Rupture of zona (top right) with partial collapse of blastocyst after 84 hours. A significant cavity (bottom) has developed in the doubtful specimen.*

properties which are expressed on a continuous scale, and (3) semiquantal measurements in which more than two possible states may occur.

Quantal observations

Quantal observations are usually concerned with morphological responses (Biggers and Brinster, 1965), and these have, until recently, constituted the main type of observation since very few of the ultramicrotechniques needed to make the other type were available. To use this procedure the criterion of response chosen should be clearly recognizable, and the effect of a particular medium measured by exposing a group of embryos to it and observing the proportion that exhibit the criterion. There is an almost infinite number of criteria which may be chosen to define a positive response. For example, an embyro may have (*a*) cleaved, (*b*) divided into more than eight cells, (*c*) developed into a morula, (*d*) developed a blastocoele, or (*e*) developed into an early blastocyst with an inner cell mass. The choice of a suitable end point is not entirely arbitrary since a criterion must be chosen which is easily recognized by the optical techniques used. One of the most satisfactory is the formation of a blastocyst since this stage can be easily seen through the dissecting microscope. The recognition of cleavage and blastocoele formation requires very careful microscopy with an inverted microscope. For example, cleavage has to be distinguished from fragmentation, which is one of the early stages of degeneration of mammalian ova; and formation of a blastocoele cavity must be distinguished from a vacuole within a blastomere (Tarkowski and Wróblewska, 1967).

If several media are compared using a quantal measure of response the data are analyzed statistically by methods of quantal analysis (Finney, 1952). For such analyses to be valid it is essential that the embryos be allotted at random to the experimental treatments. It is for this reason that the embryos from all donor mice are first collected in a tube and then randomly distributed. Quantal analysis is greatly simplified if the number of embryos per drop is kept constant. Under these conditions the angular transformation can be used. The availability of this technique allows the analysis of many experimental designs (including factorial designs) which are invaluable in the exploration of concentration-response surfaces. In experiments of this type the group size (number of embryos per drop) must be greater than ten (Claringbold, Biggers, and Emmens, 1952).

A complication arises in the use of mammalian embryos in experiments where the response is quantal. It cannot be assumed that all embryos are normal and capable of responding. Abnormal fertilization, or collection of embryos before a critical time may cause a lack of response. Biggers and Brinster (1965) have shown that provided two-cell embryos are selected at random from a common pool the probability that $0, 1, 2, \ldots, n$ embryos in a group of n embryos will develop into blastocysts follows a binomial distribution. The argument can be applied generally to the other embryonic stages, and validates the general use of the angular transformation and other methods of quantal analysis despite the occurrence of inherent nondevelopers.

Quantitative observations

Only two types of measurement have been made on single embryos. These are the volume at different stages of development (Lewis and Wright, 1935), and lactic dehydrogenase (Brinster, 1965). More recently, techniques have been developed to measure properties of relatively small groups (<100) of embryos. Several of these techniques are listed in Table 6-7. In applying them in comparative culture experiments it is essential to use embryos chosen randomly from a common pool. This not only insures the validity of the statistical analyses but distributes the number of potential nondevelopers randomly between treatments. The use of batches of embryos from individual mice on a single treatment may result in severe biases due to variation in the quality of embryos between the individuals.

The response of an embryo may also be more exhaustively measured by the simultaneous measurement of several (p) properties. The response is then represented by a vector of p-correlated measurements. The analysis of such data requires the techniques of multivariate analysis many of which are now available on high speed digital computers. An account of the use of these methods in biology is given by Seal (1964).

TABLE 6-7
Microtechniques applied in the study of mouse embryos

Techniques	Study
Dry weight	Loewenstein and Cohen (1964)
Lipid	Loewenstein and Cohen (1964)
Total protein	Loewenstein and Cohen (1964); Brinster (1967a)
Lactic dehydrogenase	Brinster (1965c)
Lactic dehydrogenase isozyme	Auerbach and Brinster (1967); Rapola and Koskimies (1967)
Malic dehydrogenase	Brinster (1966a)
Glucose–6–phosphate dehydrogenase	Brinster (1966b)
Carbon dioxide production	Brinster (1967b); Wales and Biggers (1968); Wales and Whittingham (1967)
Oxygen consumption	Mills and Brinster (1967)
Uptake labeled substrates:	
Estradiol–17β	Smith (1966, 1968)
Uridine, leucine, lysine	Monesi and Salfi (1967)
Malic acid	Wales and Biggers (1968)
Lactate and pyruvate	Wales and Whittingham (1968)
Glucose	Popp (1958); Wales and Brinster (1968)
Glycogen	Stern and Biggers (1968)

Semiquantal observations

In some situations it is possible to classify the response of the embryos into more than two states. Such responses are said to be semiquantal or polychotomous. An example is shown in Table 6-8 in which seven possible responses of two-cell mouse embryos are listed. One method of analyzing such data is to alot a numerical value (score) to each state. The response of each embryo is then measured by a score which is used in standard statistical analyses. Whitten (1957b) used such a system in his study of the response of mouse embryos to progesterone. The scoring system is shown in Table 6-8. This system assumes that each criterion is equidistant from its neighbors on a linear scale. The validity of this assumption, and estimates of improved locations for each criterion, could be obtained by discriminant analysis (Fisher, 1936). This procedure is now readily done on modern digital computers. Recently a second method of analyzing polychotomous data has been described which may be preferable to the use of a scoring system (Walker and Duncan, 1967).

TABLE 6–8
Scoring criteria for the quantitative examination of ova

Criterion	Score
Dead fragmented cells	0
Morula stage	1
Early blastula (blastocoele smaller than cell mass)	2
Small blastula (capsule not completely filled)	3
Blastula (capsule filled)	4
Large or expanded blastula (capsule distended)	5
Blastula (capsule herniated)	6

SOURCE: Whitten (1957b).

RELATION BETWEEN THE RESPONSE OF THE EMBRYOS AND THE COMPOSITION OF THE MEDIUM

In 1957, Biggers, Rinaldini and Webb discussed theoretically the responses of tissue cultures on media of varying composition. It was pointed out that a particular response could be regarded as a dependent variable subject to variations in each of $i = 1, 2, \ldots, n$ compounds in the medium. If this is visualized geometrically, the locus of all possible responses will form a concentration-response surface in i-dimensional space. Thus, in general, the practical experimental problem is to explore and map this surface. The exploration of complex concentration-response surfaces has been developed to a high degree in the field of chemistry. Further details on problems connected with setting up suitable mathematical models and experimental designs are given by Box and Wilson (1951) and Box (1954).

The simplest—but not necessarily the best—type of experimental design for the exploration of concentration-response surfaces are factorial experiments. These designs have been used in the study of the optimum combination of energy sources for the cleaving two-cell mouse embryo using a quantal response (Brinster, 1965b).

The concept of a concentration-response surface emphasizes the fact that responses may be related to an infinity of media of different composition. The magnitude of the responses obtained on these media will vary, and one may therefore speak of the medium which gives a *maximal* response. This should not be confused with an *optimal* medium, which is one that gives a response ideally suited to a particular purpose. Neither should the maximal response be identified as representing the natural degree of development which occurs *in vivo*, since conditions can be produced in chemically defined media which may lead to an excessively large response.

MISCELLANEOUS

Pharmacology of Embryos

Median effective dose (MED)

In some types of work it is necessary to obtain a measure of the activity of an agent, such as an inhibitor. Examples of this were given by Thomson and Biggers (1966) who estimated the activity of mitomycin C, actinomycin D, puromycin and fluorophenylalanine on two-cell and eight-cell mouse embryos. The technique is to use a quantal response, for example the development of the blastocyst, to measure the proportion of embryos responding at different concentrations of the drug. Such data can be used to estimate the dose of drug which causes the response in 50% of the population of embryos. Other levels of response can also be estimated for special purposes, such as the dose which blocks the development of 90% of the embryos. However, as pointed out by Trevan (1927), it is quite meaningless to attempt to measure so-called minimal and maximal effective doses.

The design of the experiments and statistical procedures for estimating the median effective dose is discussed in detail by Biggers and Brinster (1965). Because of the availability of statistical tables and computer programs, it is usual to analyze such data by probit analysis (Finney, 1952). This procedure also removes the necessity of insuring an equal number of embryos on each dose of the drug. The calculation is complicated, however, because it is necessary to allow for the presence of an unknown number of embryos which cannot develop.

Steroids

The pharmacology of the effect of steroids on mammalian eggs deserves special mention because of the interest in the relationship between embryo development and function in the presence of antifertility compounds. Some of the principles described above have been applied to the study of the

toxic effect of progesterone on mouse morulae (Whitten, 1956b, 1957b).
Daniel (1964) and Daniel and Levy (1964) observed a similar effect on
rabbit eggs and noted that this could be prevented by increasing the serum
or amino acid content of the medium. These observations suggest the need
for a concentration-response surface study which should be carried out in a
system without oil because of the oil solubility of most steroids.

The Blastocyst *in Vitro*

The culture of numbers of blastocysts from earlier stages is useful
in work in which embryos are required that have had minimal exposure to
maternal hormones. Gwatkin (1966a) used a medium resembling that in
Table 6-5 in studies of the attachment and outgrowth of mouse blastocysts
in vitro. Smith (1966, 1968) used the pyruvate/lactate medium (see Table
6-4) in studies which showed the active uptake and binding of estradiol-
17β by the mouse blastocyst, and the ability of such blastocysts to implant
in a nonestrogen-treated ovariectomized host.

Mouse blastocysts produced in the simple media shown in Tables 6-4 and
6-5 escape from the zona pellucida and then cease further development.
Gwatkin (1966b) showed that if these blastocysts were transferred to media
F10 they would spread on glass and undergo some differentiation. The
changes occurred, however, only when dialyzed serum, fetuin, or other serum
fractions were added to the medium and provided certain free amino acids
were present. These amino acids were later shown to be arginine, cystine,
histidine, leucine and threonine, and possibly some others (Gwatkin,
1966a). These observations may indicate a radical change in the nutritional
requirements of mouse blastocysts at the time of implantation.

The findings also emphasize the need for extreme caution in designing
assays with chemically defined media for the detection of specific substances
which may regulate early mammalian development. The addition of protein fractions to the medium appears necessary for other compounds in the
medium to be used. The mechanism of action is not understood but may,
among several possibilities, involve the introduction of small molecular
impurities or an interaction with the components of the medium. It is quite
possible that the addition of partially purified fractions of a biological fluid,
for example from a Sephadex column, to a medium like F10, with many
components including free amino acids, may exert effects similar to these
serum fractions. Thus they may merely be creating the conditions necessary
for the growth and differentiation of the blastocysts. Under these circumstances it is very difficult, if not impossible, to prove the fraction being
tested has a unique biological effect. The application of chemically defined
media in biological testing is relatively unstudied, and requires very critical
analysis of its limitations.

Addendum

Since this chapter was written some success with the culture of
one-cell ova from certain hybrids has been reported (Whitten and Biggers,

1968). The major change in the medium was a reduction in the sodium chloride concentration such that the molarity was about 250 milliosmols.

Subsequently, one-cell ova from several F_1 hybrids—some were inbred strains and others were from a random bred strain—have been cultured beyond the two-cell stage. The work has been done independently at The Jackson Laboratory and, in a slightly different procedure, at Johns Hopkins University.

Whitten (1969) found that all normal one-cell ova from several F_1

TABLE 6-9

Modified ovum culture media

	Whitten (1969)		Biggers, Stern, and Whittingham (unpub.)	
	g/l[1]	mM[1]	g/l[2]	mM[2]
NaCl	4.14	70.89	5.52	94.6
KCl	0.356	4.78	0.356	4.78
KH_2PO_4	0.162	1.19	0.162	1.19
$MgSO_4 \cdot 7H_2O$	0.294	1.19	0.294	1.19
Ca lactate $5H_2O$	0.527	1.71	0.527	1.71
Na-lactate (DL)	2.494[3]	22.28	2.416[4]	21.58
Na-pyruvate	0.036	0.33	0.036	0.33
$NaHCO_3$	1.900	22.61	2.106	25.07
Glucose	1.0	5.56	1.0	5.56
Crystalline bovine albumin	3.0	—	4.0	—
K-penicillin-G	0.075	—	100 I.U./ml	—
Streptomycin SO_4	0.05	—	0.05	—

[1] Gassed in 5% O_2, 5% CO_2, 90% N_2.
[2] Gassed in 5% CO_2 in air.
[3] 3.80 ml 60% syrup.
[4] 3.68 ml 60% syrup.

hybrids and many that were from inbreds developed into expanded blastocysts within four days if the oxygen concentration of the gas phase was reduced to 5% and if the bicarbonate was also reduced (see the modified formula in Table 6-9). The embryos developed both in glass tubes with rubber stoppers and in drops under oil. These results were obtained consistently at The Jackson Laboratory but were not completely confirmed at Johns Hopkins.

Biggers, Stern and Whittingham (unpublished) have also been able to culture one-cell ova to blastocysts from several F_1 hybrids and from inbred strains by using the modified formula also shown in Table 6-9. In this medium the sodium chloride and bovine plasma albumin concentrations are higher than in that used at The Jackson Laboratory, and the oxygen tension is not reduced.

The causes for the differences in the results are being investigated.

BIBLIOGRAPHY

Adams, C. E. (1965) The influence of maternal environment on preimplantation stages of pregnancy in the rabbit. In: Preimplantation Stages of Pregnancy. Ed. by G. E. W. Wolstenholme and M. O'Connor. Little, Brown, Boston, pp. 345–373.

Auerbach, S., and R. L. Brinster (1967) Lactate dehydrogenase isozymes in the early mouse embryo. Exp. Cell Res., 46:89–92.

Biggers, J. D. (1963) Studies on the development of embryonic cartilaginous long-bone rudiments *in vitro*. Nat. Cancer Inst. Monograph, 11:1–21.

———— (1965) Cartilage and Bone. In: Cells and Tissues in Culture, vol. 2, chap. 4, pp. 198–260 Ed. by E. N. Willmer. Academic Press, London.

———— and R. L. Brinster (1965) Biometrical problems in the study of early mammalian embryos *in vitro*. J. Exp. Zool., 158:39–48.

———— and R. B. L. Gwatkin (1961) The effect of bendaryl on the water content of embryonic chick tibiotarsi *in vitro*. J. Exp. Zool., 148:21–29.

————, R. B. L. Gwatkin, and R. L. Brinster (1962) Development of mouse embryos in organ culture of fallopian tubes on a chemically defined medium. Nature, 194:747–749.

————, R. B. L. Gwatkin, and S. Heyner (1961) Growth of embryonic avian and mammalian tibiae on a relatively simple chemically defined medium. Exp. Cell Res., 25:41–58.

————, B. D. Moore, and D. G. Whittingham (1965) Development of mouse embryos *in vivo* after cultivation from two-cell ova to blastocysts *in vitro*. Nature, 206:734–735.

————, L. R. Rinaldini, and M. Webb (1957) The studies of growth factors in tissue culture. Symp. Soc. Exp. Biol., 11:264–297.

————, D. G. Whittingham, and R. P. Donahue (1967) The pattern of energy metabolism in the mouse oocyte and zygote. Proc. Nat. Acad. Sci. (United States), 58:560–567.

Box, G. E. P. (1954) The exploration and exploitation of response surfaces: Some general considerations and examples. Biometrics, 10:16–60.

———— and K. B. Wilson (1951) On the experimental attainment of optimum conditions. J. Roy. Statist. Soc., Ser. B, 13:1–45.

Brachet, A. (1912) Developpement *in vitro* de blastomeres et jeunes embryons de mammiferes. C. R. Acad. Sci. (Paris) 155:1191.

Brinster, R. L. (1963) A method for *in vitro* cultivation of mouse ova from two-cell to blastocyst. Exp. Cell Res., 32:205–208.

———— (1964) Studies on the development of mouse embryos *in vitro*. Ph.D. thesis, University of Pennsylvania.

———— (1965a) Studies on the development of mouse embryos *in vitro*. II. The effect of energy source. J. Exp. Zool., 158:59–68.

———— (1965b) Studies on the development of mouse embryos *in vitro*. IV. Interaction of energy sources. J. Reprod. Fertility, 10:227–240.

———— (1965c) Lactate dehydrogenase activity in the preimplanted mouse embryo. Biochim Biophys. Acta, 110:439–41.

———— (1965d) Studies on the development of mouse embryos *in vitro*. III. The effect of fixed nitrogen source. J. Exp. Zool., 158:69–77.

———— (1966a) Malic dehydrogenase activity in the preimplantation mouse embryo. Exp. Cell Res., 43:131–135.

———— (1966b) Glucose-6-phosphate dehydrogenase activity in the preimplantation mouse embryo. Biochem. J., 101:161–163.

———— (1967a) Protein content of the mouse embryo during the first five days of development. J. Reprod. Fertility, 13:413–420.

———— (1967b) Carbon dioxide production from glucose by the preimplantation mouse embryo. Exp. Cell Res., 47:271–277.

———— (1968a) Mammalian embryo culture. In: The Mammalian Oviduct. Ed. by E. S. E. Hafez and R. J. Blandau. University of Chicago Press, pp. 419–444.

Bronson, F. H., C. P. Dagg, and G. D. Snell (1966) Reproduction. In: Biology of the Laboratory Mouse, chap. 11, pp. 187–204. Ed. by E. L. Green. McGraw-Hill, New York.

Brown, A. C. (1965) Passive and active transport. In: Physiology and Biophysics, 19th ed., chap. 43, pp. 820–842. Ed. by T. C. Ruch and H. D. Patton. Saunders, Philadelphia.

Chen, J. M. (1954) The cultivation in fluid medium of organized liver, pancreas and other tissues of foetal rats. Exp. Cell Res., 7:518–529.

Claringbold, P. J., J. D. Biggers, and C. W. Emmens (1953) The angular transformation in quantal analysis. Biometrics, 9:467–484.

Daniel, J. C. (1964) Some effects of steroids on cleavage of rabbit eggs in vitro. Endocrinology, 75:706–710.

———— (1965) Studies on the growth of five-day-old rabbit blastocysts in vitro. J. Embryol. Exp. Morphol., 13:83–95.

———— (1967) The pattern of utilization of respiratory metabolic intermediates by preimplantation rabbit embryos in vitro. Exp. Cell Res., 47:619–623.

———— and J. D. Levy (1964) Action of progesterone as a cleavage inhibitor of rabbit ova in vitro. J. Reprod. Fertility, 7:323–329.

Donahue, R. P., and S. Stern (1968) Follicular cell support of oocyte maturation: production of pyruvate in vitro. J. Reprod. Fertility, 17:395–398.

Dulbecco, R., and M. Vogt (1954) Plaque formation and isolation of pure lines with poliomyelitis viruses. J. Exp. Med., 99:167–182.

Edwards, R. G., and A. H. Gates (1959) Timing of the stages of the maturation divisions, ovulation, fertilization and the first cleavage of eggs of adult mice treated with gonadotrophin. J. Endocrinol., 18:292–304.

Finney, D. J. (1952) Probit Analysis, 2nd ed., Cambridge University Press, Cambridge.

Fisher, R. A. (1936) The use of multiple measures in taxonomic problems. Ann. Eugen., 7:179–188.

Fowler, R. E., and R. G. Edwards (1957) Induction of superovulation and pregnancy in mature mice by gonadotrophins. J. Endocrinol., 15:374–384.

Gates, A. H. (1965) Rate of ovular development as a factor in embryonic survival. In: Preimplantation Stages of Pregnancy, pp. 270–288. Ed. by G. E. W. Wolstenholme and M. O'Connor. Little, Brown, Boston.

Glick, D., and H. Holter (1961) Quantitative chemical techniques of histo- and cytochemistry. Vol. 1, pp. 173–179. Interscience, New York.

Gwatkin, R. B. L. (1964) Effect of enzymes and acidity on the zona pellucida of the mouse egg before and after fertilization. J. Reprod. Fertility, 7:99–105.

———— (1966a) Amino acid requirements for attachment and outgrowth of the mouse blastocyst in vitro. J. Cell Physiol., 68:335–343.

———— (1966b) Defined media and development of mammalian eggs in vitro. Ann. N.Y. Acad. Sci., 139:79–90.

Ham, R. G. (1963) An improved nutrient solution for diploid Chinese hamster and human cell lines. Exp. Cell Res., 29:515–526.

Hammond, J., Jr. (1949) Recovery and culture of tubal mouse ova. Nature, 163:28–29.

Harrison, R. G., (1907) Observations on the living nerve fiber. Proc. Soc. Exp. Biol. (N.Y.) , 4:140–143.

Jensen, F. C., and G. A. Castellano (1960) The long-term maintenance of tumour explants. Cancer Chemotherapy Reports, no. 8, 135–140.

Jensen, F. C., R. B. L. Gwatkin, and J. D. Biggers (1964) A simple organ culture method which allows simultaneous isolation of specific types of cells. Exp. Cell Res., 34:440–447.

Krzanowski, H. (1964) Time interval between copulation and fertilization in inbred lines of mice and their crosses. Folia Biologica, 12:231–244.

Lewis, W. H., and M. R. Lewis (1912) The cultivation of chick tissues in media of known composition. Anat. Record, 6:207–211.

Lewis, W. H., and E. S. Wright (1935) On the early development of the mouse egg. Contributions to Embryology. Carnegie Institute, Washington. 25:115–144.

Loewenstein, J. E., and A. I. Cohen (1964) Dry mass, lipid content and protein content of the intact and zona free mouse ovum. J. Embryol. Exp. Morphol., 12:113–119.

McLaren, A., and J. D. Biggers (1958) Successful development and birth of mice cultivated *in vitro* as early embryos. Nature, 182:877–878.

Mills, R. M., and R. L. Brinster (1967) Oxygen consumption of preimplantation mouse embryos. Exp. Cell Res., 47:337–344.

Mintz, B. (1962) Experimental study of the developing mammalian egg: removal of the zona pellucida. Science, 138:594–595.

——— (1964) Formation of genetically mosaic mouse embryos and early development of lethal (t^{12}/t^{12}) –normal mosaics. J. Exp. Zool., 157:273–292.

——— (1968) Mammalian embryo culture: In: Methods in Developmental Biology. Ed. by. F. Wilt and N. Wessells. Thomas and Crowell, New York.

Monesi, V., and V. Salfi (1967) Macromolecular syntheses during early development in the mouse embryo. Exp. Cell Res., 46:632–635.

Moscona, A., D. A. Trowell, and E. N. Willmer (1965) Methods. In: Cells and Tissues in Culture. Vol. 1. Ed. by E. N. Willmer. Academic Press, London and New York, pp. 19–86.

New, D. A. T. (1966) The culture of vertebrate embryos. Logos Press, London.

Onanoff, J. (1893) Recherches sur la fécondation et la gestation des mammifères. C. R. Soc. Biol. (Paris) , 45:719.

Parker, R. C. (1961) Methods of tissue culture, 3rd ed. Harper, New York.

Paul J. (1965) Cell and tissue culture. 3d ed. Williams and Wilkins, Baltimore.

Pavlok, A. (1967) Development of mouse ova in explanted oviducts: Fertilization, cultivation and transplantation. Science, 157:1457–1458.

Pincus G. (1937) The metabolism of ovarian hormones especially in relation to the growth of the fertilized ovum. Cold Spring Harbor Symposia on Quantitative Biology, 5:44–55.

Popp, R. A. (1958) Comparative metabolism of blastocysts, extra–embryonic membranes and uterine endometrium of the mouse. J. Exp. Zool., 138:1–23.

Rapola, J., and O. Koskimies (1967) Embryonic enzyme patterns: Characterization of the single lactate dehydrogenase isozyme in preimplanted mouse ova. Science, 157:1311–1312.

Schenck, S. L. (1880) Das Sangethierei Kunstlick befruchte ausserhalb des Mutter-thieres. Mitt Embr. Inst. K. K. Univ. Wien, 1:107–118.

Seal, H. L. (1964) Multivariate statistical analysis for biologists. Methuen, London.

Smith, B. D. M. (1966) Implantation in the Mouse. Ph.D. thesis, University of Pennsylvania.

———— (1968) Implantation in the mouse. The effect on implantation of treating cultured mouse blastocysts with oestrogen *in vitro* and blastocyst uptake of H^3–oestradiol. J. Endocrinol., 41:17–29.

Stern, S., and J. D. Biggers (1968) Enzymatic estimation of glycogen in the cleaving mouse embryo. J. Exp. Zool., 168:61–65.

Tarkowski, A. K. (1961) Mouse chimaeras developed from fused eggs. Nature, 190:857–60.

———— (1963) Studies on mouse chimeras developed from eggs fused *in vitro*. Nat. Cancer Inst. Monograph, 11:51–71.

———— and J. Wróblewska (1967) Development of blastomeres of mouse eggs isolated at the four- and eight-cell stage. J. Embryol. Exp. Morphol., 18:155–180.

Thomson, J. L., and J. D. Biggers (1965) Effect of inhibitors of protein synthesis on the development of preimplantation mouse embryos. Exp. Cell Res., 41:411–427.

Trevan, J. W. (1927) The error of determination of toxicity. Proc. Roy. Soc. (London), *B* 101:483–514.

Trowell, O. A. (1954) A modified technique for organ culture *in vitro*. Exp. Cell Res., 6:246–248.

———— (1959) The culture of mature organs in a synthetic medium. Exp. Cell Res., 16:118–147.

Wales, R. G., and J. D. Biggers (1968) The permeability of two- and eight-cell mouse embryos to L-malic acid. J. Reprod. Fertility, 15:103–111.

Wales, R. G., and R. L. Brinster (1968) The uptake of hexoses by mouse embryos. J. Reprod. Fertility, 15:415–422.

Wales, R. G., and D. G. Whittingham (1968) A comparison of the uptake and utilization of lactate and pyruvate by one- and two-cell mouse embryos. Biochim. Biophys. Acta., 148:703–712.

Walker, S. H., and D. B. Duncan (1967) Estimation of the probability of an event as a function of several independent variables. Biometrika, 54:167–179.

Waymouth, C. (1954) The nutrition of animal cells. Int. Rev. Cytol., 3:1–68.

———— (1965) Construction and use of synthetic media. In: Cells and Tissue Culture, vol. I, chap. 3, pp. 99–142. Ed. by E. N. Willmer, Academic Press, London.

Whitten, W. K. (1956a) Culture of tubal mouse ova. Nature, 177:96.

———— (1956b) Physiological control of population growth. Nature, 178:992.

———— (1957a) Culture of tubal ova. Nature, 179:1081–1082.

———— (1957b) The effect of progesterone on the development of mouse eggs *in vitro*. J. Endocrinol., 16:80–85.

———— (1966) Pheromones and mammalian reproduction. Advances in Reproductive Physiology, 1:155–177.

———— (1969) The effect of oxygen on cleavage of mouse eggs *in vitro*. Proc. Society for the Study of Reproduction, Second Annual Symposium, p. 29.

———— and C. P. Dagg (1961) Influence of spermatozoa on the cleavage rate of mouse eggs. J. Exp. Zool., 148:173–183.

———— and J. D. Biggers (1968) Complete culture of the preimplantation stages of the mouse *in vitro*. J. Reprod. Fertility, 17:399–401.

Whittingham, D. G. (1967) Studies on the early preimplantation stages of mammalian development. Ph.D. thesis, University of London.

———— and J. D. Biggers (1967) Fallopian tube and early cleavage in the mouse. Nature, 213:942–943.

7

Egg Storage

E. S. E. HAFEZ
Department of Gynecology-Obstetrics
School of Medicine
Wayne State University
Detroit, Michigan 48207

Several techniques have been used to preserve viable cells, tissues, and organs: refrigeration, low temperature freezing, freeze-drying, and ultrarapid freezing have been among the methods used to preserve blood cells, malignant cells, skin tissues, glands, and even intact lower forms of animal life (Belehradek, 1935, Parkes, 1956). Fertilized eggs of the ant (Pictet, 1893), fruit fly (Bach and Pemberton, 1916), and chicken (Moran, 1925) do not survive prolonged exposure to cold, but those of ascaris can be stored at 5°C for a few months and still develop normally upon rewarming. Furthermore, fertilized eggs of some species do not develop unless previously cooled (Greeley, 1903).

Delayed implantation, a fairly common phenomenon in a number of mammalian species, prompted investigation of the possibilities of storage of mammalian eggs. For practical use in egg transfer experiments, it is desirable to maintain the egg *in vitro* in a dormant state to allow accurate synchronization of egg development and endometrial differentiation. Storage of eggs also makes possible long-distance transportation of embryos (Marden and Chang, 1952; Chang and Marden, 1954; Adams *et al.*, 1961; Hunter *et al.*, 1962).

The eggs of several mammalian species have been successfully stored for a few days at subnormal temperatures. The survival rate depends on the species, the developmental stage of the egg, the physical and biochemical properties of the storage media, the storage temperature, the rate of cooling

Unpublished data in this paper are part of an investigation supported in part by the Lillian Banta Research Fund.

and rewarming the eggs, and the techniques of storage and egg transfer. Fertilized eggs are more likely to be damaged by cold storage than unfertilized eggs. Unfertilized rabbit eggs, recovered 2 hours after ovulation, were kept at 0°C for 48 to 72 hours, or at 10°C for up to 96 hours and still were fertilized after transfer (Chang, 1952, 1953, 1955). Although fertilization appeared to be normal, most of the embryos degenerated before birth.

The following discussion deals with the present techniques available for collection of mammalian eggs for storage and for transfer to recipients.

STORAGE TECHNIQUES

COLLECTION OF EGGS

Donors are superovulated by the administration of a combination of gonadotropins such as pregnant mare serum (PMS) and human chorionic gonadotropins (HCG). The specific protocol for rabbits is an intramuscular or subcutaneous injection of 150 I.U. of PMS (Hafez, 1961c), followed 66 to 72 hours later by artificial insemination or natural mating to two fertile bucks, and intravenous injection of 50 I.U. of HCG (Hafez, 1961c). At autopsy, 40 to 48 hours postcoitum, the oviducts are removed and placed in sterilized Petri dishes. The eggs can be flushed from the oviducts with a sterilized physiological saline-serum solution, or may be recovered *in situ* (see Fig. 7-1). All flushings are performed in a partially enclosed culture room without ultraviolet light (see Fig. 7-2). The eggs are collected with a glass pipette under a stereoscopic binocular microscope and

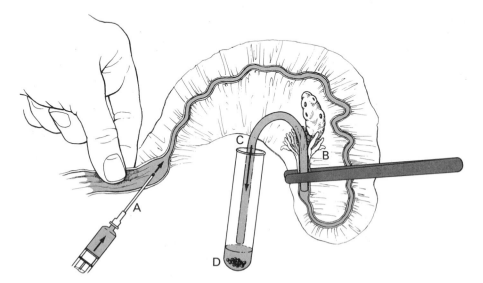

FIGURE 7-1

Collection of mammalian eggs in vivo: **(A)** *flushing medium is injected into the uterus near the uterotubal junction,* **(B)** *fimbriae,* **(C)** *polyethylene tube,* **(D)** *flushings containing the eggs.*

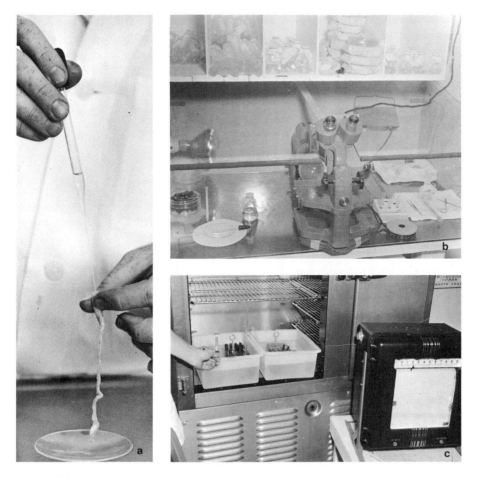

FIGURE 7-2

(a) *Collection of eggs from an oviduct using a glass pipette. The oviduct is flushed with a saline/serum mixture into a watch glass for subsequent microscopic examination.* (b) *Culture cabinet used for manipulation of eggs. A stereoscope microscope is mounted in a glass top. The cabinet is equipped with ultraviolet, infrared, and microscope lamps. Autoclaved pipettes, saline, gelatin, and watch glasses (inside Petri dishes) are on the top shelf for convenience.* (c) *Storage of eggs in a water bath inside the refrigerator at 10° C. The temperature of the water is monitored by the recorder in the foreground.*

transferred through two sterile solutions of saline to decrease any possible contamination. They are then counted ($\times 15$) and carefully examined ($\times 40$ to $\times 100$) for abnormalities.

MANIPULATION OF EGGS

Fire-polished glass pipettes having a diameter of 1 or 2 mm are used to handle the eggs. During manipulation it is desirable to have minimum exposure to visible light. For prolonged exposure, red light is preferred; short exposure to ultraviolet light may inhibit cleavage of rabbit eggs (Daniel, 1964). All equipment that is used to handle the eggs should be sterilized and held at 30° to 37°C either on a warm plate or in an incubator. In the interval between recovery and storage, the eggs can be stored in the culture cabinet at room temperature.

Holding sheep eggs for 12 to 60 minutes at room temperature before cooling at 10°C does not affect their survival when stored for two days *in vitro* or their subsequent development *in vivo* (Kardymowicz *et al.* 1966a); nor is exposure to daylight for one hour detrimental.

SELECTION OF EGGS

The integrity of an egg is important for its survival, both *in vivo* and *in vitro*. Microscopic defects may prevent the egg from implanting (Hafez, 1962a). Moreover, defective eggs that do implant may not survive during subsequent embryonic and fetal life.

Structural abnormalities in the egg may result from cytological, genetic, environmental, pathological, or artifactual conditions; the first three may contribute to the variability in successful egg storage. Such anomalies that have been described (Hafez, 1961b) include aberrations in size, shape, and degree of cytoplasmic granulation or pigmentation. Morphologically abnormal eggs may have the shape of a helmet, a kidney, an amoeba, an ovoid, or a paramecium (see Fig. 7-3). Lentil-shaped ova do not develop normally, although oval-shaped ova do (Kvasnitskii, 1956). Thus, eggs used for storage should be classified according to the stage of cleavage and examined under a high magnification (×40 to ×100) to determine any structural malformation.

In the future, more refined techniques of examining eggs (phase-contrast microscopy) seem likely. Fertilization is usually associated with the presence of the second polar body, which can be studied only through more accurate, high-power microscopes.

STORAGE IN ANOTHER SPECIES

Eggs of one species can be transferred into the oviduct or uterus of another species, and, after short-term storage, may be recovered for subsequent transfer to an appropriate recipient. Little is known, however, of the survival of fertilized eggs after interspecific egg transfer (Warwick and Berry, 1949; Briones and Beatty, 1954; Averill *et al.*, 1955). Ferret eggs can survive and develop in the rabbit oviduct, but not in the uterus. Rabbit eggs, on the contrary, cannot survive in either the ferret oviduct or uterus (Chang 1966). Following transfer to the genital tract of the rabbit, two-celled sheep eggs can survive for at least five days and can develop to the early blastocyst stage (Averill, Adams and Rowson, 1955). The survival rate of such eggs in the rabbit is better than that of eggs stored *in vitro* three days at 7°C (Averill and Rowson, 1959). Such an incubator offers ample opportunity for successful long distance transportation of sheep eggs (Adams *et al.,* 1961; Hunter *et al.,* 1962).

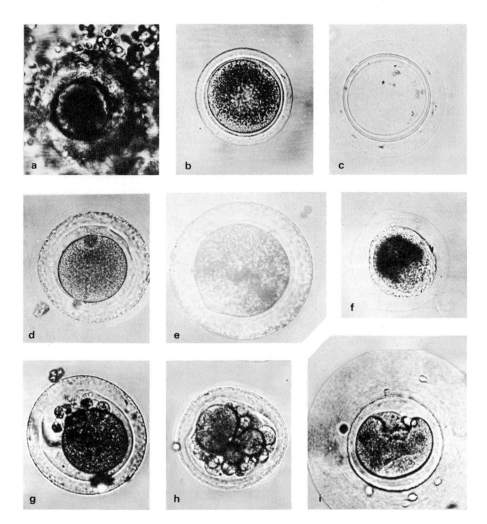

FIGURE 7-3

(a) *A normal freshly ovulated ovum surrounded by cumulus cells 17 hours postcoitum.*
×122. (b) *A normal fertilized ovum recovered 17 hours postcoitum from superovulated donor 264 after treating the egg* in vitro *with hyaluronidase.* ×122. (c) *An
ovum shell without cytoplasm recovered 97½ hours postcoitum from the uterus of
recipient 168B. Note the intact zona pellucida and regular deposition of mucin coat.*
×53. (d) *An atypical ovum recovered 24 hours postcoitum from superovulated donor
247. Note irregularity of zona pellucida.* ×122. (e) *An atypical ovum recovered 24 hours
postcoitum from superovulated donor 314. Note bulging, light granulated cytoplasm.*
×158. (f) *Unfertilized ovum recovered 25 hours after mating to a fertile buck.
Many normal fertilized ova were, however, recovered from this oviduct.* ×102.
(g) *Ovum recovered 25 hours postcoitum from superovulated donor 285. Note abnormal
granulated cytoplasm and several fragments at one pole of the ovum.* ×122. (h) *Fragmenting ovum recovered 24 hours after mating to two fertile bucks.* ×122. (i) *Abnormally shaped ova recovered from superovulated donors. Helmet-shaped ovum.*
×115. (From Hafez 1961, Intern. J. Fertility, 6:393.)

STORAGE MEDIA

Most storage media consist of salt solutions, serum, proteins and antibiotics (Paul, 1960). Physiological saline, containing 0.9% NaCl, maintains the tonicity of eggs for short periods, but it lacks both the buffering capacity and many of the essential organic ions that are necessary for long-term survival of eggs. Various preparations have been used for manipulating eggs; two of these are Krebs solution (Black *et al.*, 1951) and phosphate-buffered Ringer-Dale solution containing glucose (Chang, 1952). Omission of calcium, magnesium, potassium, or glucose from Krebs-Ringer bicarbonate prevents growth of mouse eggs, whereas omission of phosphate results in delayed development (Whitten, 1956).

Salt solution

At present, physiological solutions are used to: (1) serve as flushing fluids while maintaining tonicity within the vitellus of the eggs; (2) buffer the medium and maintain it in the physiological pH range (7.2 to 7.6); and (3) provide an aqueous ionic environment for cell metabolism (Parker, 1961). Gassing the medium with CO_2 did not improve the eggs' storage survival rate (Kardymowicz, 1961; Hafez, 1963).

Blood Serum

Most successful egg storage media contain serum; routine storage media contain 25 to 50% homologous serum (Chang, 1949; Gates and Runner, 1952; Hancock and Hovell, 1962; Hancock, 1963; Hunter *et al.*, 1962; Kvasnitskii, 1956; Pincus, 1936, 1939; Willett *et al.*, 1953). Occasionally the serum of other species can be used: horse, dog, guinea pig, rat, and pig serum can be used to store rabbit eggs, but serum from man, sheep, cattle, goats, and fowl contain a factor which is ovicidal to rabbit eggs (Chang, 1949). Substitutes for fresh serum include serum previously frozen at $-20°C$, reconstituted freeze-dried serum, blastocyst fluid, and egg-yolk citrate.

Rabbit blood is collected by heart puncture and allowed to clot for 30 minutes at room temperature prior to centrifugation for 20 minutes at 3,000 rpm; it is then decanted and recentrifuged, drawn into a 10 ml syringe, filtered through a 0.45μ millipore filter into an equal volume of sterile 0.9% physiological saline, and stored at 3° or 4°C. Freezing the serum for one month does not affect the viability of the stored rabbit embryos. Eggs do not survive when stored in chicken egg-white diluted with Tyrodes solution (Hafez, 1965).

Proteins

Gates and Runner (1952) compared Locke's solution with Ortho bovine semen diluter containing egg yolk and found the latter to be superior as a medium for egg transfer. Kiessling (1963) reports successful storage of mouse eggs for three days in Krebs-Ringer bicarbonate with one mg per ml of glucose and crystalline bovine plasma albumin added; however, no reports of chemically defined storage media are available.

Gelatin has been used in storage experiments also. This medium is prepared by dissolving 14 grams of gelatin per ml of saline and autoclaving immediately after preparation to avoid any bacterial growth. After storage in 7% gelatin, the rate of implantation of rabbit eggs was 53% after seven days, gradually declining to 3% after fourteen days; the addition of antibiotics did not prolong the survival time (Hafez, 1961c, 1962b). The addition of 7% gelatin to the storage medium is thus recommended.

Antibiotics

Smith (1949) found that a concentration of 0.02% streptomycin in a culture medium containing serum prevents bacterial growth which normally occurs even when manipulations are carried out under antiseptic conditions. The cleavage of rabbit eggs is not affected by penicillin at concentrations up to 10%; however, streptomycin at concentrations greater than 1% does inhibit cleavage, whereas concentrations of less than 1% have no effect. Acetosulphamine and sulfamerazine also tend to inhibit egg cleavage (Kurosaki *et al.,* 1954).

The survival of eggs is not affected by storage at low temperatures in concentrations of 7.5 mg of streptomycin, 4 mg chloromycetin, 6.5 mg paromomycin, or 23.9 mg of penicillin per ml of liquid medium (Hafez, 1962b). Information is available concerning the effects on embryonic survival of storage in physiological levels of antibiotics.

Storage Containers

Four ml glass tubes with an 8 mm inner diameter are used for storage. Gelatin, saline, and serum are measured with sterile syringes and transferred to the tubes. When the final volume of storage medium in the tube is reached, the medium is thoroughly mixed with a glass pipette fitted with a rubber bulb. Care is taken to avoid the formation of air bubbles. Filling the tubes with water will prevent loss from the storage medium by evaporation and condensation. Ten to one hundred eggs can be stored in one tube.

Sealed glass ampules containing 10 to 15 eggs in 0.3 ml of medium may be used. Averill and Rowson (1958) have used special dialysis tubes for storage of eggs. Subsequent findings suggest that a dialysis tube offers no advantage (Hancock and Hovell, 1961).

Storage Temperature

Several studies have been undertaken to measure the effect of cold storage on mammalian tissues. The temperature below which the cell membrane of rat diaphragm muscle is unable to maintain its cellular ionic components is 15° to 17°C (Taylor, 1956). When mammalian eggs are stored below 20°C they do not undergo further cleavage. Thus, the two-day rabbit egg, after storage, may be transplanted to a recipient doe that was injected with LH two days before egg transfer.

The optimal temperature for storage has been estimated as 10°C for the rabbit (Chang, 1947, 1948a,b), 5° to 10°C for the rat (Sugawara and Takeuchi, 1962), 5°C for the mouse (Sherman and Lin, 1959), and 7°C for the sheep (Averill and Rowson, 1959; Harper and Rowson, 1963). Two-celled rabbit eggs survive in rabbit serum at 0° to 15°C for 96 to 120 hours. At 22° to 24°C they are viable for 24 to 48 hours (Chang, 1948b).

Sudden cooling of sperm causes *temperature shock* (Hancock, 1951; Blackshaw, 1954), which can be prevented by slow cooling in an egg-yolk buffer (Philips and Lardy, 1940; Meyer and Lasley, 1945). Temperature shock also adversely affects eggs, especially at lower temperatures; it can be prevented by *acclimation* or slow cooling. To control the rate of cooling, storage tubes containing eggs are placed in a water bath at 20°C. The bath with the tubes is then placed in a cooler at 10°C (see Fig. 2c); the rate is −0.01°C/minutes.

Freezing

The discovery that glycerol can protect living cells against the effects of freezing has been successfully applied in the preservation of testes (Parkes, 1954) and ovaries (Deansley, 1957; Parrott, 1960). Normal offspring were obtained from mice with orthotopic ovarian grafts of tissue that had been frozen and stored at −79°C (Parrot and Parkes, 1960). Sherman and Lin (1958, 1959) made a thorough study of storage of unfertilized mouse eggs at temperatures below freezing.

The method of preserving eggs by freezing has had little success. Smith (1952, 1953) equilibrated rabbit eggs with 15% glycerol by 10-minute exposures to increasing concentrations of glycerol at 37°C. Few of these glycerolated eggs survived cooling to −79°C, and pregnancy occurred in but two of eight rabbits after they received two-celled eggs stored at −79°C (Ferdows *et al.*, 1958). A blood-serum–glycerol medium was superior to a physiological saline–glycerol–skim-milk medium for freezing (Ferdows, 1960). Sheep eggs did not survive storage in 12.5% glycerol in serum at −79°C (Averill and Rowson, 1959).

STORAGE PERIOD

The period for which the egg can be stored without losing its viability and without causing a harmful effect varies with the species, the stage of development, the storage medium, and the temperature.

Rabbit eggs have apparently been kept viable for longer periods than any other mammalian egg. Such eggs differ from most other species in that they become covered with a mucin coat during transportation through the oviduct. Although evidence is incomplete, this layer apparently protects the egg during manipulation and storage *in vitro*. If so, it may be possible to increase the resistance of noncoated eggs of other species by storing them in the oviducts of rabbits for one day before *in vitro* storage.

Eggs of other species that have been successfully stored include: four-to-twelve-celled eggs of sheep kept 48 hours in heat-treated sheep serum with added streptomycin (Buttle and Hancock, 1964); and sheep eggs in autologous serum, kept viable for five hours (Hancock and Hovell, 1961) to a few days (Averill and Rowson, 1959). Eighteen percent of the eggs that were stored for 5 days survived at 10°C and 63% survived storage for 3 days (Kardymowicz *et al.,* 1966b). Unfertilized mouse eggs survive up to $3\frac{1}{2}$ hours in a modified Locke's solution containing 5% glycerol at −10°C (Sherman and Lin, 1958), and up to 6 hours at 0°C (Sherman and Lin, 1959).

The effect of temperature can be seen most easily after storage for 74 hours (Chang, 1948a). A marked decline in the survival rate of rabbit eggs occurs after 5 days of storage, however some eggs have been kept viable in storage for 12 days (Hafez, 1961a, 1965).

Blastocysts are less resistant to cold storage than eggs; however, they have developed to birth after storage of only one day (Chang, 1950).

TRANSFER OF STORED EGGS

The techniques that are to be described for the transfer of stored rabbit eggs can easily be adapted to other species.

SELECTION OF EGGS

The eggs should be examined microscopically before transfer to an appropriate recipient. The reproductive cycle of the recipient should correspond to the stage of development of the eggs (Averill and Rowson, 1958; Beatty, 1959; Chang, 1950; McLaren and Michie, 1956; Noyes and Dickmann, 1960).

The storage of rabbit eggs at 0°C causes swelling of the cells and darkening or roughness of the cell membranes, but results in no appreciable deterioration of the zona pellucida (Chang, 1948a). At 10°C prolonged storage may produce granulation, loss of spheroid shape, lack of distinct blastomeres, and marked indentation in the cell mass of the egg. These changes do not necessarily indicate death. If the indentation is slight, the eggs may recover and implant, but when the depression is deep, as it is in most cases, it will not implant.

Unfertilized eggs can be stored for limited periods without loss of fertilizability. Shrinkage of the cytoplasm within the zona pellucida due to exposure to glycerol does not appear to affect fertilizability (Lin *et al.,* 1957). Chang (1952) found that some unfertilized rabbit eggs were still fertilizable following exposure at 10° to 0°C for 48 to 72 hours.

SELECTION OF RECIPIENTS

Healthy recipients with known fertility should be used when the viability of stored eggs is tested. Transfer the embryo into a recipient

TABLE 7-1

Storage media used for mammalian eggs

Medium	Study
MOUSE	
Modified Lockes solution + 5% glycerol	Sherman and Lin (1958, 1959)
Yolk citrate Lockes + 5% glycerol	Sherman and Lin (1959); Lin *et al.* (1957)
Bovine serum; serum Lockes + 5% glycerol; Illini variable temperature diluter	Sherman and Lin (1959)
Krebs-Ringer bicarbonate + 1 mg/ml of glucose + bovine plasma albumin	Kiessling (1963)
PIG	
(a) Serum of pig jugular blood heated to 56°C for 30 minutes and stored at −20°C	Hancock and Hovell (1962)
(b) Tyrode's solution with 1/g per liter bovine plasma albumin (Armour and Co.—Fraction V; Paul, 1960; Tarkowski, 1961)	
(c) Hank's solution; with 10% v/v horse serum (preservative free) (Merchant *et al.*, 1960)	
(Media (a) and (b) sterilized by Seitz filtration; all media contained 50 μg/ml penicillin and streptomycin sulphate.)	
RABBIT	
Rabbit blood serum	Chang (1948a,b; Marden and Chang (1952); Chang and Marden (1954)
Rabbit blood serum + 5% buffered Ringer solution	Chang (1953)
Phosphate buffered Ringer-Dale + 0.1% glucose + blood; centrifuged, supernatant used	Chang (1952)
Rabbit serum + Lockes solution + 0.02% streptomycin	Kardymowicz (1960, 1961, 1962)
50% rabbit serum in Ca–free Ringer solution	Ketchel and Pincus (1964)
50% rabbit serum in 0.85% normal saline	Hafez (1961, 1962a)
50% rabbit serum in normal saline plus 1, 5, or 7% gelatine	Hafez (1961)
Egg yolk + saline 8 or 15 day embryonic fluids + saline	Hafez (1963)

TABLE 7-1 (*Continued*)

Medium	Study
	RAT
Krebs-Ringer phosphate + 0.1% glucose solution; iso-serum or dilute serum	Sugawara and Takeuchi (1962)
	SHEEP
Sheep blood serum	Hancock (1963); Averill and Rowson (1959); Buttle and Hancock (1964)
Homologous blood serum + penicillin + streptomycin	Harper and Rowson (1963)
Sheep blood serum + 12.5% glycerol; Ringer solution	Averill and Rowson (1958)

SOURCE: From the literature, and literature compiled by Austin (1961) and Mauer (1966).

female, using careful and adequate controls. In the rabbit, each group of transferred eggs should have its own control within the same recipient; eggs stored in tested media are transferred to one oviduct and eggs stored in a control medium are transferred into the other (see Table 7-2). Moore and Rowson (1960) investigated the factors affecting the survival and development of transferred eggs.

Two days before the transfer of stored two-day-old eggs, the recipients are injected intravenously with 15 I.U. of HCG. The injection of HCG mimics copulation in the rabbit by inducing ovulation; thus the term postcoitum (PC) will be used for the recipients although they were not bred. The recipients are anesthetized with intravenous Nembutal. Laparotomy is performed on both flanks, and three to six eggs are transferred into each oviduct. The recipients are laparotomized eight days postcoitum to count the number of implantations and killed 15 to 29 days postcoitum, to count the number of viable fetuses.

CALCULATION AND INTERPRETATION OF RESULTS

A few investigators have judged the viability of stored eggs by their ability to grow in culture. The cleavage of eggs *in vitro* is not necessarily comparable to egg development *in vivo*. It is to be emphasized that the only valid criterion of successful egg preservation is the birth of viable young after the stored eggs have been transferred into appropriate recipients.

The effectiveness of egg storage can be evaluated by the following two indices:

Survival of stored eggs =

$$\frac{\text{number of implants, 8 days PC, in a group of recipients}}{\text{number of stored eggs transferred to a group of recipients}} \times 100$$

TABLE 7-2

Implantation and embryonic survival of rabbit eggs after storage in homologous serum containing 7% gelatine at 10°C (exp. Ia)

Storage Time (days)	No. of Recipients	Treatment Medium						Control Medium[1]					
		No. of Eggs Transferred	Embryos Implanted		Embryos Survived			No. of Eggs Transferred	Embryos Implanted		Embryos Survived		
			Total	%	Total	%			Total	%	Total	%[2]	
7	12	68	36	53	21	58		70	28	40	16	57	
8	6	33	8	2	2	25		34	11	32	3	27	
9	4	25	12	48	7	58		24	0	0	0	0	
11	5	25	10	40	7	70		30	4	13	0	0	
13	4	19	0	0				19	1	5	0		
14	12	62	2	3	1	50		65	0	0			
Total		232	68		31	50		242	44		19		

SOURCE: Hafez (1965.)
[1] Storage medium without gelatine.
[2] Total number of viable implants at 15 days postcoitum as percentage of total number of implants at 8 days postcoitum.

Prenatal survival resulting from stored eggs =

$$\frac{\text{number of viable fetuses, 15 to 29 days PC, in a group of recipients}}{\text{number of viable implants, 8 days PC, in a group of recipients}} \times 100$$

Prenatal survival of rabbit eggs stored for seven to nine days was lower than that of eggs stored for two days. This indicates that prolonged storage of eggs may have a harmful carry-over effect; that is, some of the eggs survive *in vitro* storage without losing their ability to implant but degenerate during midpregnancy. The physiological mechanism involved in this phenomenon is not yet understood.

The developmental stage of the eggs seems to have an effect on their ability to withstand storage. With rabbit eggs collected at 25, 72, and 96 hours PC, the younger embryos survived storage best (Chang, 1948b); two-to-four-celled eggs survived storage in larger numbers than eight-to-thirty-two-celled embryos (Hafez, 1961a). It is interesting to note that eggs produced by the same donor vary not only in their resistance to cold storage, but also in their resistance to antibiotics.

FUTURE RESEARCH

It is possible that the survival of eggs *in vitro* may be prolonged by increasing the osmotic pressure of the storage medium. Further studies are needed to determine the optimal physical and physiochemical properties of media required for long-term storage of eggs of a variety of species. This may involve addition of natural or synthetic proteins, inert molecules, and hormones or amino acids (Ketchel and Pincus, 1964). Attempts should be made to freeze the eggs for prolonged storage as has been done with sperm.

BIBLIOGRAPHY

Adams, C. E., L. E. A. Rowson, G. L. Hunter, and G. P. Bishop (1961) Long distance transport of sheep ova. 4th Internat. Congr. Anim. Reprod. The Hague, 11:381.

Austin, C. R. (1961) The mammalian egg. Blackwell Scientific Publications, Oxford.

Averill, R. L. W., C. E. Adams, and L. E. A. Rowson (1955) Transfer of mammalian ova between species. Nature, 176:167.

Averill, R. L. W., and L. E. A. Rowson (1958). Ovum transfer in the sheep. J. Endocrinol. 16:326.

———— (1959) Attempts at storage of sheep ova at low temperatures. J. Agric. Sci., 52:392–395.

Bach, E. A., and C. E. Pemberton (1916) Effect of cold-storage temperatures upon the Mediterranean fruit fly. U.S.D.A. J. Agr. Res., 5:657.

Beatty, R. A. (1959) Transplantation of mouse eggs. Nature, 168:995.

Belehradek, J. (1935) Temperature and Living Matter. Protoplasma-Monographien. Vol. 8. Gebruder Borntraeger, Berlin.

Black, W. G., G. Otto, and L. E. Casida (1951) Embryonic mortality in pregnancies induced in rabbits of different reproductive stages. Endocrinology, 49:237.

Blackshaw, A. W. (1954) The prevention of temperature shock of bull and ram semen. Australian J. Biol. Sci., 7:573.

Briones, H., and R. A. Beatty (1954) Interspecific transfers of rodent eggs. J. Exp. Zool., 125:99.

Buttle, H. R. L., and J. L. Hancock (1964) Birth of lambs after storage of sheep eggs *in vitro*. J. Reprod. Fertility, 7:417.

Chang, M. C. (1947) Normal development of fertilized rabbit ova stored at low temperature for several days. Nature, 159:602–603.

———— (1948a) Probability of normal development after transplantation of fertilized rabbit ova stored at different temperatures. Proc. Soc. Exp. Biol. Med. 68:680.

———— (1948b) The effects of low temperature on fertilized rabbit ova *in vitro,* and the normal development of ova kept at low temperature for several days. J. Gen. Physiol., 31:385–410.

———— (1949) Effects of heterologous sera on fertilized rabbit ova. J. Gen. Physiol., 32:291.

———— (1950) Development and fate of transferred rabbit ova or blastocysts in relation to ovulation times of recipients. J. Exp. Zool., 114:197.

———— (1952) Fertilizability of rabbit ova and the effects of temperature *in vitro* on their subsequent fertilization and activation *in vivo*. J. Exp. Zool., 121:351–381.

———— (1953) Fertilizability of rabbit germ cells. In: Mammalian Germ Cells. Ed. by G. E. W. Wolstenholme. Little, Brown, Boston.

———— (1955) Fertilization and normal development of follicular oocytes in the rabbit. Science, 121:867.

———— (1966) Reciprocal transplantation of eggs between rabbit and ferret. J. Exp. Zool., 161:297.

———— and W. G. R. Marden (1954) The aerial transport of fertilized mammalian ova. J. Heredity, 45:75.

Daniel, J. C., Jr. (1964) Cleavage of mammalian ova inhibited by visible light. Nature, 201:316–317.

Deansley, R. (1957) Egg survival in immature rat ovaries grafted after freezing and thawing. Proc. Roy. Soc. (London) Ser. B., 147:412–421.

Ferdows, M. (1960) The effects of different methods of freezing on the viability of rabbit ova. Dissert. Abstr., 21:1311–1312.

————, C. L. Moore, and A. E. Dracy (1958) Survival of rabbit ova stored at 79°C. J. Dairy Sci., 41:739 (Abstract).

Gates, A., and M. Runner (1952) Factors affecting survival of transplanted ova of the mouse. Anat. Record., 113:555.

Greeley, A. W. (1903) On the effect of variations in the temperature upon the process of artificial parthenogenesis. Biol. Bull., 4:129.

Hafez, E. S. E. (1961a) Storage of rabbit ova in gelled media at 10°C. J. Reprod. Fertility, 2:163–178.

———— (1961b) Structural and developmental anomalies of rabbit ova. Intern. J. Fertility, 6:393–407.

———— (1961c) Procedures and problems of manipulation, selection, storage, and transfer of mammalian ova. Cornell Vet., 51, 299–333.

———— (1962a) *In vitro* and *in vivo* survival of morphologically atypical embryos in rabbits. Nature, 196:1226–1227.

———— (1962b) Effects of antibiotics on viability of fertilized rabbit ova *in vitro*. Fertility Sterility, 13:583–597.

———— (1963) Storage of fertilized ova. Intern. J. Fertility, 8:459.

—— (1965) Storage media for rabbit ova. J. Appl. Physiol., 21:731–736.

Hancock, J. L. (1951) A staining technique for the study of temperature-shock in semen. Nature, 167:323.

—— (1963) Survival *in vitro* of sheep eggs. Animal Prod., 5:237.

—— and G. J. R. Hovell (1961) Transfer of sheep ova. J. Reprod. Fertility, 2:295.

—— (1962) Egg transfer in the cow. J. Reprod. Fertility, 4:195.

Harper, M. J. K., and L. E. A. Rowson (1963) Attempted storage of sheep ova at 7°C. J. Reprod. Fertility, 6:183.

Hunter, G. L., G. P. Bishop, C. E. Adams, and L. E. Rowson (1962) Successful long-distance aerial transport of fertilized sheep ova. J. Reprod. Fertility, 3:33.

Kardymowicz, O. (1961) Investigations on storage of fertilized rabbit ova. Acta. Biol. (Cracov), 4:183.

——, M. Kardymowicz, and K. Grouchowalski (1966a) A study on the effect of cooling of sheep ova to 10°C on their capability of further development. Acta Biol. (Cracov), 9:113.

——, M. Kardymowicz, and M. Kremer (1966b) Successful *in vitro* storage of sheep ova for 5 days. Acta Biol. (Cracov), 9:117.

Ketchel, M. M., and G. Pincus (1964) *In vitro* exposure of rabbit ova to estrogens. Proc. Soc. Exp. Biol. Med., 115:419.

Kiessling, J. (1963) The effect of low temperature storage on the developmental capacity of mouse zygotes. Am. Zoologist, 3:485 (Abstract).

Kurosaki, Z., Y. Sakuma, T. Sato, and A. Mori (1954) The effects of antibiotics on fertilized ova. Tohoku J. Agric. Res., 4:193.

Kvasnitskii, A. V. (1956) Generative function of ovary and reproductivity of farm animals. Proc. 3rd Internat. Congr. Anim. Reprod., Plenary Session, p. 59. Cambridge, England.

Lin, T. P., J. K. Sherman, and E. L. Willet (1957) Survival of unfertilized mouse eggs in media containing glycerol and glycine. J. Exp. Zool., 134:275.

Marden, W. G. R., and M. C. Chang (1952) The aerial transport of mammalian ova for transplantation. Science, 115:705.

Mauer, R. E. (1966) Culture and storage of tubal rabbit embryos in synthetic media. Ph.D. thesis, Washington State University.

McLaren, A., and D. Michie (1956) Studies on the transfer of fertilized mouse eggs to uterine foster mothers. J. Exp. Biol., 33:394.

Meyer, D. T., and J. F. Lasley (1945) The factor in egg yolk affecting the resistance, storage potentialities, and fertilizing capacity of mammalian spermatozoa. J. Animal Sci., 4:261.

Moore, N. W., and L. E. A. Rowson (1960) Egg transfer in sheep. Factors affecting the survival and development of transferred eggs. J. Reprod. Fertility, 1:332.

Moran, T. (1925) The effect of low temperatures on hens' eggs. Proc. Roy. Soc., (London) Ser. B., 98:436.

Noyes, R. W., and Z. Dickmann (1960) Relationship of ovular age to endometrial development. J. Reprod. Fertility, 1:186–196.

Parker, R. C. (1961) Methods of Tissue Culture. Hoeber, New York.

Parkes, A. S. (1956) Preservation of living cells and tissues at low temperatures. Proc. 3rd. Internat. Congr. Anim. Reprod., Plenary Session, p. 59. Cambridge, England.

Parrott, D. M. V. (1960) The fertility of mice with orthotopic ovarian grafts derived from frozen tissue. J. Reprod. Fertility, 1:230–241.

—— and A. S. Parkes (1960) Dynamics of the orthotopic ovarian grafts. Sex differentiation and development. Mem. Soc. Endocrinol, 7:71.

Paul, G. (1960) Cell and Tissue Culture. Livingstone, Edinburgh.

Phillips, P. H., and H. A. Lardy (1940) A yolk-buffer pabulum for the preservation of bull semen. J. Dairy Sci., 23:399.

Picket, R. (1893) De l'empoli methodique des basses temperatures en biologie. Arch. Sc. Phys. Nat. Ser. 3, 30:293.

Pincus, G. (1936) The Eggs of Mammals. Macmillan, New York.

——— (1939) The comparative behavior of mammalian eggs *in vivo* and *in vitro*. IV. The development of fertilized and artificially activated rabbit eggs. J. Exp. Zool., 82:85–129.

Sherman, J. K., and T. P. Lin (1958) Survival of unfertilized mouse eggs during freezing and thawing. Proc. Soc. Expt. Biol. Med., 98:902–905.

——— (1959) Temperature shock and cold storage of unfertilized mouse eggs. Fertility Sterility, 10:384.

Smith, A. U. (1949) Cultivation of rabbit eggs and cumuli for phase-contrast microscopy. Nature, 164:1136.

——— (1952) Behaviour of fertilized rabbit eggs exposed to glycerol and to low temperatures. Nature, 170:374.

——— (1953) *In vitro* experiments with rabbit's eggs. In: Mammalian Germ Cells, p. 217. Ed. by G. E. W. Wolstenholme. Little, Brown, Boston.

Sugawara, S., and S. Takeuchi (1962) On the respiratory activity of the rat ova stored at low temperature. Jap. J. Animal Reprod., 8:65–68.

Taylor, I. M. (1956) The effect of low temperature upon the intra-cellular potassium in isolated tissues. In: The Physiology of induced hypothermia. Ed. by R. D. Dripps. NAS-NRC Publication, 451:26–31.

Warwick, B. L., and R. O. Berry (1949) Inter-generic and intra-specific embryo transfers in sheep and goats. Heredity, 40:297.

Whitten, W. K. (1956) Culture of tubal mouse ova. Nature, 177:96.

Willett, E. L., P. J. Buckner, and G. L. Larson (1953) Three successful transplantations of fertilized bovine eggs. J. Dairy Sci., 36:520–523.

8

Egg Transfer

ZEEV DICKMANN
*Departments of Gynecology
and Obstetrics and Anatomy
University of Kansas Medical Center,
Kansas City, Kansas 66103*

"In this preliminary note I wish merely to record an experiment by which it is shown that it is possible to make use of the uterus of one variety of rabbit as a medium for the growth and complete foetal development of fertilized ova of another variety of rabbit."

This is the introductory statement from Walter Heape's paper (1890) in which he reported the first successful egg transfer experiment.

In this chapter, the term *egg* includes ovarian eggs (also known as follicular eggs and as oocytes), unfertilized ovulated eggs, and fertilized eggs from the one-cell stage through the blastocyst stage. When we speak of *egg transfer* we refer to the technique by which eggs are recovered from a *donor* and are transferred to a *recipient*. Transfers may be classified into three categories, according to *who* the recipient is: (1) autotransfer—eggs are transferred to the same individual from whom they were recovered; (2) homotransfer—eggs are transferred to another individual of the same species; and (3) heterotransfer—eggs are transferred to an animal of another species. Another criterion for classifying transfers is according to the *site* to which the eggs are transferred: (1) sites within the female reproductive tract (e.g., ovarian bursa, oviduct, and uterus); and (2) sites outside the female reproductive tract (e.g., kidney, spleen, testis). This chapter will be confined to consideration of the first category only. The reader who is

The author gratefully acknowledges financial support from the Population Council and from the Ford Foundation.

interested in transfers to sites outside the female reproductive tract should consult Kirby's review (1965) and Chapter 9 in this volume.

A logical way to explore a technique is to study it in two major sequential steps: how to do it, and what can be done with it. This article follows this approach and is divided into two main sections: (1) description of the egg transfer technique; and (2) problems that can be solved by using the egg transfer technique.

EGG TRANSFER TECHNIQUE

The egg transfer technique (ETT) has been used in the mouse (McLaren and Michie, 1956; Tarkowski, 1959), rat (Dickmann and De Feo, 1967; Yoshinaga and Adams, 1967), hamster (Blaha, 1964), rabbit (Chang, 1950; Adams, 1962; Dickmann, 1964), sheep (Moore and Rowson, 1960; Hancock and Hovell, 1961), pig (Hancock and Hovell, 1962; Dziuk, Polge and Rowson, 1964), cow (Willett, Buckner and Larson, 1953; Rowson, Moor and Lawson, 1969), and perhaps in some other species. (Only a few of many references are cited here.) The principles of the ETT are the same for all species, so that familiarity with the technique in one species provides a basis for using it in others. The rat was chosen as the model of a detailed description of the ETT because (1) the execution of the ETT in the rat is probably more difficult than, or at least as difficult as, it is in other species, and (2) I have had a great deal of experience with the ETT in the rat (Noyes and Dickmann, 1960; Dickmann and Noyes, 1960; Noyes and Dickmann, 1961; Dickmann and Noyes, 1961; Dickmann and De Feo, 1967; Dickmann, 1967).

The ETT can be conveniently divided into three phases:
(1) Recovery of eggs from the donor.
(2) Handling of eggs *in vitro*.
(3) Transfer of eggs to the recipient.

Recovery of Eggs from the Donor

The donor is anesthetized with ether and is then killed by breaking its neck. The egg-containing organ (i.e., ovary, oviduct, or uterus) is excised, rinsed in room-temperature saline (0.9% NaCl), and blotted on a paper towel in order to remove surface blood. The recovery of eggs from each of the three organs is as follows:

Recovery of ovarian eggs

The excised, cleaned ovaries are placed in a watch glass containing 37°C transfer-medium (see discussion on page 137). The harvesting of eggs from the follicles is done under a dissecting microscope. With a sharp, pointed blade, each of the large follicles is cut open and its contents are either gently squeezed or shelled into the medium. The eggs, which are imbedded in a cumulus mass, are left in the watch glass until they are transferred to a recipient.

In large and/or expensive animals in which one does not wish to kill the donor, the donor is anesthetized, the ovaries are exposed, the large follicles are punctured, and their contents are aspirated into a pipette, which is then emptied into a watch glass containing transfer medium. This procedure allows the donor to be used repeatedly.

Recovery of oviductal eggs

The procedure for recovering eggs from the oviduct varies according to the day of pregnancy on which the donor is killed. During Days 1, 2 and 3 (Day 1=spermatozoa found in the vagina), the oviduct is served from the uterus at the uterotubal junction. During the morning of Day 1 the eggs are still invested in cumulus and are located in the ampulla. There are two convenient methods by which the eggs can be obtained at this location; both methods are performed under the dissecting microscope:

Method 1: The oviduct is placed in a watch glass containing transfer medium, and the ampulla is cut with an iridectomy scissors. This procedure usually allows the cumulus mass containing the eggs to escape through the cut into the medium. Occasionally, when the escape of the cumulus does not occur, slight pressure must be applied at one or both sides of the cut.

Method 2: The oviduct is cut in half. The ovarian half is flushed with transfer medium from the uterine end toward the ovarian end, using a syringe fitted with a 30-gauge, blunted hypodermic needle. While flushing, the hypodermic needle is held in place with a fine watchmaker's forceps.

In the afternoon of Day 1 and on Days 2 and 3, the eggs are denuded (i.e., free of cumulus). They are recovered by flushing the oviduct from the ovarian end toward the uterine end. On Day 4, most eggs are located near the uterotubal junction; some, however, may already be in the oviducal end of the uterus. To obtain eggs from these locations, the oviduct is excised in a way so as to include about 3 mm of uterus. The eggs are flushed from the ovarian end toward the uterine end.

In small, inexpensive animals such as the rat, mouse, and hamster, it would be quite difficult, as well as impractical, to recover eggs from the oviduct of the living animal. On the other hand, in larger animals such as the rabbit, sheep, and pig, recovery of eggs from the living animal is easily done, but it normally requires two operators. The animal is anesthetized and the oviduct, with a portion of the uterus, is exposed. Operator-1 cannulates the oviduct at its uterine end with a hypodermic needle attached to a syringe containing the transfer medium. He holds the hypodermic needle tightly in place either with his fingers or with a pair of forceps. Operator-2 slides a bent glass tube into the fimbriated end of the oviduct, holds it in place between his thumb and index finger, and directs the other end of the glass tube into a collecting dish. Then, operator-1 flushes the oviduct from the uterine end toward the ovarian end (see Fig. 8-1). After the egg recovery is completed, the donor is sewn up and can be used again

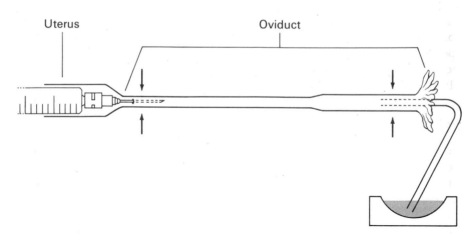

FIGURE 8-1

Diagrammatic representation of in vivo *egg recovery from the rabbit oviduct. The arrows indicate the areas where the operator's fingers are pressed against the glass tube and the hypodermic needle. (not to scale)*

at a later time. With some variations, the above technique has been used by Avis and Sawin (1951), Chang (1952), Hunter, Adams, and Rowson (1955), and others.

Recovery of uterine eggs

Fertilized eggs are recovered from the uterus on Day 5 of pregnancy by which time they are blastocysts. The excised, cleaned uteri are flushed from the oviducal toward the vaginal end, using a 23-gauge, blunted hypodermic needle fitted to a syringe containing the transfer medium. While flushing, the hypodermic needle is held in place with forceps.

Although I have used the above method in all my experiments, and intend to continue to do so, blastocysts can also be easily recovered from the uterus of the living rat. The uterus is exposed via a midventral incision. A 23-gauge hypodermic needle fitted into a syringe containing transfer medium is inserted into the uterine lumen at the oviducal end. Before flushing toward the vaginal end, a polyethylene tube (external diameter: 5.5 mm; internal diameter: 3.5 mm) is inserted into the vagina until it touches the cervix. The other end of the polyethylene tube is directed into a watch glass. The uterus is then flushed. In my experience, the number of eggs recovered in this manner is as high as when the donor is first killed and then its excised uteri are flushed. This technique was originally demonstrated to me by Dr. P. F. Kraicer.

When working with larger animals such as rabbits, sheep, and pigs, recovery of eggs without killing the donor is often preferred. The method of doing this is similar to the one used to flush the oviduct. One operator inserts a hypodermic needle attached to a syringe into the lumen of the uterus at the oviducal end. The second operator makes a small incision through the uterine wall at the vaginal end. Through this incision, he inserts into the uterine lumen one end of a bent glass tube and, while holding it in place, directs its other end toward a receiving dish. The first operator then flushes the eggs into the dish. With some variations, the above technique has been used by Venge (1952), Hunter, Adams and Rowson (1955), Hancock and Hovell (1962), and others.

Nonsurgical recovery of uterine eggs

Rowson and Dowling (1949) reported that they were able to recover eggs from cow uteri by a nonsurgical technique. They said their apparatus was "in a very experimental stage." To my knowledge, no further work has been done to streamline this technique, nor has the technique been tried in species other than the cow. It is obvious that a simplified nonsurgical technique for recovering uterine eggs would greatly facilitate egg transfer experiments in large animals.

HANDLING OF EGGS in Vitro

The recovered eggs are kept *in vitro* in the flushing medium which consists of two parts saline (0.9% NaCl) to one part rat plasma. To prevent clotting of the plasma, one drop of heparin (1000 U.S.P. units/ml) is added to every 5 ml of whole blood before it is centrifuged. Until the time of flushing, the medium is kept at 37°C, but thereafter no further heating is applied. Eggs do survive when handled in room-temperature medium. However, no control experiments have been done to determine whether or not keeping the eggs at 37°C increases the percentage of survival to term fetuses. The eggs remain *in vitro* for 10 to 20 minutes.

As to the type of medium, it was my contention that if a poor medium was used for transferring eggs, normal term fetuses could still be obtained, if the eggs were left in the medium for only a short time. To test this contention, I used tap water as the medium in an experiment in which blastocysts were transferred into the uteri of Day 4 pseudopregnant recipients. The water was not treated in any way; its *p*H was 7.5 and its temperature was 26°C. Exposure of the blastocysts to the water was timed from the moment of flushing the donor's uterus until the moment the blastocysts were deposited in the recipient's uterus. The results cannot be expressed in terms of "percentage of blastocysts which developed into term fetuses, "because the number of experiments done was too small. However, the results can be expressed qualitatively: a few live, normal, term fetuses were obtained, when the exposure time was 150 seconds or less. When the exposure time was between 170 and 200 seconds, a few implantation sites (i.e., sites at which blastocysts implanted, but where some time later the embryos died and were resorbed), but no fetuses, were obtained. These experiments show that eggs transferred in a poor medium can yield full-term fetuses if the transfer time is sufficiently short. That the water was a poor medium was further demonstrated when blastocysts were left in the water for 300 seconds or longer: gross damage to the blastocysts could be seen under the dissecting microscope. When these blastocysts were subse-

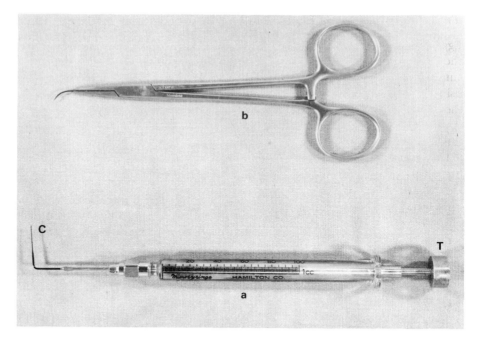

FIGURE 8-2

(a) *A 1 cc microsyringe (#0010 Hamilton Co., Whittier, California) with a threaded plunger which permits the delivery of a very small quantity of fluid by turning the thumbwheel* (T) . *The capillary pipette* (C) *is inserted and cemented into a 16-gauge hypodermic needle. For transfers into the oviduct and into the uterus, the pipette is tapered at its tip to 150 μ internal diameter (as shown in the photograph). In order for the pipette to show up better in the photograph, it was filled with dark fluid. For intra bursal transfers, the pipette is not tapered and has a 500 μ internal diameter. ×0.66.* (b) *A hemostat which serves as a convenient holder for a half curved cutting-edge stainless surgical needle (size 22, Anchor Brand). The needle is used for puncturing the uterus. In order for the needle to show up better in the photograph, it was painted with a black dye. ×0.66.*

quently fixed, stained, and examined under the compound microscope, severe cytological damage was seen.

It has been my experience that, of the various culture media used for transferring rat eggs, saline/plasma yielded the highest percentage of term fetuses. This is not to say that this medium would also be the best when used with eggs of other species. Media which have been used in various species are listed by Austin (1961).

After the flushing is collected in a watch glass, the eggs are found with the aid of a dissecting microscope; they are picked up with a micropipette (see Fig. 8-2a) and are placed in the center of the watch glass, where they remain until they are picked up again for transfer.

Transfer of Eggs to the Recipient

The transfer procedure varies according to the site to which the eggs are transferred. The sites are ovarian bursa, oviduct, and uterus.

Transfer to the ovarian bursa

Eggs that are transferred to the bursa, must be invested in cumulus cells or they will not be picked up by the oviducal fimbriae. Eggs invested in cumulus are picked up within an hour after transfer (Dickmann, unpublished). Moreover, eggs in cumulus have to be transferred into the bursa, as they cannot be transferred directly into the oviduct, because a pipette which is wide enough to accommodate the cumulus mass is too wide to be cannulated into the oviduct.

Under ether anesthesia, the ovary, oviduct, and the oviducal end of the uterus are exposed via a dorsal incision. The eggs, surrounded by cumulus along with a minimal amount of medium are picked up with a capillary pipette (diameters are: external—900μ, internal—500μ). Under a dissecting microscope, the pipette is inserted into the bursal cavity via the bursal foramen (described by Alden, 1942), and the eggs are expelled. The exposed organs (ovary, oviduct, and part of the uterus) are then lowered back into the abdominal cavity, and the animal is sutured.

Runner and Palm (1953), working with mice, used another route to reach the bursal cavity: the pipette was inserted through the fat pad adjacent to the bursa, through the bursal membrane, and into the bursal cavity.

Transfer into the oviduct

Under ether anesthesia, a dorsal incision is made through which the ovary, oviduct, and the oviducal end of the uterus are exposed. The fimbriated end of the oviduct is located with the aid of a dissecting microscope, and, using an electroscalpel to prevent bleeding, a hole, approximately one mm in diameter, is made in the bursal membrane overlying the fimbria. The eggs to be transferred, along with 0.0001 to 0.0005 ml of medium, are drawn into a micropipette, which is then introduced into the fimbriated end of the oviduct, and the eggs are gently expelled. The micropipette used is tapered towards its fire-polished tip (see Fig. 8-2a) ; the internal diameter of the tip is approximately 150μ. After the transfer, the exposed organs are lowered back into the abdominal cavity and the animal is sutured.

The smallness of the oviducal lumen (and also the uterine lumen), require that the eggs be transferred in a small volume of medium. This can be achieved by using: (1) a micropipette of the type shown in Figure 8-2a; (2) a compound mouth micropipette (see Noyes, Doyle, Gates, and Bently, 1961) ; (3) a simple mouth micropipette drawn to about 150μ internal diameter, which is controlled by air pressure from the mouth via a rubber tube connected to the pipette (H. M. Weitlauf, personal communication) ; (4) a micropipette controlled like an eye dropper by a short length of polyethylene tubing (Yoshinaga and Adams, 1966).

For larger animals, such as the rabbit, the transfer procedure is much easier than the one used with the rat. This is so because, the much larger size of the rabbit oviduct allows the operation to be performed without a dissecting microscope, the transfer pipette can be large enough to accommodate eggs embedded in cumulus, and there is no bursal membrane.

Transfer into the uterus

Under ether anesthesia, the uteri are exposed via a midventral incision. The ovarian end of the uterus is held gently between the thumb and the index finger, and the antimesometrial wall of the uterus is punctured with a fine cutting-edge surgical needle so as to create a passage into the uterine lumen (see Fig. 8-2b). The micropipette (the same size as the one used for oviducal transfers) containing the eggs is inserted through this passage, the eggs are deposited in the uterine lumen and the abdominal incision is sutured. The transfer procedure is illustrated in Figures 8-3a,b,c. An alternative route to the uterus is via a dorsal incision. The transfer procedure is the same as that described above.

When transferring early rabbit blastocysts, the uterine wall is not prepunctured; instead, the pipette containing the eggs is thrust through the uterine wall. Since the tip of the pipette has to be sharp, it is not firepolished. When transferring late rabbit blastocysts, which are 3 to 5 mm in diameter, an incision is made in the antimesometrial side of the uterine wall, and the pipette containing the eggs is inserted into the uterine lumen through this incision. The above technique has been used by Chang (1950). In other large animals, the transfer procedure is similar to that used in rabbit. It should be noted that as long as the embryo is not implanted, it can be transferred without difficulty. Thus, 14-day-old sheep blastomeric vesicles (the embryo and its membranes) have been transferred (Moor and Rowson, 1966a). These vesicles are about 8 mm in diameter and more than 100 mm in length (Rowson and Moor, 1966a).

Nonsurgical transfer into the uterus

It is obvious that a nonsurgical technique for transferring eggs into the uterus is preferable to a surgical one. Beatty (1951) and Tarkowski (1959) reported that term fetuses resulted from their nonsurgical transfers of mouse eggs, but that the percentage of successes was low. The only other species in which this technique has been successfully employed is the cow (Sugie, 1965; Rowson and Moor, 1966b). In this species, too, the percentage of successes was low. On the face of it, the nonsurgical transfer technique is very simple: the transfer pipette is inserted through the vagina and the cervix, into the uterine lumen, and the eggs are deposited. Nevertheless, for some reason not yet known, the results have been poor. One possible explanation for the failure is that uterine contractions expel the eggs via the vagina. Such contractions could be caused by oxytocin which might be released from the pituitary upon cervical stimulation (Harper, Bennet, and Rowson, 1961).

PROBLEMS THAT CAN BE SOLVED BY USING THE EGG TRANSFER TECHNIQUE (ETT)

The ETT is an absolute necessity for solving a wide variety of problems in developmental biology. Examples of such problems are listed

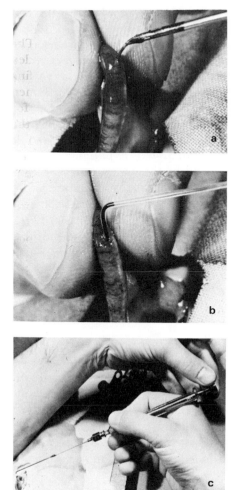

FIGURE 8-3

Three steps in the procedure of egg transfer into the rat uterus are: (a) Puncture of the antimesometrial wall of the uterus with a needle; (b) insertion of the micropipette (containing dark fluid) through the uterine wall; and (c) expulsion of the eggs into the uterine lumen by turning the thumbwheel of the microsyringe.

below. It is emphasized that the purpose here is not to explain or to discuss these problems, but merely to state them and to indicate that their solutions are dependent on the use of the ETT. For the benefit of those readers who wish to explore these problems further, a few references are cited.

Problem 1: Are ovarian eggs fertilizable?

To answer this question, ovarian eggs are obtained from a donor and are then transferred into the oviducts of a previously mated recipient. Later a determination is made whether or not fertilization has taken place. This problem has been investigated by Noyes (1952) and by Chang (1955)

Problem 2: Fertilization in vitro

Since the end of the nineteenth century, many researchers have reported achieving fertilization of mammalian eggs *in vitro*. But, the evidence for fertilization, which was based on various changes in the egg, did not satisfy the critical student of fertilization. The absolute proof of successful fertilization *in vitro* was provided by Chang (1959). He fertilized rabbit eggs *in vitro* and subsequently transferred them into recipients where they developed into normal young. Chang's evidence could not be contested by even the harshest skeptics.

Problem 3: Culture of eggs in vitro

Many studies have been conducted in which eggs were cultivated *in vitro* to determine their chemical, physical, and nutritional requirements. Such studies however, are incomplete if at the end of the experiments the eggs are not tested for their capacity to develop into normal term fetuses. The only known way by which this capacity can be tested is to transfer the eggs into appropriate recipients. A case in point is a recent study (Pavlok, 1967) in which mouse eggs, grown in explanted oviducts, developed into morulae and blastocysts which apparently looked normal. But, when these morulae and blastocysts were transferred into recipients, only 5% of them implanted. How many of the implanted embryos developed into term fetuses was not reported.

Problem 4: Study of experimental chimeras

Mintz (1965) and Tarkowski (1965) have shown that cleaving mouse eggs of two different strains can be fused *in vitro* and that the two fused embryos continue to develop *in vitro* to form a single blastocyst. This result, in itself, was a remarkable finding. However, the great reward was achieved when blastocysts so formed were transferred into the uteri of suitable recipients wherein they developed into mosaic young. The ability to produce such mosaic individuals opens the door for many studies in embryology, genetics, and immunology.

Problem 5: Egg-endometrial interrelationships

One may speak of three major events during the preimplantation period of pregnancy: development of the fertilized egg, transport of the egg through the oviduct, and changes in the uterus in preparation for implantation. Normally, these events are synchronized so that by the time the egg reaches the uterus the uterine milieu is hospitable to the egg. A basic problem is: what would happen to a pregnancy if any one of the three events was either accelerated or retarded. The best way to study this problem is by asynchronous egg transfers. Such experiments were done in the rabbit (Chang, 1950), the mouse (McLaren and Michie, 1956), the rat (Noyes and Dickmann, 1960), and the sheep (Moore and Shelton, 1964).

Problem 6: The effects of various physical and chemical agents (e.g., irradiation, temperature, drugs) on the preimplanting embryo

We may consider a case in which a pregnant female, during the preimplantation period, is subjected to a dose of irradiation which results in the death of the fetus. The questions that arise are: Does the irradiation affect the embryo directly; or does it affect the mother, who in turn affects the embryo; or does it affect both the embryo and the mother. These questions can be answered by performing the following transfer experiments: (1) A short time after irradiating a pregnant donor, her eggs are recovered and transferred into a nonirradiated recipient; (2) Eggs are recovered from a nontreated pregnant donor and are transferred into a recipient who has been irradiated; (3) Fertilized eggs are irradiated *in vitro* and are then transferred into a nonirradiated recipient. Problems of this kind were studied by Glass and McClure (1964).

Problem 7: Fecundity decline in aging females

Decline in reproductive capacity with eventual total sterility in aging females who still ovulate could be explained in terms of defects in either the (1) egg, (2) reproductive tract, (3) endocrine system, or (4) any combination of the three factors. To test the capacity of eggs from old females to develop into normal fetuses, the eggs are transferred into the reproductive tract of young recipients. To test the capacity of the uterus of old females to support a normal pregnancy, eggs recovered from young donors are transferred into the uteri of old recipients. Egg transfer experiments aimed at studying the above problems were done in hamster (Blaha, 1964), and in mouse (Talbert and Krohn, 1966).

Problem 8: Cancer studies

It is well known that the incidence of a particular tumor is high in certain strains of mice and is low in certain other strains. We may ask, is the intrauterine milieu a factor contributing to the tumor incidence? This question can be answered by performing reciprocal egg transfers between high- and low-incidence strains. Experiments of this kind were done by Fekete and Little (1942).

Conclusion

The above list of problems demonstrates that the ETT has been, and will continue to be, an indispensable tool for investigating and solving a wide variety of problems in developmental biology.

BIBLIOGRAPHY

Adams, C. E. (1962) Studies on prenatal mortality in the rabbit, oryctolagus cuniculus: The effect of transferring varying numbers of eggs. J. Endocrinol., 24:471–490.

Alden, R. H. (1942) The periovarial sac in the albino rat. Anat. Record, 83:421–435.

Austin, C. R. (1961) The Mammalian Egg. Blackwell Scientific Publications, Oxford, pp. 125–143.

Avis, F. R., and P. B. Sawin (1951) A surgical technique for the reciprocal transplantation of fertilized eggs. J. Heredity, 42:259–260.

Beatty, R. A. (1951) Transplantation of mouse eggs. Nature, 168:995.

Blaha, G. C. (1964) Effect of age of the donor and recipient on the development of transferred golden hamster ova. Anat. Record, 150:413–416.

Chang, M. C. (1950) Development and fate of transferred rabbit ova or blastocyst in relation to the ovulation time of recipients. J. Exp. Zool., 114:197–216.

——— (1952) Fertilizability of rabbit ova and the effects of temperature in vitro on their subsequent fertilization and activation in vivo. J. Exp. Zool., 121:351–381.

——— (1955) The maturation of rabbit oocytes in culture and their maturation, activation, fertilization and subsequent development in the fallopian tubes. J. Exp. Zool., 128:379–405.

——— (1959) Fertilization of rabbit ova *in vitro*. Nature, 184:466–467.

Dickmann, Z. (1961) The zona pellucida at the time of implantation. Fertility and Sterility, 12:310–318.

——— (1964) Fertilization and development of rabbit eggs following the removal of the cumulus oophorus. J. Anat., 98:397–402.

——— (1967) Hormonal requirements for the survival of blastocysts in the uterus of the rat. J. Endocrinol, 27:455–461.

——— and V. J. DeFeo (1967) The rat blastocyst during normal pregnancy and during delayed implantation, including an observation on the shedding of the zona pellucida. J. Reprod. Fertility, 13:3–9.

——— and R. W. Noyes (1960) The fate of ova transferred into the uterus of the rat. J. Reprod. Fertility, 1:197–212.

Dziuk, P. J., C. Polge, and L. E. A. Rowson (1964) Intrauterine migration and mixing of embryos in swine following egg transfer. J. Animal Sci., 23:37–42.

Fekete, E., and C. C. Little (1942). Observations on the mammary tumor incidence of mice born from transferred ova. Cancer Res., 2:525–530.

Glass, L. E., and T. R. McClure (1964) Survival of mouse eggs after *in vivo* or *in vitro* x-irradiation. Anat. Record, 148:286.

Hancock, J. L., and G. J. R. Hovell (1961) Transfer of sheep ova. J. Reprod. Fertility, 2:295–306.

——— (1962) Egg transfer in the sow. J. Reprod. Fertility, 4:195–201.

Harper, M. J. K., J. P. Bennett, and L. E. A. Rowson (1961) A possible explanation for the failure of non-surgical ovum transfer in the cow. Nature, 190:789–790.

Heape, W. (1890) Preliminary note on the transplantation and growth of mammalian ova within a uterine foster-mother. Proc. Roy. Soc. (London), 48:457–458.

Hunter, G. L., C. E. Adams, and L. E. A. Rowson (1955) Interbreed ovum transfer in sheep. J. Agric. Sci., 46:143–149.

McLaren, A., and D. Michie (1956) Studies on the transfer of fertilized mouse eggs to uterine foster-mothers. J. Exp. Biol., 33:394–416.

Mintz, B. (1965) Experimental genetic mosaicism in the mouse. In: Ciba Foundation Symposium on Preimplantation Stages of Pregnancy. Ed. by G. E. W. Wolstenholme and M. O'Conner, Churchill, London, pp. 194–207.

Moor, R. M., and L. E. A. Rowson (1966) The corpus luteum of the sheep: functional relationship between the embryo and the corpus luteum. J. Endocrinol., 34:233–239.

Moore, N. W., and L. E. A. Rowson (1960) Egg transfer in sheep: Factors affecting the survival and development of transferred eggs. J. Reprod. Fertility, 1:332–349.

Moore, N. W., and J. N. Shelton (1964) Egg transfer in sheep. Effect of degree of synchronization between donor and recipient, age of egg, and site of transfer on the survival of transferred eggs. J. Reprod. Fertility, 7:145–152.

Noyes, R. W. (1952) Fertilization of follicular ova. Fertility and Sterility, 3:1–12.

—— and Z. Dickmann (1960) Relationship of ovular age to endometrial development. J. Reprod. Fertility, 1:186–196.

—— and Z. Dickmann (1961) Survival of ova transferred into the oviduct of the rat. Fertility and Sterility, 12:67–79.

——, L. L. Doyle, A. H. Gates, and D. L. Bentley (1961) Ovular maturation and fetal development. Fertility and Sterility, 12:405–416.

Pavolk, A. (1967) Development of mouse ova in explanted oviducts: Fertilization, cultivation, and transplantation. Science, 157:1457–1458.

Rowson, L. E. A., and D. F. Dowling (1949) An apparatus for the extraction of fertilized eggs from the living cow. Vet. Rev. Annotations, 61:191.

Rowson, L. E. A., and R. M. Moor (1966a) Development of the sheep conceptus during the first fourteen days. J. Anat., 100:777–785.

—— (1966b) Non-surgical transfer of cow eggs. J. Reprod. Fertility 11:311–312.

——, R. M. Moor, and R. A. S. Lawson (1969) Fertility following egg transfer in the cow; effect of method, medium and synchronization of oestrus. J. Reprod. Fertility, 18:517–523.

Runner, M. N. and J. Palm (1953) Transplantation and survival of unfertilized ova of the mouse in relation to postovulation age. J. Exp. Zool., 124:303–316.

Sugie, T. (1965) Successful transfer of a fertilized bovine egg by non-surgical techniques. J. Reprod. Fertility, 10:197–201.

Tarkowski, A. K. (1959) Experiments on the transplantation of ova in mice. Acta Theriol., 2:251–267.

—— (1965) Embryonic and postnatal development of mouse chimeras. In Ciba Foundation Symposium on Preimplantation Stages of Pregnancy. Ed. by G. E. W. Wolstenholme and M. O'Conner, Churchill, London. pp. 183–193.

Talbert, G. B., and P. L. Krohn (1966) Effect of maternal age on viability of ova and uterine support of pregnancy in mice. J. Reprod. Fertility, 11:399–406.

Venge, O. (1952) A method for continuous chromosome control of growing rabbits. Nature, 169:590–591.

Willett, E. L., P. J. Buckner, and G. L. Larson (1953) Three successful transplantations of fertilized bovine eggs. J. Dairy Sci., 36:520–523.

Yoshinaga, K., and C. E. Adams (1966) Endocrine aspects of egg implantation in the rat. J. Reprod. Fertility, 12:583–586.

—— (1967) Luteotrophic activity of the young conceptus in the rat. J. Reprod. Fertility, 13:505–509.

9

The Transplantation of Mouse Eggs and Trophoblast to Extrauterine Sites

D. R. S. KIRBY
Department of Zoology
University of Oxford
Oxford, England

Although extrauterine pregnancies occur occasionally in humans, there are less than ten reports in the past 50 years describing spontaneous extrauterine pregnancies in laboratory or domestic animals. However, the experimental production of these abnormally located pregnancies is proving to be a valuable technique to investigate such processes as ovoimplantation, production of placental hormones, control of trophoblast invasion and general immunological problems of pregnancy.

Dickmann describes, in Chapter 8 of this book, a method for transplanting eggs to a recipient uterus. The apparatus used for the extrauterine transplantation of eggs and trophoblast makes it possible to transfer the graft in a very small quantity of carrying fluid. This overcomes the main difficulty associated with extrauterine transplantation, the dislodgement of the graft from its precarious position by the flux of excess carrying fluid.

OBTAINING THE GRAFT MATERIAL

Early cleavage eggs are obtained from the oviduct and blastocysts from the uterus by the method described by Dickmann.

Trophoblast differentiates from the cells comprising the wall of the blastocyst by two pathways. The cells of the blastocoele wall may give rise

Dr. Kirby died on November 11, 1969 from injuries sustained in an automobile accident. He was a cherished colleague and will long be remembered by the other contributors to this volume and by the scientists who will be served by it. Dr. W. David Billington has generously helped in the final preparations of this chapter.

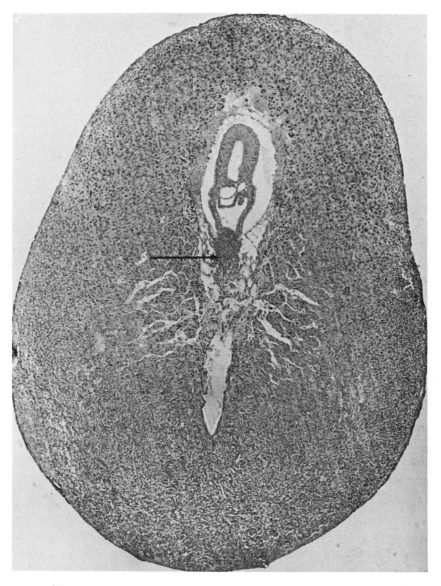

FIGURE 9-1

Section of a mouse decidual swelling removed from the uterus on Day 7 of pregnancy. Note the smooth outline of the decidua showing that it can be cleanly removed from the uterus and also that the ectoplacental cone (at arrow) is only loosely attached to the decidual tissue (×50).

directly to trophoblast giant cells [primary giant cells of Snell (1966)], whose chief function seems to be to anchor the blastocyst to the wall of the uterus. The cells overlying the inner cell mass of the blastocyst do not give rise directly to definitive trophoblast, but to a cone-shaped mass of small cells which are mitotically active. It is this structure, the ectoplacental cone or träger, the proliferative center of the placental trophoblast, which is most

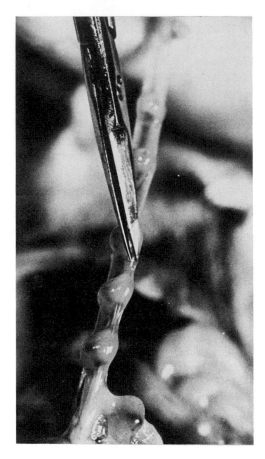

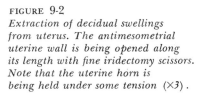

FIGURE 9-2
*Extraction of decidual swellings
from uterus. The antimesometrial
uterine wall is being opened along
its length with fine iridectomy scissors.
Note that the uterine horn is
being held under some tension* (×3).

conveniently transplanted to produce extrauterine growths of pure trophoblast, devoid of an embryo or extraembryonic membranes. The ectoplacental cone is most easily removed from the uterus on Day 7 of pregnancy (the day of sperm plug is designated Day 0) for three reasons: first, it is large enough to manipulate; second, it is loosely attached to the surrounding decidual tissue (see Fig. 9-1); and finally, it has just begun to differentiate into trophoblast and thus retains most of its developmental potential.

The ectoplacental cones are harvested as follows. The pregnant mouse is killed by cervical dislocation and pinned down with the ventral part uppermost. The abdomen is shaved and opened, and the uterine horn is grasped at the ovarian end and the adventitious mesometrium is stripped to the vagina. (It is important not to sever the uterus from the abdomen as the next stage is much easier if the uterus is firmly anchored to the carcass.) Hold the uterine horn under slight tension, making sure that the antimesometrial side is uppermost. With fine iridectomy scissors make an incision along the length of the antimesometrial wall of the uterus (see Fig. 9-2). The severed myometrium retracts, thus exposing the pear-shaped decidual swellings (see Fig. 9-3). These are loosely attached to the uterus and can be removed cleanly with fine forceps. They are then placed in physio-

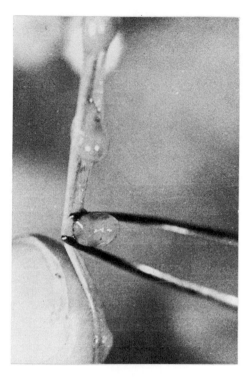

FIGURE 9-3
The cut uterine wall has retracted, leaving the ecto-placental cone prominently exposed and easy to remove with curved forceps (×3).

logical saline solution at room temperature and remain there while the ectoplacental cones are dissected under a binocular microscope (×10). The pear-shaped decidua is bissected from apex to base with fine scissors. The cut should be off-center to avoid damage to the conceptus, exposing it along its length in the larger of the two parts of the decidual shell.

The conceptus is gently teased out of the decidua with tungsten needles and fine forceps. Still immersed in saline solution, the embryo and the dark red ectoplacental cone are separated with fine scissors (see Figs. 9-4 and 9-5). With a Pasteur pipette transfer the dissected ectoplacental cones to a solid watch glass containing fresh Pannett-Compton solution, then cover with a glass slide. (Although ectoplacental cones remain viable for many hours, they tend to dissociate and become more difficult to handle).

The procedure for obtaining and transplanting eggs and trophoblast in the rat is essentially the same as for the mouse, except for a difference in the times of obtaining the grafts: eggs from the oviduct on Day 0 to Day 3 of pregnancy, blastocysts from the uterus on Day 4, and ectoplacental cones on Days 8 and 9.

TRANSPLANTATION APPARATUS

The transplantation apparatus consists of a microdissector which moves in all planes of space and gives a 5:1 reduction in movement (see Appendix, page 155). Once positioned, the counterbalance weights ensure

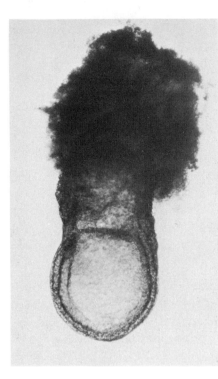

FIGURE 9-4
*The embryo and ectoplacental
cone as removed from the
decidua (×80).*

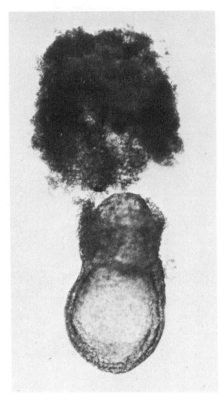

FIGURE 9-5
*The ectoplacental cone (top)
has been separated from the
embryo with fine scissors. Note
the undamaged dome shaped
chorion still attached to the
embryo (×80).*

that it will remain in place, thus freeing the hand (see Fig. 9-6).

The micropipette is made by hand-pulling soda-glass melting-point tubes. These should first be washed with acid and then rinsed to prevent capillary forces in a dirty tube from interfering with the flow of fluid. The internal diameter of the end of the pipette should be 100 to 250μ for eggs and 150 to 300μ for $7\frac{1}{2}$-day ectoplacental cones. It is important to flame-polish the tip of the pipette; a jagged or sharp tip may accidentally pierce the capsule or damage the parenchyma of the host organ. The micrometer syringe (Agla) consists of a narrow-bore all-glass hypodermic syringe attached by a rigid holder to a micrometer screw-gauge that operates the plunger. Small quantities of fluid (less than 0.01 ml) can be released with an accuracy of 0.0005 ml; for this use, the plunger of the syringe is fitted with a low-tension expansion spring which will withdraw the plunger when the pressure is released by the micrometer. The syringe is filled with liquid paraffin and connected to the pipette by a hypodermic needle and fine-bore polyethylene tubing. Before the pipette is inserted into the polyethylene tubing, a small

FIGURE 9-6
The transplantation apparatus.

quantity of saline solution is drawn into it, so that the final 5 cm of the pipette are filled with saline solution rather than liquid paraffin.

LOADING THE PIPETTE

The eggs or ectoplacental cones are usually drawn into the tip of the pipette between two small air bubbles (see Fig. 9-7). The expulsion of these bubbles, indicating ejection of the graft, can be observed when the pipette is beneath the capsule of the host organ. The pipette can, if necessary, remain loaded for several hours without harming the graft.

ANESTHESIA

Host animals are anesthetised with an intraperitoneal injection of Avertin (0.01 ml 2.5% solution/gm body weight).

TRANSPLANTATION SITES

The kidney, spleen, testis, and anterior chamber of the eye are the most commonly used host organs for extrauterine pregnancies, although eggs and trophoblast have been successfully transplanted to the brain, ovarian

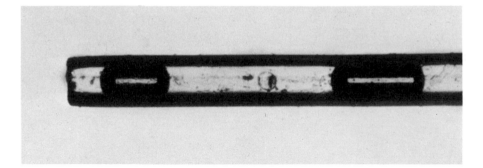

FIGURE 9-7
*Blastocyst positioned in the pipette between two "marker-bubbles."
Note that the tip of the pipette has been flame polished to prevent damage to
the host organ (×65).*

bursa, liver, wall of the alimentary canal, peritoneum, and various malignant tissues. All transplants are made with the aid of a dissecting microscope (×10). One should expect almost 100% success with transplanted
ectoplacental cones, but the proportion of transplanted blastocysts that
develop successfully varies with the host organ: testes (cryptorchid and
scrotal) 80 to 90%, kidney 80%, brain 50 to 60%, spleen and anterior
chamber of the eye 25 to 35%, other sites less than 25%.

KIDNEY

Expose the left kidney through an incision approximately 6 mm
long, 10 mm anterior to the iliac crest. Slight pressure on the abdomen
should cause the kidney to protrude and remain outside the body wall (see
Fig. 9-8). Holding the capsule with fine-pointed watchmakers forceps make
a small tear with a tungsten needle, insert the micropipette as far as the
curvature of the organ will allow, and eject the graft. It is important to keep
the surface of the kidney moist during this process to prevent the drying
capsule from adhering to the pipette. The incision in the body wall is made
to gape, allowing the kidney to withdraw to its original position. Suture or
clip the peritoneum and body wall. The right kidney is more anterior and
closer to the midline of the body than the contralateral organ. It is more
difficult to mobilize, and hence is less easy to use as a host organ. (For the
development of eggs and trophoblast on the kidney see Fawcett, 1950,
and Kirby, 1962.)

SPLEEN

Expose the spleen through an incision placed anterior to that
used to expose the left kidney. The spleen is flabby and best handled under
slight tension away from the body. The pipette is inserted through a small
tear in the capsule on the convex side and pushed along immediately below
the capsule. Fibrous trabeculae extending from the capsule into the parenchyma impede the smooth progress of the pipette. Hemorrhage resulting

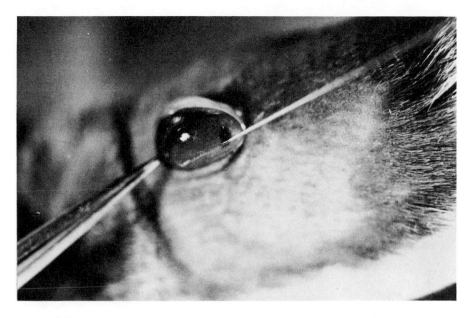

FIGURE 9-8

Pipette in position beneath the kidney capsule. The forceps are simply steadying the organ. Note that the pipette has been pushed well along under the capsule before the graft is ejected (×5).

from damage to the parenchyma can also be an annoying complication of transplantation to this organ. After the graft has been deposited, the pipette is withdrawn swiftly and its path is occluded for approximately a minute by applying gentle pressure to the surface of the spleen; this helps prevent the graft from being expelled by the spontaneous contractions of the spleen which often follow surgical interference. The spleen is a particularly favorable site for the luxuriant growth of trophoblast (Kirby, 1963a; Lichton, 1965).

SCROTAL TESTIS

An incision is made 10 mm anterior to the penis and sufficiently lateral to avoid damage to the preputial gland that lies between the skin and peritoneum. The paratesticular fat pad is withdrawn, bringing with it the testis. The testis, unlike the kidney and the spleen, is most easily handled if the capsule (tunica albuginea) is allowed to become slightly dry. The pipette is introduced through a small tear in the capsule and the graft deposited at the opposite pole. The testis is returned to the abdomen and allowed to go down the inguinal canal into the scrotum.

CRYPTORCHID TESTIS

After transplantation of the graft, the testis is prevented from returning to the scrotum by cutting the gubernaculum and suturing the portion that is still attached to the testis to the peritoneal wall (see Fig.

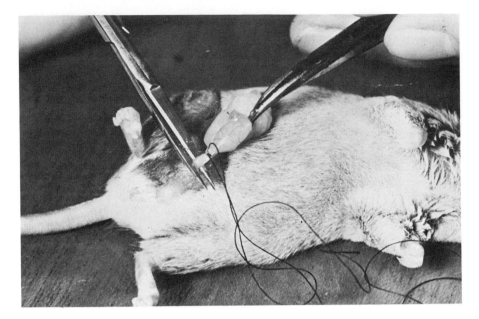

FIGURE 9-9
*Procedure for cryptorchidizing the testis. The gubernaculum is cut distal to the
attachment of the thread, leaving the blood supply to the testis intact. The thread
is sutured to the peritoneum in an anterior position in the abdomen (×1.5).*

9-9). (For descriptions of the fate of mouse egg and trophoblast grafts to
the testis see Kirby, 1963b; Billington, 1965.)

ANTERIOR CHAMBER OF THE EYE

The skin on either side of the eye is pulled tight with finger and
thumb, making the eye protrude. With a tungsten needle a small insertion
is made in the cornea close to the margin of the conjunctiva, which is the
greyish rim of the eyeball. The pipette is introduced and directed away
from the lens towards the front of the eye. Much aqueous humour is lost
when the anterior chamber is punctured, but this is replenished within a
few hours and the eye shows little sign of operative trauma. The eye is a
favorable site for the study of such problems as egg orientation and the
vascular changes associated with implantation. (See Fawcett, Wislocki, and
Waldo, 1947; Runner, 1947; Grobstein, 1949; Kirby, Potts, and Wilson,
1967.)

EXTRAUTERINE TRANSPLANTATION IN OTHER SPECIES

Extrauterine transplantation of eggs and trophoblast has been
described for two other species of rodents: the guinea pig (Bland and
Donovan, 1965) and the hamster (Billington, 1966).

APPENDIX

Agla micrometer syringe is made by Burroughs Wellcome and Co., Tne Wellcome Foundation, Euston Road, London, N.W.1.

The Oxford Manipulator made by Micro Techniques (Oxford) Ltd., 7 Little Clarendon Street., Oxford, England.

Singer-Microdissector is made by Singer Instruments, Ltd., London Road, Reading, England. Left- and right-hand models are available.

Tungsten needles are fashioned from tungsten wire that is chemically eroded in fused (heated) sodium nitrite crystals. These needles can be made very sharp and remain relatively strong.

Pannett-Compton Physiological Saline Solution.

Stock Solution A	NaCl	80. g
	KCl	0.42 g
	CaCl	0.20 g
	Distilled water	100 ml
Stock Solution B	$Na_2HPO_4 12H_2O$	0.43 g
	$NaH_2PO_4 2H_2O$	0.043 g
	Distilled water	100 ml
Final Solution	8 ml of A	
	4 ml of B	
	88 ml distilled water	

Sterilize solutions A and B and allow to cool before the final solution is made up.

BIBLIOGRAPHY

Billington, W. D. (1965) The invasiveness of transplanted mouse trophoblast and the influence of immunological factors. J. Reprod. Fertility, 10:343–352.

——— (1966) Vascular migration of transplanted mouse trophoblast in the Golden hamster. Nature, 211:988–989.

Bland, K. P., and B. T. Donovan (1965) Experimental ectopic implantation of eggs and early embryos in guinea-pigs. J. Reprod. Fertility, 10:189–196.

Fawcett, D. W. (1950) The development of mouse ova under the capsule of the kidney. Anat. Record, 108:71–91.

———, G. B. Wislocki, and C. M. Waldo (1947) The development of mouse ova in the anterior chamber of the eye and in the abdominal cavity. Am. J. Anat., 81:413–443.

Grobstein, C. (1949) Behaviour of components of the early embryo of the mouse in culture and in the anterior chamber of the eye. Anat. Record, 105:490–491.

Kirby, D. R. S. (1962) The influence of the uterine environment on the development of mouse eggs. J. Embryol. Exp. Morphol., 10:496–506.

——— (1963a) Development of the mouse blastocyst transplanted to the spleen. J. Reprod. Fertility, 5:1–12.

——— (1963b) The development of mouse blastocysts transplanted to the scrotal and cryptorchid testis. J. Anat., 97:119–130.

———, D. M. Potts, and I. B. Wilson (1967) On the orientation of the implanting blastocyst. J. Embryol. Exp. Morphol., 17:527–532.

Lichton, I. J. (1965) Survival and endocrine function of rat placenta implanted to the spleen. Endocrinology, 76: 1068–1078.

Runner, M. N. (1947) Development of mouse eggs in the anterior chamber of the eye. Anat. Record, 98:1–17.

Snell, G. D. (1941) The embryology of the mouse. In: The Biology of the Laboratory Mouse, pp. 1–54. Ed. by G. D. Snell. Blakiston, Philadelphia.

10

Egg Micromanipulation

TEH PING LIN
Department of Anatomy
School of Medicine
University of California Medical Center
San Francisco, California 94122

Micrurgical study of living cells has been practiced for more than a century (Chambers, 1940), but only recently have micrurgical techniques been applied to the study of cell differentiation and the regulation of genic activities. As a rule, such experiments have been performed on rather large somatic or germ cells, as found in acetabularia (Hammerling, 1953), amphibian embryos (King and Briggs, 1956; Markert and Ursprung, 1963), or chironomid larval glands (Kroeger, 1966).

During the past three decades, micrurgical studies have utilized living eggs of the rat (Nicholas and Hall, 1942), the human (Duryee, 1954), the mouse (Tarkowski, 1959), or the rabbit (Seidel, 1960). Mammalian eggs are difficult to inject because both their thick zona pellucida and vitelline membrane are highly elastic and resistant to penetration of microinstruments, especially in the unfertilized state (Lin, 1966). In my laboratory, an investigation of ways to surmount these obstacles was undertaken. This chapter summarizes the current stage of development of micromanipulation techniques as used for the eggs of the mouse (Lin, 1966, 1967); the methods may also be used for observing the effects of micromanipulation of the ova of other mammals.

This work is supported by grant HD-02186 from the National Institute of Child Health and Human Development, National Institutes of Health, Bethesda, Maryland. The author wishes further to express his appreciation to Dr. W. O. Reinhardt and Dr. C. W. Asling for review and suggestions made in the preparation of this manuscript. He also thanks Mrs. Joan Florence for her valuable assistance in the laboratory, and Mr. D. R. Akers and Mr. Wayne Emory for preparation of the illustrations.

INSTRUMENTS

In the micromanipulation of living cells, micrurgists use a variety of equipment and modify the procedures for different biological materials. For mammalian eggs, micromanipulators have been employed, and microtools and other facilities have been devised and prepared; a description follows.

LEITZ MICROMANIPULATOR

The micromanipulator consists of two lever-activated fine horizontal movements equipped with ball-bearing slides. There are two independent coarse adjustments for motion in the horizontal plane; vertical motion is handled by coaxial, coarse and fine controls. The fine motions in the horizontal plane are independent of the coarse adjustments and are operated by a single control lever. The movement of the operator's hand produces an eccentric motion which is transmitted to the microtool via the compound stage. The ratio of transmission between the travel of the control lever and the travel of the microtool can be varied continuously from 16:1 to 800:1. Such a ratio is very effective in minimizing vibrations transmitted from the operator's hand to the microtools.

The mountings opposite a set of two independently moving microtools is essential for the manipulation of mammalian eggs. Four micromanipulator units can be arranged in two sets for simultaneous multiple operations under a microscope (Kopac, 1961). The micromanipulator heads and the microscope are mounted upon a heavy steel plate; the entire arrangement is placed on sheets of foam rubber to reduce the effects of extraneous vibrations. For the general construction and operation of the micromanipulator, manufactured by E. Leitz, Wetzlar, Germany, refer to El-Badry (1963).

MICROPIPETTES

In the micromanipulation of mammalian eggs, micropipettes are required for injection, suction, and egg-holding. Manufacture of the microtools is the most crucial part of the entire procedure, and, because the instruments must be replaced periodically because of their delicacy, the experimenter is urged to master the technique of preparing the microtools.

Fine micropipettes may be made by hand from flint glass capillaries of about one mm outside diameter using a suitable microburner (see El-Badry, 1963) or using the flame of an air-gas blow torch (Ling and Gerard, 1949). The capillary is gradually moved from the side toward the flame; when the glass begins to soften, it should be pulled abruptly. The success of the process depends on the time and strength of the pull (Barber, 1911) : if the capillary separates with little or no feeling of resistance, the tip is likely to be too long and too flexible; if separation is accompanied by a snap, the point is usually too blunt (a blunt micropipette may, however, be used for preparing an egg-holder) ; but an intermediate resistance, felt as a slight tug as the glass separates, indicates the right amount of heating and pulling.

Injection micropipette

Micropipettes are usually drawn from glass capillaries by mechanical pulling devices, which depend upon electrical heating of the central point of a glass capillary (the heating elements are incandescent platinum loops). The softened capillary is automatically pulled apart at its hottest point, yielding two pipettes with identical microtips. The electrical and mechanical settings of these machines can be adjusted to produce pipette tips of a specified shape and size. Experimentation is necessary to determine the setting of the pipette puller, but, once determined, it can be relied upon to prepare fine micropipettes in numbers sufficient for routine experiments. The pipette puller of Du Bois (1931) can be set to produce exceedingly fine pipettes from flint glass capillaries (Aloe, V 48302, Aloe Scientific, St. Louis, Missouri).

The tip of the injecting micropipette must be fine enough to minimize injury to the vitellus of an egg and of sufficient rigidity to pierce the elastic zona pellucida and vitelline membrane. Micropipettes for microinjection of mammalian eggs are selected from those whose terminal length is tapered about 250μ; the proximal diameter is 25μ and the distal orifice is about 1μ (see Fig. 10-1). The outside diameter of the tip ($\pm 2\mu$) is about the width of the head of a mouse sperm.

Suction micropipette

The suction micropipette is used for removing material from an egg. Because the vitellar substance of the newly fertilized egg of the mouse is very viscous (Lin, 1966), and a large quantity of material cannot be sucked into a fine micropipette, the orifice of a suctioning micropipette must be wider than that of an injection micropipette; pipettes with an orifice of about 3μ are usually used for suctioning. An oblique orifice is desirable and can be prepared by breaking the tip of the terminal region of a pipette with a pair of pointed forceps under dissecting microscope. A microforge may be needed to refine the shape of the cut edge of the pipette.

Egg-holding micropipette

The egg-holder is a suction pipette, the blunt tip of which is fire-polished using a microburner. After fire-polishing, the holder pipette has an internal diameter (I.D.) of about 15μ; the entire tip of the pipette is somewhat smaller than the diameter of the mouse egg. Eggs cannot be held firmly if the pipette opening is too narrow; they can also be ruptured if the blunt end is not smooth, or if the suction is too strong.

Injection, suction, and egg-holding micropipettes are prepared from glass capillaries of about 1 mm outside diameter (O.D.); these can be inserted tightly into polyethylene tubing (Clay-Adams PE 60) cut to lengths of 12 to 18 inches. The tubing is filled with mineral oil and the micropipette is then inserted into the tubing and filled to the tip with oil; the tubing and pipette must be free of air bubbles. Afterward the pipette is wax-sealed in a

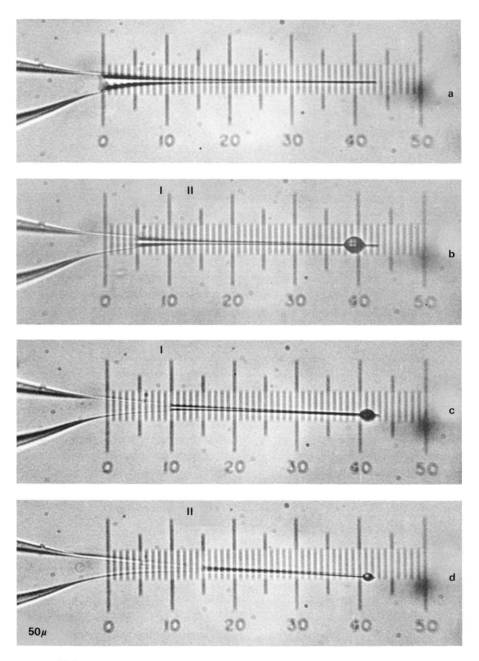

FIGURE 10-1

Terminal portion of micropipette showing the size of corresponding droplets as measured with a calibrated ocular micrometer. (a) Terminal portion of pipette filled with solution; (b) Proximal region (2730 µ³); (c) Middle region I (1340 µ³); and (d) Middle region II (180 µ³). (From: Microinjection of Mouse Eggs, by Teh Ping Lin. Science, Jan. 21, 1966. Copyright 1966 by American Association for the Advancement of Science.)

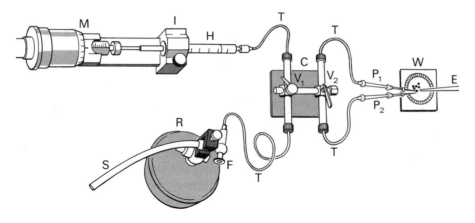

FIGURE 10-2

Diagram of the vacuum micropipetting system. (C) Plug-valve changer: (V₁) outlet valve 1 to microinjector or pressure regulator; (V₂) outlet valve 2 to micropipette P₁ or P₂; (I) Microinjector unit: (H) Hamilton Screw-Syringe; (M) Wells Microinjector; (E) Egg-holder pipette; (R) Pressure regulator: (F) finger tip control; (S) Gomco suction tube to vacuum pump; (T) Polyethylene tubing; (W) Egg-well. Details in text. (From: Micropipetting Cytoplasm from the Mouse Egg, by Teh Ping Lin. Nature, vol. 216, 1967.)

double-glass sleeve (made from a glass capillary 85 mm long and 2 mm wide), which is then inserted into a glass tube, 50 mm long and 4 mm wide. The capillary and the glass tube are sealed with the capillary protruding 10 mm at one end and 25 mm at the other. The PE 60 tubing connected to the micropipette is inserted into the long end, and the tubing is then pulled through until the shank of the micropipette can be sealed in the double-glass sleeve. Thus, the oil-filled micropipette is held in a glass sleeve which in turn is held by the clamp of the micromanipulator. The free end of the PE 60 tubing of the injection, suction, or egg-holding micropipette is then tightly connected, via a 20 gauge hypodermic needle, to a Hamilton Screw-Syringe (Hamilton Co., Whittier, California), which is held in the microinjector unit (see Fig. 10-2) or to the pressure regulating syringe for holding the egg.

Micropipettes for injection and suction may be siliconized according to directions by Siliclad (Clay-Adams Co., New York). Siliconized micropipettes have a water-repellent coating that improves the precision of delivery of materials.

MICROINJECTOR UNIT

The microinjection apparatus together with a micromanipulator is used to inject materials into the mammalian eggs. The Wells microinjector (Mechanical Developments Co., South Gate, California) used (see Fig. 10-2) is made of aluminum, has a thimble control mounted on ball bearings, and is graduated (20 divisions) to regulate the quantity of injection. A Hamilton Screw-Syringe of 1 ml capacity is held by the injector; injection may also be regulated by the screw. Thus, the complete microinjector unit has two controls: the thimble control in the Wells injector regulates delivery of a large volume of injection medium and the screw control in the Hamilton syringe regulates delivery of a small volume. The syringe with a hypodermic needle is inserted into the polyethylene tubing, the other end of which is connected to the micropipette. The entire system from the syringe to the microtip of the pipette is completely filled with mineral oil. The extraordinary accuracy of the device results from the great resistance to the flow of fluid produced in the tubing that varies from 0.1 mm in diameter in the body to about 1μ at the tip.

Vacuum Micropipetting System

Removal of cellular material from the mammalian egg by micropipetting can be difficult because the cellular substance is quite viscous, especially during the fertilized, pronuclear stage. To overcome this difficulty, a vacuum pipetting system was devised.

The vacuum micropipetting system (see Fig. 10-2) consists of (1) a small vacuum pump, (2) a pressure regulator, and (3) a plug-valve changer. Negative pressure can be regulated using a Gomco surgical suction tube with a fingertip control at one end; the other end of the suction tube is attached to a vacuum bottle which, in turn, connects to the vacuum pump. The plug-valve changer has two interconnected valves each of which has two outlets. The outlets of one valve are connected by polyethylene tubing to the microinjector and the vacuum pump; the outlets of the other valve connect with two micropipettes. Using this device the function of a pipette can be switched quickly from suction to injection or vice versa.

Egg-well

Most micrurgical experiments with attached living cells are performed in hanging-drop preparations to permit the highest possible magnification. But the unattached specimen usually does not come to rest on a solid support and is not easily held in position with a micropipette. Mammalian eggs are large cells and can be observed distinctly in a drop of medium on a glass surface. Therefore, a well to contain the eggs is prepared. The egg-well is made with nonreactive Lubriseal (A. H. Thomas Co., Philadelphia, Pennsylvania), a vaseline-like grease, placed in a 1.25 inch-diameter circle on a 2 inch-square glass slide. A small drop (± 3 mm diameter) of injection medium is put in the center of the well which is then filled with mineral oil to prevent evaporation. Eggs are pipetted through the mineral oil into the medium in the well (see Fig. 10-3).

OPERATING PROCEDURES

The procedures for the micromanipulation of fertilized mouse eggs do not require unique talents, but do require careful practice and observation. Some of the steps were learned by trial-and-error, therefore the

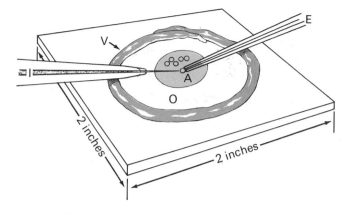

FIGURE 10-3

Diagram of the egg well with micropipettes in operation. (A) Aqueous drop containing six eggs; (E) Egg-holder pipette; (I) Injection pipette; (O) Mineral oil (covering the drop of egg medium); (V) Vaseline ring. Details in text.

beginner is advised to acquire facility in the handling of mammalian eggs before undertaking micrurgical experiments.

PREPARATION OF EGGS AND EGG MEDIA

Female mice of black-agouti strain (C3H/HeJ) are induced to ovulate by intraperitoneal injection of pregnant mare's serum and human chorionic gonadotrophin (Lin and Bailey, 1965; see Chap. 4 by Gates for details of this method.) At the scheduled time of hormonally induced ovulation, about 14 hours after the injection of human chorionic gonadotrophin, the female mice are individually caged with fertile males of albino strain (BALB/c) for mating. Between 12 and 18 hours after mating, the oviducts are removed from mated females. A blunted 30-gauge needle attached to a small syringe is inserted into the fimbrial end of the oviduct; the eggs are flushed out of each oviduct (by injection of three drops of egg medium) into a depression slide; and the eggs are pipetted into a drop of medium in an egg-well covered by mineral oil. The fertilized eggs are then identified by the presence of the second polar body or by observation of pronuclei. As many as twenty eggs may be pooled and maintained in a small drop of egg medium in the center of the well for micromanipulation.

If unfertilized eggs are to be used, they can be obtained from the tubal ampulla of the superovulated unmated female 14 to 16 hours after HCG injection; the eggs may be treated with a medium containing hyaluronidase, 12.5 U/ml, to disassociate follicular cells and to soften the zona pellucida. Such treated eggs are still fertilizable (Chang and Bedford, 1962).

Globulin citrate-Locke's solution (GCL) or phosphate buffer saline (PBS) may be used for suspending the eggs. The globulin citrate-Locke's solution consists of three parts of modified Locke's solution (distilled water, 250 ml; NaCl, 1.69 gm; CaCl$_2$, 0.06 gm; KCl, 0.08 gm; NaHCO$_3$, 0.03 gm;

dextrose, 0.19 gm) plus one part 2.9% sodium citrate dihydrate; 10 to 25 mg of bovine gamma globulin are added per milliliter of solution as a macromolecular substance to be microinjected into the eggs. The viscosity of the GCL preparation may reach 1.1591 centipoise at room temperature; the pH is maintained at 7.2

Microinjection of deoxyribonuclease (DNase) into the eggs will require a solution which can enhance the activities of the enzyme. Phosphate buffer saline (distilled water, 200 ml; NaCl, 1.75 gm; NaH_2PO_4, 0.17 gm; Na_2HPO_4, 0.40 gm; $MgCl_2 \cdot 6H_2O$, 0.07 gm; dextrose, 0.25 gm; bovine crystalline albumin, 10 mg/ml) contains Mg^{++} which may activate DNase. Ten mg of DNase (containing 900,000 Dornase units of activity) are dissolved in each milliliter of PBS for the microinjection. The pH of PBS is 6.8. Penicillin (100 U/ml) and streptomycin (50 μg/ml) are added to each of the egg media.

Positioning of Micropipettes in the Egg-Well

The egg-holder and injection micropipettes are held separately by the left and the right unit of the micromanipulators. Both pipettes are pointed towards each other in a straight line. Before placing the micropipettes into the egg-well, the eggs and the tips of the pipettes must be properly arranged in the microscope field. Eggs to be micromanipulated should be located in the upper half of the field above the micropipettes (see Fig. 10-3) leaving the area below the micropipettes empty for placing the eggs after micromanipulation (see Fig. 10-4). The egg-holder and injection micropipettes are placed first in contact with the surface of the mineral oil in the egg-well by using the coarse vertical adjustment of the micromanipulators. The holder is maintained in the mineral oil above the drop of egg medium while the tip of the injecting or suction micropipette slowly enters the drop. The injecting or suction micropipette is gradually lowered by the fine vertical adjustment to the level of the eggs in the drop. Then the holder pipette can be lowered into the drop to hold the eggs for operation. The drop should remain round, the eggs not being disturbed or moved. Micromanipulation can be carried out under a light microscope at a magnification of ×100 or ×200.

Calibration of Micropipette

The difficulty of manipulating an exact quantity of material into and out of a living cell by a micropipette has been only partially solved. A precise method is still needed for the quantitative calibration of a specific micropipette. The micropipette for the micromanipulative experiment with the mammalian eggs requires a short, stiff terminal. The terminal extends from the capillary shank and consists of tip, shaftlet, and taper; the diameter of the bore of each of these divisions is not uniform; thus, the terminal of the pipette is arbitrarily divided into four regions: proximal, two middle regions, and the tip end; each region may be subdivided as shown in Figure 10-1.

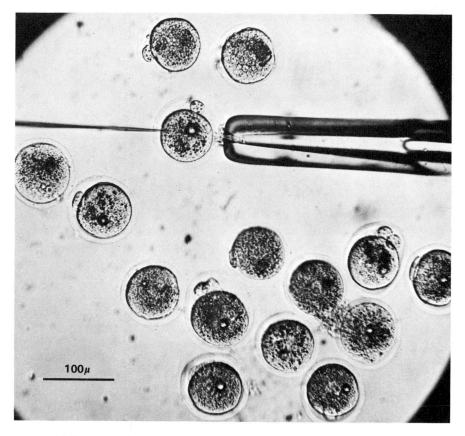

100μ

FIGURE 10-4

A group of 15 newly fertilized mouse eggs in process of microinjection. Thirteen eggs have been injected with an oil droplet; above the micropipettes are two uninjected eggs. (From: Microinjection of Mouse Eggs, by Teh Ping Lin. Science, Jan. 21, 1966. Copyright 1966 by American Association for the Advancement of Science.)

The bore diameter of the shaftlet is determined with a calibrated ocular micrometer while the shaftlet is in the plane of focus, immersed in mineral oil in the egg-well. The oil has nearly the same refractive index as glass, hence it eliminates the lens action of the glass body of the shaftlet, and the use of mineral oil prevents exaggeration of the bore image. The quantity of liquid dispensed by a micropipette can be measured with a calibrated eyepiece micrometer by the displacement of the meniscus in the shaftlet of the micropipette; the displacement corresponds to the size of the drop accumulated near the tip of the micropipette. The shape of the drop is not quite that of a sphere, but is assumed to be one for a quantitative approximation. The volume (V) of the microdrop is calculated from the measurement of the radius (r) of the drop: $V = 4/3\pi r^3$.

As shown in Figure 10-1, the holding capacity (a) of a pipette terminus is about $5000\mu^3$; the proximal region (b) contains about $2730 \pm 570\mu^3$; middle region I (c), about $1340 \pm 340\mu^3$; and middle region (d), $180 \pm 80\mu^3$.

For each volume the known amount of a substance dissolved in the solution can be calculated. For example, if 25 mg of BGG per milliliter of the citrate-Locke's solution were used, microinjection of $180\mu^3$ contains approximately 5 picograms (pg) of BGG; injection of $770\mu^3$ contains 20 pg of BGG; or injection of $2730\mu^3$ contains 68 pg of BGG ($1\mu^3 = 10^{-12}$ ml; 1 pg $= 10^{-9}$ mg).

Microinjection of Eggs

Penetration by and withdrawal of the tip of a micropipette from a living cell are two steps essential to a successful microinjection; therefore we will discuss these steps in the microinjection of the mouse egg.

Healthy pronuclear eggs remain spherical in the egg-well whereas unhealthy eggs usually flatten under the cover of mineral oil. Healthy eggs are mounted in citrate-Locke's solution plus BGG or in phosphate buffer saline plus DNase for the microinjection. By capillary action the egg-holder and the injecting pipette can be filled with the test solution from the egg-well. The pipettes can be refilled in this way as many times as required for injecting many eggs. The egg is maintained in position by a negative pressure in the holder pipette and is released by increasing the pressure in the egg-holding syringe. During the period of micromanipulation, mouse eggs maintained *in vitro* at room temperature for 30 to 40 minutes may shrink within their zonae pellucidae; the shrunken eggs usually survive and are regarded as normal.

The zona pellucida and vitelline membrane of unfertilized eggs are elastic, and the ooplasm is watery; in contrast, the coverings of a fertilized pronuclear egg are less elastic and the vitellar substance is more viscous: thus microinjection is easier at the pronuclear stage than at the unfertilized stage. The membranes of the unfertilized eggs may be softened for easier microinjection by hyaluronidase treatment. When the fine tip of a pipette passes through the zona pellucida, but does not penetrate the vitellus, a deep indentation of the vitelline membrane is formed and injected solution flows not into the vitellus, but into the perivitelline space. Injection into the vitellus requires that the tip of the micropipette pass through the vitelline membrane; immediately afterward the zona pellucida and the vitelline membrane resume their normal contour and the egg regains its original appearance (see Fig. 10-5). If the tip of a micropipette is not sharp, penetration of the vitellus is difficult and is likely to injure the egg, causing eventual disorganization of the vitellus.

In general, living cells can tolerate the thrust of a fine micropipette without visible sign of injury (Chambers and Chambers, 1961). Cells can be punctured without damage if the successive movements of the pipette are performed slowly; sudden penetration may cause disruption of the membrane surrounding the protoplasm. The ooplasm of the pronuclear eggs of the mouse is quite viscous, especially in its cortical layer, and is filled with large yolk particulates; these eggs may also be penetrated safely by fine microtips. For successful microinjection experiments, the experimenter must be able to control the delivery of amounts of solution which are

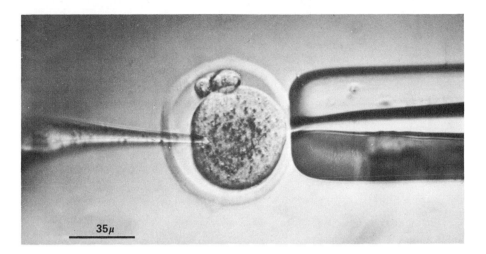

35μ

FIGURE 10-5

A pronuclear mouse egg held for microinjection. Note the normal appearance of the egg with vitelline membrane bordering on the zona pellucida. (From: Microinjection of Mouse Eggs, by Teh Ping Lin. Science, Jan. 21, 1966. Copyright 1966 by American Association for the Advancement of Science.)

minute compared to the volume of the egg being injected. It is necessary to inject substances gradually, for even minute amounts may produce cytolysis if introduced suddenly; however the protoplasm of many cells will accept, by gradual introduction, surprisingly large amounts of fluid, even of water, without destruction (Chambers, 1940).

The volume of a spherical mouse egg 70μ in diameter is calculated to be about $180,000\mu^3$; injection of fluid in a volume of approximately $1000\mu^3$ did not cause any detectable distortion of the vitellus (see Fig. 10-5). Gradual injection of approximately $2700\mu^3$ of the fluid, however, distended the egg; the vitellus expanded, pressing the polar bodies closely against the zona pellucida, but no egg damage was observed (see Fig. 10-6).

Microinjection is also feasible in other preimplantation stages of the egg of the mouse. The blastomere of the four-cell egg was slightly distended by an injection of about $800\mu^3$, but it did not rupture. On the contrary, injection into the cavity of full-grown blastocysts causes the embryo to contract within the zona pellucida, and, in many cases, close up the entire blastocoele of the embryo. The techniques of injection and microsurgery of cells into and from the blastocysts were used to study the capacity of regulation (Gardner, 1968; Lin, 1969a).

A micrurgical procedure was developed to explore the possibility of delivery of mouse sperm into unfertilized egg (Lin, 1969b).

REMOVAL OF CYTOPLASM FROM THE EGGS

The exchange of cytoplasm or nucleus between cells is of particular interest in the study of cell differentiation and interactions between cell organelles. Using a vacuum micropipetting system, the possibility of the

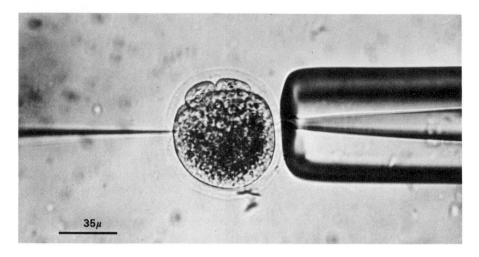

35μ

FIGURE 10-6

The same egg as in Fig. 10-5 showing expansion of the vitellus by injection of gamma globulin citrate-Locke's solution; the polar bodies are flattened against the zona pellucida. (From: Microinjection of Mouse Eggs, by Teh Ping Lin. Science, Jan. 21, 1966. Copyright 1966 by American Association for the Advancement of Science.)

exchange of cytoplasm in the eggs of different strains of mice has been explored in this laboratory. The removal of egg cytoplasm is the first step in this exploratory study. In Figure 10-7, a living newly fertilized mouse egg, shown at the pronuclear stage, was maintained in a drop of phosphate-buffered saline plus bovine plasma albumin (10 mg/ml). The egg was held by an egg-holder pipette, and a microsuctioning pipette was operated from the opposite side.

In microsuctioning, the operator's thumb is placed over the fingertip control of the Gomco surgical suction tube and the vacuum pump is turned on (see Fig. 10-2). Suction of the newly fertilized denuded egg starts when the vacuum gauge reaches a negative pressure of 125 to 250 mm Hg, and the flow of cytoplasmic materials from the egg into the micropipette can be maintained up to a negative pressure of 505 to 635 mm Hg, by on-and-off action of the fingertip control. Negative pressure of 710 mm Hg can be produced at the tip of the micropipette, but this increases the danger of cellular destruction.

The size of the micropipette lumen is critical in determining the rate of flow of materials into the pipette; a high negative pressure may be required if the pipette orifice is too narrow; if the orifice is too small the particulate materials in the egg may block the flow. For these reasons, a pipette with an orifice of about 3μ is employed. Figure 10-7 illustrates the removal of cytoplasm from a pronuclear egg. The vitelline membrane of this egg originally bordered the zona pellucida (see Lin, 1967); the volume of the egg was estimated to be $176,000\mu^3$; and the volume of its vitellus, about $97,000\mu^3$. After suctioning, the egg vitellus became smaller, about $54,000\mu^3$, and a large perivitelline space was present. Although half the cytoplasm was removed, and the vitelline membrane was shrunken; the egg was intact.

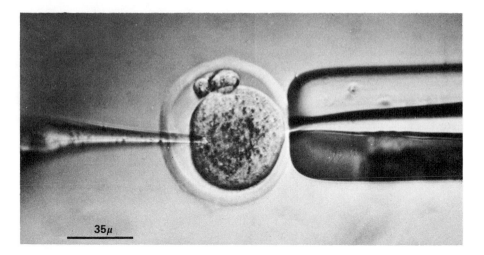

FIGURE 10-7

A pronuclear mouse egg in the process of removal of cytoplasm by microsuction. Note decrease in size of vitellus and enlargement of perivitelline space. (From: Micropipetting Cytoplasm from the Mouse Egg, by Teh Ping Lin. Nature, vol. 216, 1967.)

TRANSPLANTATION OF POSTOPERATIVE EGGS

The test for a successful operation is to transplant the micromanipulated eggs into a foster mother. After microinjection or removal of cytoplasm, the operated and fertilized eggs of black-agouti strain (C3H/HeJ) can be transplanted into mated females of albino strain (BALB/c) for observation of development. The donor and the recipient must be at the same reproductive stage. The micromanipulated eggs are taken out of the egg-well by a mouth pipette prepared according to the design of Runner and Gates (1954). Another mouth pipette, curved about 135° at its tip, is used to transfer the eggs into the oviduct through a small opening made in the membrane of the ovarian bursa of the recipient; a small air bubble in the pipette is used as a marker to indicate when all the operated eggs are released into the infundibulum. The entire procedure, from the time experimental eggs are flushed from the donor oviduct to the time of their transplantation into the recipient, requires 30 to 40 minutes.

The early embryonic development of the micromanipulated eggs can be followed by radioautography if the transferred eggs are previously radiolabelled (Lin, 1956). The fetuses developing from the transplanted eggs can be identified in the uterus of albino recipients by fetal eye pigment, and later also by coat color.

The viability of the manipulated eggs varies: 71 pronuclear eggs injected with $180 \pm 80\mu^3$ of citrate-Locke's solution containing BGG (25 mg/ml) showed no more observable damage than did those eggs which were punctured but uninjected; after transplantation into 15 recipients, 11 of the injected eggs (15%) developed into living fetuses which appeared normal when autopsied on Day 17 of gestation. Of a group of 57 eggs, each injected with $770 \pm 150\mu^3$ of GCL, 12% survived to near term; when injected with $2730 \pm 570\mu^3$, some eggs became slightly distended and the survival rate dropped; only 5 living fetuses (7%) resulted from 74 eggs transplanted in this way into females.

Pronuclear mouse eggs have also survived the injection of PBS containing DNase (10 mg/ml). The enzyme is presumed to be able to degrade parts of the substrate DNA in cellular organelles (Mirsky, 1964; Lima-de-Faria, 1965); some eggs maintained in the medium became swollen even before injection. Eighty-two eggs received the DNase injection in amounts of 1340 to $2730\mu^3$ and nine living fetuses (11%) were obtained.

The microsuctioning of cytoplasm from the pronuclear mouse eggs is in the initial stage of investigation.

CONCLUDING REMARKS

Procedures were developed for the micromanipulation of the one-celled egg of the mouse. The survival of manipulated pronuclear eggs varies according to the type and degree of microinjection. The techniques may also be used for the micromanipulation of other preimplantation stages of mammalian eggs. Injection of the early cleavage eggs suggests a possible method for cytochemico-embryological mapping, or differentiation, of mammalian blastomeres by macromolecular substances.

At present, our technique for the removal of a quantity of cytoplasm from an egg is quite promising. The plug-valve switching device of the vacuum micropipetting system, which permits microsuction followed by injection, will facilitate the exchange of cytoplasm between the eggs of different mouse strains. Microsuction by the vacuum pipetting system may ultimately be useful for the transplantation of a minute subnuclear structure, such as a chromosome (see McClendon, 1907; Kopac, 1961) into a fertilized egg of the mouse.

BIBLIOGRAPHY

Barber, M. A. (1911) A technic for the inoculation of bacteria and other substances into living cells. J. Inf. Diseases, 8:348–360.

Chambers, R. (1940) The micromanipulation of living cells. In: The Cell and Protoplasm, pp. 20–30. Ed. by F. R. Moulton. The Science Press, Washington, D.C.

———— and E. L. Chambers (1961) Explorations into the nature of the living cell. Harvard Univ. Press, Cambridge, Mass.

Chang, M. C., and J. M. Bedford (1962) Fertilizability of rabbit ova after removal of the corona radiata. Fertility Sterility, 13:421–425.

Du Bois, D. (1931) A machine for pulling glass micropipettes and needles. Science, 73:344–345.

Duryee, W. R. (1954) Microdissection studies on human ovarian eggs. Trans. N.Y. Acad. Sci. II, 17:103–108.

El-Badry, H. M. (1963) Micromanipulators and micromanipulation. Academic Press, New York.

Gardner, R. L. (1968) Mouse chimaeras obtained by the injection of cells into the blastocysts. Nature, 220:596–597.

Hämmerling, J. (1953) Nucleo-cytoplasmic relationships in the development of Acetabularia. Intern. Rev. Cytol., 2:475–498.

King, J. T., and R. Briggs (1956) Serial transplantation of embryonic nuclei. Cold Spr. Harb. Symp. Quant. Biol., 21:271–290.

Kopac, M. J. (1961) Exploring living cells by microsurgery. Trans. N.Y. Acad. Sci. II, 23:200–214.

Kroeger, H. (1966) Micrurgy on cells with polytene chromosomes. Methods in Cell Physiol., 2:61–92.

Lima-de-Faria, A. (1965) Labelling of the cytoplasm and the meiotic chromosomes of Agapanthus with H³-thymidine. Hereditas, 53:1–11.

Lin, T. P. (1956) DL-Methionine (Sulphur-35) for labelling unfertilized mouse eggs in transplantation. Nature, 178:1175–1176.

———— (1966) Microinjection of mouse eggs. Science, 151:333–337.

———— (1967) Micropipetting cytoplasm from the mouse eggs. Nature, 216:162–163.

———— (1969a) Microsurgery of inner cell mass of mouse blastocysts. Nature, 222:480–481.

———— (1969b) Development of a micrurgical procedure for delivery of mouse sperm into unfertilized egg. Soc. Study of Reprod., Sec. Ann. Meeting, Abstracts, no. 38, p. 19.

———— and D. W. Bailey (1965) Difference between two inbred strains of mice in ovulatory response to repeated administration of gonadotrophins. J. Reprod. Fertility, 10:253–259.

Ling, G., and R. W. Gerard (1949) The normal membrane potential of frog sartorius fibers. J. Cellular Comp. Physiol., 34:383–396.

Markert, C. L., and H. Ursprung (1963) Production of replicable persistent changes in zygote chromosomes of *Rana pipiens* by injected proteins from adult liver nuclei. Develop. Biol., 7:560–577.

McClendon, J. F. (1907) Experiments on the eggs of Chaetopterus and Asterias in which the chromatin was removed. Biol. Bull., 12:141–145.

Mirsky, A. E. (1964) Regulation of genetic expression. General survey. J. Exp. Zool., 157:45–48.

Nicholas, J. S., and B. V. Hall (1942) Experiments on developing rats. II. The development of isolated blastomeres and fused eggs. J. Exp. Zool., 90:441–458.

Runner, M. N., and A. Gates (1954) Sterile, obese mothers. J. Heredity, 45:51–55.

Seidel, F. (1960) Die Entwicklungsfähigkeiten isolierter Furchungszellen aus dem Ei des Kaninchens *Oryctolagus ciniculus*. Arch. f. Entwicklungsmech., 152:43–130.

Tarkowski, A. K. (1959) Experimental studies on regulation in the development of isolated blastomeres of mouse eggs. Acta Theriologica, 3:191–267.

11

Development of
Single Blastomeres

A. K. TARKOWSKI
Department of Embryology
University of Warsaw
Warsaw 64, Poland

Studies of the development of single blastomeres, first undertaken on oviparous vertebrates and invertebrates in the last century, made it possible to acquire knowledge of the organization of animal eggs and their developmental mechanisms. Despite the great technical advances made in experimental embryology since that time, this simple study has lost nothing of its usefulness.

Viviparity presents a serious obstacle in these studies in mammals but this obstacle can be overcome partly by performing a short-term operation on embryos *in vitro* and transplanting single blastomeres back to the reproductive tract of recipient females. This was done in the rat by Nicholas and Hall (1942), in the rabbit by Seidel (1952, 1956, 1960) and by Moore *et al.* (1968), and in the mouse by Tarkowski (1959b,c). Recent progress in culture methods enabled Mulnard (1965a,b), and Tarkowski and Wróblewska (1967) to follow the development of single blastomeres *in vitro* to the blastocyst stage, but transfer to recipient females is indispensable if later stages of development of blastomere-embryos are to be studied.

The developmental capacity of a single mammalian blastomere can first be determined at the end of its preimplantation development at which time it becomes known if the embryo was able to blastulate (cavitate) and, if so, if the inner cell mass was formed with the trophoblast. It is already known that in both the rabbit and the mouse some of the single blastomeres give rise to trophoblastic vesicles that are devoid of the inner mass (Seidel, 1956, 1960; Tarkowski, 1959b,c; Mulnard, 1965a,b; Tarkowski and Wróblewska, 1967), and their development thus comes to a halt. A primary condition for survival of the blastomere embryo beyond implantation is, therefore, formation of a normally formed blastocyst. If the over-all size of such a blastocyst and of its inner mass in particular does not decline below a "reasonable" limit, normal development during the uterine period and even birth of a viable individual can be expected. This was proved in the mouse (Tarkowski, 1959a,b) with single blastomeres of the two-cell egg, and in the rabbit with single blastomeres of two-, four-, and eight-cell eggs (Seidel, 1952, 1960; Moore *et al.* 1968). Species can differ in regard to the stage from which a single blastomere can still develop into a viable individual. The blastomere-embryo has to regulate its size and it may be that the longer the pregnancy the greater the chance for this regulation to be successful.

Since the majority of the techniques employed in research on single blastomeres are not specific to this type of experimenting and are described in detail in other chapters of this book, this chapter presents only a general outline of the experiment. Emphasis is placed on the mouse which has been the object of intensive investigations and with which I am most familiar. Although the techniques were elaborated in investigations on only three species (mouse, rabbit and rat), it seems that they can be adapted to other species without introducing major modifications.

TERMINOLOGY

Despite its obvious inaccuracy, the term *cleaving egg* is persistently used by the majority of embryologists, and its continued use in embryological terminology seems assured. To please both the linguistic purists and those, myself included, who cannot refrain from calling a cleaving embryo a cleaving egg, both terms will be used in this chapter (e.g., two-cell embryo—two-cell egg).

Single blastomeres can be obtained either by killing the remaining ones *in situ* or by separating sister blastomeres from each other. These two methods are not, however, discriminated terminologically. The term *isolated blastomere,* although generally accepted, is misleading because it implies an isolation that is not always the case. The neutral term *single blastomere* is most accurate and will therefore be used throughout this chapter.

DESCRIPTION OF METHODS

OBTAINING AND HANDLING OF CLEAVING EGGS

The two-cell stage is obviously the earliest convenient stage for the experiment; the last stage is probably the eight-cell stage. Although it is possible to disassemble advanced embryos (such as the morula or the blastocyst) into individual cells (see Chap. 34 by Cole), these cells are no longer able to produce organized entities comparable to the embryo—they proliferate but they do not develop. Interesting as they are, such experiments are not considered in this chapter.

As donors of cleaving embryos both the females that had ovulated naturally and those which had been induced to ovulate by administration of gonadotropins can be used (see Chap. 4 by Gates). Eggs are removed from the oviduct either by cutting it into pieces in saline (mouse, rat) or by flushing it (rabbit, rat, mouse). With training, the flushing method is superior because of the quickness and ease with which the eggs can be collected from the drop of clear saline solution. The minute oviduct of the mouse has to be manipulated under the dissecting microscope; the procedure is as follows: The oviduct is placed on a watch glass and a short micropipette with a rubber nipple (or one that is mouth-controlled) is inserted into the infundibulum. We found it more convenient to carry out this procedure dry than under saline. During the flushing the pipette has to be held in place by squeezing it through the wall of the oviduct with forceps. A strong point-illumination directed onto the oviduct is very helpful.

Any of the basic salt solutions (e.g., Tyrode, Locke, Hanks, Ringer, etc.) or any of the media devised for egg culture are suitable for washing and storing the eggs (see Chap. 6 by Biggers, Whitten, and Whittingham, and Chap. 7 by Hafez). It is recommended that these be supplemented with protein—crystalline bovine serum albumin is now commonly used (although the concentration is not critical over a wide range, a 0.1% concentration is recommended). Solutions rich in bicarbonate should be used with caution because exposure to air increases their alkalinity to a point which may be harmful for eggs. During manipulations, which rarely extend over an hour, the eggs and blastomeres can be kept safely at room temperature.

OBTAINING SINGLE BLASTOMERES

There are two ways to obtain single blastomeres; each has its advantages and disadvantages (see Figs. 11-1 and 11-2). The choice of procedure depends on the aims of the experiment.

By elimination of sister blastomeres inside the zona pellucida

The advantage of this method is that the blastomere remains inside its natural shielding which is its zona pellucida. The main disadvantage of this procedure is that the existence of possible differences in the potency of sister blastomeres can never be proved or disproved directly, and can be deduced only indirectly by examining a large series of embryos. Blastomeres can be killed either surgically, by puncturing them with a glass microneedle, or with microbeam irradiation. The usefulness of the latter method is anticipated rather than proved since the only irradiation attempt was a pilot experiment by Daniel and Takahashi (1965), who used laser beam

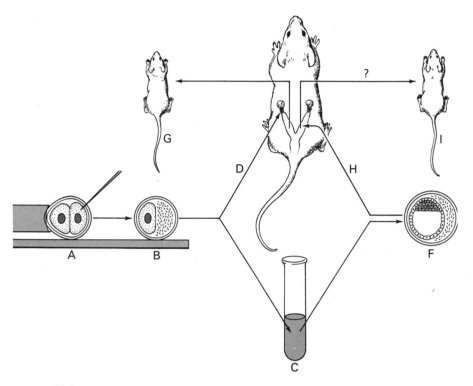

FIGURE 11-1

A scheme for experimentation with mouse half-blastomeres with zona pellucida intact. (A) A two-cell egg is immobilized by suction at the mouth of a micropipette and one blastomere is punctured with a needle. The single blastomere thus obtained (B) can either be cultured in vitro *(C) to the blastocyst stage (F), or immediately transferred into the oviduct of a recipient (D), in which it can develop till term (G). It is believed that the blastocyst developed from the half-blastomere in vitro (F), when transferred to the uterus of a recipient (H) will also develop till term (I). (Drawing by J. Modlinski.)*

irradiation to destroy blastomeres in two-, four-, eight-, and sixteen-cell rabbit eggs. Only the surgical method will be described here.

The surgical manipulations are performed under a dissecting microscope. After being placed in a drop of medium on the slide, the egg which is to be operated upon must first be immobilized. This can be accomplished by holding the egg by suction at the mouth of a micropipette (see Figs. 11-3a,b) of an internal diameter one-fourth to one-half the size of the egg (Nicholas and Hall, 1942; Tarkowski, 1959b and c). The pipette is held in a joint clamp and is connected by rubber tubing to a compressible rubber bulb. Suction pressure can be effectively regulated by manipulating the rubber bulb. With its tip immersed in the medium, the pipette is secured and the eggs are directed to its mouth, if necessary by moving the slide. Suction keeps the egg firmly in place. Another method of immobilizing the egg is to wedge it in the angular crevice between two cover slips attached to the bottom of the dish (Seidel, 1960).

The blastomere is punctured through the zona pellucida with a glass

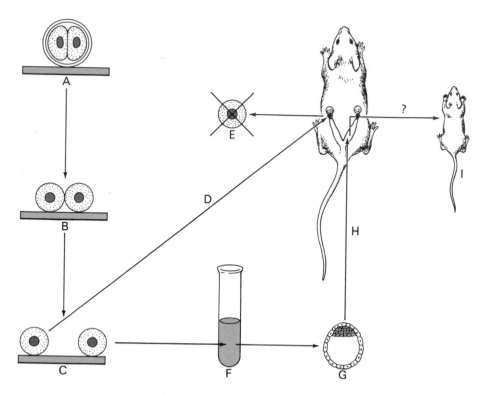

FIGURE 11-2

A scheme of experimenting with naked mouse half-blastomeres. From a two-cell egg (A)
zona pellucida is removed (B) *and the blastomeres are isolated from each other* (C) .
Transplantation of naked blastomeres to the oviduct of a recipient (D) *will not result
in further development, since the blastomeres do not survive in this environment* (E) .
If, however, the blastomere is cultured in vitro (F) , *it can develop into a blastocyst* (G) .
It is believed that such a blastocyst, if transferred into the uterus of a recipient (H) ,
will develop till term (I) . (*Drawing by J. Modlinski.*)

microneedle. The needle can be handled by any micromanipulator; those
constructed for low-power magnification are suitable. A very simple micromanipulator invented by Goldacre (1954) has been used routinely in our
laboratory and has been found to be convenient for operating upon two-cell
eggs, although for later stages a more subtle apparatus is necessary. Micro-
needles can be drawn in an ordinary microburner; a microforge is not
necessary. The puncture should be as delicate as possible and should not
result in serious damage to the zona pellucida. A severely punctured blastomere degenerates in a few minutes: it first becomes opaque and either
remains as a compact body for a long time, or soon falls into a loose detritus
(see Figs. 11-3c,d) .

By isolating them from each other

The major advantage of this method over that previously described is that
it makes possible a complete representation of developmental potency of all
cells constituting the "maternal" embryo. Technically this approach is more

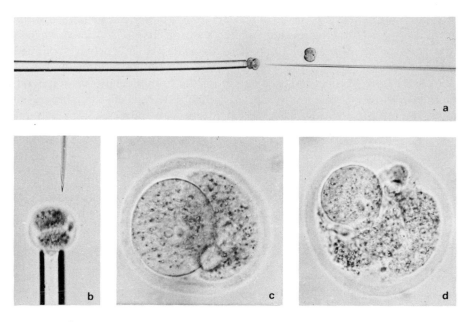

FIGURE 11-3

(a) *A two-cell mouse egg immobilized by suction at the mouth of a micropipette. On the right is a glass microneedle used to puncture blastomeres. ×50. (b) A view of the immobilized egg under higher magnification. The figure shows orientation of the egg most convenient for performing the operation or for making a puncture. A part of one blastomere together with the overlying portion of zona pellucida has been drawn into the lumen of the pipette. ×150. (c) A single half-blastomere of the mouse obtained by destroying the sister cell in situ. The punctured blastomere has lost its regular shape and its cytoplasm has become opaque. ×400. (d) A single quarter-blastomere of the mouse obtained by destroying the remaining three blastomeres in situ. The punctured blastomeres have disintegrated into structureless debris. ×400. (Photo d, Tarkowski, 1959b.)*

difficult because of the necessity for removing the zona pellucida and for handling minute and delicate blastomeres. Moreover, development of naked blastomeres to the blastocyst stage can proceed only *in vitro*.

REMOVAL OF ZONA PELLUCIDA. Removal of the zona pellucida can be accomplished either mechanically or enzymatically. The first method devised for mouse eggs, consists in rapid suction of the cleaving egg into a micropipette of an internal diameter of about 65 microns (Tarkowski, 1959c, 1961). The pipette is connected by rubber tubing to a compressed rubber bulb which creates a negative pressure inside. The bulb regulates the suction pressure of the pipette. The equipment is similar to that used to immobilize the eggs (see previous section), the only difference is that the pipette is not held motionless but is freely manipulated by hand. The internal diameter of the mouth of the pipette and the suction pressure applied are critical for the success of this operation; slow suction of the egg will not rupture the zona, which is sufficiently flexible to accommodate itself even in a very narrow pipette. Once the pipette is selected and adequate skill is acquired, the technique is very effective. The advantages of this technique are: simplicity of technical devices, rapidity, immediate disintegration of those blastomeres that suffered damage. Its main disadvantage is that survival of all embryo blastomeres is achieved only when "rounded out" morulae are used. In two-cell eggs one blastomere is often destroyed (rarely are both of them lost) ; in eggs from the four- to eight-cell stage, the number of lost blastomeres varies. Although I succeeded many times in removing zona in this way from eggs at the two- to eight-cell stage without damaging any blastomeres, the method is certainly not recommended if survival of all blastomeres of the cleaving egg is essential.

Enzymatic digestion of zona pellucida by pronase was discovered by Mintz (see Chap. 12) in 1962; the usefulness of this technique has been confirmed by several authors. In our laboratory we have used it successfully in experiments designed to obtain complete sets of blastomeres from eggs at the four- to eight-cell stage (Tarkowski and Wróblewska, 1967) .

ISOLATION OF BLASTOMERES. Adhesion of sister blastomeres to each other changes considerably during the intermitotic period. Blastomeres which have just undergone cleavage are firmly connected and isolation may lead to their disruption. Conversely, when the next cleavage approaches, they separate easily. Pipetting is sometimes sufficient, but separation by chemical means may yet be necessary. The naked, cleaving embryos are placed in calcium- and magnesium-free BSS supplemented with EDTA (Versene) in concentration of 0.02%. Delicate pipetting of embryos hastens isolation which is usually accomplished after a few minutes (see Figs. 11-4a,b,c) . The isolated blastomeres are washed in normal BSS and placed in culture medium. If protein is not included in BSS, naked blastomeres tend to adhere strongly to glass and can be damaged easily. This obstacle can also be overcome by using a dish coated with a layer of agar gel (1% in 0.9% sodium chloride) , to which naked blastomeres do not stick (Mulnard 1965b) .

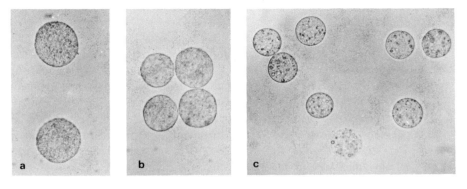

FIGURE 11-4

Naked single blastomeres of the mouse isolated at the two-, four-, and eight-cell stage, respectively: ×240. (a) Two sister half-blastomeres. (b) Four sister quarter-blastomeres. (c) Eight sister blastomeres. The blastomere lowest in the figure has degenerated during manipulation.

PREIMPLANTATION DEVELOPMENT OF SINGLE BLASTOMERES

Studies of the development of single blastomeres are often confined to the preimplantation period only—a period in which much important information about developmental potency of blastomeres can be obtained. Preimplantation development of single blastomeres with zona pellucida intact can be investigated either *in vitro* or *in vivo*. In the latter case the blastomeres are immediately transferred to the oviduct of a recipient female and, depending on the time spent in the reproductive tract, recovered for examination either from the oviduct or the uterus. Naked mouse eggs at stages from one- to eight-cell, as well as single blastomeres originating from such eggs, do not survive when transferred to the oviduct (Tarkowski, unpublished observations; Modliński, 1970). Similar observations were reported in the rabbit by Moore *et al.* (1968); none of the 62 naked blastomeres obtained from two- and four-cell eggs developed into blastocysts *in vivo*. Thus culture *in vitro* is the only possible way of following development of naked single blastomeres to the blastocyst stage.

Culture in vitro *of single blastomeres*

There is no doubt that the conditions to which eggs are exposed *in vitro* only approximate, at best, those prevailing *in vivo*. It is conceivable, therefore, that cultivation *per se* may modify the development of eggs and, in consequence, influence the morphology of resulting embryos. Although it is relatively simple to discriminate between normal and abnormal development of whole eggs, it is not simple with single blastomeres for which a canon of normalcy is not *a priori* available. Morphology of the blastomere embryo is the only criterion of the prospective potency of the given blastomere. However, the question of how much the embryo has been shaped by internal factors and how much by external factors is not easy to resolve. It is therefore extremely important that in each culture single blastomeres be accompanied by whole eggs, and that the behavior of the latter be analyzed with the same scrutiny as that of blastomeres.

BLASTOMERES IN ZONA PELLUCIDA. Culture can be carried out in any medium and by any technique recommended for whole eggs at the corresponding stage (see Chap. 6 by Biggers, Whitten, and Whittingham). In our laboratory "half" blastomeres of the mouse were successfully cultured (unpublished observations) in tightly corked test tubes, about 10 cm by 1 cm, under 5% CO_2, in 0.5 to 1 ml of Brinster's medium (Brinster, 1963) with lactate alone or with lactate and pyruvate (Brinster, 1965). Culture in drops of medium submerged in liquid paraffin is also possible (see below). A technique elaborated by Mulnard (1964, 1965a,b, 1967) has also been found to be very efficient and is highly recommended for cinematographic studies.

NAKED BLASTOMERES. Choosing the appropriate method of culturing depends to a degree on the size of the blastomeres, i.e. the stage at which they were isolated, and if it is essential to recover all blastomeres of the embryo at the end of the culturing. Blastomeres originating from two- and four-cell eggs can be cultured in test tubes but recovery of embryos is inefficient, possibly because of their adhesion to glass. Microtubes devised by Mulnard (*ibid.*) appear more suitable than big tubes because after culturing the blastomere embryos can be localized easily and then withdrawn under constant observation. In our attempts to culture sets of blastomeres origi- nating from four- to eight-cell mouse eggs we tested different techniques and found it most convenient to culture blastomeres in drops of medium placed on the bottom of a dish and covered with liquid paraffin (Tarkowski and Wróblewska, 1967). The technique was developed for culturing pairs of fused eight-cell eggs and early morulae (Tarkowski, 1961), but the medium used (Krebs-Ringer bicarbonate with glucose) in the original experiment is unsuitable for cultivating either eggs or single blastomeres explanted at earlier stages. However, starting with a late two-cell stage, eggs will develop to blastocysts if cultured in the medium with lactate or with lactate and pyruvate (Brinster, 1963, 1965; see also Chap. 6 by Biggers, Whitten, and Whittingham). According to our latest evidence, early cleav- age stages develop much better in this system if Mulnard's medium (Mulnard, 1967, formula KRb) is used instead of Brinster's media (Brin- ster, 1963, 1965). The liquid paraffin method is especially suitable for culti- vating minute naked blastomeres because the blastomeres remain undis- turbed and can be easily spotted during the culture period.

In this procedure (Tarkowski and Wróblewska, 1967) plastic or glass Petri dishes are half filled with liquid paraffin, then small drops of sterilized Brinster's or Mulnard's medium are placed on the bottom. If glass dishes are used they must be coated with silicone (*Repelcote*) to prevent the drops from spreading. Although the size of the drops is not critical, we prefer drops not exceeding 5 mm in diameter. The dish is subsequently placed in a dessicating chamber with 5% CO_2 in air or with alveolar air. The culture dishes should be prepared a few hours in advance since it takes time to adjust the pH of the medium to a proper level. This adjustment can be hastened if the gas mixture is bubbled through the liquid paraffin before it is poured into the dishes. Single blastomeres are first washed in a drop of medium and then each is placed in a separate drop. At the end of the culture period the blastomere-embryos adhere to the bottom of the dish and care is necessary in detaching them—the best procedure is to direct a stream of fluid on them from the pipette.

Methods of studying morphology of preimplantation blastomere-embryos

Inspection of embryos in the living state (see Figs. 11-5a,b,c) does not allow for a detailed quantitative and qualitative analysis of their structure. Mak- ing permanent preparations is, therefore, indispensable. Whole mounts are best; they retain the spatial organization of the embryo and there is thus no need to reconstruct it from sections. In my opinion the best method of making whole mounts from mammalian eggs and early embryos is the one developed by Dalcq (1952) and Mulnard (1955). The essential step of this method is the mounting of the previously fixed eggs in a drop of albumin on a slide and fixing the albumin in alcohol vapor. After this has been done

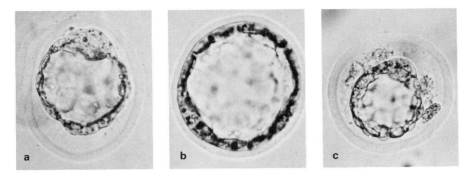

FIGURE 11-5

(a) *A living mouse blastocyst developed* in vivo *from a half-blastomere similar to that shown in Fig. 11-3, c. ×350. (Tarkowski, 1959c)* (b) *A living trophoblastic vesicle developed* in vivo *from a half-blastomere similar to that shown in Fig. 11-3, c. ×350. (Tarkowski, 1959b)* (c) *A living mouse blastocyst developed* in vivo *from a quarter-blastomere similar to that shown in Fig. 11-3, d. ×350. (Tarkowski, 1959b.)*

the slide can be processed like an ordinary histological preparation, and egg manipulations are restricted to a minimum. The procedure described below presents our modification of the original method of Dalcq and Mulnard. It is now used routinely in our laboratory for normal mouse eggs (one-cell to blastocyst) and for forms developed from single blastomeres.

1. Fix eggs in Heidenchein fixative for 5 minutes or longer (saturated aqueous solution of mercuric chloride, two parts; distilled water, two parts; formaline alkalinized with calcium carbonate, one part; filter before use). The solution from which the eggs are taken must be protein-free.

2. Immerse the eggs in aqueous Lugol solution for 5 minutes.

3. Place them in distilled water for 1 to 2 minutes.

4. Place a drop of egg albumin 2 to 3 mm in diameter on the slide and introduce the eggs into it with minimal amount of water. Excess albumin should be withdrawn. (To prepare albumin, three parts of thin egg albumin are mixed with one part of distilled water and then filtered and stored in a refrigerator).

The manipulations described under points 1 to 4 are carried out under a dissecting microscope. The solutions are kept either in embryological watch glasses or as drops on a sheet of plexiglas, and the eggs are transferred from one to another with the help of a mouth-controlled pipette.

5. Place the slide on the embryological watch glass containing 96% ethyl alcohol to fix the albumin in alcohol vapor. After this step the slide can be handled like an ordinary histological preparation.

6. Immerse first in 50% alcohol for 5 minutes (if necessary the slide can be stored in it overnight); then immerse successively in haematoxylin—50% alcohol with few drops of 1N HCl—until the albumin is bleached;

tap water for 5 minutes; alcohols 50, 80, 90, and 96%—one minute each; absolute alcohol—two changes of 5 minutes; xylen—two changes of 5 minutes. Mount in balsam, DPX, or other synthetic resin.

Preparations produced in the above way permit evaluation of the structure of the embryo, number of cells, presence of mitotic figures and details of cell structure (see Fig. 11-6). In multicellular embryos precise determina-

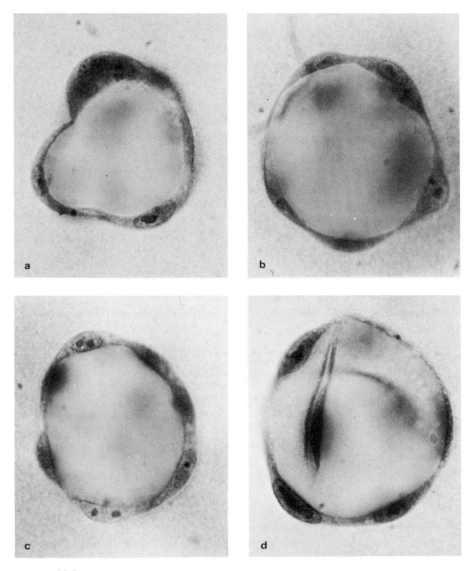

FIGURE 11-6

Four sister forms developed in vitro *from a seven-cell mouse egg that had been separated into three pairs of eighth-blastomeres and one quarter-blastomere. Only the form shown in* (a) *is a blastocyst, the three others are trophoblastic vesicles. Photomicrographs taken from whole mounts prepared by the method described in the text. ×900. (Tarkowski and Wróblewska, 1967.)*

tion of the cell number is difficult and if this information is important, air-dried preparations are recommended (Tarkowski, 1966) : with them it is possible to obtain a scattering of nuclei and metaphase plates in one layer, thus making the counting of cells a very simple task.

The method of making whole mounts is also very valuable in cytochemical studies. The only reports of such studies of blastomere embryos are the papers by Mulnard (1965a and b) that describe his investigation of the presence and distribution of acid phosphatase in mouse embryos developed from "half" blastomeres.

TRANSPLANTATION OF SINGLE BLASTOMERES AND BLASTOMERE-EMBRYOS OF THE MOUSE

A single blastomere develops at the same rate as the whole egg and consecutive stages are achieved after the same periods of time, though in blastomere-embryos it is on the basis of an appropriately smaller number of cells. The rules of transplantation of single blastomeres, or embryos developed from them, are thus the same as for normal embryos of the corresponding stage. As a detailed account of egg transfer techniques is presented in Chap. 8 by Dickmann and Chap. 21 by Staples, this section will be confined to remarks on transplantation in the mouse.

Transfer of single blastomeres in zona pellucida

As mentioned above, only blastomeres with an intact zona pellucida can be transferred to the reproductive tract. If the investigator is interested in postimplantation development of single blastomeres, they must be transferred to the oviduct on the first day of pregnancy or pseudopregnancy of a recipient (first day being the day when the vaginal plug is found). Eggs transplanted on the second day do not always pass into the uterus; those transferred on the third day and following are regularly tube-locked and remain there as blastocysts for several days (Tarkowski, unpublished observations). When blastomeres are transplanted on the first day donors are always in more advanced pregnancy than the recipients; this is advantageous because the development of the transplanted eggs is retarded for several hours by operational shock. Although blastomere-embryos can easily be distinguished from normal embryos during preimplantation development, and use of genetic markers makes it possible to discriminate between native and alien embryos at later stages, pseudopregnant females should be used as recipients.

Following is the technique of transferring mouse eggs to the oviduct (Tarkowski, 1959a) :

1. After administering ether or nembutal anesthesia make a small longitudinal incision in the skin and muscular layer above the ovary.
2. Pull out the ovary and oviduct and hold them in place by grasping the fat pad adjoining the ovary with a Peane's clamp.

3. With fine scissors make an incision in the distal (aboviducal) half of the ovarian bursa.

4. Hold the proximal portion of the dissected bursa with fine forceps and insert the pipette into the oviduct through the infundibulum to expel the eggs.

5. Replace the upper portion of the reproductive tract in the body cavity and suture the muscular layer and the skin separately.

Steps 3 and 4 should be performed under a dissecting microscope; point-illumination of the operation field is necessary. If there is bleeding from the fat adjoining the ovary or from the bursa, try to remove the blood before inserting the pipette into the oviduct. Rounded edges of the tip of the pipette facilitate its insertion into the oviduct and prevent its damaging oviducal epithelium. A mouth-controlled pipette guarantees adequate control when expelling eggs; the amount of fluid introduced with eggs should be as small as possible. A small air bubble formed in the pipette subsequent to introduction of eggs is not harmful and the bubble often serves as a sign of successful egg transfer. Transfers to the left oviduct are much easier than to the right because of the natural position of the infundibulum.

Transfer of embryos developed in vitro *from single blastomeres*

Although transfer of embryos (morulae, blastocysts) developed *in vitro* from single blastomeres has not been attempted, there are reasons for believing that such experiments will result in implantation and further uterine development. The following recommendations may be helpful for those interested in undertaking these investigations.

Blastomere-morulae or blastocysts, whether in zona or naked, can be transferred either into the oviduct by the method previously described or directly into the uterus. Transfer to the uterus is recommended because of its simplicity and high effectiveness. The operation itself consists in making a puncture with a hypodermic needle on the antimesometrial side of the horn, close to the tube-uterine junction, and inserting through it a mouth-controlled pipette. The eggs should be drawn into the narrow part of the pipette and should be expelled without introducing air into the uterus. According to our experience with transplanting normal morulae and blastocysts, best results are obtained when embryos are transferred during the period lasting from the late afternoon hours of the third day to noon on the fourth day (optimal time limits vary slightly among various strains).

BIBLIOGRAPHY

Brinster, R. L. (1963) A method for *in vitro* cultivation of mouse ova from two-cell to blastocyst. Exp. Cell Res., 32:205–208.

——— (1965) Studies on the development of mouse embryos *in vitro*. IV. Interaction of energy sources. J. Reprod. Fertility, 10:227–240.

Dalcq, A. M. (1952) L'oeuf des Mammiféres comme objet cytologique (avec une technique de montage in toto et ses premiers résultats). Bull. Acad. r. Med. Belg., 6ᵉ Serie, 17:236–264.

Daniel, J. C., and K. Takahashi (1965) Selective laser destruction of rabbit blastomeres and continued cleavage of survivors in vitro. Exp. Cell Res., 39:475–482.

Goldacre, R. J. (1954) A simplified micromanipulator. Nature (London), 173:45.

Mintz, B. (1962) Experimental study of the developing mammalian egg: removal of the zona pellucida. Science, 138:594–595.

Modliński, J. A. (1970) The role of the zona pellucida in the development of mouse eggs in vivo. J. Embryol. Exp. Morphol., 23:539–547.

Moore, N. W., C. E. Adams, and L. E. A. Rowson (1968) Developmental potential of single blastomeres of the rabbit egg. J. Reprod. Fertility, 17:527–531.

Mulnard, J. G. (1955) Contribution à la connaissance des enzymes dans l'ontogénèse. Les phosphomonoestérases acide et alcaline dans le développement du Rat et de la Souris. Arch. Biol., Liège, 66:525–685.

——— (1964) Obtention in vitro du développement continu de l'oeuf de Souris du stade II au stade du blastocyste. C. R. Acad. Sc. Paris, 258:6228–6229.

——— (1965a) Aspects cytochemiques de la régulation in vitro de l'oeuf de souris après destruction d'un des blastomères du stade II. 1. La phosphomonoestérase acide. Bull. Acad. r. Med. Belg., 2ᵉ Série, 5:31–67.

——— (1965b) Studies of regulation of mouse ova in vitro. In: Preimplantation Stages of Pregnancy. A Ciba Foundation Symposium. Ed. by G. E. W. Wolstenholme and M. O'Connor, pp. 123–138. Churchill, London.

——— (1967) Analyse microcinématographique du développement de l'oeuf de souris du stade II au blastocyste. Arch. Biol. (Liège), 78:107–138.

Nicholas, J. S., and B. V. Hall (1942) Experiments on developing rats. II. The development of isolated blastomeres and fused eggs. J. Exp. Zool., 90: 441–459.

Seidel, F. (1952) Die Entwicklungspotenzen einen isolierten Blastomere des Zweizellenstadiums im Säugetierei. Naturwissenschaften, 39:355–356.

——— (1956) Nachweis eines Zentrums zur Bildung der Keimscheibe im Säugetierei. Naturwissenschaften, 43:306–307.

——— (1960) Die Entwicklungsfähigkeiten isolierter Furchungszellen aus dem Ei des Kaninchens Oryctolagus cuniculus. Roux Arch. EntwMech., 152:43–130.

Tarkowski, A. K. (1959a) Experiments on the transplantation of ova in mice. Acta Theriologica, 2:251–267.

——— (1959b) Experiments on the development of isolated blastomeres of mouse eggs. Nature (London), 184:1286–1287.

——— (1959c) Experimental studies on regulation in the development of isolated blastomeres of mouse eggs. Acta Theriologica, 3:191–267.

——— (1961) Mouse chimaeras developed from fused eggs. Nature (London), 190:857–860.

——— (1966) An air-drying method for chromosome preparations from mouse eggs. Cytogenetics, 5:394–400.

——— and Wróblewska, J. (1967) Development of blastomeres of mouse eggs isolated at the 4- and 8-cell stage. J. Embryol. Exp. Morphol., 18:155–180.

12

Allophenic Mice of Multi-Embryo Origin

BEATRICE MINTZ
The Institute for Cancer Research
Fox Chase, Philadelphia, Pennsylvania 19111

Differentiation may be described as the acquisition of a phenotype. The individual's genetic complement, established at fertilization, does not necessarily come into expression during embryonic life: some parts of the genome may be expressed before birth, others are expressed postnatally. In the broadest sense, any actualization of genetic potential is an example of differentiation; this includes genetically dictated events in embryogenesis, in many diseases, in aging, and even in behavior. Thus the mechanisms of differentiation, which have remained obscure, offer a challenge of considerable scope.

In multicellular species, differentiation occurs at several levels of organization. At the cellular level, phenotypic diversification arises despite the genotypic identity of the cells; in tissues and organs, orderly associations are formed and interactions occur among diversely specialized cells. The total phenotype is a genotype-specific complex of all these phenomena.

A new approach to the problem was formulated in our laboratory for the purpose of analyzing mammalian differentiation within the framework of the intact animal, where biological organization would be preserved. The plan was to introduce different cellular genotypes into a single individ-

These investigations were supported by United States Public Health Service grants No. HD-01646, CA-06927, and FR-05539, and by an appropriation from the Commonwealth of Pennsylvania.

ual at a sufficiently early developmental stage so that all tissues and organs could eventually contain two genetic subpopulations of cells. The allelic differences between the subpopulations would then provide the kinds of cellular markers that might be capable of revealing the sequence and nature of gene-controlled developmental events. At the time this plan was conceived, genetic mosaicism had already become a major experimental tool in Drosophila, and Owen's work (1945) on the erythropoietic system of fraternal cattle twins had just opened the way to a wide range of important investigations based on mosaicism in mammals. In no species had there been any early experimental mosaicism in which the genotypes could be chosen without restriction.

The studies of the classical embryologists on blastomere recombinations in invertebrate and amphibian embryos (reviewed by Morgan, 1927) suggested to me a practical means of realizing this possibility in a mammal. The projected procedure was to explant cleavage-stage embryos from genetically dissimilar donors, to remove the enveloping zona pellucida, and to aggregate the blastomeres from two (or more) embryos into one group. After the cells had had a chance to reconstitute an integrated spherical embryo during a brief culture period, the genetic composite was to be surgically transferred to a pseudopregnant "incubator mother" for further development. The embryo would of course have to be developmentally labile at the stage of rearrangement to obtain viable mosaics. The mouse appeared to be the most favorable species because of the many loci with known allelic alternatives and the availability of homozygous inbred strains. Experiments summarized elsewhere (Mintz, 1964a, 1965b) demonstrated that mouse embryo cells are indeed labile, even as late as blastocyst cavitation, and that cell rearrangements were compatible with normalcy (see Fig. 9-12, p. 210).

Our technical explorations, undertaken in 1960, were aimed at replacing the laborious mechanical procedures for egg membrane removal and blastomere aggregation used in the classical studies with techniques better suited to large-scale work and to embryos of warm-blooded species. These explorations yielded methods (see Figs. 12-1,2) with a high incidence of success. During the *in vitro* culture period, 95% or more of paired midcleavage-stage mouse embryos became unified single blastocysts or late morulae after one day; following surgical transfer, 29% of all transferred embryos survived to birth and beyond, but survival was as high as 48% in some large recent series.

The first demonstrably mosaic and fully viable mice "synthesized" from two embryos were born New Year's Day, 1965 (Mintz, 1965a,b). Since then, we have produced over 1,000 adult mice comprising 38 different paired combinations of cellular genotypes. Two genetic populations have been detected in all tissues that have thus far been examined, including skin and hair follicles (Mintz, 1965a,b; Mintz, 1969b; Mintz, 1970a,c,d; Mintz and Silvers, 1967, 1970), melanocytes (Mintz, 1967a, 1969a,b, 1970c,d; and see Fig. 12-2), erythrocytes (Mintz and Palm, 1965, 1969), leucocytes (Mintz, 1970f), 7S2a γ-globulin serum allotypes (Weiler and Mintz, data cited in Mintz and Palm, 1969), spleen, thymus, and lymph nodes (Mintz, 1970a), liver (Mintz and Baker, 1967; Mintz, 1970a), skeletal muscle and cardiac muscle (Mintz and Baker, 1967; Baker and Mintz, 1969), germ cells and gonadal soma (Mintz, 1965b, 1968, 1969a), mammary gland (Mintz and Slemmer, 1969), visual retina and pigmented retina (Mintz and Sanyal, 1970), and many other tissues.

The inclusion of two genetic cell populations within each histological cell type has led to the proposal that each specialized kind of cell is multi-clonal in origin (Mintz, 1969a, 1970d). That is, each kind of cell is developmentally derived from at least two genetically determined cells, each of which proliferates to form a clone in which the tissue-specific functional genetic specialization is maintained. The actual clonal number is often greater than two, however. For example, all coat melanoblasts appear to be derived from 34 clonal initiator cells (Mintz, 1967a), hair follicles from 150 to 200 cells (Mintz, 1969b), and the photoreceptor layer of the retina from 20 cells (Mintz and Sanyal, 1970). These genetic mosaic mice have thus made it possible to analyze mammalian development experimentally as a clonal process and to learn ways in which development is genetically controlled at the clonal level (Mintz, 1970d).

When the two cell strains present in one individual differ in their immunogenetic constitution (e.g., at the *histocompatibility-2*-locus), there is permanent immunological tolerance of both types and normal immune competence to reject a foreign type (Mintz and Palm, 1965, 1969; Mintz and Silvers, 1967). The tolerance that these animals display has been designated *intrinsic* (as opposed to experimentally *acquired* tolerance), since it presumably reflects the normal developmental process of recognition of ontogenetically new proteins (Mintz and Silvers, 1967). Thus there is no immunological restriction on the choice of genotypes to be combined. Even cells from a lethal genotype (e.g., W/W) can be used to produce a mosaic in which lethal and normal cells are admixed (Mintz, 1967a. Thus the multinucleate fibers of allophenics are heterokaryons.

In most tissues, the cells do not actually fuse (Mintz, 1970b). A notable exception is skeletal muscle (Mintz and Baker, 1967; Baker and Mintz, 1969).

Each animal obtained from two embryos can be considered "quadri-parental" (Mintz, 1965a), and a larger number of embryos, and therefore of parents, might contribute. The term *allophenic mice* (Mintz, 1967a) has been coined for all these animals of multi-embryo origin, and designates their chief feature: presence of cells of different phenotypes, ascribable to genotypic differences, as integrated components in any or all tissues throughout development. Allophenic mice are unique among experimental mammalian mosaics not only in their origin, but also in the tissue distributions of their genotypically different components. They do not in any way resemble the well-known experimental "chimeras" made by introducing a limited tissue graft into a host of a relatively advanced stage. The use of "allophenic mice" for our animals avoids confusion with these chimeras. (It may be added that the mythological derivation of the word *chimera* signifies a monstrous agglomeration of unassimilated parts; an allophenic animal, quite to the contrary, appears entirely normal and has joint participation of the genetically diverse cells in its ontogeny.) The component genotypes are shown with the special connecting symbol ↔, which distinguishes this association from genetic crosses, cell mixtures, etc. (Mintz, 1964a).

The term "chimera" has been arbitrarily used by some workers to signify genotypic cell admixtures derived from separate individuals, in juxtaposition to use of "mosaic" for mixtures arising by genetic, mitotic, or other errors among cells within a single individual. Insistence upon this distinction has the disadvantage that it presupposes knowledge of the origin of the condition. In point of fact, the origin may be unknown or debatable, as in many human clinical cases, and a forced choice between these alternatives invites error. The usage is therefore of questionable value. I suggest that the term "genetic mosaic" be employed impartially as a simple, general descriptive term in all cases of genotypic cell mixtures, without prejudgement as to their provenance. Specialized technical terms (e.g., "allophenic mice") can then also be coined to designate genetic mosaics of certain kinds, as a further unambiguous convenience.

There have been only two other attempts to produce early mammalian embryos having two genotypes; both relied on the classical approach of blastomere recombination. The first efforts, by Nicholas and Hall (1942) with rat eggs, yielded one dead fetus with no identifiable mosaicism. Experiments on aggregation of mouse blastomeres were undertaken by Tarkowski (1961, 1963) concurrently with and independently of our own and, not surprisingly, resulted in methods different from ours. He ruptured the zona pellucida by rapid suction of the egg into a narrow pipette, and then squeezed the donor blastomeres together in a microdrop of (non-serum) medium under paraffin oil. Confinement in the microdrop was carried out at room temperature and the gas phase was not controlled during culturing. Tarkowski's techniques did, in fact, lead to formation of some fetuses and neonates that were visibly mosaic. Further viability appeared to be limited, however; there were only two postnatal survivors, neither of which had evidence of mosaicism. (The experimenter should expect to obtain some non-mosaic mice with our methods as well; the reasons for complete loss of mosaicism are discussed in a later section.) The first adult mosaics that were reported from Tarkowski's laboratory were nine animals described in 1968 (Mystkowska and Tarkowski, 1968); they were all produced by our basic techniques introduced earlier (Mintz, 1962b, 1964a).

These techniques will be given in detail here, along with ancillary methods for the study of early development and for histological processing of preimplantation embryos. Our procedures for making allophenic mice revolve chiefly around the discovery that the zona pellucida can be enzymatically lysed with pronase without harming the embryo (Mintz, 1962a), and that blastomere adhesion is temperature-dependent (Mintz, 1962b,c, 1964a) and is easily inducible by placing the cells in contact at 37°C without mechanical confinement. Environmental fluctuations (of temperature, pH, gas phase, etc.) must be rigorously prevented during the *in vitro* period. In presenting the ways in which we maintain environmental control, simple and inexpensive alternatives are given for any equipment requiring custom shopwork.

OBTAINING DONOR EMBRYOS

Mouse embryos may be obtained in several ways: (1) from sexually mature females, which normally ovulate spontaneously at estrus even if they do not mate; (2) from immature females that can be induced by hormone treatment to ovulate precociously; and (3) from mature females in which ovulation can be induced by hormone treatment, irrespective of phase of the estrous cycle.

Spontaneous ovulation and mating generally occur at night at or near the period of estrus. Mating usually precedes ovulation by 1 to 5 hours. It is desirable to exclude all outdoor light from the animal colony and to use an automatic time switch for maintenance of a constant 24-hour ratio of light and dark. In our colony, the light period is from 6 A.M. to 6 P.M. throughout the year. Shorter dark periods, down to 4 hours, can be used to reduce the variability in ovulation time. Estrous females are identified by vaginal smears, and are then individually paired with males. Or three or four randomly chosen females can be placed with each male. Females are transferred to the male's cage near the end of the light period. The presence of a vaginal plug (detectable with a dissecting probe) on the next morning indicates that mating has occurred. Females should not be stored for long periods without matings, and postpartum matings should not be used for timed embryo stages because the mating time is then relatively variable and implantation is delayed. Further suggestions (e.g., caging arrangements, etc.) for maximizing the incidence of spontaneous matings are reviewed and summarized elsewhere in a chapter on methods for mammalian embryo culture (Mintz, 1967b).

Hormone-induced ovulation of prepuberal females yields more matings and more eggs per mated female than does spontaneous ovulation of adults. We therefore use superovulated prepuberal females as egg donors whenever possible. Pregnant mare serum (PMS) is given to stimulate follicular growth, then human chorionic gonadotropin (HCG) is administered to stimulate ovulation (Runner and Gates, 1954). Ovulation generally follows HCG by 11 to 14 hours. There are strain differences in optimal age of treatment and dose of each hormone; these differences can be quite marked and it is worthwhile to plot the optima if only a few strains are involved. Since we work with many strains, we have found it impractical to obtain control data for all of them, and have adopted the following uniform schedule: (1) Females are weaned at 21 to 25 days of age; (2) At 25 to 30 days, a single injection of 4 International Units (I.U.) of PMS (Equinex, Ayerst Laboratories, New York) in .04 ml of .9% sodium chloride is given intraperitoneally at 5 P.M. (or at noon) and is followed 40 (or 45) hours later, respectively, by an intraperitoneal injection of the same dose of HCG (A.P.L., Ayerst Laboratories); (3) Females are housed individually or in pairs with one mature, untreated male in the male's cage, at the time HCG is given; (4) The check for vaginal plugs is made the next day. Hormone-primed prepuberal mice tend to produce more abnormal eggs than are commonly found in adult spontaneous matings; a reduction in PMS dose may reduce the incidence of abnormal eggs but it may also reduce egg number.

Females can be superovulated between weaning and sexual maturity (at 1 to 2 months of age), and mature females have also been superovulated with doses ranging from 2.5 to 15 I.U. of each hormone. Here also the animals are caged together at the second injection. We have had lower yields of plugs and of eggs from superovulated females at these ages than at prepuberal ages. Injected females (prepuberal or mature) that have not mated may be reinjected 2 or 3 more times at biweekly intervals.

The date of plug detection is designated Day 0 in the developmental timetable. Roughly 24 hours elapse until the first cleavage, and about 12 hours separate the succeeding cleavages. Cleavages are not synchronous among the cells of one embryo. There are strain differences in developmental rates and, if hormones are administered, the time of day that HCG is given will also influence developmental staging on subsequent days. The embryo has 2 cells on Day 1, 3 to 16 cells on Day 2, and 16 or more cells (late morula or blastocyst) on Day 3.

EXPLANTATION PROCEDURES
AND CULTURE MEDIUM

Embryos can be successfully aggregated to form a unified composite (see Figs. 12-1,2) if they are united at any time during cleavage, from the two-cell stage through the early blastocyst (Mintz, 1964a). The eight-to-twelve-cell stage is optimal because at that time the blastomeres are sticky and ameboid at 37°C and the aggregates quickly form one sphere. Prior to the eight-cell stage, the blastomeres have not developed their maximum stickiness and motility and are less readily induced to adhere when they are placed in contact. Late-stage aggregations that are near the blastocyst stage leave little *in vitro* time for cell rearrangements prior to surgical transfer, since the blastocyst is the terminal culture stage, and may yield conjoined

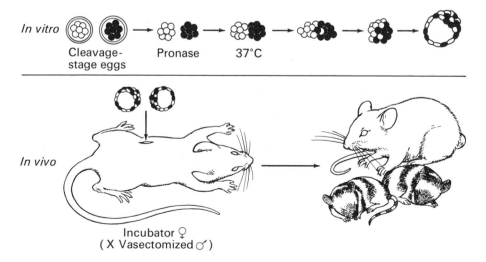

FIGURE 12-1

Diagram of procedures for obtaining allophenic mice from aggregated eggs.

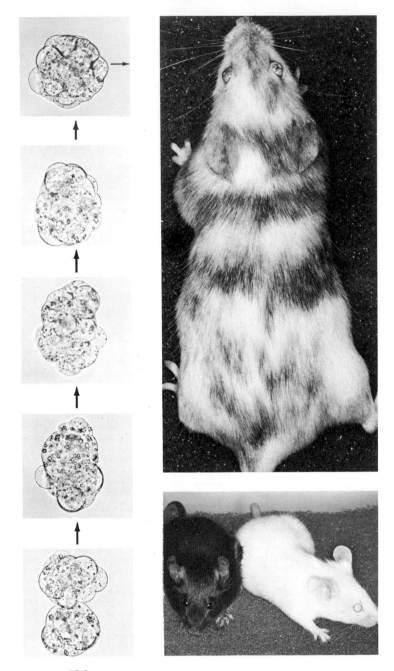

FIGURE 12-2

Lower left: Two living cleavage-stage embryos of pigmented (CC) and albino (cc) genotypes, aggregated in vitro *after zona pellucida removal with pronase. Successive photographs show formation of one spherical embryo (CC ↔ cc) from all the blastomeres of two eggs. One of the resultant allophenic mice from these paired genotypes is at the upper right; note transverse clones of black and white in the coat and radiating clones in the eyes. The allophenic mouse was a germinal mosaic female; at the lower right are two of her offspring, one from a genetically pigmented and one from an albino germ cell, after mating with a pure-strain albino male.*

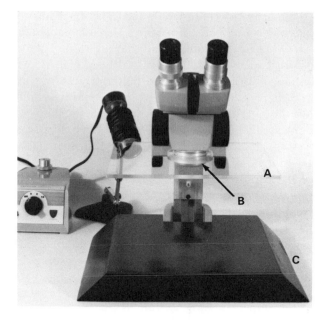

FIGURE 12-3

The microscope for dissection of the egg donor is fitted with a protective horizontal Lucite shield (A) supported from below by a Lucite ring (B). A hollow wooden base (C) over the microscope stage increases working space.

parabiotic embryos in culture rather than a single unified sphere. The embryos that are aggregated should be at approximately the same developmental stage though they need not have the same cell number. If the partners are markedly dissimilar in stage, a parabiotic pair is formed with each member on its own developmental schedule (Mintz, 1964a, 1965b).

To perform aggregations at the eight-to-twelve-cell stage, the egg donors of the different genotypes are killed on Day 2, in succession rather than simultaneously. The animal is killed by pinching apart the cervical vertebrae or by stretching the spinal cord; it is briefly dipped in 95% alcohol and is placed on its back in a shallow pan made from aluminum foil. Dissection should be performed either in a work area where there are no strong air currents or in a laminar flow sterile hood. Without a hood, adequate protection against vertical drop of airborne contaminants is afforded by the shield shown in Figure 12-3. The dissecting microscope (with internal rather than projecting objectives) is fitted with a horizontal transparent Lucite plate (20 cm square) with a central hole surrounding the objective lens. The plate is supported from below by a Lucite ring (with set screws) which encircles the objective housing, or by the threaded lens shield (Mintz, 1967b). A similar shield can be hand-cut from a thin sheet of any stiff, clear plastic. A hollow base (of wood or other material) over the microscope stage serves to increase the working space, or a swinging-arm microscope can be used without a stage and the work is then done directly on the table surface.

The embryo donor is dissected by reflected light. Instruments need not be sterilized in advance. They are dipped in alcohol, flamed with a micro-burner (Touch-O-Matic), and used when they are cool. They are repeatedly cleaned and sterilized as needed. The ventral body wall is cut with scissors in a wide V-shaped incision from the posterior midline anteriorly and laterally and the flap is deflected forward. Each oviduct is removed with a short piece of the uterine horn since some of the embryos may have traversed the uterotubal junction on Day 2. With fine scissors and #5 watchmaker forceps, most of the mesentery and fat can be cut away. The pair of ducts is immersed in 37°C culture medium in one concavity of a warm triple-well clear Pyrex spot test plate (85 × 34 mm, A. H. Thomas Co., Philadelphia). (Warming and dispensing of medium will be discussed later.) The spot test plate is enclosed in a Petri dish for sterility and convenience. The corners of the plate may need to be ground down slightly to fit in the dish; if a grinder is not available, a double-well 75 × 25 mm microslide, a single-well Maximow slide, or an embryological watch glass will fit into the Petri dish.

The same kind of medium that is to be employed for final culture should also be used in flushing the embryos out of the ducts and in rinsing and examining them. Theoretically, any culture medium is satisfactory if con-trol embryos, after explantation in it at the same initial stage as the allophenic embryos, can reach the blastocyst stage by the normal (*in vivo*) time. The medium must, however, be tested under conditions that permit the embryos to remain at the 37°C *in vivo* temperature throughout *in vitro* handling, inasmuch as cooling causes a slowing down of development. Although several media can fulfill the requirements, we have found that a high-serum medium is optimal for making allophenic embryos because of good pH retention during handling; in addition, the proteins afford rapid blocking of pronase action after the zona pellucida is removed (see the fol-lowing section). The quality of different batches of serum from a given source may vary and each batch should be tested by explanting controls in it.

The medium in which we have secured an extremely high survival of allophenics (Mintz, 1964a) contains 50% fetal calf serum (the non-hemo-lysed type, which we presently obtain from Grand Island Biological Co.) and is at pH 7. It is composed of equal volumes of fetal calf serum and Earle's balanced salt solution, to which are added 0.1% lactic acid (which may be added from a commercial 40% solution of L(+)–lactic acid) and .002% phenol red; the pH is adjusted with sodium bicarbonate (7.5%) under 5% CO_2 in air. (Use of lactate was based on Whitten's [1957] earlier observations of its beneficial effects.) Inclusion of phenol red is necessary for continuous monitoring of the pH; excessive or sustained loss of CO_2 at any time during the *in vitro* portion of this work will appreciably reduce the frequency of success. No antibiotics are included; if sterile precautions are taken, contamination of cultures is very rare.

The culture medium should be stored under 5% CO_2 in air. This is easily accomplished by using sealed serum bottles with rubber puncture caps. The screw-cap type of bottle with a rubber diaphragm in the cap (Aloe Scien-tific) is adequate. Before introducing the medium, the bottle is quickly gassed and closed. Each component of the medium is then added from a hypodermic syringe with a needle. During the large-volume additions (of serum and Earle's solution), a separate hypodermic needle is temporarily inserted in the cap as a vent for gas replacement; it is withdrawn before the remaining solutions are introduced, so that the complete medium is under a slight positive pressure. A sample for pH check is withdrawn with a hypodermic syringe and needle. Medium used for embryo work is always withdrawn in a syringe and the syringe is subsequently kept at 37°C in the warmed hood or incubator; liquid is dispensed into dishes whenever it is needed. This arrangement is superior to the dispensing of medium from pipettes since it permits retention of CO_2 in a closed system. The bottle of medium can be stored in the refrigerator (not frozen) for a few weeks.

We now return to the explantation of embryos. The Petri dish and depression slide containing medium, with the pair of ducts from the first donor, is kept on a warm stage (see Fig. 12-4; p. 199) or is moved to a warmed hood (see Fig. 12-5; p. 200) at 37°C. All further handling of the embryos is performed at this temperature. The construction of the warm stage and hood, as well as a number of suggested substitutes for them, will be described later in connection with the aggregation of blastomeres.

The eggs are flushed from each duct from the infundibular end with prewarmed culture medium (usually less than 0.1 ml) from a 1- or 2-ml short syringe fitted with a 30-gauge short needle. The bevel is first cut off the needle, leaving about 6 mm, and the tip is polished smooth. The infundibulum of the oviduct is held in place around the needle with #5 watchmaker forceps. For this and all later operations, transmitted light is preferable to reflected light and total magnifications of ×20 to ×80 are adequate. The eggs are picked up in a fine pipette and transferred into an additional rinse of medium before further processing.

Although various rotating-knob micrometer types of attachments are available for operating pipettes when small volumes are involved, we have found that an ordinary rubber bulb (of slightly aged, thin amber rubber) is faster and easier to operate. A mouth-held rubber tube is also acceptable but it is less convenient in a hood. A short pipette of about 7 cm total length offers good leverage during egg handling. Our pipettes are made by pulling Pyrex tubing in a hot flame (e.g., from a compressed air-gas blast burner) and repulling once or twice more in an ordinary flame (from a microburner) in the region of gradual taper of the glass. A long (2 to 2.5 cm) uniformly narrow and thin-walled tip is essential for fine control. The internal diameter of the tip should be only slightly larger than the embryo; a series of embryos can then be rapidly picked up like a string of beads and a minimal amount of fluid carried along with them into the next solution. The tip is broken off so as to form a slant bevel, by holding the capillary under slight finger pressure at right angles to the sharp edge of a hard Arkansas stone, scratching one side lightly, and progressively increasing the pressure toward that side. The bevel is fire-polished by brief passage at the edge of a microburner flame. With this beveled tip, the pipette can be conveniently held at an angle and the oval opening quickly cupped over each egg in turn. Pipettes need not be siliconed, even when they are used with zona-free sticky eggs, if they are kept clean. A cotton filter may be omitted, as the fibers tend to fall into the medium and entrap the eggs.

We have designed a rack for sterilizing these pipettes; it is shown here as item (I) in Figure 12-5 and is diagrammed elsewhere (Mintz, 1967b). The rack is made of stainless steel or aluminum. Each of nine pipettes stands in a hollow plug support affixed to a solid base and is covered by its own test-tube sheath, allowing removal of one pipette without contaminating the others. A satisfactory substitute is to sterilize pipettes individually in steel-capped test tubes, with the cap down and the fine tip pointing upward.

We have also designed a drying rack (see the diagram in Mintz, 1967b) for fine-bore delicate pipettes; the rack consists of a steel or aluminum block with cylindrical depressions for the broad ends of the pipettes and a narrow hole drilled through the floor of each depression for warm air circulation. A substitute requiring no machine shop work is to partition a wire-screen pipette basket or test tube basket into individual compartments for free-standing pipettes, by stringing wires at right angles to each other in each of two horizontal layers.

PRONASE METHOD FOR REMOVAL
OF THE ZONA PELLUCIDA

Removal of the egg's enveloping membrane, or zona pellucida, is necessary before the blastomeres of genetically different embryos can be aggregated to form allophenics. Mechanical methods for denuding mammalian eggs are relatively hazardous and are not applicable to the processing of large number of eggs. The discovery (Mintz, 1962a,b,c) that the proteolytic enzyme *pronase* could digest the zona without injuring the egg permitted these difficulties to be surmounted. Pronase treatment has since become the preferred method for denuding eggs or embryos at any stage, in a wide variety of species. It also digests the matrix of the follicle (cumulus and corona) cells that surround uncleaved eggs (Mintz, 1962a). The enzyme is a product of *Streptomyces griseus* (Nomoto and Narahashi, 1959; Satake *et al.*, 1961); the impure preparation (Calbiochem, B-grade) is satisfactory.

A 0.5% solution of pronase is prepared in bicarbonate-free Hanks' balanced salt solution with phenol red. Use of a phosphate-buffered saline without bicarbonate is permissible for the brief time (3 min) involved, and has the advantage that the dissolution of the zona can be monitored in an open dish by direct observation without pH change. Many other balanced salt solutions that are commonly used in mammalian cell culture work can be substituted for Hanks'. Pronase is the only protein in the medium used during zona removal and no other macromolecular additives are needed. With appropriate handling, no fatalities result.

In our standard procedure, the 0.5% pronase solution is first cleared by passage through a fluted paper filter and is then adjusted to pH 7 with $0.3N$ NaOH. A Millipore filter in a pressure (rather than suction) type of sterile filtration apparatus is washed through by filtration of approximately 200 ml of Hanks' solution before the pronase is filtered; the first fraction of pronase will then be dilute and may be discarded. The filtered enzyme solution can be stored indefinitely if it is kept frozen at −20°C. If it is tubed in 1 ml aliquots in sterile plastic test tubes (Falcon Plastics), each batch of eggs can be treated with a separate thawed preparation. The frozen-thawed sample has a precipitate and should be centrifuged before being used. About 0.4 to 1 ml of supernatant liquid is pipetted into a culture dish, a spot test plate, a watch glass, or a depression slide, and is warmed to 37°C in the incubator or hood. The warming should not be prolonged unnecessarily as a precipitate will form and enzyme activity will decline.

After the explanted embryos have been rinsed in culture medium, they are transferred directly into pronase and are mixed with the enzyme by quickly pipetting them to a fresh area while using the pipette as a stirring rod. Lysis of the zona pellucida can be seen by tilting the substage mirror so as to darken the field slightly. Complete lysis requires about 3 minutes. The embryos should be picked up just before the last vestige of the zona has disappeared (lysis will be completed inside the pipette); they are then transferred through three or four changes of 37°C culture medium (preferably in triple-well Pyrex spot test plates), with two rinses in each. The zona-free embryos are sticky at this temperature and should be scattered to prevent them from sticking together. If accidental adhesions occur between like-genotype eggs, the eggs can be separated by squirting a jet of medium between them or by quickly pipetting them up and down once. The post-pronase rinse dishes are not siliconed; we have had no difficulty with adhesions to the glass during the brief rinse periods.

The quality of pronase lots may vary. If a test sample gives suboptimal results, an alternative procedure for digestion of contaminating nuclease in the preparation (Gillespie and Spiegelman, 1965) may be introduced. After filtration through paper, the solution is incubated for 2 hours at 37°C and is then dialyzed against Hanks' solution overnight at 3°C. Enzyme activity is reduced after this treatment and the time required for zona lysis is slightly prolonged.

AGGREGATION OF BLASTOMERES
TO FORM ALLOPHENIC EMBRYOS

Monozygotic mammalian twins are not miniatures. It therefore seemed reasonable to suppose that genetic mechanisms control embryo size in some way and that they would also restore giant embryos to normal size. The prediction was borne out: aggregation of two entire embryos results in a single giant blastocyst *in vitro* but size regulation occurs *in vivo* during the implantation period (Mintz, 1967a). Thus, in the following description, whole embryos rather than halves are conjoined in pairs and the added work and risk required for separating the cells is avoided. More than two whole embryos can in fact be combined (see Fig. 12-9c, p. 210); as many as sixteen have been successfully united to form one enormous blastocyst in culture (Mintz, 1965b). Although such blastocysts may well regulate to normal embryo size once they have been transferred *in vivo*, we cannot assume that all the cellular genotypes would be represented in the resultant embryo. It is well known that the embryo proper is derived from some (not necessarily all) cells of the inner cell mass, with extraembryonic tissues coming from other parts. Our analyses of the genotypic composition of allophenic mice from paired embryos reveal that some completely non-mosaic individuals are obtained (Mintz, 1967a, 1970a). On the basis of the frequency of the mosaics in a large experiment, we have proposed that the mouse "embryo," in the strict sense of that term, may originate from a very small number (possibly as few as three) of cells in the inner cell mass (Mintz, 1970d).

Since the central principle in constructing allophenic embryos is to induce adhesion between unrelated donors by the use of physiological temperature (Mintz, 1962b, 1964a), the *in vitro* environment should optimally be kept at 37°C during aggregation and the embryos should remain motionless under the microscope. Carbon dioxide would ordinarily be lost from the medium, causing a *p*H rise, during that period; bicarbonate-containing medium is required for development, so this problem cannot be circumvented. The procedures for combining embryos therefore depend upon simultaneous maintenance of 37°C temperature and of gas phase constancy. High incidence of mosaicism and high viability are obtainable only when temperature and *p*H fluctuations are prevented in all the rest of the *in vitro* handling and the incubation as well. The following methods for environmental control are applicable not only to allophenic embryo work but also to many other kinds of *in vitro* experimentation with mammalian embryos, organ rudiments, tissues, and cell cultures.

The 37°C temperature essential for inducing blastomere adhesion is most accurately and conveniently achieved by carrying out the aggregations either on a thermostatically controlled transmitted-light type of warm stage or in a thermostatically controlled hood surrounding the microscope. With either of these methods, pairs of denuded embryos remain conjoined and continue their *in vitro* development normally and without developmental delay in more than 95% of all cases.

The warm stage in Figure 12-4 is a circular model with a center hole for transmitting light upward from the microscope substage mirror, and with a thermostatic control and pilot light and a thermometer. This kind of stage is ideally suited to the purpose and can readily be made by adapting a well-type rectangular warm stage (Incu-Stage incubator, 115 volts, Labline or Chicago Surgical and Electrical Co.). (The prototype circular model, made by the same company, has recently been discontinued.) The Incu-Stage (preferably the 95 × 75 mm size) is adapted by first cutting a center hole in the floor, removing the rectangular grill, bending the grill to a round shape, and reinserting it around the hole. The heating element is then lifted up, a sheet of asbestos is introduced under it (clearing the center hole), and the heating element is replaced and is spread over the surface of the asbestos. A screw-down circular top plate of aluminum is finally added, directly over the heating element. The size of the plate is most conveniently between 100 and 125 mm diameter and the center hole in the plate should be about 20 to 30 mm diameter.

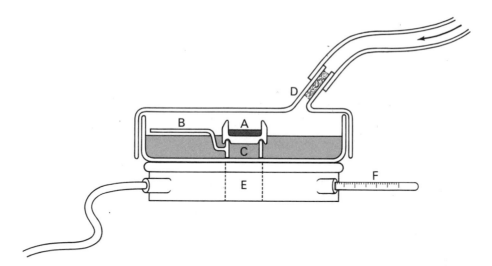

FIGURE 12-4

Egg-handling chamber for embryo aggregation at 37° C under 5% CO_2 in air. The culture dish (A) on a stainless steel rack (B) is partly immersed in sterile water (C) in a Petri dish. A glass tubulation (D) fused to the lid contains a cotton plug and is attached by rubber tubing to the gas inlet line (arrow). The water bath rests on a warm stage with a center hole (E) for transmitting light and a thermometer (F) for temperature monitoring.

A heated microscope hood or a warm stage are equally effective alternatives for formation of allophenic embryos, but the hood affords greater convenience and efficiency of manipulation. The hood of our design is shown in Figure 12-5. The frame is made of wood at the back, sides, and top, with a piano hinge across the top for servicing the interior. A clear Lucite slant-front panel is set into a groove in the frame on each side. Below the Lucite panel hangs a pliable plastic drape, which is open across the lower edge; hands and forearms are inserted under this drape. Hood size depends on work space; our model in Figure 12-5 is 4 feet wide, 2 feet deep, 2 feet high at the back, and 9 inches high at the front. There are three outside control switches on the right side (not visible in the photograph). One operates an ultraviolet germicidal lamp located just under the roof of the hood; the lamp should be off during handling of cultures. Another switch operates the heating system, and a third controls the air circulation system. Heat is supplied by four ceramic-coated Ohmite resistors (150 ohms, 100 watts) mounted across the lower back panel on an aluminum reflector. A Fenwal thermostat in the 37°C range is at the upper right; its adjustment screw is externally located, and a 110-volt pilot light is nearby. A blower (15 cu ft/min capacity) pulls the warmed air out at the upper right and into a hose (vacuum cleaner type) that passes around in back of the hood and reenters at the lower left through a single opening in the side wall. A second wooden wall just median and parallel to the first is perforated with holes and acts as an air diffuser. Three small openings near the base of the

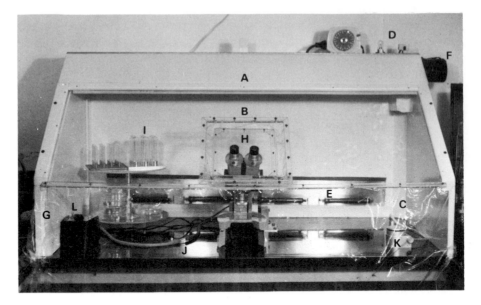

FIGURE 12-5

Temperature-controlled microscope hood. The hood has a wooden frame (A) , *slant-front Lucite panel* (B) , *and pliable plastic drape* (C) *open across the lower edge. A thermostat with pilot light* (D) *controls four resistors mounted on an aluminum reflector* (E) *along the lower rear wall, and a blower* (F) *circulates the warmed air. Air passes out of the hood at the upper right into a hose which re-enters at the lower left through a double wall with an inner perforated baffle* (G) . *A movable Lucite plate* (H) *rests on rings encircling the microscope eyepieces and is attached to the fixed hood surface by pliable plastic sheeting. A double-tiered "lazy Susan" carries pipette racks* (I) . *The CO_2-and-air line* (J) *enters through a hole in the back wall at the lower left, the gas line for a burner* (K) *through a hole at the lower right, and the electrical cord* (L) *for the microscope lamp at the center.*

back panel provide access for a tube from the CO_2-and-air source, a tube from the gas line for a burner (Touch-O-Matic, with small pilot flame) used in flaming instruments, and an electrical cord for the microscope lamp.

With this hood, excellent visibility is maintained over the working area by means of a "floating" Lucite plate that rides up and down with the microscope during focusing. Only the oculars of the microscope are outside the hood. The plate ($7\frac{1}{2}$ inches wide $\times$ $8\frac{1}{2}$ inches high) is parallel to the slant hood surface and is bolted to a pair of Lucite rings that encircle the microscope eyepieces; the rings are held in place with set screws. There is a two-inch space around the edge of the plate, and a loosely fitting frame of pliable plastic is fastened by overlying Lucite strips and screws to the plate and the fixed hood surface. A foot-focus attachment for the microscope is very convenient but not essential.

The actual pairing of denuded eggs is carried out in a simple chamber of our design (see Fig. 12-4) . When the source of heat is a warm stage, the chamber contains water to hold the heat; the water may be omitted if egg aggregations are performed in a heated hood. In either case, this chamber permits a controlled gas atmosphere to be introduced over the embryos and

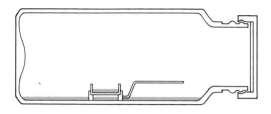

FIGURE 12-6

An 8-oz individual culture chamber for embryos, with culture dish on a stainless steel rack. The gassed French square bottle has a screw-cap (rubber-lined) closure. Tyrode's solution on the floor of the bottle maintains high humidity and has phenol red as a pH indicator.

allows the pH of the culture medium to be kept constant, not only while egg pairs are being adhered, but also at any other time during *in vitro* handling of embryos. This "egg-handling chamber" is made from a Petri dish (100×20 mm) with a glass tubulation fused at an angle onto the top of the lid near the margin. A slow stream of 5% CO_2 in air passes from the supply tank through two successive gas washing bottles, enters the lid sidearm (containing a sterile cotton plug), flows over the culture, and leaves the chamber between the loose-fitting upper and lower halves. (A substitute for this type of gassing lid, if glass-blowing facilities are not available, is an ordinary funnel with a sterile cotton plug in the stem.) The zona-free eggs should be paired in the same dish in which they will later be incubated; thus the egg-handling chamber can contain either a small culture dish on a rack (as in Figs. 12-4 and 12-6) or a small (60 mm diameter) Petri dish (as in Fig. 12-7) on the floor, without a rack. If a Petri dish, whether small or large, is to serve later as the culture dish, the lid that normally accompanies it can be fitted with a glass sidearm for gassing purposes so that no additional surrounding chamber need be used during egg aggregation. For standard Petri dish cultures, the gassing lid is then later replaced with a conventional lid.

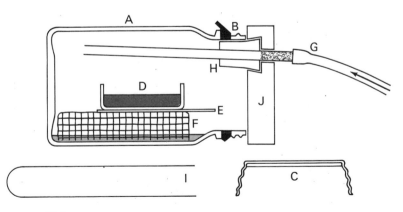

FIGURE 12-7

A one-pint individual culture chamber for larger volume cultures in a Petri dish, with gassing apparatus in use. The chamber is a flat-sided Mason jar (A) with a rubber ring (B) and a glass-lined screw-cap (C). The culture (D) is in a Petri dish half on a glass slide carrier (E) resting on a wire screen table (F). The gassing apparatus (G) is a cotton-plugged glass tube with a one-hole rubber stopper (H) for sterile storage in a test tube (I). The incoming gas line (arrow) passes through a Lucite shield (J).

If there is no machine shop for adapting an Incu-Stage or other warm stage, or for constructing a hood, there are many other satisfactory heating arrangements that require only moderately priced commercially available equipment or simple laboratory improvisations. Success in each case depends on the accuracy of temperature control since the embryo pairs tend to come apart when the temperature is less than 37°C and temperatures above that level are increasingly lethal. Examples of other culture-warming devices are the following: (1) a light-heat box (such as the one from Fisher Scientific Co.) with an opal glass insert plate in the top and two 7-watt bulbs inside, supplemented by a microscope lamp for top lighting during the light-off period of the box; (2) a warming plate (e.g., Gentl-Therm, from Will Scientific, Inc.) on which aggregations are to be conducted by reflected rather than transmitted light; (3) a heating tape mantle wound around a Petri dish or other type of water bath; (4) an immersion heater in the water bath; (5) a liquid circulating stage (like the model from A. H. Thomas Co.) fed with prewarmed water from a thermostatically controlled water bath or from a sink with a temperature-controlling faucet; (6) a hair-dryer type of heater-blower directed at the bottom of the egg-handling chamber; (7) a bunsen burner flame for occasional warming of the water bath or of the nearby air (though this is less subject to control and is likely to give a lower yield of viable mosaics than the preceding arrangements) .

If no warming devices are used and the eggs are paired at room temperature, some will successfully adhere after the culture has been incubated, since the adhesion of blastomeres is temperature-dependent (Mintz, 1964a). In this case, it is advisable to use prewarmed culture medium during the handling, to work rapidly, and to withdraw and recheck the incubated culture after a period of time, in order to reassemble separated members of pairs. However, the *in vitro* yield of allophenic embryos is likely to be substantially less than with the temperature-controlled conditions described above. In addition, development is slowed and cells are less motile; subsequent *in vivo* development and retention of mosaicism may be impaired.

If the eggs are aggregated in a small culture dish (see the next section for a description of culture dishes) , only 0.8 to 1 ml of medium is needed. If a Petri dish is used, a larger amount of medium is required, depending on the size of the dish. In that case, medium can be economized by using drops that are dispensed with a syringe under a relatively light paraffin oil in the manner used by Brinster (1963) for explanting unaggregated eggs. The drops also help to isolate the pairs from each other and to locate them at the end of incubation if the dish is large. A drop may contain roughly .05 to .1 ml of medium and does not physically constrain the eggs. The oven-sterilized oil (viscosity 125/135, Fisher Scientific Co.) is equilibrated with the medium a few days in advance by mixing approximately 95 ml with 5 ml of medium, gassing with 5% CO_2 in air, and incubating at 37°C under the same gas mixture.

The rinsed, zona-free embryos of the first contributing genotype are pipetted into the prewarmed culture dish with medium at 37°C, in the egg-handling chamber, and are spread wide apart over the bottom of the dish. In the dishes shown in Figures 12-4 and 12-6, we introduce 10 to 12 of these embryos for later pairing. The dishes must be siliconed (e.g., Beckman Desicote) to prevent embryos from sticking to the floor. The chamber with the gassing lid is carefully removed from the microscope and is left in the hood or on a warm stage with a slow stream of 5% CO_2 in air passing through it.

Another female is then killed and dissected and the eggs of the second contributing genotype are flushed from the ducts, washed, pronased, and rinsed 3 or 4 times in the same manner as the first set, at 37°C. The second group is picked up in a pipette, the chamber with the first set is returned to the microscope, and one embryo is deposited alongside each that is already in the chamber. With a little practice, the pipetting of the second member can be done so that it contacts the first one in a jet upon leaving the pipette; in that case, there is a good chance that the two will adhere immediately (see Fig. 12-2). (If there is any tendency for the eggs to stick together while they are in the pipette, they may all be deposited first in the center of the dish and then picked up singly and moved to their proper partners. This extra step can be eliminated, and adequate speed and efficiency of the initial pipetting can be gained, with further practice.) A finely drawn glass rod with a beaded tip slightly larger than the embryos is used next to push one embryo of each pair firmly and forcibly against the other. Broad contacts between the members, in which several cells are apposed, are more effective than tangential ones. The gassing lid (see Fig. 12-4) is replaced and the chamber is left undisturbed with continuous, slow gassing. After 10 to 15 minutes the lid is removed, any pairs that have come apart are again pushed together with the beaded rod, the lid is replaced, and the gassing is resumed for another 10 to 15 minutes. The unions generally become firm by then. At no time should the pH of the culture be allowed to rise appreciably (as judged from the phenol red color) before gassing. If the embryos do not adhere easily in one or two rounds of being prodded together, they are likely to have been injured by overpronasing or in some other way, and are not promising material for formation of allophenics. The final check is made before incubation; nonaggregated pairs should be removed, but such failures should be very rare.

CULTURE OF EMBRYOS AND TYPES OF CULTURE CHAMBERS

The culture dish is now ready to be moved to the incubator, where gas phase control must be continued. Many commercial 37°C high-humidity incubators can be flushed with a continuous supply of 5% CO_2 in air; Petri dish cultures can be used in them. We have found, however, that this arrangement gives relatively poor development of mammalian embryos. After the incubator door has been opened, replacement of the gas phase is slow and the resultant disturbances of pH tend to impair development. Parodoxically, as efficiency in the aggregation of the embryos increases and more cultures are explanted, the survival rate declines because of frequent opening of the incubator to add new cultures. We have therefore devised two types of individual, closed culture chambers that are pregassed and humidified; these have given excellent results and are inexpensive. Many separate containers can be placed in one incubator; the incubator is not gassed (Mintz, 1964a, 1967b). One type of chamber is the French square bottle in which a small dish of about 1 ml capacity is accommodated (two dish styles are seen in Figs. 12-4 and 12-6). Another type of chamber is the more versatile Mason jar (see Fig. 12-7), which can accommodate either a small culture dish or a larger dish.

The Mason jar chamber used is the one-pint size, and is wide-mouthed and fairly flat-sided, with a glass-lined screw cap and rubber compression ring. It may contain either a Petri dish up to 60×15 mm, a long cut strip of tissue culture plate with a series of separate wells (Falcon Plastics), a 1 ml embryological watch glass (see Fig. 12-4) available from A. H. Thomas Co., or a custom-made low-walled (25×7 mm) Pyrex cylinder (Bellco Glass, Inc.). In this jar, the culture dish is supported at mid-height by a table made from bent wire screening; a $1'' \times 3''$ glass microscope slide between the dish and the supporting screen functions as a carrier to insert or remove the dish. Culture volumes of 15 ml are possible in the Petri dish; hence the chamber is also usable for organ cultures. The Petri dish lid may be left off. As already pointed out, small drops of medium under paraffin oil may also be placed in the Petri dish. On the floor of the chamber is bicarbonate-containing balanced saline (e.g., Tyrode's, Earle's, or Hanks' solution) for humidification and pH indication with .002% phenol red and 10,000 units each of penicillin and streptomycin per 100 ml of saline. The salt solution is introduced, the bottle is gassed with 5% CO_2 in air, and the closed chamber is warmed to 37°C. The chamber is then reopened, the culture is added, there is a final gassing for a few minutes, and the bottle is closed and incubated.

The gassing apparatus (see Fig. 12-7) is a sterile cotton-plugged glass tube with a one-hole rubber stopper for sterile storage in a test tube. To avoid excessive gas loss from the wide mouth of the culture bottle, a Lucite block is used as a free-standing loose shield which is placed against the mouth during gassing. A satisfactory shield can also be cut from a thin stiff plastic sheet.

We usually use the French square bottle (A. H. Thomas Co.) for culturing allophenic embryos because we prefer the small culture dishes (.8 to 1 ml) for performing the aggregations, and the 8-oz bottle is adequate to house these. The bottle is flat-sided with a wide mouth (35 mm I.D.) and a gas-tight, rubber-lined screw cap. The culture dish is either a commercially available embryological watch glass (A. H. Thomas Co.), seen in Figure 12-4, or a custom-made Pyrex cylinder (25 × 7 mm, Bellco Glass, Inc.), shown in Figure 12-6. The watch glass does not have a perfectly flat bottom and is optically inferior to the other dish, but we have successfully assembled over 300 viable allophenics with ease in the watch glasses. Either kind of dish can be moved in and out of the bottle on a $1'' \times 3''$ glass slide; the slide is supported on a wire screen table in the bottle, in the manner already described for the Mason jar chamber that is shown in Figure 12-7. If a glass slide carrier is used for the dish during incubation, the dish can be supported during the (previous) embryo aggregation on a commercial glass ring of slightly smaller diameter than the dish itself. (The ring raises the level of the dish, thereby permitting more water to be contained in the water bath (see Fig. 12-4). If no water is included, e.g., if aggregation is conducted in a warm hood, the culture dish can be seated on the glass slide throughout handling.)

We have designed a stainless steel rack for each type of small dish; the rack is slightly more convenient than the glass slide carrier. Each rack is an open ring (for viewing of the culture in transmitted light) with a wire handle. The one used with the watch glass (see Fig. 12-4) fits into a circular groove on the bottom of the dish; the one used with the custom-made dish (see Fig. 12-6) has an inside shelf in which the dish is seated. The French square bottle is handled like the Mason jar chamber. It is humidified with bicarbonate-containing balanced salt solution (with phenol red, penicillin, and streptomycin), and is gassed and warmed before use, and then regassed briefly after the dish has been introduced.

We have also used a smaller (4 oz) French square bottle for allophenic embryos, with a smaller, custom-made (Bellco Glass, Inc.) Pyrex dish on a steel rack of the same design as the larger one in Figure 12-6. Alternatively, a commercial Pyrex microdish with attached handle (A. H. Thomas Co.) can be housed in the 4-oz bottle. It needs no rack and is placed on a wire screen table in the bottle (see Fig. 6, Mintz, 1967b; this was mistakenly printed upside down). With this dish, an angle bend in the pipette tip is needed because of the high walls and small diameter.

Dessicators or anaerobic culture jars can also be employed as closed chambers, but they are relatively expensive and unnecessarily large for single cultures. The Dri-Jar (Scientific Glass Apparatus Co.) is one of the least expensive dessicators and has a convenient shelf rack on which the Petri or other type of dish can be suspended from the lid.

During incubation of the paired embryos, the blastomeres migrate within each aggregate. With time-lapse photography under high-power magnification at 37°C, we have observed the formation of pseudopodia and what was apparently random cell movement. Each mass of cells changes progressively from an elongate to a spherical shape within a few hours after aggregation (see Fig. 12-2), and development continues as though no experimental intervention had occurred. Under the conditions described here, most allophenic and control embryos explanted on Day 2 reach the blastocyst stage after one day in culture; the remaining embryos are usually advanced morulae, thus maintaining the normal developmental rate. The cultures are best terminated at this time. The blastocysts do not continue their development normally *in vitro,* but finally collapse onto the glass and, in our medium, the cells attach to the substratum and form cultures with some giant cells and droplets of extracellular material. These cultures, first described elsewhere (Mintz, 1964a), are presumably trophoblastic and may survive for long periods.

SURGICAL TRANSFER OF EMBRYOS

The allophenic embryos must now be returned to an *in vivo* environment. A pseudopregnant, rather than a pregnant, recipient is used in order to avoid having to distinguish native from alien offspring. The female is not artificially primed with hormones. She is mated to a vasectomized male of known sterility who is also not hormone-treated. Females that have previously had one litter may be superior recipients. We have found it somewhat impractical to have an adequate constant supply of uniparous females and we have generally used nulliparous ones, with excellent results. Of all our allophenic embryos thus far transferred to pseudopregnant hosts, 1,014 have survived to birth and beyond. Viability has ranged from a low of 13% in a few genotypic combinations to much higher levels in most of the genotypic series, with the best yields close to 50% in each of several large, recent series. These figures are not corrected for incidence of infertile matings in control females of each strain; corrected survival rates of allophenic embryos would of course be higher than those stated here, and are therefore now over 50%.

Any strain of female can be employed as an incubator host; we prefer ICR randombred females because they normally produce, and successfully raise, large litters of their own. Some special strain of host may, however, be required for a particular investigation. Thus in some of our experiments on gene control of neoplasias caused by maternally transmitted viruses, the allophenic embryos with cells of both the susceptible and resistant strains were transferred to a virus-positive incubator mother of the susceptible strain (Mintz, 1970a).

The males may all be of a single strain; we use ICR males, though other strains with normally good breeding performance are satisfactory. The male is vasectomized through a small longitudinal incision on each side of the posterior ventral abdominal wall. The entire complex of testis and associated ducts is drawn through the incision and two silk ligatures are placed around each vas deferens, one near the cranial and one near the caudal level of the adjacent epididymis. A section of the vas is then excised between the ligatures and the cut ends are turned slightly away from each other. The muscular body wall is closed with sutures and the skin is closed with wound clips (Mintz, 1967b). The males may be test-mated to verify sterility. We have never encountered any instances of functional regeneration of ducts.

Embryos explanted on Day 2, which become blastocysts or morulae after one day in culture under the conditions described here, should then be transferred to the uterus of a Day 2 pseudopregnant recipient. This slight disparity in the age of the embryos and host allows for any developmental lag that may result from the inevitable cooling of the embryos during transfer. As already pointed out in a detailed discussion of appropriate recipient stages and transfer sites in relation to various embryo stages (Mintz, 1967b), developmental synchrony between host and donor is the ideal arrangement but this requires empirical tests under actual conditions. We usually transfer into one recipient approximately 1 to $1\frac{1}{2}$ times the number of embryos born, on the average, to that host strain. (Our ICR subline has an average litter size of eleven.) Both uterine horns of the host are used, with about half the allophenic embryos put into each horn.

The prospective recipient is anesthetized, e.g., with intraperitoneally injected sodium pentobarbital (Diabutal, Diamond Laboratories, Inc., Des Moines), or with ether or some other commonly used anesthetic. She need not be tied to an operating board; shaving the operative site is also unnecessary. The dorsal approach is used. The operative area is swabbed

FIGURE 12-8

*Pipette for surgical egg transfer. The hollow Chambers holder
(A) has a conical threaded nut (B) containing a one-hole
rubber compression ring (C). A fine capillary with an angle
bend (D) is tapered at the upper end. The one-hole stopper
(E) holds the transfer pipette in a test tube for storage.*

with 70% alcohol and the wet hairs are evenly parted. A short longitudinal
incision is made through the skin over one ovary at a time. The ovary can
be recognized through the overlying muscular body wall. The wall is then
cut over the ovary, the ovarian fat pad is grasped with blunt forceps, and
the attached oviduct and anterior uterine horn are drawn out without
twisting them. They are laid carefully on the surface, to which they adhere
adequately to keep them in place. The thicker part of the mesentery at the
uterotubal junction is held with fine watchmaker forceps and a small hole is
made in the anterior end of the uterus with a sharply pointed, short-beveled
needle. The embryos are introduced into the uterine horn in a small
amount of culture fluid, without air bubbles, from a pipette placed in the
hole, pointing posteriorly. The tract on that side is then carefully eased
back into its original position by pulling the sides of the incision up over it;
the muscle wall is closed with silk sutures; and the skin is closed with
Michel-type wound clips (e.g., Autoclips, 9 mm, with automatic dispenser,
from Clay-Adams, Inc.). No sterile precautions need be taken during the
surgery; we have never encountered any subsequent infections.

Our transfer pipette (see Fig. 12-8) is based on the principle of Holter's
(1943) braking pipette. A thin-walled capillary is pulled from Pyrex glass;
a gradual taper at the upper end serves as a braking mechanism and permits
very small changes in fluid volume. The capillary is about 9 cm long, with
an internal diameter slightly larger than an allophenic blastocyst, and the
lower end has an angle bend and a straight, fire-polished tip. The capillary
is housed in a Chambers instrument holder with a removable head (Brink-
mann Instruments). A small rubber compression ring inside the conical
threaded nut of the holder keeps the capillary in place without air leakage.
The pipette is operated by means of a rubber mouth tube. A one-hole
rubber stopper holds the pipette in a test tube for storage.

Allophenic mice are born at the normal time and are almost always well
cared for by their foster mothers. Neonatal deaths or lack of maternal care
have been rare in our experience and have occurred only in some of the
small litters; inadequate maternal lactation may be involved. On Day 19 of
pregnancy, if the foster mother looks as though she has no more than two
fetuses, she is either caged with another pregnant female or the young are
delivered by Caesarean section and transferred to a mother with a litter
born the same or the preceding day. The young animals in one cage should
of course be of the same genotypic provenance (e.g., the same allophenic
combination), or readily distinguishable (e.g., by coat color), or else toe-

marked for identification. For Caesarean delivery, the host is killed and not anesthetized. The uterus is removed and cut open and each fetus is eased out onto moist paper toweling. The umbilical cord is pinched between fine forceps held close to the body and is then cut distal to the forceps; the amniotic covering is removed. The animal should not be handled directly; it should be rolled promptly onto a dry sheet (e.g., of Saran) under a warm lamp. It is then intermittently rolled back and forth by cradling the sheet, until breathing is normal and the skin is bright red.

Two points must be strongly emphasized in all allophenic work. The first, mentioned earlier, is that in any experiment, regardless of the genotypes initially combined, some non-mosaics are obtained. Extensive tissue tests with many genetic markers have confirmed this (Mintz, 1970a). Positive evidence of mosaicism must therefore be sought in each case. This is a complicating factor in many types of experiments. The reason for non-mosaicism may be chiefly the derivation of the embryo itself from only a few cells of the blastocyst (Mintz, 1970d); the haphazard distribution of the component genotypes (Mintz, 1964a) may cause some individuals to arise entirely from genetically identical cells. The other noteworthy point is that in the true mosaics the proportions and tissue distributions of the two cellular genotypes vary widely from one allophenic individual to another of the same paired combination (Mintz, 1967a, 1970a; Mintz and Palm, 1969). As a result, a broad spectrum of permutations and combinations is produced and these offer entirely novel experimental possibilities for the investigation of many basic biological problems.

A final comment which may be of general interest concerns the question whether allophenic individuals can ever be formed spontaneously *in vivo*. The answer appears to be clearly in the affirmative. A normal uterine factor capable of lysing the zona pellucida has been discovered in the mouse, and an *in vivo* bioassay has demonstrated that it reaches peak activity at the time of implantation, acting as an "implantation initiating factor" (Mintz, 1970e). In most instances, embryos are in the blastocyst stage when the product begins to have an appreciable effect on the zona pellucida. However, these studies have shown that in certain genetic strains (e.g., our Balb/c subline) a number of embryos are still in the morula stage when the uterine enzyme lyses their zonas. In such strains, there is a statistically predictable frequency with which two denuded morulae in the uterine fluid might stick to each other and remain together. The resultant allophenic embryo, in an inbred strain, might have sex chromosomal mosaicism, as in some experimental allophenics (Mintz, 1968). It seems possible that a comparable uterine factor may exist in the human and may have caused some of the clinical cases of genetic mosaicism in man (Mintz, 1970e).

ADAPTATIONS OF ALLOPHENIC EMBRYO TECHNIQUES FOR STUDY OF EARLY DEVELOPMENTAL CAPACITIES

As indicated above, rearranged blastomeres from two entire embryos can form one normal mouse with two genotypes of cells, even when the cells are rearranged as late as the inception of blastocyst cavitation. Such facts appear to demonstrate that the cells remain developmentally labile during this period (Mintz, 1962b,c; 1964a,b; 1965a,b), despite distinct progressive (non-regional) changes that are detectable in their nucleic acid and protein metabolism (Mintz, 1962b, 1964b, 1965c). Although previous workers had suggested that early mammalian blastomeres diverge in developmental capacities, their conclusion was based on negative evidence, namely, the non-development of some of the blastomeres that they isolated (Seidel, 1952, 1960; Tarkowski, 1959), or the failure of late-cleavage-stage aggregated egg pairs to become unified embryos (Tarkowski, 1961, 1963). In both instances, the results are attributable to technical difficulties, such as injury in the first example or the use of room temperature instead of 37°C in the latter.

Apart from our positive results in operational tests (the production of known mosaic adults from giant groups of artificially arranged blastomeres), we have conducted a variety of other experiments that directly demonstrate that blastomeres can be rearranged and that this nevertheless is followed by the formation of morphologically normal, albeit abnormal size, blastocysts. The experimental methods, which involve simple adaptations of the basic allophenic embryo techniques, fall into three general classes.

In the first, blastomeres that are radioactively labeled can be admixed with unlabeled blastomeres and the resultant blastocysts can be examined autoradiographically to determine if the components have come to occupy any consistent topographic location (see Fig. 12-9). Many different arrangements have in fact been seen in such experiments, indicating that the cells have no predetermined positions, and yet development continues normally. In some cases, all labeled cells, from one of the original embryo donors, ended up either outside or inside the composite blastocyst. Labeling is accomplished by incubating embryos in medium containing tritiated thymidine during the S-period, until all nuclei have their DNA labeled sufficiently to retain the label through mitoses into the blastocyst stage. The embryos are washed with unlabeled thymidine in the medium and are pronased; each is then adhered at 37°C to a denuded embryo without label. After the aggregates have become blastocysts *in vitro*, they are fixed, and paraffin sections are cut at 4μ. Sections are placed onto subbed slides which are deparaffinized in toluene, rinsed in absolute alcohol, and air dried. The slides are dipped in liquid emulsion (Ilford L4) and are exposed in the dark. Exposed slides are developed in amidol, rinsed, fixed in sodium hyposulfite, and washed in water; the emulsion is then hardened overnight in 70% ethyl alcohol. After rinsing, they can be stained in azure B, run quickly through water, and air dried. Further details of histological fixing and handling procedures adapted for mammalian embryos are given in the final section of this paper. For autoradiography, a number of alternative methods can be followed; those mentioned above, as well as other possible methods, can be found in Chapters 15 through 18 of Prescott (1964).

The second type of experiment involves a cytochemical rather than an isotopic marker. In the lethal mutant t^{12}/t^{12} homozygote, developmental arrest and death occur in the late morula stage, at a time when the cells are

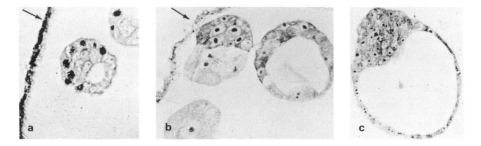

FIGURE 12-9

(a) *Allophenic blastocyst formed from unlabeled blastomeres combined with blastomeres labeled in the nuclei with* ^3H*-thymidine. Note ant pupa case wall (arrow)*. (b) *Allophenic* $t^{12}/t^{12} \leftrightarrow +/+$ *morula and blastocyst, with palely stained (azure B) mutant cells. (Arrow points to pupa case wall.)* (c) *Giant allophenic blastocyst from nine entire embryos, formed by aggregating three separate trios and then combining them into a single group.*

exceptionally low in cytoplasmic basophilia (Smith, 1956; Mintz, 1963, 1964c). Allophenic $t^{12}/t^{12} \leftrightarrow +/+$ embryos can be assembled and may become blastocysts; the mutant cells are clearly identifiable in stained sections (see Fig. 12-9b) and appear to be located in variable positions (Mintz, 1964a).

The third class of experiments involves the aggregation of normal cleavage-stage embryos in unusual arrangements and in increasing numbers. Specific experiments, detailed elsewhere (Mintz, 1965b), include using linear rather than spherical alignments of embryos, performing aggregations in two stages instead of one, and employing increasingly large numbers of embryos in a cluster. Huge blastocysts were obtained from masses of as many as sixteen entire embryos, and development occurred at the normal rate. The blastocyst shown in Figure 12-9c was formed from all the cells of nine mid-cleavage eggs, in two stages. They were first arranged in three trios, and, after each of the three had become a rounded mass, they were brought together into one group. These experiments testify dramatically to the developmental lability of the early mammalian embryo.

FIXATION AND HISTOLOGICAL PROCESSING OF PREIMPLANTATION EMBRYOS

Special fixing and handling methods are required for preimplantation mammalian embryos because of their smallness and fragility. Known procedures were found to be inadequate in many respects, and certain methods that have been introduced in our laboratory are useful not only for the experiments just outlined, but also for many other kinds of studies. The techniques (previously unpublished) are simple and rapid; they rely only on readily available materials, and give uniformly excellent results.

Beginning with the premise that there should be a free choice of fixatives (the choice being determined by the cytological problem at hand), the greatest need seemed to be an egg container that would be inert and yield good sections. Metz (1921) used the bag-like transparent covering of the abdomen of Drosophila midpupae as a holder. A large supply of Drosophila is required and preparation of the sacs is slow. More importantly, our results with such sacs were variable because the exchange of fluids through the bag occurs too slowly. In a similar approach, first pointed out to me by Christina M. Richards, Cooper and Ris (1943) used ant cocoons. These pupal cases are spun of fine silk that forms a microscopic mesh which is freely permeable, and the sacs occur in a great variety of sizes (see Fig. 12-9a,b,c). We have introduced a method in which ant cocoons are used and the procedures are modified to fit the special needs of mammalian embryo material.

Pupae are either purchased or collected in the field. A can of several thousand is sold cheaply in most pet shops under the misnomer "dried ant eggs" (as turtle, fish, and bird food). These are European species; smaller cocoons of even finer silk are to be found in many regions of the United States, and we have collected especially good quality sacs in sandy soil under stones and wood in New Jersey. They were identified by Marion R. Smith of the U.S. Department of Agriculture as the species *Acanthomyops claviger* (Rog.). Dirt is sieved out, the caudal tip of the cocoon is snipped off with scissors, and the pupa is pulled out with watchmaker forceps. The empty cases are washed in 70% alcohol; it is essential to inspect them carefully and to remove any attached glass-like particles. Preparation is rapid; the cleaned cases can be stored in alcohol indefinitely. Just before use, a sac, after a water rinse, is immersed in protein-containing culture medium and is agitated to replace the water in it with medium. Eggs in medium are pipetted into the sac and it is then lifted into the fixative by holding the lip with fine forceps. It is quickly but gently agitated in the fixative; most of the protein is shaken off, and the eggs are entrapped without distortion in a fine film of coagulated protein; thus, no closure of the sac is required. In all further handling, it is transferred with forceps held over the lip. We have also similarly processed a group of sacs of graduated sizes that telescoped into each other; experimental and control eggs were separated into the compartments thus formed. Thus, in a single slide of sections, it is possible to compare results of parts of an entire experiment within concentric pupa case rings with the assurance that all have been handled identically.

For fixation, we generally use cold glacial acetic-absolute alcohol (1:3) for about 10 minutes, followed by two changes of cold absolute alcohol (for a total of 30 minutes), toluene at room temperature (3 minutes), and two changes of paraffin (30 minutes' total for both) followed by embedding in paraffin.

BIBLIOGRAPHY

Baker, W. W., and B. Mintz (1969) Subunit structure and gene control of mouse NADP-malate dehydrogenase. Biochem. Genet., 2:351–360.

Brinster, R. L. (1963) A Method for *in vitro* cultivation of mouse ova from two-cell to blastocyst. Exp. Cell Res., 32:205–208.

Cooper, K. W., and H. Ris (1943) Handling small objects in bulk for cytological preparations. Stain Tech., 18:175–176.

Gillespie, D., and S. Spiegelman (1965) A quantitative assay for DNA-RNA hybrids with DNA immobilized on a membrane. J. Molec. Biol., 12:829–842.

Holter, H. (1943) Technique of the Cartesian diver. C. R. Trav. Lab. Carlsberg, Sér. Chim., 24:399–478.

Metz, C. W. (1921) A simple method for handling small objects in making microscopic preparations. Anat. Record, 21:373–374.

Mintz, B (1962a) Experimental study of the developing mammalian egg: removal of the zona pellucida. Science, 138:594–595.

——— (1962b) Formation of genotypically mosaic mouse embryos. Am. Zool., 2:432.

——— (1962c) Experimental recombination of cells in the developing mouse egg: normal and lethal mutant genotypes. Am. Zool., 2:541–542.

——— (1962d) Incorporation of nucleic acid and protein precursors by developing mouse eggs. Am. Zool., 2:432.

——— (1963) Growth *in vitro* of t^{12}/t^{12} lethal mutant mouse eggs. Am. Zool., 3:550–551.

——— (1964a) Formation of genetically mosaic mouse embryos, and early development of "lethal (t^{12}/t^{12})-normal" mosaics. J. Exp. Zool., 157:273–292.

——— (1964b) Synthetic processes and early development in the mammalian egg. J. Exp. Zool., 157:85–100.

——— (1964c) Gene expression in the morula stage of mouse embryos, as observed during development of t^{12}/t^{12} lethal mutants *in vitro*. J. Exp. Zool., 157:267–272.

——— (1965a) Genetic mosaicism in adult mice of quadriparental lineage. Science, 148:1232–1233.

——— (1965b) Experimental genetic mosaicism in the mouse. In: Ciba Foundation Symposium on Preimplantation Stages of Pregnancy. Ed. by G. E. W. Wolstenholme and M. O'Connor. Churchill, London, pp. 194–207.

——— (1965c) Nucleic acid and protein synthesis in the developing mouse embryo. In: Ciba Foundation Symposium on Preimplantation Stages of Pregnancy. Ed. by G. E. W. Wolstenholme and M. O'Connor. Churchill, London, pp. 145–155.

——— (1967a) Gene control of mammalian pigmentary differentiation. I. Clonal origin of melanocytes. Proc. Nat. Acad. Sci., 58: 344–351.

——— (1967b) Mammalian embryo culture. In: Methods in Developmental Biology. Ed. by F. Wilt and N. Wessells. Crowell, New York, pp. 379–400.

——— (1968) Hermaphroditism, sex chromosomal mosaicism, and germ cell selection in allophenic mice. J. Animal Sci., 27 (Suppl. I) :51–60.

——— (1969a) Developmental mechanisms found in allophenic mice with sex chromosomal and pigmentary mosaicism. Birth Defects: Original Article Series, 5:11–22.

——— (1969b) Gene control of the mouse pigmentary system. Genetics, 61 (Suppl.) :41.

——— (1970a) Neoplasia and gene activity in allophenic mice. In: Genetic Concepts and Neoplasia, 23rd Annual Symposium on Fundamental Cancer Research, University of Texas M. D. Anderson Hospital and Tumor Institute. Williams and Wilkins Co., Baltimore, pp. 477–517.

——— (1970b) Do cells fuse *in vivo*? Advances in Tissue Culture, "In Vitro," 5:40–47.

———— (1970c) Gene control of differentiation of the mouse melanocyte system. J. Investig. Dermatol., 54:93.

———— (1970d) Gene expression in allophenic mice. In: Control Mechanisms in Expression of Cellular Phenotypes, Symposium of the International Society for Cell Biology. Ed. by H. Padykula. Academic Press, New York. Vol. 9:15–42.

———— (1970e) Control of embryo implantation and survival. In: Schering Symposium on Intrinsic and Extrinsic Factors in Early Mammalian Development. Ed. by G. Raspé, Pergamon Press, Oxford. In Press.

———— (1970f) Genetic mosaicism in vivo: Development and disease in allophenic mice. Federation Proceedings. In press.

———— and W. W. Baker (1967) Normal mammalian muscle differentiation and gene control of isocitrate dehydrogenase synthesis. Proc. Nat. Acad. Sci., 58:592–598.

———— and J. Palm (1965) Erythrocyte mosaicism and immunological tolerance in mice from aggregated eggs. J. Cell Biol., 27:66A.

———— and J. Palm (1969) Gene control of hematopoiesis. I. Erythrocyte mosaicism and permanent immunological tolerance in allophenic mice. J. Exp. Med., 129:1013–1027.

———— and S. Sanyal (1970) Clonal origin of the mouse visual retina mapped from genetically mosaic eyes. Genetics. 64 (Suppl.):43–44.

———— and W. K. Silvers (1967) "Intrinsic" immunological tolerance in allophenic mice. Science, 158:1484–1487.

———— and W. K. Silvers (1970) Histocompatibility antigens on melanoblasts and hair-follicle cells: Cell-localized homograft rejection in allophenic skin grafts. Transplantation, 9:497–505.

———— and G. Slemmer (1969) Gene control of neoplasia. I. Genotypic mosaicism in normal and preneoplastic mammary glands of allophenic mice. J. Nat. Canc. Inst., 43:87–95.

Morgan, T. H. (1927) Experimental Embryology. Columbia University Press, New York, 766p.

Mystkowska, E. T., and A. K. Tarkowski (1968) Observations on CBA-p/CBA-T6T6 mouse chimeras. J. Embryol. Exp. Morphol., 20:33–52.

Nicholas, J. S., and B. V. Hall (1942) Experiments on developing rats. II. The development of isolated blastomeres and fused eggs. J. Exp. Zool., 90:441–459.

Nomoto, M., and Y. Narahashi (1959) A proteolytic enzyme of Streptomyces griseus. I. Purification of a protease of Streptomyces griseus. J. Biochem. (Japan), 46:653–667.

Noyes, R. W., L. L. Doyle, A. H. Gates, and D. L. Bentley (1961) Ovular maturation and fetal development. Fertility and Sterility, 12:405–416.

Owen, R. D. (1945) Immunogenetic consequences of vascular anastomoses between bovine twins. Science, 102:400–401.

Prescott, D. M. (Ed.) (1964) Methods in cell physiology. Vol. 1, pp. 305–379. Academic Press, New York.

Runner, M. N., and A. Gates (1954) Conception in prepuberal mice following artificially induced ovulation and mating. Nature, 174:222–223.

Satake, K., S. Kurioka, and T. Neyasaki (1961) A limited proteolysis of native ovalbumin with a proteinase from Streptomyces griseus. J. Biochem. (Japan). 50:95–101.

Seidel, F. (1952) Die Entwicklungspotenzen einer isolierten Blastomere des zweizellstadiums in Säugetierei. Naturwiss., 39:355–356.

———— (1960) Die Entwicklungsfähigkeiten isolierter Furchungszellen aus dem Ei des Kaninchens Oryctolagus cuniculus. Roux' Arch. f. Entw.-Mech., 152:43–130.

Smith, L. J. (1956) A morphological and histochemical investigation of a preimplantation lethal (t^{12}) in the house mouse. J. Exp. Zool., 132:51–83.

Tarkowski, A. K. (1959) Experimental studies on regulation in the development of isolated blastomeres of mouse eggs. Acta Theriolog., 3:191–267.

—— (1961) Mouse chimaeras developed from fused eggs. Nature, 190:857–860.

—— (1963) Studies on mouse chimeras developed from eggs fused *in vitro*. Nat. Canc. Inst. Monogr., 11:51–67.

Whitten, W. K. (1957) Culture of tubal ova. Nature, 179:1081–1082.

13

Measuring Embryonic Enzyme Activity

RALPH L. BRINSTER
Laboratory of Reproductive Physiology
Department of Animal Biology
School of Veterinary Medicine
University of Pennsylvania
Philadelphia, Pennsylvania 19104

Enzyme activity measurements have been made only recently on preimplantation mammalian embryos. The small amount of data thus obtained stands in marked contrast to the wealth of information available on enzyme activity in the embryos of invertebrates and amphibians. The lack of information about mammalian embryos stems from the almost miniscule amount of embryonic tissue that is available for analysis. In many invertebrate species it is possible to collect several grams of ova from a single individual, but most mammals produce only a few ova at each ovulation. Although it is possible to superovulate mammals, the maximum number of ova obtained from one individual is between 100 and 200, and the average from each animal is generally 30 or lower.

The small numbers of embryos, combined with the small size of the individual embryo, provide very little tissue for the enzyme assay. The zona-free one-cell ovum of the mouse has a wet weight of 178 mμg (Loewenstein and Cohen, 1964) and a protein content of 27.77 mμg (Brinster, 1967a); the protein content is between 20 and 28 mμg during the entire preimplantation period. Although the mouse ovum is one of the smallest of the mammalian ova, the differences among the eutherian mammals are small. However, in some species, such as the rabbit, blastocyst formation is

The work discussed here has been supported in part by the National Institutes of Health (HD-03071), the National Science Foundation (GB-4465) and the Pennsylvania Department of Agriculture (ME-8).

followed by a rapid and marked increase in size. In the six-day rabbit embryo, for example, the wet weight just prior to implantation, may reach 50 to 100 mg (Lutwak-Mann, 1962). In these species enzyme determinations using later stages, in which large quantities of tissue are available, are straightforward and offer few problems.

PREPARATION OF EMBRYOS

Cumulus Removal

Embryos for enzyme analyses are obtained by flushing them from the reproductive tract (Brinster, 1967a, 1968c); oocytes are obtained by dissection of mature ovarian follicles (Brinster, 1968a). One-cell embryos and oocytes are surrounded by cumulus cells and corona radiata cells, which must be removed if accurate activity measurements are to be made. Cumulus cells are easily removed by hyaluronidase at a concentration of 300 units/ml (Rowlands, 1944; Brinster, 1965a). The hyaluronidase can be dissolved in phosphate-buffered saline (Parker, 1961) or in Brinster's Media for Ovum Culture, shown in Table 13-1 (Brinster, 1965b).

Hyaluronidase will remove the corona cells as well as the cumulus of some ova. However, in some species, such as the rabbit, the corona cells do not come off the zona pellucida with hyaluronidase treatment. In these

TABLE 13-1

Brinster's medium for ovum culture plus glucose

Component	mM	g/l	ml of 0.154M stock in 13 ml
NaCl	94.88	5.546	5.90
Na-Lactate	25.00	2.253[1]	2.10
Na-Pyruvate	0.25	0.028	2.10[2]
KCl	4.78	0.356	0.40
CaCl$_2$	1.71	0.189	0.20[3]
KH$_2$PO$_4$	1.19	0.162	0.10
MgSO$_4 \cdot$ 7H$_2$O	1.19	0.294	0.10
NaHCO$_3$	25.00	2.106	2.10
Penicillin/strep	100 I.U./ml of pen, 50 μg/ml of strep		—
Bovine serum albumin	1 mg/ml	1.000	—
Glucose	1 mg/ml	—	—

Source: Brinster (1965).
[1] NaLac added as liquid prepared as follows: Add 1.82 ml of concentrated lactic acid (85 to 90%) to 200 ml of double distilled H$_2$O. Neutralize to pH 7.4 (about 15 to 20 ml of 1N NaOH). This is enough lactic acid to make one liter of medium.
[2] NaPyr stock is 0.00154M pyruvate in 0.154M NaCl (17 mg/100 ml).
[3] CaCl$_2$ stock is 0.11M.

cases, it is necessary to employ mechanical means such as shaking the embryos in a serological test tube (Bedford and Chang, 1962) or drawing the embryos back and forth in a pipette with a diameter slightly larger than the ovum (Brinster, 1968a).

Embryo Coverings

To remove the zona pellucida of the ovum or embryo, use pronase at a concentration of 2.5mg/ml (Mintz, 1962; Brinster, 1965a; Chap. 12 above by Mintz). Removal of the zona pellucida had no effect on the activity of lactate dehydrogenase, malate dehydrogenase, and glucose-6-phosphate dehydrogenase in the mouse ovum, and of lactate dehydrogenase in the rabbit ovum (Brinster, 1965a; 1966a, b; 1967b).

In the rabbit, the embryo is covered with a thick layer of mucin in the fallopian tube. This can be removed by pronase solution also. Lactate dehydrogenase activity in the rabbit embryo does not appear to be affected by removal of the mucin layer (Brinster, 1967c), but, if the embryonic cells are incorporated into the mucin layer as it is formed, then the mucin must be removed to obtain accurate measurements.

Washing the Embryos

Before enzyme measurements can be made, all reproductive tract fluids and cellular debris must be removed from the medium containing the embryos, and, if the embryos have been exposed to enzyme treatment, the enzymes must also be removed. The embryos are usually washed at least three times before being placed in tubes for analyses. The wash medium, which contains no energy substrates, is shown in Table 13-2.

To wash the embryos, pick them up in as small a volume as possible (several hundred mouse embryos can be picked up in 20 to 30 μl) and place

TABLE 13-2
Medium for washing embryos

Component	g/l	mM
NaCl	6.975	119.32
KCl	0.356	4.78
CaCl$_2$	0.189	1.71
KH$_2$PO$_4$	0.162	1.19
MgSO$_4$ · 7H$_2$O	0.294	1.19
NaHCO$_3$	2.106	25.07
Penicillin-G (potassium)	100 U/ml	—
Streptomycin sulfate	50 μg/ml	—
Crystalline bovine serum albumin	1.000	—

them in 3 ml of the wash medium. The embryos are picked up with finely-drawn Pasteur pipettes with a tip diameter slightly larger than the diameter of the embryo. The tip should be flat, at a 90° angle with the stem, in order to pick up the embryos in a minimum of fluid. After each transfer, the embryos and transferred fluid are dispersed to assure mixing, and the embryos are then reaggregated in the center for the next transfer. Washing is performed at a temperature of 32° to 37°C. After the last wash, the embryos are picked up in 3 to 5 μl of fluid and are placed in the bottom of a 6×50 mm tube. After each transfer, the pipette is examined to make sure that all the embryos have been transferred to the tube.

Freezing the Embryos

After the embryos have been placed in the tube, it is stored at $-70°$C until the assay. Since some enzymes lose activity when stored, it is important to determine this loss for each enzyme. Lactate dehydrogenase in the mouse embryo is stable for at least a month. In general, the activity is measured as soon as possible after freezing.

A second function of freezing is the disruption of cell membranes, thus allowing the escape of cell enzymes. Enzymes are bound in various ways within the cell and hence are liberated into solution under different conditions. One might expect that alternate freezing and thawing would increase the disruption of the cell and thereby increase enzyme activity, but freezing and thawing the embryos as many as five times did not increase either lactate or malate dehydrogenase activity any more than a single freezing and thawing (Brinster, 1965a, 1966a). However, the multiple freezing and thawing did decrease the activity of glucose-6-phosphate dehydrogenase (Brinster, 1966b); this decrease probably resulted from denaturing of the enzyme. Protection from loss of enzyme activity is important not only during the manipulations of the living egg but also during storage.

SPECTROPHOTOMETRY

General Assay Procedures

Many methods can be used to assay enzyme activity, but the two most versatile, and the two which provide the most accurate measurement of small amounts of tissue, are spectrophotometry and fluorometry. Most enzyme activities can be measured by determining changes in the level of one of the pyridine nucleotides NAD, NADH, NADP, or NADPH. These nucleotides may be directly involved in the reaction catalyzed by the enzyme, or the enzyme-catalyzed reaction is coupled to a pyridine nucleotide-dependent reaction. For example, hexokinase catalyzes a reaction not involving a pyridine nucleotide:

$$\text{Glucose} + \text{ATP} \xrightarrow[\text{Hexokinase}]{\text{Mg}^{++}} \text{Glucose-6-phosphate} + \text{ADP}$$

The activity of hexokinase can be determined by measuring the amount of NADPH formed (this is equivalent per molecule to the glucose-6-phosphate formed) when the following reaction is coupled to the hexokinase reaction.

Glucose-6-phosphate	Mg^{++}	6-phosphogluconate
+	$\xrightarrow{\hspace{2cm}}$ $\xleftarrow{\hspace{2cm}}$	+
NADP	Glucose-6-phosphate dehydrogenase	NADPH

To perform the assay, add everything but the hexokinase, Glucose-6-phosphate, and the NADPH to the ovum, then measure the NADPH produced per unit of time. This particular assay will be discussed in greater detail later.

In spectrophotometric measurements of enzyme activity in mammalian embryos, the amount of NADH or NADPH oxidized or the amount of NAD or NADP reduced may be determined. In either case, the optical density at a wave length of 340 mμ determines the quantity of reduced pyridine nucleotide present at any given time. Since the extinction coefficient is 6.22×10^3 cm^2/m mole, the limit of sensitivity, using 1-ml cuvettes with a 1-cm light path, is approximately 10^{-8} moles. This corresponds to an optical density reading of 0.062.

Because one is interested in measuring the initial, maximum velocity, it is essential that the changes in optical density with time be linear and that they be taken from the first part of the reaction rate curve. The time during which the reaction rate will be linear varies with the enzyme and assay conditions, but it is important to determine this time in preliminary experiments, because it will fix the upper limit of the time over which changes in optical density may be used to calculate activity. It is best to use the shortest possible time. Most enzyme assay procedures have been performed on several tissues, and the optimum assay conditions are recorded in the literature. Since these conditions may vary slightly among different tissues and among different species, optimum assay conditions should be ascertained for each enzyme to be studied in the mammalian embryo. When assaying a new tissue, the best method is to start with the known optimum conditions for a similar tissue and vary each reactant until maximum rates are achieved.

Two of the most important assay variables are temperature and pH. If the temperature of the reaction mixture can be kept at 37°C rather than 22° or 23°C (room temperature), the reaction rate is doubled for most enzymes; this increases the sensitivity two-fold, and the assay is performed at a physiological temperature. Although the physiological pH range is usually 7.4 to 7.5, it is often found that maximum rates for a given assay are achieved at another pH (Long, 1961). Thus the optimum pH for maximum activity rate should be determined experimentally for each enzyme. In those reactions where hydrogen ions enter into the reaction as a reactant or product, pH will have a direct and marked effect on the reaction rate.

MACROASSAY METHOD

Lactate dehydrogenase (LDH) is an important enzyme in the early mouse embryo (Brinster, 1965a). It catalyzes

$$\text{Pyruvate} + \text{NADH} \underset{\text{LDH}}{\overset{\longrightarrow}{\longleftarrow}} \text{Lactate} + \text{NAD} + \text{H}^+$$

The activity of LDH can be measured by adding pyruvate and NADH to a buffered solution (pH 7.5) containing the dispersed embryo and observing the decrease in optical density due to the oxidation of NADH. Alternatively lactate and NAD may be added to a buffered solution (pH 11) containing the embryo and the reduction of NAD is measured. Both methods have been used for the mouse embryo, but the second method gives slightly lower activities (Brinster, unpublished).

The optimum method for assaying LDH activity spectrophotometrically is from the rate of oxidation of NADH at 37°C. Optical density readings are taken at 3 minute intervals following a 15 minute preincubation of the mixture of embryo, buffer, and NADH. The preincubation assures equilibration at 37°C and indicates the rate of NADH oxidation in the absence of pyruvate. The reaction mixture consists of 1 ml of 50 mM phosphate buffer, 0.1 μmole of NADH, and 1.0 μmole of sodium pyruvate. The reaction is initiated by the addition of pyruvate and the decrease in absorbence due to NADH oxidation during the first 15 minutes is used to calculate the activity of the enzyme. Table 13-3 shows a typical assay.

If the net optical density (O.D.) change is less than 0.045 after the initial 15 minutes, even though the change is linear, it is advisable to increase the O.D. change per unit time either by increasing the number of embryos per assay cuvette or by decreasing the assay volume. The limited number of mammalian embryos available sets restrictive limits for the first of these alternatives. However, it is possible to reduce the reaction mixture volume to 50 μl with only moderate modifications; the volume may be decreased even further with more elaborate modification. Glick (1963) has reviewed a number of methods for increasing the sensitivity of spectrophotometric methods.

MICROASSAY METHOD

To determine LDH activity in volumes down to 50 μl, with a 1-cm light path, optical density readings are made in my laboratory with a Beckman model DU spectrophotometer, which is modified with a precision cell positioner and microcells (Greenberg and Rodder, 1964). Although most spectrophotometers can be modified to make readings on 50 μl with a 1-cm light path, the most critical requirement is accurate and repeatable sample positioning. The reaction mixture consists of 50 μl of 50 mM phosphate buffer which contains 0.5 mg/ml crystalline bovine serum albumin, 5 mμmole of NADH and 50 mμmole of sodium pyruvate. The reaction is initiated by the addition of pyruvate, and the preincubation and calcula-

TABLE 13-3

Spectrophotometer readings in assay for lactate dehydrogenase activity

Cuvette Contents	Preincubation[1]		Assay Incubation[1]						Optical Density Change[2]	
	0	900	0	180	360	540	720	900	Total	Net
1. Phosphate buffer (PB)	0	0	0	0	0	0	0	0	0	—
2. PB + embryo + NADH	0.675	0.673	0.680	0.665	0.650	0.635	0.625	0.610	−0.070	−0.068
3. PB + embryo + NADH	0.680	0.675	0.680	0.665	0.650	0.635	0.620	0.605	−0.075	−0.073
4. PB + embryo + NADH	0.655	0.655	0.660	0.660	0.659	0.659	0.659	0.658	−0.002	—

[1] All times are in seconds.
[2] Numbers are optical density at 340 mμ. Readings were made on a Beckman model DU spectrophotometer. The light path was 1 cm; slit width, 0.6 mm. Pyruvate is added to cuvettes 1 to 3 after 15 minutes preincubation. Cuvette 4 is the control.

tion of activity is the same as described previously. Using the 50 μl method, changes of the order of (3 to 5) $\times$ 10^{-10} moles of reduced nucleotide per 15 minutes can be determined. Most mammalian embryos thus examined have LDH activity greater than this.

Either the macro (1 ml) or micro (50 to 100 μl) method can be used to assay enzyme activity in the mammalian embryo. So far, however, only LDH activity in mouse embryos has been sufficiently high to make measurement by the macromethod feasible. LDH in other species and the other enzymes in the mouse which have been measured have been determined using the micromethod. The only exception to this has been hexokinase, which was analyzed fluorometrically.

FLUOROMETRY

GENERAL ASSAY PROCEDURES

By using fluorometric techniques it is possible to extend the range of sensitivity appreciably beyond that of spectrophotometry. Whereas the present practical limit of sensitivity of spectrophotometric measurement of NADH or NADPH is of the order of 10^{-5} moles per liter or 5×10^{-10} moles total, the present practical limit of fluorometric measurement of reduced or oxidized pyridine nucleotide is of the order of 10^{-8} moles per liter, or 10^{-12} moles total. Thus, there is considerable advantage in using fluorometry to measure enzyme activity in mammalian embryos.

The fluorometric technique is quite simple; the principles on which the analyses are based follow. Di- and triphosphopyridine nucleotide (NAD and NADP) do not fluoresce and are destroyed by weak alkali, but with strong alkali they develop marked fluorescence. NADH and NADPH do fluoresce and are destroyed by hydrochloric acid, but their native fluorescence is only one-tenth that resulting from their oxidation to NAD and NADP by treatment with strong alkali. The various methods used have been described by a number of workers (Lowry, Roberts, and Kapphahn, 1957; Lowry, Passonneau, and Rock, 1961; Glick, 1961); only a brief account will be presented here.

To analyze for NAD or NADP, the sample is made 0.2N in HCl for at least 30 seconds to destroy any reduced nucleotide. Strong NaOH is added to give a final concentration of 6N and the sample is heated to 60°C for 15 minutes. It is then diluted at least five-fold with water and read in a fluorometer: the nucleotides have an excitation maxima at 340 mμ and a fluorescence maxima at 456 mμ (Udenfriend, 1962). The final nucleotide solution may be from 10^{-8} to 10^{-5} M. Above 10^{-5} M, the fluorescence is no longer proportional to concentration, but these higher concentrations can be read in the spectrophotometer or diluted.

To analyze the native fluorescence of NADH or NADPH is quite simple. The sample is diluted to a concentration below 10^{-5} M in a buffered solution having a pH between 8 and 10. If Mg^{++} is present in a concentration greater than 10^{-5} M and the pH is greater than 11, the magnesium should be removed with EDTA (ethylenediaminetetraacetate).

To analyze for NADH or NADPH indirectly, by the strong alkali method, the sample is made 0.02 to 0.04N in NaOH and heated for 15 minutes at 60°C to destroy any NAD or NADP that may be present. NaOH and H_2O_2 are then added to give a final concentration of 6N NaOH and 0.03% H_2O_2. The alkaline peroxide is prepared from aqueous 3% H_2O_2, which is prepared each day from 30% H_2O_2. The alkaline peroxide solution is not stable and must be prepared within an hour of its use. After the alkali has been added, the sample is heated to 60°C for 15 minutes. It is then diluted at least five-fold and read in the fluorometer.

The volumes used in the preceding fluorometric procedures are arbitrary and depend on the experimental enzyme assay conditions which precede the analysis for the nucleotide. It is essential that blanks and standards be carried not only through all steps of the enzyme reaction, but also through all steps of the analysis. One of the problems encountered in fluorescence work is the high degree of background fluorescence of reagents and enzyme preparations. It is essential to minimize this background fluorescence, and consequently, good quality double-distilled H_2O, clean glassware, good quality reagents, and extreme care in procedures are desirable.

DIRECT ASSAY METHOD

Table 13-4 is an example of a typical analysis for glucose-6-phosphate dehydrogenase activity in a mammalian embryo. In the example only

TABLE 13-4

Fluorometer readings in assay for
glucose-6-phosphate dehydrogenase activity

Tube Contents	Fluorometer Reading[1]	
	×3 Sensitivity	×10 Sensitivity
1. Opaque blank	0	—
2. H_2O	16	—
3. Tris buffer	17	—
4. Tris + NADP	41	—
5. Tris + NADP + G–6–P	42	0
6. Tris + NADP + ½ mouse embryo	42	0
7. #5 + 1.25 × 10^{-10} moles NADPH	—	21
8. #5 + 2.50 × 10^{-10} moles NADPH	—	44
9. #5 + 5.00 × 10^{-10} moles NADPH	—	82
10. #5 + ¼ mouse embryo	—	25
11. #5 + ½ mouse embryo	—	51

[1] Readings for blanks were made at ×3 sensitivity set, then the reagent blank was set to zero at ×10 sensitivity to read standards and embryos. This increases the separation of the standards and increases the accuracy of readings.

two samples contain embryos; generally a larger number are used during the analysis. In fact, if different developmental stages are being compared, it is desirable to run all the stages in each assay group. The incubation mixture consists of 50 μl of Tris buffer, 50 mM, pH 7.8, containing 50 mμmoles of NADP. To this is added 50 mμmoles, in 5 μl, of glucose-6-phosphate (G-6-P). All tubes are kept in an ice bath and, immediately after the addition of the G-6-P to the tubes containing embryos, all the tubes are transferred to a water bath at 37°C for 15 minutes. The important blanks are number 4 (Tris + NADP + G-6-P) and number 5 (Tris + NADP + embryo). The standards should cover the expected range of NADPH production by the embryos in steps of two-fold concentration increase. Natural fluorescence of NADPH was read in a Turner model 111 fluorometer and the final volume in the fluorometer tube was 4 ml. The sensitivity settings are made in arbitrary units.

Assay after Enzymatic Cycling

The sensitivity of the assay may be increased considerably by enzymatic cycling of the pyridine nucleotide (Lowry, Passonneau, Schulz, and Rock, 1961). In this method, the nucleotide produced by the enzyme reaction is multiplied by a cyclic process. The system used for NADP or NADPH measurement uses glucose-6-phosphate dehydrogenase and glutamate dehydrogenase:

$$\text{NADPH} + \alpha\text{-ketoglutarate} + \text{NH}_4^+ \rightarrow \text{NADP} + \text{glutamate}$$

$$\text{NADP} + \text{glucose-6-P} \rightarrow \text{NADPH} + \text{6-phosphogluconate}$$

Each molecule of NADP catalyzes the formation of 5000 to 10,000 molecules of 6-phosphogluconate in 30 minutes. The 6-phosphogluconate is then measured in a second incubation with 6-phosphogluconate dehydrogenase and extra NADP. The NADPH formed by this last reaction is determined fluorometrically, either by native fluorescence or after treatment with alkaline peroxide as described previously.

When the original enzyme reaction in the embryo tissue produces NAD or NADH, the nucleotide is multiplied with lactate dehydrogenase and glutamate dehydrogenase:

$$\text{NADH} + \alpha\text{-ketoglutarate} + \text{NH}_4^+ \rightarrow \text{NAD} + \text{glutamate}$$

$$\text{NAD} + \text{lactate} \rightarrow \text{NADH} + \text{pyruvate}$$

Pyruvate is produced in a 2500-fold yield in 30 minutes and is measured in a second step with added NADH and lactate dehydrogenase. The resulting NAD is measured fluorometrically.

The cycling method can be used to measure either the oxidized or the reduced form of the nucleotide in concentrations as low as 10^{-9} M and in amounts as small as 10^{-15} moles. In measuring NAD or NADP, any NADH and NADPH, can be destroyed by a brief acid treatment (as described above, see p. 222). In measuring NADH or NADPH, mild alkaline treatment is used to destroy NAD and NADP. Interfering enzymes can be destroyed by acid treatment at 100°C. Recently these methods have been used to measure hexokinase in the preimplantation mouse embryo (Brinster, 1968b). The procedure used will be described as an example of the technique; the coupling of the hexokinase reaction to the glucose-6-phosphate dehydrogenase reaction for analyses has been described earlier. In the following, no glucose-6-phosphate dehydrogenase is added to the reaction mixture since the embryo contains high levels of this enzyme (Brinster, 1966b).

Place 50 μl of 50 mM Tris buffer (pH 7.7) containing 20 mM MgCl$_2$, 1 mM glucose, 1 mM ATP, 0.1 mM NAD, and 0.5 mg/ml bovine serum albumin into a 6 × 60 mm tube containing the embryos, in an ice bath. After the addition of this reaction mixture, place the tubes in a water bath at 37°C for 15 minutes, return it to the ice bath; then add 100 μl of 0.16 N NaOH to stop the reaction and destroy the NAD. The NAD was destroyed by heating the alkaline mixture to 60°C for 15 minutes. Twenty μl of this mixture were placed in 100 μl of the cycling reagent described by Lowry et al. (1961) for NADH determination, and incubated as described, at 38°C for 30 minutes.

The cycling mixture is made in 0.1 M Tris buffer, pH 8.0, and contains 5 mM α-ketoglutarate, 1 mM glucose-6-phosphate, 0.1 mM ADP, 0.025 M ammonium acetate, 0.2 mg per ml of bovine plasma albumin, 0.2 mg per ml of crystalline beef liver glutamate dehydrogenase and 0.05 mg per ml of glucose-6-phosphate dehydrogenase. With optimal substrate and NADPH concentrations, the glutamate dehydrogenase concentration should give a calculated velocity of approximately 0.4 mole per liter per hour at 25°C. The glucose-6-phosphate dehydrogenase should give a velocity of 0.6 mole per liter per hour at 25°C. The enzymes are resuspended in 2 M ammonium acetate before addition to the cycling mixture. This eliminates the sulfate ion, which is inhibitory, and provides the necessary NH$_4^+$ in the mixture.

Cycling was stopped by heating to 100°C for two minutes; the cycling

TABLE 13-5

Hexokinase activity in the preimplantation stages of the mouse embryo[1]

Stage of Development	Approximate Hours after Ovulation	Number of Determinations	Total Activity ($\mu\mu$moles of NADP reduced/ hour/embryo)	Specific Activity (mμmoles of NADP reduced/min/mg of protein)
Unfertilized	12	16	1.23 ± 0.12	0.738 ± 0.072
Fertilized	12	16	1.76 ± 0.15	1.056 ± 0.090
Two–cell	36	16	1.70 ± 0.13	1.086 ± 0.083
Eight–cell	60	16	2.24 ± 0.18	1.596 ± 0.128
Morula	84	16	5.63 ± 0.63	4.559 ± 0.510
Blastocyst	84	16	7.94 ± 0.80	5.549 ± 0.559
Late blastocyst	108	16	9.40 ± 0.86	7.170 ± 0.656

[1] The values are means ± S. E. The protein content was determined by the method of Lowry et al. (1951) and has been reported previously (Brinster, 1967b).

mixture was then transferred to a 5 ml fluorometer tube containing 1 ml of 20 mM Tris buffer (*p*H 8.0), with 0.2 mM NAD, 0.1 mM EDTA, and enough 6-phosphogluconate dehydrogenase to oxidize 0.005 mM 6-phosphogluconate in three minutes or less. After 30 minutes at room temperature, 3 ml of H_2O were added and the fluorescence of each sample was measured, as were the standards and blanks carried through the entire procedure. The multiplication of NADH during cycling was 5000- to 10,000-fold.

Each experimental run contained blanks and standards as well as embryos representative of each developmental stage. Table 13-5 shows the results obtained from these experiments on hexokinase activity in the mouse embryo. The method has great sensitivity; with additional effort, it is possible to repeat the cycling process with an overall multiplication of 10^7 to 10^8.

CONCLUSIONS

During the past ten years there have been advances in techniques that allow us to measure enzyme activity at very low concentrations. Lowry *et al.* (1961) state that it is possible to measure one molecule of any enzyme which forms a product that can be led to a pyridine nucleotide system. This is indeed a powerful tool for the mammalian embryologist. We have found that enzyme activity can be measured spectrophotometrically, in some cases, in single mammalian eggs (Brinster, 1965a), and that fluorometrically many enzymes can be examined at very low concentrations.

Enzyme activity measurements of early mammalian embryos should provide us with valuable information about the metabolism of the embryo and its nutritional needs. For example, the high lactate dehydrogenase activity in the early mouse embryo (ten times as high as that in any other tissue in the mouse body) appears to be associated with a metabolic requirement of the oocyte of this species. Perhaps all the relationships may not be so striking, and perhaps we will need to look at changes in more than one enzyme and relate these to other metabolic information in order to obtain a complete picture. However, there seems little doubt that determining the activities of key enzymes in the embryo, and ascertaining the changes these activities undergo with development, can provide us with valuable information about the metabolism of the embryo as well as about the embryo's progressive differentiation and the activity of specific genes.

BIBLIOGRAPHY

Bedford, J. M., and M. C. Chang (1962) Fertilization of rabbit ova *in vitro*. Nature, 193:898.

Brinster, R. L. (1965a) Lactic dehydrogenase activity in the preimplanted mouse embryo. Biochim. Biophys. Acta, 10:439–441.

——— (1965b) Studies on the development of mouse embryos *in vitro*. IV. Interaction of energy sources. J. Reprod. Fertility, 10:227–240.

——— (1966a) Malic dehydrogenase activity in the preimplanted mouse embryo. Exp. Cell Res., 43:131–135.

——— (1966b) Glucose-6-phosphate dehydrogenase activity in the preimplantation mouse embryo. Biochem. J., 101:161–163.

——— (1967a) Protein content of the mouse embryo during the first five days of development. J. Reprod. Fertility, 13:413–420.

——— (1967b) Lactate dehydrogenase activity in the preimplantation rabbit embryo. Biochim. Biophys. Acta, 148:298–300.

——— (1968a) Lactate dehydrogenase activity in the oocytes of mammals. J. Reprod. Fertility, 17:139–146.

——— (1968b) Hexokinase activity in the preimplantation mouse embryo. Enzymologia, 34:304–308.

——— (1968c) Carbon dioxide production from glucose by the preimplantation rabbit embryo. Exp. Cell Res., 51:330–334.

Glick, D. (1961) Quantitative chemical techniques of histo- and cytochemistry. Vol. I. Wiley, New York.

——— (1963) Quantitative chemical techniques of histo- and cytochemistry. Vol. II. Wiley, New York.

Greenberg, L. J., and J. A. Rodder (1964) New precision microcell positioner and cell holder for the Beckman model DU Spectrophotometer. Anal. Biochem., 8:137–141.

Loewenstein, J. E., and A. I. Cohen (1964) Dry mass, lipid content and protein content of the intact and zona-free mouse ovum. J. Embryol. Exp. Morphol, 12:113–121.

Long, C. (Ed.) (1961) Biochemist's handbook. 1st ed. D. Van Nostrand Company, Princeton, New Jersey.

Lowry, O. H., J. V. Passonneau, and M. K. Rock (1961) The stability of pyridine nucleotides. J. Biol. Chem., 236:2756–2759.

Lowry, O. H., J. V. Passonneau, D. W. Schulz, and M. K. Rock (1961) The measurement of pyridine nucleotides by enzymatic cycling. J. Biol. Chem., 236:2746–2755.

Lowry, O. H., N. R. Roberts, and J. I. Kapphahn (1957) The fluorometric measurement of pyridine nucleotides. J. Biol. Chem., 224:1047–1064.

Lowry, O. H., N. J. Rosebrough, A. L. Farr, and R. J. Randall (1951) Protein measurement with the Folin phenol reagent. J. Biol. Chem., 193:265–275.

Lutwak-Mann, C. (1962) Glucose, lactic acid and bicarbonate in rabbit blastocyst fluid. Nature, 193:653–654.

Mintz, B. (1962) Experimental study of developing mammalian eggs. Removal of the zona pellucida (mice). Science, 138:594–595.

Parker, R. C. (1961) Methods of tissue culture. 3rd ed. Paul B. Hoeber Inc., New York.

Rowlands, I. W. (1944) Capacity of hyaluronidase to increase the fertilizing power of sperm. Nature, 154:332.

Udenfriend, S. (1962) Fluorescence assay in biology and medicine. Academic Press, New York.

14

Studying the Effect of Viruses on Eggs

R. B. L. GWATKIN
Merck Institute for
Therapeutic Research
Rahway, New Jersey 07067

The interaction of the mammalian egg with viruses is an important area of investigation because viruses are known to exist in the genital tract (Christian *et al.*, 1965; Virgilio *et al.*, 1965) and infections occurring in the egg could result in defects at a later stage of development. A number of viruses are known to produce congenital malformations (see Blattner and Heys, 1961; Evans and Brown, 1963; Medearis, 1964; Gordon *et al.*, 1964; Williamson *et al.*, 1965; Ferm and Kilham, 1965; Gordon *et al.*, 1965; Hardy, 1965; Stoller and Collmann, 1965) and thus the response of the egg as a host may lead to a better understanding of the factors that affect virus infections in general, and may also reveal characteristics of mammalian eggs that are difficult to detect by other means.

Recently, the infection of mouse eggs *in vitro* with Mengo virus was proposed as a model system for such studies (Gwatkin, 1963, 1966). This appears to be the only virus infection of a mammalian egg which has been demonstrated experimentally. The techniques employed are summarized in this chapter.

COLLECTION AND CULTURE OF OVA

Methods for the collection and *in vitro* cultivation of mammalian eggs are given in other chapters. The eggs at various stages of preimplanta-

The techniques reviewed in this chapter were developed while the author was on the Faculty of the University of Pennsylvania and held a Career Development Award (K3-HD21) from the U.S. Public Health Service. Grants GB 617 and GB 4053 from the National Science Foundation are gratefully acknowledged.

tion development are removed from the oviducts and uteri of superovulated 6 to 8 week old Swiss mice and cultured in droplets of lactate-pyruvate medium (Brinster, 1965) under light paraffin oil (No. 0-119, Fisher Scientific Co.) in 60 mm plastic Petri dishes. These cultures are incubated at 37°C in an atmosphere of water-saturated 5% CO_2 in air and examined under a Wild M5 stereoscopic microscope for inhibition of cleavage and pathological changes.

The eggs are handled with micropipettes prepared from disposable Pasteur pipettes (length 23 cm long, no. 8216-D, A.H.Thomas Co., Philadelphia, Pa.) by drawing the capillary portion over the pilot light of an ordinary bunsen burner. The micropipette is controlled by a rubber tube whose end is attached to a plastic mouthpiece.

REMOVAL OF THE CUMULUS OOPHORUS AND ZONA PELLUCIDA

These structures are usually removed, since by hindering virus adsorption and the washing away of unadsorbed virus they make precise work impossible. The cumulus oophorus, present before fertilization and briefly thereafter, is digested by placing the eggs in medium containing hyaluronidase (300 U.S.P. units per ml) for about 5 minutes at 37°C. (See Chapter 13 by Brinster for details of the method). This treatment leaves the zona pellucida intact. If the zona pellucida is to be removed with the cumulus oophorus, a proteolytic enzyme is used. The enzyme most suitable, since its pH optimum lies close to neutrality, is a crude protease from *Streptomyces griseus* (pronase): this was first employed by Mintz (1962) and is described in Chapter 12 of this book.

The eggs are placed in one ml of the enzyme solution in an embryological watch glass (No. 9844, A. H. Thomas Co.) and observed at room temperature under a microscope until the zonae pellucidae have disappeared; this requires only a few minutes. The eggs are washed immediately by transferring them with a micropipette through three 0.5 ml volumes of medium contained in the depressions of a glass spot-test plate (No. 8289-C, A. H. Thomas Co.). Because of the tendency of the eggs to stick to glass after the zona pellucida is removed, the micropipettes, spot-test plates and watch glasses are siliconized. This is done prior to sterilization by immersing them for an hour in 1% aqueous Siliclad (Clay Adams Inc.), then rinsing them in distilled water. Pronase could conceivably destroy the cell-surface receptors required by viruses for infection, so it may be advisable to remove the zona pellucida by other means. Tarkowski (1961) has been able to do so manually by expelling the eggs from a micropipette with an orifice 65μ in diameter. Forcing the egg through the orifice cracks open the zona pellucida, but if the operation is done with care the vitellus is left intact. Papain and ficin will also dissolve the zona pellucida (Gwatkin, 1964) but the effects of these enzymes on subsequent development has not been studied.

EFFECT OF MENGO VIRUS ON MOUSE EGGS

The 37A heat stable strain of Mengo virus (Brownstein and Graham, 1961) is used in all the experiments. Figure 14-1 shows the effect of

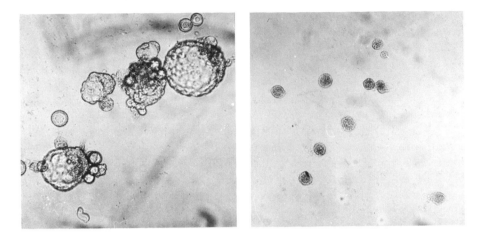

FIGURE 14-1

Effect of Mengo virus on the development of zona-free two-cell mouse eggs in vitro.
*Left: Uninfected controls. Separation of blastomeres and fusion have given rise to
miniature and giant blastocysts. Right: Infected eggs that have become necrotic. (Gwatkin,
1963.)*

adding it to cultures of zona-free eggs at the two-cell stage. Eggs exposed to
the virus fail to develop further and eventually become necrotic. Control
eggs develop into blastocysts that are characterized by a fluid-filled cavity
and by a unilateral thickening, the so-called inner cell mass. Without their
zonae pellucidae the eggs lose blastomeres easily and adjacent eggs fuse;
miniature and giant, as well as normal, blastocysts are thus formed. These
changes may be followed by examination of the droplets with an inverted
microscope equipped with a long-distance phase condenser. The Wild M40
is a satisfactory instrument.

NET VIRUS SYNTHESIS

To establish beyond a doubt that virus production has occurred,
it is necessary to show that more virus is present in the cultures than was
added initially. This is done by diluting virus with medium to about 10,000
plaque-forming units (PFU) /ml, and preparing a series of 0.01 ml droplets
of this suspension under oil: one-half of the series is used as a control; each
of the remaining series receive ten zona-free ova. After incubation for 24 or
48 hours each drop is transferred to 1 ml of medium in a 12×75 mm capped
plastic tube. These tubes are frozen and thawed three times to release
intraovular virus; their contents are then assayed by a standard plaque
count on mouse embryo monolayers. With such a low concentration no
virus is produced by the eggs after 24 hours of incubation; this is presumably because virus adsorption under these conditions takes place slowly.
However, after 48 hours about one-third of the drops are found to contain
relatively large numbers of infectious virus particles (see Fig. 14-2), thus
proving conclusively that the eggs are capable of supporting the synthesis of
new virus.

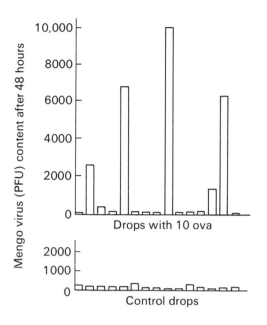

Drops with 10 ova

Control drops

FIGURE 14-2

Synthesis of Mengo virus by groups of zona-free two-cell mouse eggs added to droplets containing small numbers of virus particles. Assays were made 48 hours after adding the eggs. (Gwatkin and Auerbach, 1966.)

Penetration of virus through the zona pellucida

Mengo virus particles are 27 or 28 mμ in diameter and readily pass across the zona pellucida to infect the vitellus; their entry is presumably made through pores left when the cellular processes of the corona radiata are withdrawn, and it may be that these channels are not wide enough for larger viruses to pass through them.

Proof that Mengo virus can traverse the zona pellucida is obtained by exposing eggs for one hour to a concentrated suspension of the virus (10^9 PFU/ml), conditions shown previously to result in the infection of all zona-free two-cell eggs (see below). The eggs are washed five times to remove unadsorbed virus and are then placed individually in droplets of medium under mineral oil. Some of these droplets are assayed immediately to detect any carry over of virus. There should be none from the naked eggs, but when the zona pellucida is present some virus will become trapped in the perivitelline space. The remaining droplet cultures are incubated overnight at 37°C to permit virus multiplication. Formation of plaques in excess of the small number carried over in the wash indicates infection of the egg. Table 14-1 gives the results of two such experiments, in which it was demonstrated that the zona pellucida of the mouse egg is permeable to Mengo virus at both the two-cell and morula stages of development. In some experiments not all the eggs produce virus; this is probably because the many manipulations which the egg undergoes may cause lethal damage before virus infection can be completed. (Measurement of the virus penetration rate through the zona pellucida is described below.)

Virus concentration and duration of exposure

For quantitative studies it is important to determine the virus concentration that is required for adsorption of virus to the eggs within a convenient

TABLE 14-1

Mengovirus infection of early mouse embryos through the zona pellucida at the two–cell and morula stages of development

Stage	Zona Pellucida	Number of Embryos Infected out of 15 Exposed	
		Exposure 1	Exposure 2
Two–cell	+	15 (100.0%)	14 (93.3%)
	−	15 (100.0%)	14 (93.3%)
Morula	+	7 (46.6%)	15 (100.0%)
	−	13 (86.6%)	14 (93.3%)

SOURCE: Gwatkin (1967).

time. To do so, groups of zona-free eggs are exposed for this period to varying concentrations of virus in 0.05 ml droplet cultures. The unadsorbed virus is washed away and the eggs are distributed by micropipette to 0.01 ml droplets under paraffin oil. As in the experiments described previously, these droplet cultures are incubated overnight to allow the infected eggs to produce virus progeny, which are then assayed by plaque count. Figure 14-3 shows that 5×10^8 PFU per ml are required to infect eggs reliably within the one hour exposure period. This is approximately 100 PFU per egg volume, a surprisingly high figure, suggesting that the mouse egg may be more difficult to infect than some other types of cells which have been studied. It may turn out that pinocytosis, the probable means by which animal viruses gain entry to cells, is relatively sluggish in the egg.

Once conditions are established for infecting all the eggs, the proportion which have become infected at any given time can be studied. To do this, eggs are placed in a series of Mengo virus drops containing at least 5×10^8 PFU/ml. The droplet cultures are incubated under standard conditions. At appropriate intervals cultures are removed from the incubator, the eggs are

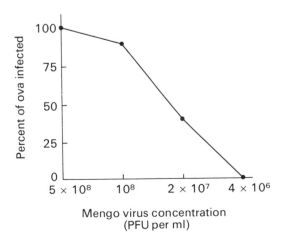

FIGURE 14-3

Proportion of two-cell eggs infected, plotted as a function of Mengo virus concentration. Each point is derived from 20 eggs. (Gwatkin, 1966.)

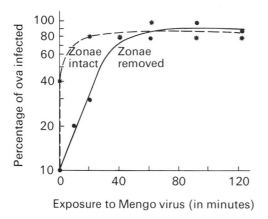

FIGURE 14-4

Proportion of eggs infected, plotted as a function of time of exposure to Mengo virus at a concentration of 10^9 PFU/ml. Each point is derived from twenty eggs. (Gwatkin, 1966.)

washed five times and placed singly in other droplets. These in turn are incubated overnight and assayed for the yield of infectious virus. In this way it is possible to measure the proportion of eggs which have become infected during each period. About 60 minutes are needed to infect all zona-free two-cell eggs under the conditions of such an experiment (see Fig. 14-4). Paradoxically, when the zona is intact the eggs appear to become infected more rapidly; this is not due to a more rapid adsorption of virus onto the zona pellucida than onto the vitellus, but simply to an accumulation of virus in the perivitelline space, from which it is not easily removed, even by repeated washing. This can be demonstrated by an experiment which measures the approximate time required for the virus to pass across the zona pellucida.

Time required for passage of virus across the zona pellucida

Eggs are pipetted rapidly back and forth with a micropipette, the orifice of which is small enough to so distort the eggs that the zona pellucida is broken open and the vitellus ruptured. The empty zonae pellucidae are collected, washed, and exposed to a concentrated suspension of Mengo virus (10^9 PFU/ml) for varying periods of time. The eggs are then washed five times and groups of ten are added to one ml of medium, which is frozen and thawed three times and assayed by plaque count. Eggs, with or without the zona pellucida, are given identical treatment: exposed to virus, washed, assayed. No virus is recovered from the zona-free eggs after exposure to the virus for 20 minutes, and the amount of virus associated with the empty zonae pellucidae, presumably in the intrazonal channels, remains constant with time. However, the virus in the intact eggs increases (see Fig. 14-5) in the minimum time required to perform the manipulations: thus the virus must pass through the zona pellucida into the perivitelline space within 10 minutes. This type of experiment may also be used to determine whether penetration of the zona pellucida has occurred, without requiring actual infection of the vitellus as a criterion of viral entry.

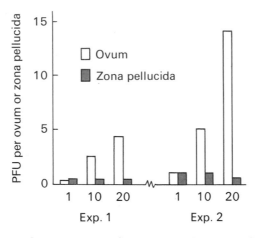

FIGURE 14-5

Mengo virus recovered after exposure of intact two-cell eggs and isolated broken zonae pellucidae after exposure to Mengo virus (10^9 PFU/ml) for brief periods. Each point is derived from twenty eggs or twenty zonae pellucidae. (Gwatkin, 1966.)

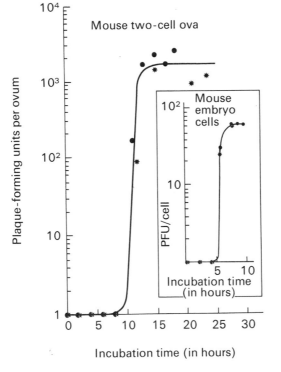

FIGURE 14-6

Growth curve of Mengo virus in two-cell mouse eggs. The insert shows the curve obtained with mouse embryo cells (predominantly fibroblasts). (Gwatkin and Auerbach, 1966.)

VIRUS GROWTH CURVE

Growth curves for viruses in mammalian eggs may be determined after conditions are established for reliably infecting the eggs within a relatively short period. It is best to do this work with zona-free eggs. Although leaving the zona pellucida intact makes it easier to handle the eggs, a small but variable number of unadsorbed virus particles remain in the perivitelline space and must be subtracted from the final titers.

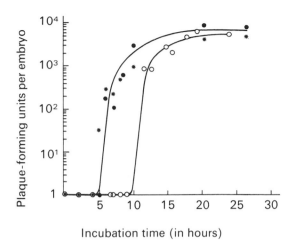

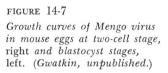

FIGURE 14-7

Growth curves of Mengo virus in mouse eggs at two-cell stage, right *and blastocyst stages,* left. (*Gwatkin, unpublished.*)

Because the growth of Mengo virus in two-cell mouse eggs extends over a relatively longer time than in other cells which have been studied, it is convenient to do the growth curve experiment in two parts of 12 hours each. Superovulatory injections are timed to yield eggs of the same developmental stage for each phase. After adsorbing virus in the manner already described, groups of ten infected eggs are placed under oil in a series of 0.01 ml droplets (three per Petri dish). These cultures are incubated and at intervals a dish is removed from the incubator; each drop is picked up with a micropipette and transferred to one-ml aliquots of medium. These samples are frozen and thawed three times and assayed. Figure 14-6 shows a growth curve obtained for two-cell eggs. Such curves differ in two respects from those obtained for primary mouse embryo cells (predominantly fibroblasts) and L-cells: (1) the period before new virus is produced is about twice as long in the two-cell egg as in these somatic cells, and (2) the amount of virus produced by the egg, although much greater than the amount produced by these cells, is not nearly what would be expected from its size. [Actually, on a unit volume basis, the egg produces only 1 or 2% of the virus yield of the cells (Gwatkin and Auerbach, 1966).] As the egg develops into a blastocyst the latent period becomes shorter but the virus yield remains essentially unchanged (see Fig. 14-7). The reasons for these responses are not known.

CONCLUDING REMARKS

As was mentioned earlier, Mengo virus is the only one known to multiply in a mammalian egg; several others have been investigated (Gwatkin, unpublished). The virus of Western Equine Encephalomyelitis caused two-cell mouse eggs to degenerate, but virus replication was not studied. The viruses of West Nile, Herpes Simplex, Rubella, Vaccinia, Cytomegalic Incluson Disease, and Polyoma all failed to produce detectable changes in mouse two-cell ova *in vitro,* or to block their development into blastocysts. Plaque assays, made on all but the Polyoma-treated eggs, failed to detect virus multiplication, but whether this means that the mouse egg is resistant to infection by these viruses, or only that the right conditions were not provided, is not yet known. In what appears to be the only other investigation of the effect of viruses on mammalian eggs, Berkovich (1963) noted that primary human amnion cells infected with Herpes Simplex, Vaccinia, and Myxomatosis did not affect the development of rabbit eggs cultured with them. However, the rabbit egg is surrounded by a thick mucin coat which may be impervious to viruses.

Obviously, a great deal of work remains to be done to define the characteristics of mammalian eggs as hosts for viruses. The methods described in this chapter should make it possible to extend these studies to any eggs that can be grown *in vitro* and any virus for which there is a sensitive plaque assay. The effect of virus infection on later development may be determined by transfer of the infected eggs to uterine or extrauterine sites (see Chap. 8 by Dickmann and Chap. 9 by Kirby). It may also be possible to study viral infections in postblastocyst cultures in which the trophoblast cells grow in a monolayer (see Gwatkin, 1966b).

BIBLIOGRAPHY

Berkovich, S. (1963) An attempt to infect with virus the early rabbit embryo maintained *in vitro*. Abstracts of 33rd Annual Meeting of the Society for Pediatric Research, p. 81.

Blattner, R. J., and F. M. Heys (1961) Role of viruses in the etiology of congenital malformations. Progr. Med. Virol., 3:311–362.

Brinster, R. B. L. (1965) Studies on the development of mouse embryos *in vitro*. IV. Interaction of energy sources. J. Reprod. Fertility, 10:227–240.

Brownstein, B., and A. F. Graham (1961) Interaction of Mengo virus with L cells. Virology, 14:303–311.

Christian, R. T., P. P. Ludovici, N. F. Miller, and M. R. Gardner (1965) Viral studies of the female reproductive tract. Am. J. Obstet. Gynecol., 91:430–436.

Dulbecco, R., and M. Vogt (1954) Plaque formation and isolation of pure lines with poliomyelitis viruses. J. Exp. Med., 99:167–182.

Evans, T. N., and G. C. Brown (1963) Congenital anomalies and virus infections. Am. J. Obstet. Gynecol., 87:749–758.

Ferm, V. H., and L. Kilham (1965) Histopathological basis of the teratogenic effects of H-1 virus on hamster embryos. J. Embryol. Exp. Morphol., 13:151–158.

Gordon, R. G., A. P. Williamson, and R. J. Blattner (1964) Origin and development of lens cataracts in mumps-infected chick embryos. Am. J. Anat., 115:473–486.

Gwatkin, R. B. L. (1963) Effect of viruses on early mammalian development. I. Action of Mengo encephalitis virus on mouse ova cultivated *in vitro*. Proc. Nat. Acad. Sci., 50:576–581.

———— (1964) Effect of enzymes and acidity on the zona pellucida of the mouse egg before and after fertilization. J. Reprod. Fertility, 7:99–105.

———— (1966) Effect of viruses on early mammalian development, III. Further studies concerning the interaction of Mengo encephalitis virus with mouse ova. Fertility Sterility, 17:411–420.

———— (1966a) Defined media and development of mammalian eggs *in vitro*. Annals. N.Y. Acad. Sci., 139:79–90.

———— (1966b) Amino acid requirements for attachment and outgrowth of the mouse blastocyst *in vitro*. J. Cell Physiol., 68:335–344.

———— (1967) Passage of Mengo virus through the zona pellucida of the mouse morula. J. Reprod. Fertility, 13:577–578.

———— and S. Auerbach (1966) Synthesis of a ribonucleic acid virus by the mammalian ovum. Nature, 209:993–994.

Hardy, J. B. (1965) Viral infections in pregnancy: a review. Am. J. Obstet. Gynecol., 93:1052–1065.

Medearis, D. N. (1964) Viral infections during pregnancy and abnormal human development. Am. J. Obstet. Gynecol., 90:1140–1148.

Mintz, B. (1962) Experimental study of the developing mammalian egg: removal of the zona pellucida. Science, 138:594.

Stoller, A., and R. D. Collmann (1965) Virus aetiology for Down's syndrome (Mongolism). Nature, 208:903–904.

Tarkowski, A. K. (1961) Mouse chimaeras developed from fused eggs. Nature, 190:103–128.

Virgilio, G. D., N. Lavenda, and D. Siegel (1965) Viruses, vaginitis-cervicitis complex and cervical cancer in man. Am. J. Obstet. Gynecol., 93:491–501.

Williamson, A. P., R. J. Blattner, and G. G. Robertson (1965) The relationship of viral antigen to virus-induced defects in chick embryos. Newcastle disease virus. Develop. Biol., 12:498–519.

15

Methods for Histologic and Autoradiographic Analysis of the Early Mouse Embryo

RICHARD G. SKALKO
Department of Anatomy
Cornell University Medical College
New York, New York 10021

Renewed interest in early mammalian morphogenesis is, for the most part, attributable to the development of experimental procedures to evaluate critically the various hormonal, biochemical, and genetic mechanisms that are operative throughout the preimplantation period. The major impetus for these studies is the standardization of conditions which permit development to proceed *in vitro* at a rate that approximates *in vivo* conditions (Whitten, 1956; Brinster, 1963; Mintz, 1964).

To the experimental teratologist and the developmental geneticist, these procedures are a useful means of utilizing the preimplantation embryo as a "model system" to evaluate the mechanism of action of drugs on a developing mammal whose environment can be precisely controlled (Skalko and Morse, 1969). Although this approach to the evaluation of teratogens has been viewed with reservations for postimplantation embryos (Beck and Lloyd, 1966), the fact that preimplantation embryos, which have been maintained *in vitro,* are capable of normal development when transplanted into foster mothers (Tarkowski, 1963; Biggers *et al.,* 1962; Mintz, 1965a) makes them ideally suited for such studies. This chapter provides an experimental protocol to use for *in vitro* cultivation as well as histologic and autoradiographic analysis of the early mouse embryo.

The expert technical assistance of Miss Gail M. Carmody during this study is gratefully acknowledged. The original work reported here was supported by grant number H.D. 01693 from the National Institute of Child Health and Human Development.

CULTURE METHODS

Since the demonstration by Hammond (1949) that glucose would not support the initiation of second cleavage *in vitro,* it has been known that the intermediary metabolism of carbohydrates is unique in the early mouse embryo. Later stages (four-cell to morula) are capable of developing in a variety of culture media, whereas the two-cell embryo requires an organic acid as an energy source. Originally lactate or lactic acid (but more recently pyruvate and oxaloacetate) have been shown to be ideal substitutes for glucose (Brinster, 1965). The experiments performed in my laboratory

TABLE 15-1
Medium for embryo cultures

Basic Salt Solution (50%)[1]	
Salt	ml of Isotonic Solution
NaCl	18.30
KCl	0.80
CaCl$_2$	0.40
KH$_2$PO$_4$	0.20
MgSO$_4$ · 7H$_2$O	0.20
NaHCO$_3$	4.20
Na-pyruvate	1.00
Phenol red	
Horse Serum (50%)	

[1] Salt solution sterilized by millipore filtration. Horse serum added aseptically. Stored in sterile vaccine bottles at 4°C.

have been restricted to embryos removed from oviducts at times after second cleavage because their use produces a high yield of blastocysts (about 85 to 100%) and permits a degree of confidence in evaluating drug action and isotope incorporation. The medium that is used is based on the one originally developed by Brinster (1963) for two-cell embryos and its composition is shown in Table 15-1.

To obtain a high yield of embryos, the technique of superovulation (Smithberg and Runner, 1956) is used. Virgin female mice (ICR strain) are given an intraperitoneal injection of 10 to 15 I.U. of PMS (Equinex, Ayerst) which is followed by a similar injection of HCG (A.P.L., Ayerst) 44 hours later. Ovulation under these conditions occurs within 11 to 14 hours (Edwards and Gates, 1959; Braden, 1962). Females with vaginal plugs after overnight exposure to fertile males are isolated and killed at appropriate times after ovulation (which is assumed to be 12 hours after

HCG injection). Details of embryo harvesting are discussed by Gates (see Chap. 4). Embryos removed from each oviduct are pooled, scored for viability (Whitten and Dagg, 1961), and then transferred to a culture dish. The culture dish that is used for all studies is a commercially obtainable plastic moist-chamber supplied by Falcon Plastics (catalog 3011). Either sterile saline or distilled water is added to the outer ring and the embryos are placed in microdrops under sterile paraffin oil in the center well. The chambers are then placed inside a desiccator that has tap water and phenol red in the base. The desiccator is gassed with a mixture of 5% CO_2 in air, sealed, and placed in a 37°C incubator.

HISTOLOGIC METHODS

The small size of early mammalian embryos, especially that of the mouse, has made them unsuitable for routine cytological study. However, the response of large numbers of these embryos to drugs and teratogenic chemicals can be conveniently evaluated using *in vitro* culture (Thomson and Biggers, 1966; Skalko and Morse, 1969). There are two standard procedures which have been used but neither of them is ideally suited to rapid processing: they are: (1) the highly accurate celloidin techniques of Heuser (Heard, 1957), and (2) the agar-paraffin procedures of Green *et al.* (1937) and Samuel (1944). The technique used in this study is a modification of the agar-paraffin methods presently available. It has the advantage of handling many embryos per block (up to five conveniently). All steps in this procedure are performed under a dissecting microscope.

Fixation

1. Place the embryos in a sealable glass chamber.[1]
2. Add fixative (Bouin's fluid; Neutral Buffered Formalin; Carnoy's Fixative) and seal the chamber with stopcock grease.
3. Fix the embryos for varying lengths of time depending on the nature of the fixative.
   Recommended times:
   Bouin's—24 to 48 hours
   Formalin—30 minutes
   Carnoy's—15 minutes

Preembedding

1. Place a drop of molten 4% agar (Harris, 1965) on a slide that has been treated with a mixture of chrome alum and gelatin, and was kept on a slide warming table at 55°C.

---

[1] Bellco Glass Company, Vineland, New Jersey.

2. Remove variable numbers of embryos from the fixative with a finely drawn Pasteur pipette that has been fitted with a hematological mouthpiece.

3. Place the embryos within the molten agar, taking care not to touch the slide with the pipette.

4. Allow the agar to gel. Cut blocks containing the embryos from the slide with a sharp scalpel and place them in a plastic embedding capsule.

Clearing and embedding

1. Dehydrate through graded alcohols.

2. Place the capsules in a mixture of 50% absolute alcohol (eosin saturated) and 50% methyl benzoate for one hour.

3. Transfer to 100% methyl benzoate for one hour.

4. Place in 50% benzene-50% paraplast for 30 minutes.

5. Infiltrate in paraplast for one hour, then embed in paraplast.[2]

Sectioning and staining

Blocks are sectioned serially at 4 microns on a rotary microtome. Sections containing the eosin-stained embryos are mounted on slides pretreated with chrome alum and gelatin. Staining follows routine procedures.

Embryos removed from pregnant females at 12-hour intervals following induced ovulation were processed according to the above procedure in order to determine cleavage rate and cytology. Because of the implication of altered nucleolar morphogenesis on the phenotypic expression of t^{12}/t^{12} homozygotes (Smith, 1956), particular attention was paid to nucleolar morphology during cleavage. Representative sections are shown in Figures 15-1a through f.

AUTORADIOGRAPHIC METHODS

Studies on the kinetics of ribonucleic acid metabolism in the preimplantation mouse embryo using radioactive precursors *in vitro* have consistently demonstrated that the pattern of isotope incorporation is quite different from that of the more extensively studied nonmammalian forms (see Brachet, 1967, for review). These differences are produced by a number of factors, all of which are intimately related to the developmental cytology of the early embryo. The first, and most important, is the early appearance of the "true" nucleolus in the four-cell embryo (Austin and Braden, 1953; Mintz, 1964, 1965b) with a corresponding increase in ^{3}H-uridine incorporation and RNA content. The reappearance of this organelle is associated with the onset of ribosomal RNA synthesis (Woodland and

[2] Arthur H. Thomas Company, Philadelphia, Pennsylvania.

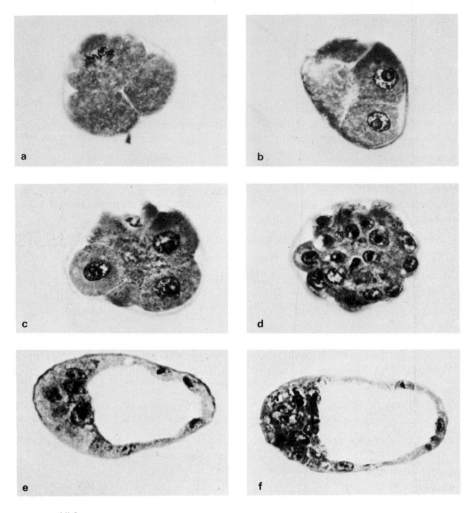

FIGURE 15-1

(a) *Three-cell embryo with one blastomere in mitosis. Sixty hours old.* (b) *Four-cell embryo showing characteristic shell type nucleoli. Sixty hours old.* (c) *Five-cell embryo showing nucleoli with increasing central basophilia. Sixty hours old.* (d) *Advanced morula with peripheral primitive trophoblast cells and intense nucleolar basophilia. Eighty-four hours old.* (e) *Blastocyst with its zona pellucida still intact. Nucleoli are now polymorphous and multiple. Eighty-four hours old.* (f) *Blastocyst without its zona pellucida. This is a littermate of the embryo shown in* (e). *Mitoses occurring in both the trophoblast and the inner cell mass. Eighty-four hours old. All sections are stained with H and E. ×284.*

Graham, 1969), and this finding is consistent with studies in other forms (Brown, 1965). That the nucleolus is the site of ribosomal RNA synthesis is, at present, well documented (Perry, 1967). When these nucleoli are in the shell configuration (see Fig. 15-1b), the nucleolus consists of a perinucleolar ring of heterochromatin which also possesses RNA and basic protein (Braden and Austin, 1953). The second factor is the presence of a preimplantation lethal mutation in the mouse, t^{12}/t^{12}, which is also related to nucleolar formation and maturation. These homozygotes never form blastocysts (Smith, 1956) and, at a time when heterozygous littermates are rapidly increasing their rate of RNA synthesis and cytoplasmic basophilia, they maintain a low level of activity until eventual death (Mintz, 1964). That these events are related to a lack of nucleolar maturation emphasizes the role that the nucleolus plays in early mammalian development.

To evaluate these phenomena under controlled conditions, embryos were removed from oviducts at 55-, 60-, and 65-hour intervals after ovulation and were cultured in the basic medium (described above), supplemented with ^{3}H-uridine (5 μc/ml, specific activity 8100 millicuries/millimole) for one hour. Following this pulse period, the embryos were either washed in saline solution and fixed in Bouin's fluid, or were washed three times in basic medium containing a high excess of unlabeled uridine (20 mg/ml), and then cultured in this new medium for up to five hours. Embryos processed by the second procedure were then washed and fixed in Bouin's. Embryos from both groups were embedded in paraplast, sectioned at 4 microns, and processed for autoradiography according to the following (This procedure is a modification of the now classical liquid emulsion technique described by Kopriwa and LeBlond, 1962.) :

Materials needed

    Slides containing mounted embryos
    Water bath
    Liquid photographic emulsion, Ilford (K-5 or L-4)
    Humidifier
    Bakelite slide boxes (equipped with Drierite)
    Air blower
    Electrician's tape

Procedure

1. Two hours prior to the processing, turn on the humidifier and allow the relative humidity in the darkroom to reach 80%.

2. In complete darkness, place the bottle of emulsion in the water bath (40° to 45°C) and allow it to melt. Once it is melted, prepare a 1:1 dilution of the emulsion with twice-distilled water. Place the diluted emulsion in a coplin jar and set it in the water bath.

3. Remove the slides from xylene (if they have been prestained), allow them to dry, then dip them in the emulsion. Wipe the back of each slide and place it on a rack in front of the blower. When all the slides are dry, place them in bakelite slide boxes with the emulsion side up. Seal the box with electrician's tape and store it for the desired exposure time (one week to two months) at 4°C.

4. To develop the slides, use the following chemicals in sequence:

D-19 developer —	3 minutes
Stop bath —	30 seconds
Hypo —	5 minutes

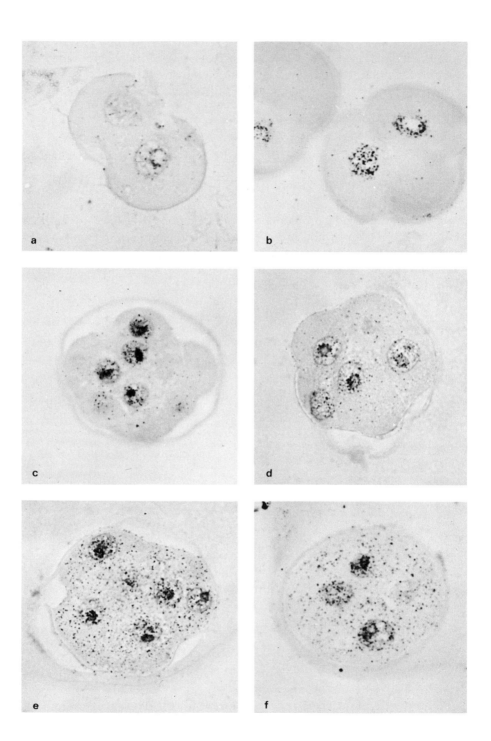

FIGURE 15-2

(a) *Four-cell embryo, 55 hours old. One hour pulse with ³H-uridine. Shell nucleolus labeled around the heterochromatic rim as is the non-nucleolar part of the nucleus. Seven-day exposure.* (b) *Four-cell embryo, 55 hours old. Littermate of the embryo shown in* (a) *but exposed for 14 days. The nucleolar core is devoid of label at this stage.* (c) *Morula, 60 hours old. One hour pulse with ³H-uridine. The nucleolus is intensely basophilic at this stage and is uniformly labeled. Seven-day exposure.* (d) *Morula, 60 hours old. One hour pulse with ³H-uridine followed by a three-hour chase. Cytoplasmic label is now significant. Seven-day exposure.* (e) *Morula, 60 hours old. One hour pulse with ³H-uridine followed by a five-hour chase. Fourteen-day exposure.* (f) *Eight-cell embryo, 60 hours old. One-hour pulse with ³H-uridine followed by a three-hour chase. Fourteen-day exposure. All sections were coated with K-5 emulsion, ×304.*

Then place the slides in running water for five minutes, follow this with a wash in distilled water for five minutes. The slides are dehydrated by xylene (or run through an appropriate staining procedure if this was not previously done), mounted in DPX (Gurr), and coverslipped.

Following this procedure, it has been possible to get reproducible autoradiographic localization of ³H-uridine in the early embryo without high levels of background. The defects that have been found are usually the result of batch variations in the emulsion; these have been discussed by Caro (1964). Autoradiographs from representative experiments which demonstrate the effectiveness of this procedure are shown in Figures 15-2a to f.

BIBLIOGRAPHY

Austin, C. R., and A. W. H. Braden (1953) The distribution of nucleic acids in rat eggs in fertilization and early segmentation. I. Studies on living eggs by ultraviolet microscopy. Australian J. Biol. Sci., 6:324–333.

Beck, F., and J. B. Lloyd (1966) The teratogenic effects of azo dyes. Adv. in Teratol., 1:131–193.

Biggers, J. D., R. B. L. Gwatkin, and R. L. Brinster (1962) Development of mouse embryos in organ cultures of fallopian tubes on a chemically defined medium. Nature, 194:747–749.

Brachet, J. (1967) Behavior of nucleic acids during early development. In: Comprehensive Biochemistry. Ed. by M. Florkin and E. H. Stotz, vol. 28: Morphogenesis, Differentiation and Development, Elsevier, Amsterdam.

Braden, A. W. H. (1962) Spermatozoan penetration and fertilization in the mouse. Symp. Genet. Biol. Ital., 9:1–8.

───── and C. R. Austin (1953) The distribution of nucleic acids in rat eggs in fertilization and early segmentation. II. Histochemical studies. Australian J. Biol. Sci., 6:665–673.

Brinster, R. L. (1963) A method for *in vitro* cultivation of mouse ova from two-cell to blastocyst. Exp. Cell Res., 32:205–208.

───── (1965) Studies on the development of mouse embryos *in vitro*. II. The effect of energy source. J. Exp. Zool., 158:59–68.

Brown, D. D. (1965) RNA synthesis during early development. In: Developmental and Metabolic Control Mechanisms and Neoplasia. Williams and Wilkins, Baltimore.

Caro, L. G. (1964) High-resolution autoradiography. In: Methods in Cell Physiology. Vol. 1. Ed. by D. M. Prescott. Academic Press, New York.

Edwards, R. G., and A. H. Gates (1959) Timing of the stages of the maturation divisions, ovulation, fertilization and the first cleavage of eggs of adult mice treated with gonadotrophins. J. Endocrinol., 18:292–304.

Green, W. W., C. Barrett, and L. M. Winters (1937) A technic for the sectioning of mammalian ova and blastocysts. Stain Technol., 12:43–47.

Hammond, J. (1949) Recovery and culture of tubal mouse ova. Nature, 163:28–29.

Harris, H. H. (1965) Simple technique for embedding and supporting delicate biological specimens. Nature, 208:199.

Heard, O. O. (1957) Methods used by C. H. Heuser in preparing and sectioning early embryos. Contrib. to Embryol., 36:1–18.

Kopriwa, B. M., and C. P. LeBlond (1962) Improvements in the coating technique of radioautography. J. Histochem. Cytochem., 10:269–284.

Mintz, B. (1964) Synthetic processes and early development in the mammalian egg. J. Exp. Zool., 157:85–100.

——— (1965a) Genetic mosaicism in adult mice of quadriparental lineage. Science, 148:1232–1233.

——— (1965b) Nucleic acid and protein synthesis in the developing mouse embryo. In: Ciba Foundation Symposium on Preimplantation Stages of Pregnancy. Ed. by G. E. W. Wolstenholme and M. O'Connor. Little, Brown, Boston.

Perry, R. P. (1967) The nucleolus and the synthesis of ribosomes. In: Progress in Nucleic Acid Research and Molecular Biology, 6:219–257.

Samuel, D. M. (1944) The use of an agar gel in the sectioning of mammalian eggs. J. Anat., 78:173–175.

Skalko, R. G. and J. M. D. Morse (1969) The differential response of the early mouse embryo to actinomycin D treatment *in vitro*. Teratology, 2:47–54.

Smith, L. J. (1956) A morphological and histochemical investigation of a preimplantation lethal (t^{12}) in the house mouse. J. Exp. Zool., 132:51–83.

Smithberg, M., and M. N. Runner (1956) The induction and maintenance of pregnancy in prepuberal mice. J. Exp. Zool., 133:441–454.

Tarkowski, A. K. (1963) Studies on mouse chimeras developed from eggs fused *in vitro*. In: Symposium on Organ Culture. National Cancer Institute Monograph, no. 11, pp. 51–67.

Thomson, J. L. and J. D. Biggers (1966) Effect of inhibitors of protein synthesis on the development of preimplantation mouse embryos. Exp. Cell Res., 41:411–427.

Whitten, W. K. (1956) Culture of tubal mouse ova. Nature, 177:96.

——— and C. P. Dagg (1961) Influence of spermatozoa on the cleavage rate of mouse eggs. J. Exp. Zool., 148:173–183.

Woodland, H. R. and C. F. Graham (1969) RNA synthesis during early development of the mouse. Nature, 221:327–332.

16

Techniques for Studying Fine Structure of Cleavage, Blastocyst, and Early Implantation Stages

ALLEN C. ENDERS
Washington University School of Medicine
St. Louis, Missouri 63110

In almost every generation since Cruickshank (1797) first collected eggs from the tubes of the rabbit, embryologists have been studying the early cleavage and implantation stages of mammals. Techniques as divergent as the celebrated lifting of the ovum from a follicle on the point of a knife by Von Baer (1827; see translation by Corner, 1933, and the historical discussion by Austin, 1961), and the microchemical methods of Loewenstein and Cohen (1964) have been used effectively. Some methods, such as the use of flat-mounts of blastocysts (Van Beneden, 1880) disappear from the literature, only to reappear (Moog and Lutwak-Mann, 1958) and be useful again in a somewhat different context. Indeed, Van Beneden used not only flat-mounts but also osmium fixation and, in some instances, a postfixation technique.[1]

Although some of these methods of handling tissues are notable, in essence each individual working with these relatively minute stages had designed methods to suit both his dexterity and the specific objectives of his study. The methods which we are presenting here are merely those which we have found simplest for handling and preparing cleavage and early implantation stages for study by electron microscopy. In general we have followed the premise that the simpler the method, the easier it is to apply consistently. Although in most instances we have tried methods which *a priori* appeared rational, the final test has always been the empirical one of the condition of the tissue when seen in the final micrographs.

The work in this chapter was supported by grant GB 5024 from the National Science Foundation.

[1] I am indebted to Dr. C. Hugh Tyndale-Biscoe for first bringing these papers to my attention.

It should be noted at the outset that this chapter is not intended to be a dissertation on fixation methods for electron microscopy *per se,* but is concerned rather with the special difficulties that handling of the early developmental stages presents. (For a more thorough discussion of the problems of the preservation of embryonic tissue for electron microscopy, see Szollosi, 1967.) Anyone who uses electron microscopy as a method of studying preimplantation stages should be aware not only of the numerous specific fixation techniques, but also of the information indicating the effects of buffering and osmolarity on tissue fixation (Maser *et al.,* 1967; Wood and Luft, 1967) and should realize that about 12% of the protein is usually thus extracted (Luft and Wood, 1963), and an even higher amount of lipid is lost (Morgan and Huber, 1967; Cope and Williams, 1968).

COLLECTION OF STAGES

The basic principle in collecting preimplantation stages is to keep the apparatus and method simple, and the collecting fluid clean.

Oviductal Stages

Stages which are located in the oviducts can be obtained in three ways: fixation *in toto,* and either mincing or flushing of the oviduct. With small animals (e.g., the deer mouse), the entire oviduct can be fixed with some hope of good preservation of the contained preimplantation stages. Two methods deserve particular attention. One is perfusion of the oviduct by injection of fixative directly into the blood-vascular system. This method, used by Nilsson and Reinius (1969), has the advantage of preserving the oviduct and the cleavage stages in good condition, and makes subsequent handling easier. One disadvantage is that it tends to dilate the vessels. In many instances the cleavage stages probably are not moved from their initial position, but the surge of fixative in the vessels and the variable rate of muscle fixation might result in small changes in location. Nevertheless, the stages should be located more accurately by this method than by removal of the oviducts and subsequent fixation. Another possible disadvantage of this method is that the fixative reaches the blastomeres only after penetrating the intervening maternal tissues, and consequently may be altered in composition. In the smallest oviducts such alteration is probably minimal, but as the size of the oviduct increases the change could be detrimental.

In theory, freeze-substitution offers the best possibility for localization of contained stages within the oviduct. However, in addition to the difficulties of preventing ice crystal damage during either the initial quick-freezing or subsequent substitution, the number of fixing solutions is limited by the necessity of dissolving them in low freezing-point, organic solvents. We have not yet fully evaluated this method.

Oviducts of small mammals can also be cut into small pieces within a depression slide containing the fixative and some of the contained cleavage stages can be recovered. This method has the advantage of being extremely simple, but a disadvantage is that only a small percentage of the ova are generally recovered, due in part to the tremendous amount of debris present and due to the retention of these stages within folds of the oviduct. A further disadvantage is the deterioration of the fixative which may be caused by the blood and cellular debris. However, if just a few representative stages are desired, the ova can be obtained in this fashion, and following that they should be rapidly transferred to a clean fixative for subsequent processing.

The most common method of obtaining cleavage stages is to flush them from the oviduct. Details of a number of the different methods used in flushing may be obtained in other chapters of this book. In general, for small mammals such as the mouse and the rat, a 30-gauge hypodermic needle is blunted before use and introduced into the infundibulum after the oviduct has been removed from the uterus by section of the uterotubal junction. Generally less than 0.1 ml is necessary to flush the eggs into a dish. Care is taken so that the tubal end of the oviduct is directed towards the bottom of the dish and so that fat droplets are not introduced.

With larger mammals, such as the rabbit or armadillo, the removed oviduct is usually trimmed and flushed from the uterotubal junction toward the fimbria. Again, care should be taken that as little blood as possible enters the receptacle with the flushing fluid, and that the total amount of fluid is not in excess of the capacity of the receptacle.

Choice of flushing fluid is partially a matter of convenience. It might be thought that the best method is to flush with the initial fixative. This method however has two drawbacks: first, since the fixative will harden the epithelium of the oviduct, many of the contained cleavage stages are likely to be trapped; and second, the fixative, in its initial surge, may meet the ova irregularly or may be slightly altered by substances within the oviduct. An alternate procedure is to flush with saline (0.9% NaCl), Hank's solution, or an appropriate tissue culture medium. Since the deleterious effects of saline can be seen even in the dissecting microscope, most investigators prefer a more balanced medium (see Szollosi, 1967). In my laboratory, both direct flushing with the fixative and flushing with a tissue culture medium have been used satisfactorily.

As an initial receptacle for recovering the flushings, use a vessel whose entire width can be scanned in a dissecting microscope. For ova and smaller blastocysts, we have found that straight-walled depression slides are most satisfactory (see Fig. 16-1). The flat bottom on the depression slide prevents distortion, and the straight side-walled well holds enough fluid to allow ample time for getting a cover slip over the preparation without undue haste. At all times, the cover slip should be slid on and off the well of the depression slide to prevent displacement of the fluid.

The best magnification for observing these slides in a dissecting microscope depends in part on the size of the ovum. For small ova such as those of the mouse and rat, magnifications of 25× to 50× are appropriate. Using depression slides, the flushings can also be examined with the 10× objective of most compound microscopes. This permits selection of appropriate stages and photography of these stages if desired.

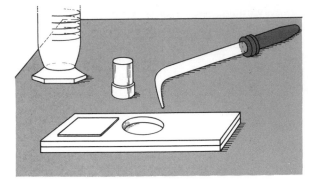

FIGURE 16-1

Sketch of the depression slide, pipette, and plastic capsule used during fixation, dehydration, and embedding of ova and blastocysts. Although special containers have been designed (Buchanan, 1965) we have found that this simple apparatus can be used effectively with a dissecting microscope.

INTRAUTERINE STAGES

Preimplantation stages can be readily obtained from the uteri of most mammals by most simple flushing techniques. As mentioned above, care should be taken to keep fat and blood from entering the chamber into which the blastocysts are flushed. The blastocysts are usually in the first drops of fluid passing through the uterus. Consequently, care must be taken that these drops land in the recipient container rather than on your lab coat.

In the armadillo, in which the blastocyst is situated in the fundic tip of the uterus simplex beneath folds of endometrium, it is necessary to evert the uterus by slicing the mesometrial margins up to the oviducts and depressing the fundic tip with a finger. In this fashion, the everted fundic tip can be dipped repeatedly in the collecting fluid.

FIXATION

Because of the smallness of the ova and the blastocysts, there should be no problem posed by penetration of fixative, extensive dilution of fixative, or damage due to manipulation or cutting into small pieces. It was therefore with some surprise that we found in our early studies that these stages were preserved very poorly when fixed and dehydrated routinely for electron microscopic study. Most of the early material, fixed in Palade's (1952) or Caulfield's (1957) solutions, showed extensive leaching. With these fixatives, it was found in a number of laboratories that in normal tissue blocks there was a zone of good preservation a short distance beneath the surface. Two general types of explanation were commonly heard concerning the lack of preservation of the surface of the block: one explanation was that the damage caused by cutting, and exposure to air and to blood gave the unsatisfactory results; the second explanation was that overlying cells have a mediating effect on the rigors of fixation and dehydration. Hence, the superficial cell layer could not be expected to be fixed well. Blastocysts placed in Caulfield's solution did not appear to alter in structure during fixation, and darkened only slowly, yet the results from these observed blastocysts and from others fixed at different intervals from five minutes to several hours were equally poor. We were forced to conclude from the broken membranes and leached cytoplasm that extensive extraction of the cytoplasm occurred either during initial fixation or more probably during dehydration and embedding. Potassium permanganate fixation (1.2% in 0.9% sodium chloride) fixed cleavage stages without the signs of excess leaching and in a fashion similar to other tissues fixed with this procedure. However, ribonucleoproteins and other proteins were clearly not preserved and the swelling so typical of permanganate fixation was produced. A further disadvantage was the opacity of the fluid, the tendency to form a precipitate when exposed to air, and the formation of a precipitate when the solution was introduced with the contained stages into alcohol. Nevertheless, tissues fixed in this fluid could be successfully examined and many of the features of cellular structure were determined (see Fig. 16-2), (Enders, 1962; Enders and Schlafke, 1962, 1965; Schlafke and Enders, 1963).

Introduction of the method of initial fixation by aldehydes followed by postfixation has been a real boon for the study of cleavage and blastocyst stages. Micrographs of tissue fixed by this procedure show greater preservation of cytoplasmic constituents than was previously possible (see Figs. 16-2 and 16-3). Several secondary boons also occur. Tissues in the initial aldehyde fixation show relatively little alteration from the living state. Because of the clarity of these solutions, the stages can be readily studied and photographed (see Fig. 16-4). Also, allowing the blastocyst to remain for periods longer than an established minimum fixation time seems to have relatively little effect on preservation *per se*. In a number of instances satisfactory results have been obtained from blastocysts kept in a buffered glutaraldehyde fixative for several days. The temperature of initial fixation does not seem to be crucial either. There will undoubtedly be a number of refinements in this two-stage fixation procedure within the next few years. The procedure and comments that follow represent simply a method that has been useful for a large number of stages in an extensive series of different mammals. Some of the modifications necessary in dealing with implantation stages will be mentioned after the basic methods.

Whether the stages have been flushed from the reproductive tract in fixative or in a culture medium, it is desirable to switch them rapidly to a fresh container of fixative. For this, and for all other transfers, the simplest device is a standard glass eyedropper drawn to a fine tip at a 45° angle to the long axis (see Fig. 16-1). It is preferable to have the opening in the pipette only slightly larger than the cleavage stage or blastocyst being transferred. Such an arrangement assures that relatively little excess fluid is transferred, and keeping the blastocysts in the small portion prevents accidental deposit

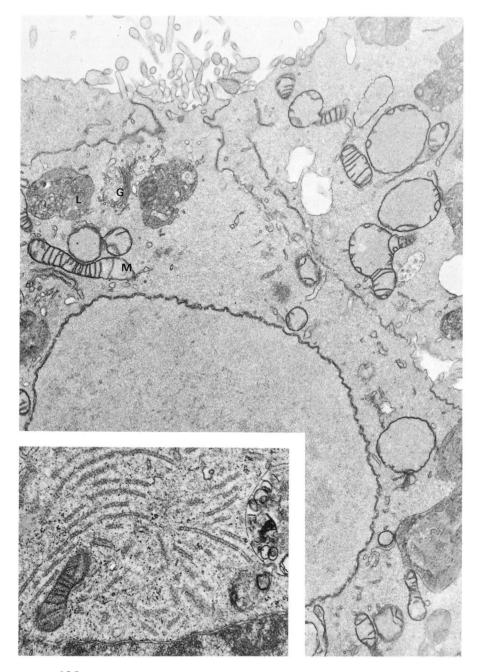

FIGURE 16-2

Potassium permanganate-fixed hamster morula. The Golgi (G), mitochondria (M) and cell membranes are clearly visible, as are some of the complex lipoidal inclusions (L). However, the bilaminar plaques so typical of the cytoplasm of hamster oocytes (Enders and Schlafke, 1965; Weakley, 1967) cannot be seen, and the chromatin within the nucleus is extremely light. Ribosomes are also absent. In the insert, from a hamster morula fixed by the glutaraldehyde–osmium method, the bilaminar plaques are clearly visible in a curved pattern, and clumps of chromatin can be seen just beneath the nuclear membrane at the bottom of the picture.

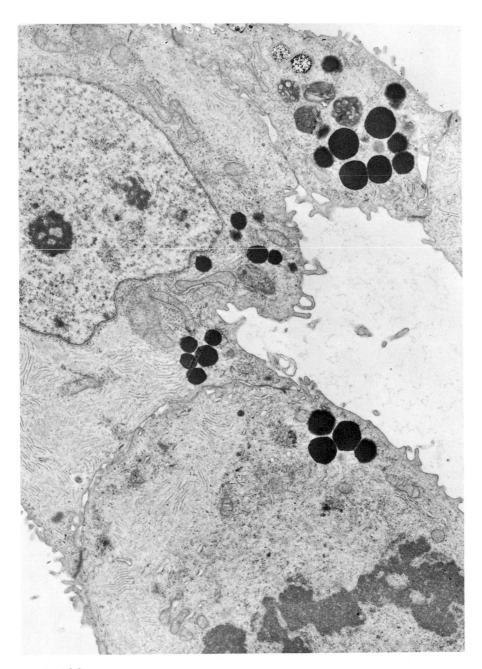

FIGURE 16-3

*Rat blastocyst fixed by the glutaraldehyde–osmium method. Note the abundance of differ-
ent materials in the cytoplasm preserved by this method. The regions of protein (?)
plaques contrast with the regions rich in ribosomes. Lipid droplets, degradation bodies,
and the usual organelles are well preserved. In the cell at the bottom of the micrograph
the chromosomes of an early metaphase are seen. The extremely dense particles in
vesicles in the cell at the top of the picture are colloidal particles of thorotrast ingested
by the trophoblast cells.*

FIGURE 16-4

Ferret blastocysts photographed during dehydration after the glutaraldehyde–osmium procedure. In this and other carnivores, lipid droplets ring the nuclei of the trophoblast cells, and the embryonic cell mass is a highly distinct group of cells. Note the absence of blastocyst distortion and that the trophoblast cells are so closely associated with the zona pellucida that it is difficult to distinguish the latter structure.

of them on the inner walls of the pipette. A large number of clean pipettes should be available; they not only make observation of the blastocysts easier, but also remove the possibility of blastocysts sticking to any dirt.

When preimplantation stages are first placed in 3% glutaraldehyde (in 0.1 M phosphate buffer, pH 7.3) they may float briefly, but will gradually sink unless they adhere to the surface film. For this reason they should be placed beneath the surface. It is better to place only two or three preimplantation stages in a single depression slide; otherwise they can not be observed simultaneously. (This is important in the dehydration stages.) Initial fixation should not be shortened just because of the small size of the object being fixed. (It may take some time for the aldehyde groups to properly cross link the free amino groups of the proteins; in our experience initial fixation times of less than one-half hour are unsatisfactory.)

Although it is customary to follow glutaraldehyde fixation with a buffer rinse, this has not been necessary when the initial fixation time does not exceed two hours. However, passage through a phosphate buffer rinse prevents contamination of the subsequent fixative from the glutaraldehyde, and it may be convenient to leave the tissues in this stage for several minutes to a few hours. Postfixation is accomplished by placing the blasto-cysts in still another dish containing 2% phosphate buffered osmium tetroxide. This process usually lasts one or two hours, during which time the dish is refrigerated. During transfers which involve osmium tetroxide or propylene oxide, a downdraft hood helps to protect the observer from eye irritation.

DEHYDRATION, EMBEDDING, AND ORIENTATION

During the dehydration, the blastocysts are kept in the original dish, and the fluid is removed with one pipette and added with another. Thus, while observing the blastocysts with the dissecting microscope, most of the fixing fluid is drawn off, and 10% alcohol is added drop by drop. The reason for withdrawing the fluid and for adding the alcohol rather than moving the blastocysts is that surface tension forces and heat of solution produce currents within the dish that otherwise make it difficult to keep track of the blastocysts. Successive changes of alcohol are made every 10 minutes by introducing 10, 30, 50, 70, 95, and 100% alcohol sequentially into the container. During this process the alcohols are usually kept cold. Under these circumstances the alcohols of the middle range (20 to 60%) are highly viscous (e.g., 45% alcohol—5.59 centipoises at 5°C, as compared with 1.62 centipoises for 100% alcohol at the same temperature). Larger blastocysts may collapse slightly in the lower alcohol concentrations, but they usually recover in 100% alcohol.

When the tissues reach 95% alcohol, they are allowed to come to room temperature. Generally three changes of 100% alcohol are used and the last change is made by transferring the blastocysts to a clean dish containing 100% alcohol.

Alcohol and araldite are slightly miscible. It is therefore possible to go directly from alcohol to the prepared monomer (Szollosi, 1967). However, we have had more consistent results by passing the tissues through an organic solvent that is more readily miscible with the araldite. The solvent that we usually use is propylene oxide, the use of which requires patience since it has an extremely high vapor pressure, evaporates rapidly, and causes a veritable turmoil of currents in the depression slide when added to alcohol. Consequently we use the following procedure: most of the alcohol is removed from the depression slide; propylene oxide is taken up in a fresh pipette, then expelled, a fresh supply is taken up and is introduced onto the slide. (The vapor pressure of the propylene oxide is such that when it is first taken into a pipette it tends to be spontaneously expelled. Hence the prefilling process.)

In this fashion two changes of propylene oxide are placed in the dish for 5 to 10 minutes. A cover slip must be slid quickly over the preparation and it must be carefully watched during this stage. A 50% mixture of propylene oxide and plastic is then introduced onto the slide (without removal of all the propylene oxide). Many large blastocysts will partially collapse at this stage. It is possible to make an extensive series of mixtures of propylene oxide in plastic, but the possibility of losing the blastocyst is increased and only rarely is the eventual collapse of one side of the blastocyst prevented.

After 15 to 30 minutes, the tissue can be readily transferred to the prepared plastic in capsules. Although it is possible to transfer blastocysts directly from propylene oxide to the prepared plastic in capsules, this procedure is not recommended because blastocysts are extremely difficult to transfer in pure propylene oxide.

The embedding capsule selected is, in part, a matter of personal preference. In our hands, plastic capsules (BEEM) with the tips cut off and inverted so that the cap forms the flat base of the capsule have proved particularly useful. When these capsules are filled with the bubble-free araldite mixture, the blastocyst can be seen clearly in the plastic. In addition, polymerization of the plastic seems to occur more uniformly in these capsules than in gelatin capsules, which have the disadvantage of being hydroscopic. The capsule is placed in a 60°C oven for polymerization. If the blastocyst has been placed beneath the surface, somewhere near the middle of the capsule, it will sink to the bottom and will tend to become so oriented that its broadest area is at the base of the capsule.

Although we usually use araldite (Durcupan—5 ml resin, 5 ml hardener, 0.2 ml accelerator, 0.05 ml plasticiser) in our laboratory other mixtures of epoxy resins would be equally suitable. When polymerization is completed (after about four days) the capsule can be cut off the block, and the contained blastocyst or egg is clearly visible in a dissecting microscope. Once it has been located, the spot can be followed with the naked eye. The block face can then be cut with a dental drill and disc, and prepared for sectioning. If the orientation is not suitable, the block can be cut out and reembedded, or glued to another block.

The collection of early implantation stages poses a special problem, one which is somewhat different in each animal. In the case of the rat and mouse, where the uterus is relatively small, a method of perfusion with glutaraldehyde has been particularly satisfactory (Enders and Schlafke, 1967, 1969; Reinius, 1967). The advantage of this method is that it hardens the muscle of the uterus so that when the tissue is cut, it is not distorted (when a fresh uterus is cut, the smooth muscle contracts, squeezing out the adjacent endometrium like toothpaste from a tube). Localization of the implantation sites can be made in a number of ways. If implantation has proceeded to the extent that there is a localized vascular permeability, injection of Pontamine Blue will indicate the site. Consequently, after perfusion the segments of uterus can be carefully and gently sliced with a razor blade so that a millimeter thick slice in the region of the blue spot can be obtained through the entire uterus. In this fashion, orientation of the specimen can be achieved and ample penetration of both the initial fixative and subsequent fixative is facilitated. If even earlier stages are desired, segments of the uterus can be embedded longitudinally and then sectioned sagittally. Thick sections can then be examined until the blastocysts are found.

For species with larger uteri or different modes of implantation, different fixation procedures must be worked out. However, initial perfusion to harden the uterus is still useful, since the common problem is to prevent distortion of the implantation site by the manipulation procedures used to remove it.

SECTIONING

The sectioning for electron microscopy of ova and blastocysts can be handled in the same fashion as more conventional tissues. The smallness of the object being sectioned occasionally makes it difficult to tell the exact depth of the ova or blastocyst in the block, and the cut face of these objects does not present a very clear image. Consequently, it is desirable to take thick sections (2 or 3μ) that can either be viewed by a phase microscope or, preferably, can be stained with Azure B. Such sections, when covered, form excellent study material for the light microscope (see Fig. 16-5). A wide variety of microtomes, grids and staining methods may be used. In our work, sections to be examined with the electron microscope are made using diamond knives on a Porter-Blum MT-2 microtome; they are then placed on slotted copper grids and are stained with a lead citrate method (Reynolds, 1963).

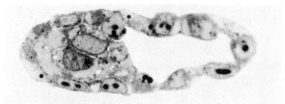

FIGURE 16-5

Light micrograph of a section from a plastic block containing a rat blastocyst prepared by the the glutaraldehyde–osmium method. The preparation has been stained with Azure B, and photographed with a yellow filter. Note that in addition to the nuclei and nucleoli, the basophilic areas of the cytoplasm are more pronounced than the less basophilic regions (those areas that are filled with plaques in electron micrographs).

CONCLUSIONS

The handling of eggs, blastocysts, and implantation stages for examination by electron microscopy is not very difficult. However, since all of the manipulations take place under a dissecting microscope, it usually requires a little practice to avoid occasional loss of an ovum. The greatly increased amount of information that one can get from these stages, when compared to conventional histological observation, makes it well worth the effort. Indeed, one can sympathize with Doyle *et al.* (1966) in their desire to prepare human ova in this fashion. We recommend, however, that ova or blastocysts from readily available species should be run as a trial procedure to overcome the initial awkwardness that dealing with such small objects produces in most of us.

BIBLIOGRAPHY

Austin, C. R. (1961) The Mammalian Egg. Thomas, Springfield, Ill.

Buchanan, G. D. (1965) A plastic microstrainer for handling mammalian blastocysts. Stain Technol., 40:177–179.

Caulfield, J. B. (1957) Effects of varying the vehicle for OsO_4 in tissue fixation. J. Biophysic. Biochem. Cytol., 3:827–829.

Cope, G. H., and M. A. Williams (1968) Quantitative studies on neutral lipid preservation in electron microscopy. J. Roy. Microscop. Soc., 88:259–277.

Corner, G. W. (1933) The discovery of the mammalian ovum. Mayo Foundation Lectures, Philadelphia.

Cruickshank, W. (1797) Experiments in which, on the third day after impregnation, the ova of rabbits were found in the Fallopian tubes; and on the fourth day after impregnation in the uterus itself; with the first appearance of the foetus. Phil. Trans., pt. 1, p. 197.

Doyle, L. L., J. Lippes, H. S. Winters, and A. J. Margolis (1966) Human ova in the Fallopian tube. Am. J. Obstet. Gynecol., 95:115–117.

Enders, A. C. (1962) The structure of the armadillo blastocyst. J. Anat., 96:39–48.

———— and S. J. Schlafke (1962) The fine structure of the blastocyst. Anat. Record, 142:338.

———— and S. J. Schlafke (1965) The fine structure of the blastocyst: some comparative studies. In: Ciba Foundation Symposium on the Preimplantation Stages of Pregnancy, pp. 29–59.

———— and S. J. Schlafke (1967) A morphological analysis of the early implantation stages in the rat. Am. J. Anat., 120:185–226.

———— and S. J. Schlafke (1969) Cytological aspects of trophoblast-uterine interaction in early implantation. Am. J. Anat., 125:1–30.

Loewenstein, J. E., and A. I. Cohen (1964) Dry mass, lipid content and protein content of the intact and zona-free mouse ovum. J. Embryol. Exp. Morph., 12:113–121.

Luft, J., and R. L. Wood (1963) The extraction of tissue protein during and after fixation with osmium tetroxide in various buffer systems. J. Cell Biol., 19:46A.

Maser, M. D., T. E. Powell, III, and C. W. Philpott (1967) Relationships among pH, osmolarity and concentration of fixative solutions. Stain Technol., 42:175–182.

Moog, F., and C. Lutwak-Mann (1958) Observations on rabbit blastocysts prepared as flat mounts. J. Embryol. Exp. Morphol., 6:57–67.

Morgan, T. E., and G. L. Huber (1967) Loss of lipid during fixation for electron microscopy. J. Cell Biol., 32:757–759.

Nilsson, O., and S. Reinius (1969) Light and electron microscopic structure of the oviduct. In: The Mammalian Oviduct. Ed. by E. S. E. Hafez and R. J. Blandau. Univ. of Chicago Press, Chicago.

Palade, G. E. (1952) A study of fixation for electron microscopy. J. Exp. Med., 95:285–298.

Reinius, S. (1967) Ultrastructure of blastocyst attachment in the mouse. Z. Zell-forsch., 77:257–266.

Reynolds, E. S. (1963) The use of lead citrate at high pH as an electron-opaque stain in electron microscopy. J. Cell Biol., 17:208–212.

Schlafke, S. J., and A. C. Enders (1963) Observations on the fine structure of the rat blastocyst. J. Anat., 97:353–360.

Szollosi, D. (1967) Fixation procedures of embryonal tissues for electron micros-copy. In: Methods of Developmental Biology. Ed. by F. Wilt and N. Wessels. Crowell, New York.

Van Benden, E. (1880) Recherches sur l'embryologie des Mammiferes. La forma-tion des feuillets chez le Lapin. Arch. Biol., 1:137–224.

Weakley, B. S. (1967) Investigations into the structure and fixation properties of cytoplasmic lamellae in the hamster oocyte. Z. Zellforsch., 81:91–99.

Wood, R. L., and J. H. Luft (1967) The influence of buffer systems on fixation with osmium tetroxide. J. Ultrastruct. Res., 12:22–45.

17

Preparation of Chromosomes and Interphase Nuclei of Early Embryos for Autoradiography

JANICE DURR KINSEY
Department of Natural Sciences and Mathematics
Avila College
Kansas City, Missouri 64145

This chapter describes methods used for mounting early mammalian embryos directly on slides for use in light-microscope cytological and autoradiographic studies of metaphase chromosomes and nuclei in the other stages of the division cycle. By making flat spreads of the embryos on slides, one may bypass the procedures involved in making paraffin sections, and thus obtain better preparations.

Other workers have reported methods for studying early embryos (Jagiello, 1965; Melander, 1962; Moog and Lutwak-Mann, 1958; Sato, 1965; Tarkowski, 1966; McFeely, 1966). Where applicable, an attempt will be made to describe briefly their techniques, as there is no single method which is best for all species, stages, and purposes. The methods reported here have been developed using rabbit and mouse embryos; it is hoped that the readers will adapt these techniques to their own uses.

MATERIALS

To examine embryos during treatment one needs a dissecting microscope with a magnification range of about 10× to 30× and a light source with substage mirror or reflector. A compound microscope is used after the embryos have been put on slides.

This work was supported by National Institutes of Health Postdoctoral Fellowship 5 F2 HD-31, 986-02. The author wishes to thank Professor Stanley M. Gartler, in whose laboratory the work was carried out, for his support and help.

Micropipettes of various bores for handling embryos in various stages are made from glass and rubber tubing as described in New (1966).

For fixation of embryos, disposable plastic Petri dishes (Falcon Plastics) are used in two sizes, 60×15 mm and 35×10 mm. Maximov tissue culture slides are used to hold the various solutions through which the embryos are passed. One may also use multiple-concavity slides which are very useful if the embryos are to be passed through a series of solutions in rapid succession. All glassware is coated with Siliclad (Clay-Adams). During treatment prior to fixation the embryos, and the solutions through which they are passed, are kept warm on a 37°C warming plate.

Fine-drawn glass needles and standard histological materials such as slides and cover slips, Coplon jars, mounting medium, alcohol series, and xylol are used. Solutions used include tissue culture medium (type will vary according to the embryos used), 0.5% pronase (Calbiochem) in Hank's BSS without bicarbonate, physiological saline, hypotonic solution (0.8 or 0.9% sodium citrate), fresh Carnoy's fixative (three parts absolute methanol: one part glacial acetic acid), and a mitotic arrestant for accumulation of metaphases [Velban (Vinblastine sulfate), Eli Lilly and Company]. Straining is done with 2% acetic orcein and McNeill's tetrachrome, or any other stain.

METHODS

Obtaining and Treating Embryos Prior to Fixation

Methods for superovulation, recovery of embryos at the various stages, and *in vitro* culture of embryos may be found in other parts of this book and elsewhere (New, 1966).

Many of the methods used for the study of *in vitro* cell cultures may be applied to early embryos. Such procedures as karyotype analysis, isotopic chromosome labeling, labeling of other cellular components, estimation of the lengths of the parts of the cell cycle, and cytological studies of heterochromatin and nucleoli may be carried out using early mammalian embryos.

If metaphase chromosomes are desired, the embryos are flushed from the reproductive tract of the mother with the medium that is to be used for culture; they are then transferred to a test tube containing another 1 or 2 ml of the medium to which a mitotic arrestant has been added. I have used Velban (kept as a stock solution of 1 gamma/ml), but colchicine and Colcemid are also effective. The optimum concentration of mitotic inhibitor varies with the exposure time and the species used; for a 7 or 8 hour exposure to Velban, I use 0.02 gamma/ml for mouse embryos, and 0.03 to 0.04 gamma/ml for rabbit embryos. Tarkowski (1966) and Sundell (1962), using the mouse and the golden hamster respectively, obtained metaphases by injecting Colcemid and colchicine into the mother prior to removal of the embryos. If autoradiography is to be done, standard radioactive labeling procedures are used, preferably *in vitro*. For most purposes it is not necessary to remove the embryonic membranes prior to incubation as the labeled compounds and mitotic arrestants penetrate them.

General Handling Procedures

Pipettes should be made in a variety of bore sizes. The bore of the pipette used should be slightly greater than the diameter of the desired embryo; too large a pipette results in too great a volume of fluid being carried from one solution to the next along with the embryos. The pipettes and depression slides should be siliconized to prevent sticking.

For best results, embryos are kept as close as possible to 37°C for all steps except fixation. A warming plate kept at 37°C may easily be kept near the microscope so that depression slides containing various solutions and embryos may be kept warm; a heated microscope stage is unnecessary.

Surface tension and convection currents are hazardous to early embryos, particularly if the membrane has been removed. Therefore, when transferring embryos from one solution to another, a small amount of air is expelled from the pipette before the tip is put into the solution, thereby avoiding the expulsion of air bubbles into the fluid. If several embryos are to be picked up at one time, the slide is swirled gently, causing the embryos to collect at the bottom of the depression. The embryos are picked up in a minimum of fluid and the tip of the pipette is placed in the next solution so that its tip rests at an angle against the substrate. The embryos are expelled slowly from the pipette, care being taken to avoid expelling air bubbles. To avoid air pockets in the pipette, one should keep the tip submerged once it is placed in the fluid.

Enzymatic removal of the zona pellucida and other egg coats may be necessary in embryos that are to be used for chromosome preparations. Pronase digestion of the membrane is described elsewhere in this book and in New (1966).

Preparation of Cells in Stages Other than Metaphase

If other than metaphase nuclei are to be labeled, embryos are incubated in medium and then rinsed several times in saline; if labeling is not done, the embryos are fixed immediately. Normally the fixative removes the zona pellucida, but in the rabbit, pronase digestion may be necessary to remove the albuminous layer, particularly in earlier embryos. The fixative used is Carnoy's, freshly made prior to use, which is easiest to handle if it is used to fill a small disposable plastic Petri dish which has been set in the lid of a larger disposable plastic Petri dish. By touching only the large Petri lid, one may move the dish of fixative while examining it under a dissecting microscope and without having the fixative touch the fingers. The arrangement may be stored in the refrigerator overnight. When covered with the bottom of the large dish, the double cover helps retard evaporation of the fixative. Since the alcohol evaporates much faster than the acid, this is important in maintaining the composition of the fixative.

I routinely allow the embryos to remain in the fixative overnight in the refrigerator to allow for complete removal of the zona pellucida and for thorough fixation. After the embryos have been fixed, they are located in the dish of fixative from which they are removed with a micropipette. If, in transferring embryos to the fixative, one is careful to expel the embryos directly onto the bottom of the dish and to allow the dish to sit undisturbed for a few minutes before moving it, the embryos tend to adhere to the plastic bottom of the dish near the spot where they were expelled. This facilitates their location. Any embryos that are stuck to the bottom of the dish are easily dislodged by a nudge with the pipette tip accompanied by a gentle suction from the pipette. Some embryos may settle at the edge of the dish at the junction of the side and the bottom, and may be extremely difficult to see. Also, some embryos usually remain free-floating in the fixative and the slightest convection current may cause them to swirl rapidly; when this happens, a low power of magnification and up-and-down focusing is used. Embryos of different species have different optical properties when fixed: early rabbit embryos become almost completely transparent in fixative, whereas mouse embryos become darker and are more easily seen. Embryo visibility can be changed considerably by altering the angle of the substage mirror or reflector. This can be especially useful in searching for embryos at the edge of the dish. With practice one can practically eliminate loss of embryos in the fixative.

If large blastocysts (e.g., four- to six-day rabbit embryos) are to be mounted for the study of structures other than metaphase chromosomes, one may follow the method described by Moog and Lutwak-Mann (1958). Basically, the technique consists of placing the embryo on a cover slip shallowly submerged in fixative and dissecting the spherical embryo with fine glass needles to form a star-shaped preparation. The preparation is then air dried and may be stained or prepared for autoradiography.

If cleavage stage embryos, small blastocysts, or large blastocysts are to be prepared (see Figs. 17-1a, b) the embryo is picked up in a pipette with a minium of fixative, and expelled directly onto a clean slide, where it may be seen as a tiny lump as the fixative spreads out; a small drop of 2% acetic orcein is placed immediately on the embryo, and a cover slip is applied carefully. [This method is essentially the one used by Melander (1962) in a study of chromatin body formation in early rabbit embryos.] The slide is then examined under a dissecting microscope to locate the embryo, and its location is marked on the underside of the slide with a felt marking pen. The slide is then examined under the compound microscope, and, if squashing is needed, gentle pressure may be applied to the cover slip until the desired degree of spreading is achieved. Usually, the use of a minimal amount of fixative obviates further squashing because the weight of the cover slip is sufficient to flatten the embryo. To remove the cover slip before either permanent mounting or autoradiography, the slide is placed on a piece of dry ice until the stain is frozen. The cover slip is quickly flicked off with a razor blade, and the slide is immediately lowered into a Coplon jar of fresh fixative in which it is allowed to remain until the stain has been removed; a minute or two is usually enough for destaining. If acid hydrolysis is needed, the slide may be transferred to an acid bath, rinsed in water,

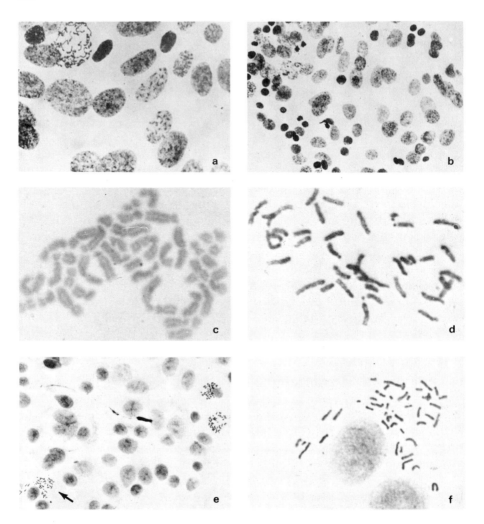

FIGURE 17-1

(a) *Two and one-half-day mouse embryo (morula)*, *fixed immediately after removal from mother, placed on slide, covered with a drop of 2% acetic orcein, and coverslip added with no additional squashing. Coverslip removed by freezing, slide rinsed in fixative, air dried, and restained in Tetrachrome for 30 seconds.* ×950. (b) *Four-day mouse embryo prepared in same way as that shown in* (a). ×380. (c) *Two-day (sixteen-cell) rabbit embryo metaphase prepared for autoradiography by six hours continuous label in 1 μc/ml tritiated thymidine and 0.03 gamma/ml Velban. Preparation is made by dropping the fixed embryo from a height of 2 or 3 inches onto a wet slide and then air-drying. Photograph made after autoradiography, following removal of silver grains from photographic emulsion. Tetrachrome stain was used.* ×3420. (d) *Three-day mouse embryo metaphase prepared for autoradiography by a 15-minute pulse label with 1 μc/ml tritiated thymidine and a 30-minute exposure to 0.04 gamma/ml Velban six hours after pulse. After 15 minute hypotonic treatment and overnight fixation, embryo was put on slide, covered with small drop of 2% orcein, and coverslip applied with no additional squashing. Photographed after autoradiography, following removal of silver grains from emulsion. Tetrachrome stain.* ×912. (e) *Four-day rabbit blastocyst incubated in 0.03 gamma/ml Velban for three hours. Zona pellucida and mucous coat removed by pronase and dissection. Embryo treated with hypotonic solution for 20 minutes, fixed overnight, placed on slide, covered with drop of 2% acetic orcein, and coverslip added with light squashing. Coverslip removed, slide rinsed in fixative, air-dried, restained with Tetrachrome, and coverslip permanently mounted.* ×456. (f) *Enlargement of metaphase (arrow) shown in* (e). ×2014.

and air dried. Otherwise, the slide is air dried immediately after its removal from the fixative and is ready for autoradiography or staining.

PREPARATION OF METAPHASES

Several workers have reported methods for the preparation of metaphase chromosomes from early mammalian embryos. An attempt will be made here to mention their various methods so that others may adapt the requisite steps to their own uses.

After incubation and rinsing, embryos with thick membranes (e.g., rabbit embryos) are denuded with pronase, repeated pipetting, and dissection. For mouse embryos, removal of the zona pellucida may be unnecessary. The next step produces swelling of the cells by treatment with a hypotonic solution, as in the preparation of metaphase chromosomes from tissue culture cells. The embryos are transferred from the saline rinse, following either incubation or pronase treatment, to a concavity slide containing a hypotonic solution. I use 0.8% sodium citrate for most cells, but tonicity may be varied according to the requirements of the experimenter. At 37°C, early mouse and rabbit embryos in cleavage stages are treated for 10 to 15 minutes. Again, the length of hypotonic treatment varies according to the species and stage used. One can observe the swelling of the cells with the aid of the dissecting microscope and, with practice, it is possible to judge accurately by observation the degree of swelling. Also, some cells, particularly in the earlier cleavage stages, tend to dissociate in the hypotonic solution. When this is observed, one should transfer the dissociating embryos to fixative that makes the cells more cohesive. Tarkowski (1966) describes a method of chromosome preparation for early mouse embryos: basically, a hypotonic treatment followed by transfer of the embryo in a tiny drop of the hypotonic solution to a slide. The cells are then spread and fixed by letting individual drops of the fixative fall directly onto the embryo, allowing each drop to spread and begin to dry before addition of the next. After air drying, the slide is ready for staining or autoradiography. Sato (1966) bypasses the fixation, putting the embryos onto a slide from the hypotonic solution, blotting most of the hypotonic solution, adding a drop of acetic dahlia solution and a cover slip, and squashing. Jagiello (1965) has prepared mouse ovum meiotic chromosomes by fixing and squashing eggs in 50% acetic acid, removing the cover slip by freezing, hydrolyzing in acid, and staining in Giemsa. McFeely (1966) has prepared metaphase chromosomes from large pig blastocysts by treating them essentially as one would tissue culture cells; the pig blastocysts are larger than the embryos considered here, however.

As shown in Figure 17-2, embryos may be transferred to the previously described dish of fixative from hypotonic solution, and may be stored in fixative for up to one day in the refrigerator. After fixation, one of two courses may be followed. The first, which was used in a study of 16- to 32-cell rabbit embryos (Kinsey, 1967), entails picking up the cells in a pipette and dropping them from a height of several inches directly onto a clean slide covered by a film of cold water, a procedure similar to that used

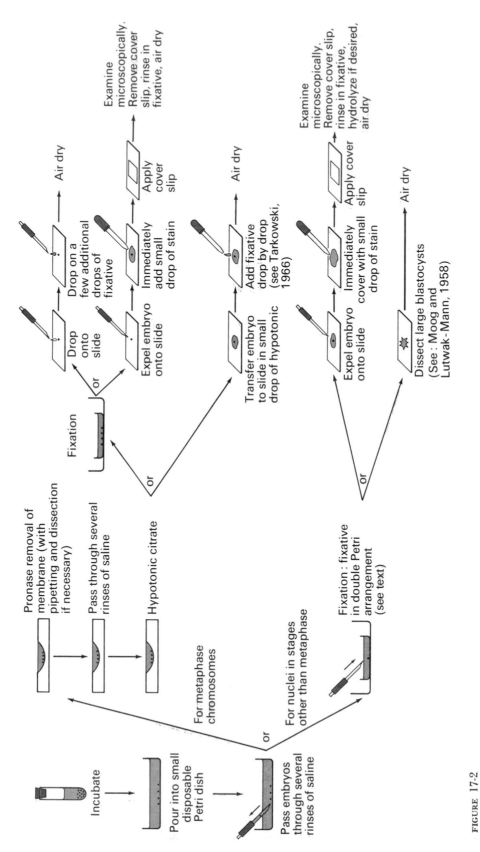

FIGURE 17-2
Diagrammatic representation of methods used in the preparation of embryos for various purposes.

in the preparation of tissue-culture cell metaphases. The slide is held at an angle of about 45° so that the drops of fixative spread when they hit the slide. The water beads at the edge of the slide, and is carefully blotted. The slide is then air dried (see Fig. 17-1c). Another method, which is suitable for later embryos as well as for cleavage stages, involves placing the embryo in a drop of the fixative on a clean slide. Immediately before the fixative dries, a small drop of stain is placed on the embryo and a cover slip is carefully applied. The embryo is found at low magnification under the dissecting microscope and its location is marked on the underside of the slide. The length of hypotonic treatment should be adjusted so that squashing, other than that provided by the weight of the cover slip, is unnecessary (see Figs. 17-1d, e, f).

The stain that has proved most satisfactory for the squashing procedure is 2% acetic orcein. The slide, prepared as described above, may be examined microscopically to determine its suitability for autoradiography or permanent mounting. If suitable, the slide is frozen on dry ice, the cover slip is flicked off, and the slide is rinsed in fresh fixative and air dried. Any further staining or autoradiography may then be done. These methods are summarized in Figure 17-2.

The methods described in this chapter have been found effective for a range of purposes that is fairly limited considering the number of possible uses. It is hoped that others may find them useful as guidelines, and will expand and improve on them for their own purposes.

BIBLIOGRAPHY

Jagiello, G. M. (1965) A method for meiotic preparations of mammalian ova. Cytogenetics, 4:245–250.

Kinsey, J. D. (1967) X-chromosome replication in early rabbit embryos. Genetics, 55:337–343.

McFeely, R. A. (1966) A direct method for the display of chromosomes from early pig embryos. J. Reprod. Fertility, 11:161–163.

Melander, Y. (1962) Chromosomal behaviour during the origin of sex chromatin in the rabbit. Hereditas, 48: 645–661.

Moog, F., and C. Lutwak-Mann (1958) Observations on rabbit blastocysts prepared as flat mounts. J. Embryol. Exp. Morphol., 6:57–67.

New, D. A. T. (1966) Culture of vertebrate embryos. Academic Press, New York.

Sato, A. (1965) A method for the demonstration of chromosomes in the cleaving mouse egg. Cytologia, 30:462–464.

Sundell, G. (1962) The sex ratio before uterine implantation in the golden hamster. J. Embryol. Exp. Morphol., 10:58–63.

Tarkowski, A. K. (1966) An air-drying method for chromosome preparations from mouse eggs. Cytogenetics, 5:394–400.

18

Studies of Metabolic Processes in the Preimplantation Conceptus

LOUIS FRIDHANDLER

Department of Medicine
University of California, Irvine
Irvine, California 92664

MICROMANOMETRIC TECHNIQUES

CARTESIAN DIVER

One of the earliest and still useful modes of investigation of metabolism is to examine one of the indexes of the overall metabolic rate: for example, the rate of oxygen uptake, CO_2 production, or water production. For a number of technical reasons it is most practical to measure oxygen uptake. A simple device for manometric measurement of gas exchange, commonly called the Warburg (Burk and Hobby, 1954), is suitable for tissue preparations of the order of 100 mg wet weight. The Warburg method depends on the principle of measurement at constant volume of the change of gas pressure within the vessel containing the tissue. The same principle is used in the method of the Cartesian diver (Holter, 1943); however, the volume of the gas space in the Warburg vessel is about 1000 times that in the Cartesian diver, and the sensitivity of the Cartesian diver method is therefore at least 1000 times that of the Warburg. This great sensitivity is enough for measurement of oxygen uptake of about 10 rabbit ova (during the cleavage stages) or of three Day 4 blastocysts, or one Day 5 blastocyst. A Day 6 blastocyst is too large for the ordinary Cartesian diver and therefore must be divided into two. Adequate directions for the diver method are given in a summary by Glick (1949). Equipment and accessories for the Cartesian diver procedure can be obtained from Ole Dich Instrument Makers, 18 Holmevej, Hvidovre, Denmark.

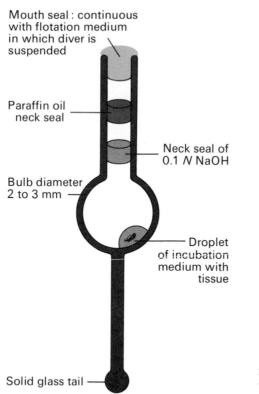

Mouth seal: continuous with flotation medium in which diver is suspended

Paraffin oil neck seal

Neck seal of 0.1 N NaOH

Bulb diameter 2 to 3 mm

Droplet of incubation medium with tissue

Solid glass tail

FIGURE 18-1
The Cartesian Diver.

Although the principle is the same, the method for achieving constant volume in the diver is far different from that used in the Warburg method. The gas space in the Warburg vessel is connected directly to the gas space over the liquid in the capillary in one arm of a U-shaped manometer. This manometer is manipulated during the experiment so as to keep the volume in the vessel at a constant level. Any change in pressure is noted by reading the level of the other arm of the manometer.

The measurement of pressure at constant volume in the Cartesian diver is achieved by a flotation method. Figure 18-1 illustrates a diver loaded with tissue and suspended in the flotation medium. The flotation vessel, which contains the medium, is connected to a manometer with which pressure in that vessel can be manipulated and measured. The manometer is not directly connected with the gas space in the diver, but it is connected to the gas space over the flotation medium and any change of pressure in the flotation vessel obviously results in a similar change of pressure in the diver gas space. Reduction of pressure will cause the diver gas space to expand and, thereby, reduce over-all density of the diver. As pressure is reduced the diver will begin to float and rise at an accelerating rate in the medium; pressure must then be increased to increase diver density and make it drop. This cycle is then repeated, but with oscillations of ever-decreasing amplitude. Within a minute or two the oscillations will practically cease, thereby signaling the achievement of an unstable equilibrium. Under that condition the density of the diver as a whole is exactly equal to the density of the flotation medium. The manometric pressure is recorded and is called the *equilibrium pressure*. Equilibrium pressure readings are taken at appropriate intervals and variations with time should quantitatively reflect the rate of gas exchange. If the density of the diver as a whole is brought to the same value at each equilibrium pressure reading, one is effectively reading pressure at constant volume. Once the diver is loaded and is in the flotation medium, the only factor that can be varied to change the diver density is the volume of gas space. Since density is brought to the same value each time, volume is the same at each reading.

As oxygen is removed by the tissue from the gas space in the diver the buoyancy of the diver will tend to decrease. Therefore, it will be necessary to reduce pressure to bring volume back up to the value required to attain the equilibrium pressure. That is, equilibrium pressure falls with time in proportion to the rate of gas uptake.

There are factors (other than gas space) controlling diver density that are constant throughout any one experiment, and can be measured with great precision. These factors are: (1) volume and density of neck seals and aqueous incubation medium; (2) density of the glass from which the diver is made; (3) weight of the diver; and (4) density of the flotation medium. The volume of gas space at equilibrium pressure is mathematically related to the factors listed above. A relatively simple equation thus permits calculation of gas space with great accuracy (Glick, 1949).

THE BRAKING PIPETTE

An important equipment accessory in diver technique is the braking pipette. (Glick, 1949, p. 359; Claff, 1947). It is a capillary pipette of special construction which prevents the uncontrollably rapid ingress (by capillary action) of aqueous medium: the medium enters spontaneously but very slowly. This allows accurate control of uptake and discharge of medium and ova by means of suction and pressure by mouth. This type of pipette is exceedingly useful in picking up and transferring very small bits of tissue with an absolute minimum of accompanying aqueous medium, and it can be accurately calibrated. A brief description follows, but for detailed instructions see the references quoted above.

The essential part of the braking pipette consists of a thin-walled glass capillary (wall thickness approx. 0.05 mm; I.D., 0.3 to 0.4 mm). One end (distal to the ova being picked up) is drawn in a microflame to an exceedingly fine capillary 2 to 3μ I.D. This narrow capillary end acts as a brake to slow the flow of air through it and thereby slow the rate of inflow or outflow of aqueous medium when sucking up or discharging the sample or specimen. The capillary braking pipette is mounted in glass tubing through an air-tight cork. The narrower end is inside the glass tubing. The broader end into which the sample is sucked extends outside the tube one or one-and-a-half inches beyond the cork. A rubber tube is fitted to the glass tube in which the braking pipette is mounted. Negative and positive pressure applied by mouth via the rubber tube will allow very fine control of pick up and discharge of ova or other tissue in accurately controlled volumes of medium. A dissecting microscope facilitates this.

Rate of oxygen uptake can be studied most conveniently with air as the gas phase, however, it is possible to replace air with a gas mixture of another oxygen concentration, or even with an anerobic atmosphere. Glick (1949) describes a suitable method, but a more convenient, simple and effective method is given by Waterlow and Borrow (1949). In their method the diver is filled with water, submerged in water, and then the water inside is replaced by the desired gas atmosphere. In my experience it is wise to apply a silicone surface to the diver with Desicote (Beckman) or Siliclad (Clay-Adams).

Anerobic glycolysis of preimplantation conceptus can be measured mano-metrically in the Cartesian diver by incubating in Ringer Bicarbonate pH 7.4 with a gas phase consisting of 95% nitrogen and 5% carbon dioxide (Fridhandler, 1961). The method is based on the fact that as lactic acid is produced the acid reacts with bicarbonate. This results in the generation of carbon dioxide which increases the equilibrium pressure. Glick (1949) provides examples of how micromanometric methods may be designed to measure a wide variety of enzymes and substrates, as well as respiration and anerobic glycolysis.

The diver can also be used to detect and measure the capacity of the conceptus to convert C^{14} labelled substrates to $C^{14}O_2$ (Fridhandler, 1961). The NaOH neck seal used to absorb CO_2 during respiration measurement may be recovered at the end of an experiment. Radioactivity in the sodium hydroxide can be assayed by planchet counting or by liquid scintillation spectrometry. Representative analogous studies with other methods were recently published by Brinster (1968).

Rabbit ova, morulae or blastocysts can be obtained very simply by selecting a doe in estrus and placing her in the cage of a fertile male. Mating should normally take place within a few minutes, and the time of coitus is recorded. Ovulation occurs in response to the mating stimulus about 10 hours later; fertilization occurs a few hours after that. The day after mating, ova can be recovered by excising the oviducts and flushing with medium suited to the experiment to be conducted. If the day of mating is called Day 0 then ova are recoverable from the oviduct on Days 1, 2, and 3. On Day 3 the ovum has developed to the morula stage and is about to enter the uterus and become a blastocyst. Free-living unattached blasto-cysts can be recovered easily by flushing the excised uterus on Days 4, 5, and 6. On Day 7, the tough mucin coat which surrounds the blastocyst begins to soften and may be ruptured during the flushing procedure. Attachment to the uterine wall will sometimes be observed on Day 7; and on Day 8, when implantation is underway, the whole blastocyst cannot be flushed out. Typical blastocyst diameters are as follows: fourth day: 0.3 mm; fifth day; 1.0 mm; sixth day: 3.0 mm. The great expansion in blastocyst volume is due almost entirely to blastocyst fluid increase.

Tubal ova and the Day 4 blastocysts can be added to the Cartesian diver directly with the braking pipette. Most Day 5 and all Day 6 blastocysts must be punctured with glass needles under a dissecting microscope, and the tissue is dissected away from the mucin coat and blastocyst fluid. This must be done since their outer diameters, in the intact state, are too wide to enter the diver. The blastocyst can be dissected after puncturing in a watch glass under the incubation medium. The isolated tissue can then easily be drawn into the braking pipette and transferred several times through fresh medium to wash away the uterine debris, mucin coat, and the blastocyst fluid. Then the tissue can be transferred to a diver in a measured volume of incubation medium (about 1 μl). The advantage of removing the blastocyst fluid is that it permits complete control of the experimental incubation environment.

I have had no extensive experience with ova of other species, but have had enough to know that the appropriate techniques and their attendant difficulties vary greatly among them.

The ova of rats and mice, for example, are smaller than those of rabbits, and the delicacy and convoluted anatomy of the rat or mouse oviduct calls for special skills in the flushing procedures. Rat and mouse blastocysts normally implant shortly after arrival in the uterus, and this, coupled with the impossibility of determining their ovulation time accurately, makes it difficult to obtain blastocysts directly from the rat or mouse uterus. Brinster and Thomson, 1966; Brinster, 1965; Sugawara and Umezu, 1961; and Boell and Nicholas, 1948 have successfully used rat and mouse ova for metabolic studies.

STUDIES OF INTERMEDIARY METABOLISM

Uptake and production of specific organic substrates can be studied directly under suitable conditions. Substrate-uptake studies become possible with microquantities of tissue only if the reservoir of substrate to be taken up by that tissue from the incubation medium is relatively small. If that reservoir is small, then the small amounts of tissue may subtract a large enough proportion of the substrate to be detected by a sensitive and precise method. Examples of such studies are given in recent papers (Fridhandler, *et al.* 1967; Fridhandler, 1968). The method involves the incubation of rabbit blastocysts in a very small volume (40 microliters) of medium containing the substrate. The small volume allows use of a sufficiently high concentration of substrate with maintenance of a high enough proportion of tissue mass (about 100 micrograms of protein) to substrate mass (of the order of 200 mμmoles). A large volume of medium (1 or 2 ml) with the same substrate concentration would result in presenting to the small amount of tissue an overwhelmingly large reservoir of substrate, which the tissue could not appreciably reduce.

Enough tissue for a single such incubation can be obtained from three to five Day 6 blastocysts, or about ten Day 5 blastocysts. It is important to have enough tissue to cause measurable changes in the medium, but too high a concentration of tissue would be undesirable since factors such as diffusion of oxygen, substrates, or products, may become limiting and distort the observations. The tissue quantities recommended appear suitable since

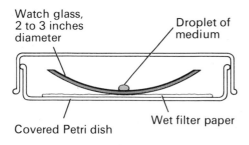

Watch glass,
2 to 3 inches
diameter

Droplet of
medium

Covered Petri dish

Wet filter paper

FIGURE 18-2

A suitable arrangement for incubation of microquantities of tissue in air in a small droplet of medium. This is adaptable to studies of substrate disappearance and production with both labelled or unlabelled compounds. This technique can also be used to study uptake of labelled precursors into tissue constituents such as fatty acids, DNA, RNA, and protein. The watchglass is silicone-coated so that the droplet of medium does not spread unpredictably. The coating can be applied with Desicote (Beckman Industries), or Siliclad (Clay-Adams, Inc., New York). The wet filter paper prevents evaporation when the whole assembly is placed in an incubator at 37°C.

there is no obvious overcrowding. The droplet is small and the oxygen needs to diffuse but a short distance through the liquid to reach any part of the tissue.

The blastocyst must be punctured and the tissue dissected as described above, and the tissue then transferred (by means of the braking pipette) to the droplet of incubation medium (see Fig. 18-2).

If desired, one can start with a 50-microliter droplet and, after stirring, take two 5 microliter samples to carry out zero-time measurements of product and substrate. A one- or two-hour incubation is sufficient for quantitative measurement by enzymatic methods of glucose disappearance, or of lactate production as described by Fridhandler *et al.* (1967). To insure that results are comparable from vessel to vessel it is important to arrange the groups of blastocysts so that each group contains blastocysts of approximately the same size range and distribution as the comparable group in the same experiment.

It may be desired to study the capacity of the preimplantation conceptus to convert labelled small molecule precursors to various intermediates. Information on the intermediary metabolic paths available to the conceptus under experimental conditions can thus be obtained. This can be done by carrying out the incubations as above for one or two hours in the presence of glucose-C^{14}. At the end of the incubation, the medium and/or tissue contents are analyzed by paper chromatography. The location and intensity of the radioactive spots give qualitative and quantitative data on disappearance of precursor and appearance of accumulated products. A specific example of the study of the conversion of glucose-U-C^{14} to amino acids is given here; Popp (1958) carried out analogous studies with mouse blastocysts.

At the end of incubation with glucose-U-C^{14}, the medium and tissue are placed in 80% aqueous ethanol and heated for 10 to 15 minutes at approximately 60°C. The resultant aqueous ethanol extract is divided into two equal parts and each is evaporated to a small volume in a 500-microliter polyethylene centrifuge tube (Beckman/Spinco). The extract is then placed on a spot on a square of filter paper (Whatman No. I, 8" x 8"). Chromatography is then carried out in two dimensions by capillary ascent. For the separation of amino acids a suitable developing solvent in the first dimension is made of 60% butanol; 15% acetic acid; and 25% water. A suitable second dimension solvent consists of 90 milliliters of 88% phenol; 10 milliliters of water; and 0.5 milliliters of 28% ammonia. Whether the tissue has formed new labelled metabolites from the original glucose is revealed by exposing the chromatogram to a sheet of X-ray film (Kodak Blue Brand), thus producing a radioautogram. Before exposure to the film, the chromatogram is marked with radioactive dye in three of the corners. This will enable precise positioning of X-ray picture and chromatogram against each other. The suspected identity of a newly formed metabolite is tested by addition of authentic crystalline nonradioactive carrier amino acid whose position on the chromatogram can be revealed by ninhydrin. Perfect correspondence as to position and shape between radioactive spots and carrier spots (chemically detected by ninhydrin) provides evidence of the identity of the spots.

Only one of the pair of parallel chromatograms obtained from each run is stained with ninhydrin. Ninhydrin causes loss of carboxyl groups in amino acids and, therefore, renders the stained chromatogram useful only for qualitative identification purposes. The relevant, identified spots in the other parallel unstained chromatogram can be eluted with water. The eluate is evaporated to about 0.2 milliliters and then put into solution in 12 milliliters of the scintillator solvent Liquifluor (Nuclear Chicago) with 2.8 milliliters of absolute ethanol. The ethanol is required to bring the aqueous extract into solution in the toluene-based scintillator solvent. Alternatively, scintillator "cocktails" may be used which accept aqueous solutions directly. The utilization of C^{14}-labelled glucose and the production of labelled pools of glutamate, aspartate, alanine, and glutamine were studied by such means (Fridhandler, 1968).

Fatty acid synthesis may be studied by incubating as above with acetate-1-C^{14} as substrate (Fridhandler, 1968). The fatty acids can then be isolated and their radioactivity measured by the method of Emerson and Van Bruggen (1958). The microquantities of tissues require that the extractions be carried out in microcentrifuge tubes (Beckman/Spinco) rather than in separatory funnels.

MACROMOLECULE BIOSYNTHESIS

Assaying Radioactivity in RNA, DNA, and Protein

The study of macromolecule synthesis can be carried out by incubating, as described above, with appropriate radioactive precursors. The tissue is then fractionated by perchloric acid extraction to separate and isolate, as well as possible, the RNA, DNA, and protein fractions. Combination of this technique, with the use of specifically labelled precursors, allows one to attain considerable specificity and quantitative reproducibility (Fridhandler *et al.*, 1967; and Palmer and Fridhandler, 1968). Tritiated thymidine for DNA, tritiated uridine for RNA, and C^{14}-labelled leucine for protein are the subtrates of choice for measurement of each of the types of macromolecules discussed here.

The method used is a modification of that of Ogur and Rosen (1950). At the end of the incubation, the tissue is drawn into the capillary, braking pipette and transferred to 500 microliter polyethylene microcentrifuge tube containing approximately 400 microliters of 70% aqueous ethanol. The tubes are vibrated to mix for approximately 20 seconds and centrifuged in a microcentrifuge (microfuge: Beckman/Spinco) for one minute. The supernates are discarded. The remaining steps in the procedure are as follows (approximately 400 microliters are used in each wash or extraction):

1. Wash with 70% aqueous ethanol containing 0.1% perchloric acid, discard supernatant.

2. Add absolute ethanol : diethylether (3:1), mix, heat for 3 minutes at 65°C, centrifuge and discard supernatant.

3. Repeat step 2.

4. Wash twice with cold 0.2 molar perchloric acid, discard supernatant.

5. Extract with 1.0 molar perchloric acid at 4°C for 12 to 18 hours.

6. Wash with cold 1.0 molar perchloric acid, add supernatant from step 5 and retain as RNA-containing extract.

7. Extract twice with 1.0 molar perchloric acid for 30 minutes at 80°C; combine supernatants as DNA-containing extract.

8. Wash remaining residue with absolute ethanol and with diethylether; evaporate residue to dryness. This residue is the perchloric acid-insoluble protein. The protein can be dissolved with Nuclear Chicago Solubilizer and added to Liquifluor (see below) for radioactive measurement.

The RNA- and DNA-containing extracts are neutralized with 1.0 molar Hyamine hydroxide, and then enough absolute ethanol is added so that the water in the extracts is brought into solution with the toluene scintillator solution. The liquid scintillation counting solution is prepared by adding 160 ml of concentrated Liquifluor (New England Nuclear Corporation) to 8 pints of toluene. Radioactivity is measured with a Nuclear Chicago Model 723 ambient temperature liquid scintillation spectrometer or other similar instruments. Efficiency is determined by the channels ratio procedure. C^{14} can be counted under these conditions at an efficiency of approximately 45% in the RNA- and DNA-containing extracts and about 70% in the protein samples. The corresponding tritium efficiencies are approximately 8 and 15% respectively.

The procedure outlined above is applicable to the Day 6 blastocysts, as well as the Day 5 blastocyst. The Day 4 blastocyst is too small to dissect and therefore the whole blastocyst will have to be added to the 70% ethanol. There should be no essential difficulty introduced by this difference in mode of handling.

To test the effects on uptake of a metabolite, drug, or analogue, the incubation medium will contain not only radioactive precursor but the appropriate metabolite, drug or analogue at the desired concentration.

This method of *in vitro* incubation offers the advantage of great control over the incubation medium. Compared to radioautographic methods, the quantitative data are obtainable much more quickly. Quantitation of the data is usually more readily attainable by means of an extraction method to isolate the relevant macromolecular fraction. Combined use of the extraction methods and a specific precursor enhances reliability of identification of the macromolecules.

Some of these studies can be carried out using the whole intact blastocyst rather than the dissected isolated tissue: for example, the uptake of precursor into the fatty acids can be studied in this way as well as the synthesis of DNA, RNA, and protein from the appropriate precursor. A sacrifice of incubation environment control is necessary but one gains the advantage of working with a more physiological preparation. Rabbit blastocysts will expand during culture in a medium such as Ham's F-10 (Ham, 1963) fortified with 5% rabbit serum. Daniel (1965) has modified the F-10 medium for optimum growth of Day 5 rabbit blastocysts. Both the rabbit serum and the enclosed blastocyst fluid in an intact blastocyst are not under the experimenter's control and contain many factors, many of them undefined. However, study of the action of metabolites, drugs, and analogues on macromolecule biosynthesis are still feasible under such conditions. Of course, at the end of incubation of intact blastocysts, it is necessary to dissect out the tissue to subject it to extraction with the micromethod described above.

BIBLIOGRAPHY

Boell, E. J., and J. S. Nicholas (1948) Respiratory metabolism of the mammalian egg. J. Exp. Zool., 109:267–281.

Brinster, R. L. (1965) Studies on the development of mouse embryos *in vitro*. IV. Interaction of Energy Sources. J. Reprod. Fertility, 10:227–240.

——— (1965) Carbon dioxide production from glucose by the preimplantation rabbit embryo. Exp. Cell Res., 51:330–334.

——— and J. L. Thomson (1966) Development of eight-cell mouse embryos *in vitro*. Exp. Cell Res., 42:308–315.

Burk, D., and G. Hobby (1954) Hydraulic-leverage principles for magnification of sensitivity of gas change in free and fixed-volume manometry. Science, 120:640–648.

Claff, C. L. (1947) Braking pipettes. Science, 105:103–104.

Daniel, J. C. (1965) Studies on the growth of five-day old rabbit blastocysts *in vitro*. Embryol. Exp. Morphol., 13:83–95.

Emerson, R. J., and J. T. Van Bruggen (1958) Acetate metabolism: effects of tracer concentration. Arch. Biochem. Biophys., 77:467–477.

Fridhandler, L. (1961) Pathways of glucose metabolism in fertilized rabbit ova at various preimplantation stages. Exp. Cell Res., 22:303–316.

——— (1968) Intermediary metabolic pathways in preimplantation rabbit blastocysts. Fertility Sterility, 19:424–434.

Fridhandler, L., W. B. Wastila, and W. M. Palmer (1967) Role of glucose in metabolism of the developing mammalian preimplantation conceptus. Fertility Sterility, 18:819–830.

Glick, D. (1949) Techniques of histo-and cyto-chemistry. Interscience, New York.

Ham, R. G. (1963) An improved nutrient solution for diploid Chinese hamster and human cell lines. Exp. Cell Res., 29:515–526.

Holter, H. (1943) Technique of the cartesian diver. Compt. Rend. Trav. Lab. Ser. Chim., 24:399–478.

Ogur, M., and G. Rosen (1950) The Nucleic Acids of Plant Tissues. I. The extraction and estimation of DNA and RNA. Arch. Biochem., 25:262–276.

Palmer, W. M., and L. Fridhandler (1968) Effects of growth-inhibiting antibiotics on macromolecule biosynthesis in preimplantation rabbit conceptus. Fertility Sterility, 19:273–285.

Popp, R. A. (1958) Comparative metabolism of blastocysts, extraembryonic membranes, and uterine endometrium of the mouse. J. Exp. Zool., 138:1–23.

Sugawara, S., and M. Umezu (1961) Studies on the metabolism of the mammalian ova II oxygen consumption of the cleaved ova of the rat. Tohoku Journal of Agricultural Research, 12:17–28.

Waterlow, J. C., and A. Borrow (1949) Observations on cartesian diver technique. Compt. Rend. Trav. Lab. Carlsberg, Ser. Chim., 27:93–123.

19

Preparation of Preimplantation Embryos for Autoradiography

HARRY M. WEITLAUF
AND GILBERT S. GREENWALD
Departments of Obstetrics,
Gynecology, and Anatomy
University of Kansas Medical Center
Kansas City, Kansas 66103

Metabolic substrates tagged with radioisotopes have been widely used to examine chemical events in early embryonic development. The bulk of the work has been done with embryos from oviparous species because of the ease with which they can be collected, experimentally manipulated, and evaluated for incorporation of tracers (See Weber, 1965; Monroy, 1965; and Gross, 1967 for reviews). The use of radioactive tracers to study metabolic changes in embryos from viviparous species has been hindered by the complexity of the fetal-maternal relationship and the relatively small number of embryos in each female. Furthermore, the study of embryos prior to implantation is complicated by the difficulties involved in manipulating the eggs for evaluation of tracer incorporation.

In some cases the radioactive materials can be detected and quantified with counting devices, however, the small amounts of radioactivity and the limited sample size in experiments with preimplantation embryos may necessitate the use of autoradiography. Therefore, the following discussion is presented to outline several methods which have been useful in preparing preimplantation embryos for autoradiography (Greenwald and Everett, 1959; Weitlauf and Greenwald, 1965, 1967).

Harry M. Weitlauf was supported in this work as a United States Public Health Service postdoctoral fellow. The research was supported by grants from NIH (HD-00596) and The Ford Foundation.

AGAR-GEL PARAFFIN DOUBLE EMBEDDING

This technique allows the removal of eggs from the reproductive tract, therefore eliminating background radiation from maternal tissues and permits the orientation of eggs for sectioning in any plane (Samuel, 1942). It can be used to prepare eggs for autoradiography by any laboratory equipped for routine histological work.

Recovery of eggs from small laboratory animals is easily accomplished (see Chap. 4 by Gates and Chap. 8 by Dickmann). Once eggs have been recovered from the reproductive tract (or from tissue culture medium in *in vitro* experiments) it is necessary to wash away excess maternal fluids and blood which may be radioactive and thus contribute to background artifacts. Washing the eggs can be tedious because they are easily lost. To reduce this potential loss it is helpful to perform all steps of washing under direct observation at ×10 to ×15 with a dissecting microscope, using indirect lighting.

First remove the eggs from the flushing solution and from the debris collected from the reproductive tract. A fine glass micropipette (inside diameter 90 to 100μ) controlled by oral air pressure through a length of rubber tubing is used to collect the eggs (singly or in groups of two or three) and to transfer them to a fresh rinse solution of normal saline. The eggs are then rinsed and agitated with a stream of saline from the pipette, re-collected in a small volume of rinse solution, subjected to a second rinse, and finally a third. In each rinse the same washing procedure is followed.

During the rinsing the eggs tend to become sticky and can adhere to the glassware. This can be overcome by two methods: Either use a drop of wetting agent in the wash water, e.g., Teepol 1:10,000, (Braden, 1952) or simply coat the depression slide with agar gel. The agar is made up in a 1.3% concentration, melted, placed in a depression slide, spread evenly, and chilled. Saline for washing the eggs is then gently poured on the gel coating. Eggs will not stick to the agar and repeated washing can be done without egg loss.

In some cases a rinse fluid other than saline may be necessary; it is best to avoid one containing serum or albumin because protein from these solutions will coagulate during fixation, covering the egg with debris and obscuring the final preparation.

The eggs are fixed after adequate rinsing. The fixative used should not interfere with the experiment, e.g., alcoholic fixatives should be avoided if the tagged compound is a steroid which is alcohol-soluble. The fixative may, however, be helpful in removing unwanted radioactivity, e.g., in experiments that use tagged amino acids to evaluate protein synthesis, Bouin's solution will remove free amino acids and will leave only those which are incorporated into protein (Droz and Warshawsky, 1963).

Fixing, like washing, is performed under direct observation. The eggs are deposited in 1 or 2 ml of fixative in a depression slide and are then covered to prevent evaporation. One to two hours is enough to fix mouse eggs when Bouin's solution is used; for other fixatives and eggs from other species it

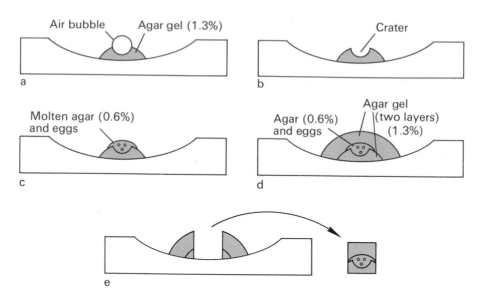

FIGURE 19-1

Steps in the formation of an agar block to contain preimplantation eggs for dehydration, embedding and sectioning: (a) Molten agar (1.3%) is placed in a depression slide with a small air bubble on the surface. (b) After the agar has cooled the bubble is broken, leaving a small crater in the agar. (c) Eggs are deposited in the crater with molten agar (0.6%, see text). (d) A second layer of agar is added to cover the crater. (e) A block of agar containing the eggs is cut from the gel sandwich and removed for processing.

may be necessary to conduct preliminary tests to determine the appropriate time. Overfixation tends to collapse blastocysts and to shrink blastomeres, thus distorting the embryo. After fixation is completed the eggs are washed several times to remove free radioactivity in the fixative; they are then ready for embedding. Once the eggs are fixed they are, of course, nonvital, and the saline rinse at this stage may remove water-soluble compounds such as glycogen. If the tagged molecule is to be incorporated into a soluble compound, this wash may be modified by using a different solution, such as, fresh fixative or 50% alcohol.

The next step is to embed the egg in an agar block. This can be difficult because each step must follow quickly and it is necessary to have everything ready before starting. The agar is made from BACTO agar (DIFCO) in two strengths: 1.3% and 0.6% by weight. Agar can be kept conveniently in a large covered test tube in the refrigerator for two to three weeks without spoiling. While the eggs are being fixed, the agar is melted in a water bath at 60°C. To prepare the embedding block, 1 to 2 ml of melted agar (1.3%) is dropped into a depression slide and a small air bubble is left at the top (see Fig. 19-1a). When this has cooled the bubble is opened with a needle and a crater is left in the agar (see Fig. 19-1b). Difficulty can be encountered in the next step which must be carried out rapidly. The washed eggs are drawn into the micropipette in a minimal amount of fluid; the slide with the agar crater is then placed on the stage of a dissecting microscope and a drop of melted agar (0.6%) is placed in the crater; before the drop cools, the pipette is inserted into the liquid agar and the eggs are expelled gently under direct observation (see Fig. 19-1c). If this is done before the agar has hardened, the eggs will lie together in the bottom of the depression and can be oriented with a glass needle before they become trapped in the hardening gel. When the agar has cooled, a second application of agar (1.3%) is spread on top and the agar sandwich is chilled in the refrigerator for several minutes (see Fig. 19-1d). A block containing the eggs is then cut from the gel for dehydration and clearing (see Fig. 19-1e).

Dehydration and clearing are performed in disposable plastic tissue capsules (Lab Tek) which protect the blocks from damage. It is important to avoid rapid changes in alcohol concentrations; these tend to collapse blastocysts and disrupt the normal morphology, especially of large blastocysts. Concentration changes that are made in 10% increments in the range between 30% and 90% are satisfactory. The blocks stay in each alcohol concentration from 1 to 1.5 hours. Finally absolute alcohol, 95% and 100% concentrations, are used in duplicate; after the second absolute alcohol change the blocks are placed in methyl benzoate and saturated alcohol eosin (1:1) for 15 minutes. Light staining with eosin facilitates recognition of the eggs once they are sectioned. They are then cleared in methyl benzoate for 24 hours or more. Blocks can be stored in methyl benzoate for several days without damage. The blocks are next placed in benzene (three changes of a half-hour each), then put in benzene paraffin (1:1 at 58°C) for 1 hour, and finally are embedded in regular paraffin. The paraffin-agar blocks are cooled in ice water and are then sectioned serially at 6 to 10μ. Serial sections are scanned with a dissecting microscope for those which contain eggs, and the appropriate sections are mounted on albumin-coated slides. After drying 24 hours at 40°C in a dust free atmosphere the slides are deparaffinized by two changes of xylene (2 to 3 minutes), two changes of absolute alcohol (3 minutes each), and then air dried.

The next step in the preparation of autoradiographs is to apply photographic emulsion to the slide. There are several liquid emulsions available and the selection of the one best suited to a particular experiment will depend on the tracer to the used and the amount of radioactivity in the eggs (see Pelc et al., 1965). Liquid emulsions may be applied either by dipping the slides into it or by applying it with a fine brush. The emulsion is allowed to dry before the slides are placed in light-tight plastic slide boxes which are then sealed with black plastic tape. It is helpful to include a small cheesecloth sack of desiccant to prevent latent-image fading (Elias, 1964). The slides are stored at 10°C for exposure. The best exposure for a particular experiment will have to be determined empirically; it will vary with the amount of radioactivity in the eggs and the type of emulsion used. (Instructions for developing the autoradiographs are provided with the film; the development presents no problem unique to egg preparations.)

The eosin that has been used to stain the eggs may be removed by the solutions used in the remainder of the procedure so a counterstain is also used. Specific stains may be desired but one must be cautious in his choice, some stains will chemically reduce the emulsion and will give false positive results if they are applied before the emulsion is developed, whereas other stains may make the emulsion so dark that the eggs cannot be seen against the background (Thurston and Joftes, 1963). For general purposes we find that hematoxylin (1:2 with distilled H_2O), applied for 2 minutes with $2\frac{1}{2}$ minutes in tap water for bluing, stains the eggs and does not interfere with interpretation of the finished autoradiograph. After staining, the slides are run through a series of alcohol solutions (starting at 50%), and are then put in xylene and cover slips are applied.

The method described yields serial sections in which the morphology of the blastocyst is preserved and subcellular localization of certain radioactive compounds is made possible. It is, however, a tedious procedure requiring hourly changes in alcohol concentrations and four or five days are spent before the slides can be coated with emulsion. In addition, the dehydration with alcohol may make the method unsuitable for studies using tagged steroids. The whole-mount method may be employed to give good quality autoradiographs under certain conditions without the use of alcohol dehydration, and it requires only one or two days to prepare the slides for coating with emulsion.

WHOLE-MOUNT PREPARATIONS

Eggs are recovered, washed, and fixed (as described previously), and are then placed on albumin-coated slides in a drop of saline and are oriented with a glass needle. Excess saline is carefully removed and the eggs are allowed to dry overnight on a warming tray at 40°C. Mark the location of the eggs on the slide so that they can be easily found. The eggs are stained lightly with eosin or with another suitable stain, and then rinsed and dried. Emulsion is applied in the same way as to sectioned preparations.

Although the whole-mount method is quick and relatively simple, it does have disadvantages. The surface of the dried egg is uneven and the thickness of the emulsion may not be uniform (this is most pronounced when eggs in cleavage stages are studied), cellular detail is poor and subcellular localization of tracers is not possible, preparations of dried eggs are thicker than those cut at 6 to 10μ, and self-absorption by the egg may reduce the effectiveness of the autoradiograph.

Whole-mounts are therefore especially useful in preliminary experiments or in those in which an all-or-none response is expected (i.e., uptake or no uptake); they are less useful when graded uptakes are expected or when their purpose is to determine subcellular localization.

SERIAL SECTIONS

The two methods described above require manipulation of the eggs outside the reproductive tract; a third method of preparing preimplantation eggs for autoradiography is to prepare all or part of the female reproductive tract as if it were for a routine histology. After serial sections of the oviduct or uterus have been cut they can be examined with a dissecting microscope and those containing eggs are mounted. The preparation of these sections for radioautography is similar to that for any other tissue.

This process is simple and does not involve the handling of eggs. The major drawback is that the maternal tissues and the fluid in the lumen of the tract may be labelled and the subsequent high background can obscure the detection of radioactivity within the egg. This is most evident when high energy tracers are used and when radiation scatter may be extensive.

INTERPRETATION OF AUTORADIOGRAPHS

Interpretation of autoradiographs can be difficult, especially if comparisons are being made among eggs of different ages or from different treatment-groups. Subjective evaluations can be made when eggs are strongly positive or strongly negative, but will be difficult to grade when responses are between these two extremes. Grain counts, made with an ocular micrometer in a consistent region such as the center section of the inner cell mass, give objective results and are thus preferable.

BIBLIOGRAPHY

Braden, A. H. (1952) Properties of the membranes of rat and rabbit eggs. Australian J. Sci., 5:460–471.

Droz, B., and H. Warshawsky (1963) Reliability of the radioautographic technique for the detection of newly synthesized protein. J. Histochem. Cytochem., 11:426–435.

Elias, J. M. (1964) A desiccant holding slide box for radioautographic exposure. Stain Technol., 39:235–236.

Greenwald, G. S., and N. B. Everett (1959) The incorporation of S_{35} methionine by the uterus and ova of the mouse. Anat. Record, 134:171–184.

Gross, P. R. (1967) The control of protein synthesis in embryonic development and differentiation. In: Current Topics in Developmental Biology. vol. 2. Ed. by A. Monroy and A. A. Moscona Academic Press, New York.

Monroy, A. (1965) Chemistry and physiology of fertilization. Holt, Rinehart, and Winston, New York.

Pelc, S. R., T. C. Appleton, and M. E. Welton (1965) State of light autoradiography. In: The Use of Radioautography in Investigating Protein Synthesis. Ed. by C. P. LeBlond and K. B. Warren, Academic Press, New York.

Samuel, D. M. (1942) The use of an agar gel in the sectioning of mammalian eggs. J. Anat., 78:173–175.

Thurston, J. M., and D. L. Joftes (1963) Stains compatible with dipping radioautography. Stain Technol., 38:231–235.

Weber, R., ed. (1965) The biochemistry of animal development. Vols. 1 and 2. Academic Press, New York.

Weitlauf, H., and G. S. Greenwald (1965) A comparison of ^{35}S methionine incorporation by the blastocysts of normal and delayed implanting mice. J. Reprod. Fertility, 10:203–208.

——— (1967) A comparison of the in vivo incorporation of S_{35} methionine by two-celled mouse eggs and blastocysts. Anat. Record, 159:249–254.

20

Culture of the Rabbit Blastocyst Across the Implantation Period

JOSEPH C. DANIEL, JR.
Department of Molecular, Cellular, and Developmental Biology
University of Colorado
Boulder, Colorado 80302

Organogenesis in the mammal may be said to begin with the formation of the embryonic shield from the inner cell mass. Because implantation, which renders the embryo relatively inaccessible, follows shortly thereafter, it has been difficult to study the major events that accompany and constitute early organogenesis. Thus, it has been desirable to develop a system for growing this stage embryo *in vitro* for both observation and experimentation.

In the rabbit the embryonic shield may be found on Day 7 (i.e., six to seven days *post coitum;* see Fig. 20-1) and implantation begins about seven and a half days *post coitum.* Blastocysts older than seven days generally cannot be removed intact from the uterus because their size packs them tightly in the lumen and because they are attaching to the endometrium. They can, however, be removed six hours earlier and, by the culture technique to be described in this chapter, be grown for two days or more, thus, enabling them to be investigated.

Cleavage and midblastocyst stages of the rabbit are relatively easy to grow *in vitro* (Purshottam and Pincus, 1961; Huff, 1962; Daniel, 1963, 1965). Older blastocysts are more difficult to handle because of their size and the

The work leading to the development of the methods described here was supported by N.S.F. Grant GB-4401 and A.E.C. contract AT (11-1)-1597.

mechanical requirements of their manipulation and support; yet one of the earliest attempts at mammalian embryo culture was done on six- and seven-day rabbit blastocysts (Brachet, 1913). Huff and Eik-Nes (1966) have grown six-day blastocysts in synthetic medium for as long as 96 hours and Lutwak-Mann *et al.* (1962) have grown the same stage, in serum, to the primitive streak stage. Ogawa and Imagawa (1969) have cultured six-day embryos until pulsating heart areas developed, and Cole and Paul (1965) reported early embryo formation with beating hearts and closure of the amniotic folds *in vitro* after starting on Day 6. Waddington and Waterman (1933) grew the explanted embryonic shields of seven-day blastocysts on plasma clots until neural folding had been completed and Glenister (1961) has shown that rabbit blastocysts will initiate implantation in uterine endometrium *in vitro*. Dwinnel (1939) cultured nine-day embryos in hanging drops for ten hours or less to study heart formation. Work with later stage preimplantation embryos has been somewhat more successful in other species: this was evidenced by the work of Nicholas and Rudnick (1934, 1938) and Turbow (1965) with the rat, and by that of Jolly and Lieure (1938) with the guinea pig. New and Stein (1963, 1964) have grown post implantation mouse and rat embryos *in vitro* until brain, eyes, kidney, limbs, and circulatory systems had begun to form. In circulating medium even better growth of rodent embryos is possible (New, 1967; Givelber and DiPaolo, 1968; New and Daniel, 1969). Thus, the method described here is not conceptually original but does represent some technological improvement over existing methods for culturing the implantation-stage rabbit embryo.

THE METHOD

Does in the seventh day of pregnancy (preferably between 160 and 165 hours *post coitum*) are killed by cervical dislocation and the uterine

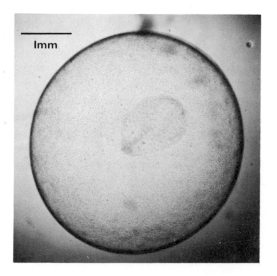

FIGURE 20-1

Typical seven-day rabbit blastocyst showing embryonic shield.

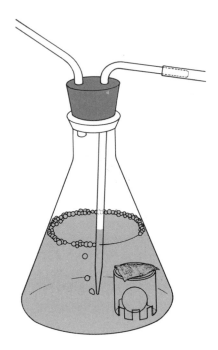

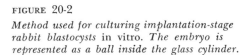

FIGURE 20-2

Method used for culturing implantation-stage rabbit blastocysts in vitro. *The embryo is represented as a ball inside the glass cylinder.*

horns are removed under sterile conditions. These uteri are carefully picked open with medium-point, angle-tipped forceps until the blastocysts are exposed; the blastocysts are lifted in the angle of the forceps or scooped into small watch glasses (10 mm diameter) for transfer to the culture chamber.

The culture chamber consists of a 125- to 150-ml Erlenmeyer flask containing 100 ml of medium (see Fig. 20-2). The medium may be any of Ham's F7-F12 series (Ham, 1962, 1963, 1965) or modifications (Daniel, 1965) with 15% to 25% maternal serum. One-tenth of a milliliter of methyl silicone is overlaid on the surface of this medium to act as an antifoaming agent. The blastocysts are protected and supported within glass cylinders that are submerged in the medium and are topped with a small stainless steel grid. These cylinders are 25 mm high by 15 mm inside diameter and have cuts in the bottom edge to permit free circulation of the medium. A mixture of 5% carbon dioxide, 10% oxygen, and 85% nitrogen is slowly, but continously, bubbled through the medium and the entire unit is retained in a water bath at 39°C. To prevent contamination, cooling, and to reduce evaporation of the medium, the gas mixture is passed through two Erlenmeyer flasks in series; each flask should contain 100 ml of sterile distilled water at incubation temperature (39°C) before entering the culture flask. Growth is improved when the culture medium is changed after 24 hours.

With this method one regularly gets, after two days *in vitro,* embryos that have closed neural tubes with optic vesicles forming, eight to eleven somites, a formed amnion, blood islands and beating hearts (see Fig. 20-3); the embryonic area appears normal at a stage approximating that of the $8\frac{1}{2}$ to

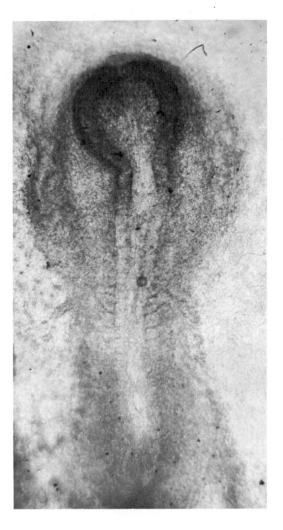

FIGURE 20-3

Embryonic area of a rabbit blastocyst that was explanted from the mother at 162 hours postcoitum and grown in vitro *for 48 hours.*

9 day embryo grown *in vivo*. After this stage is reached the development is aberrant; at this point the embryo becomes stunted or misshapen. The overall growth rates of blastocysts under these conditions, as determined by measurements made of their diameters, show that they parallel the *in vivo* growth rate for approximately 48 hours after which time a slowdown is observed. Figure 20-4 illustrates this growth rate.

ADDENDUM

A modification of New's technique for culturing rodent embryos (see Chap. 22) has now been used successfully with rabbit embryos. By this method, embryos of 18 to 20 somites can be reliably produced (Daniel, 1970; Nature, 225:193).

288

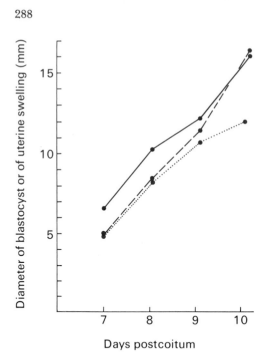

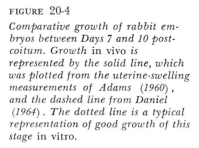

FIGURE 20-4

Comparative growth of rabbit embryos between Days 7 and 10 postcoitum. Growth in vivo *is represented by the solid line, which was plotted from the uterine-swelling measurements of Adams (1960), and the dashed line from Daniel (1964). The dotted line is a typical representation of good growth of this stage* in vitro.

Diameter of blastocyst or of uterine swelling (mm)

Days postcoitum

BIBLIOGRAPHY

Adams, C. E. (1960) Studies on prenatal mortality in the rabbit *Oryctolagus cuniculus:* The amount and distribution of loss before and after implantation. J. Endocrinol, 19:325–344.

Brachet, A. (1913) Development *in vitro* de jeunes vesicules blastodermiques du lapin. Arch. de Biol. T., 28:447–504.

Cole, R. J., and J. Paul (1965) Properties of cultured preimplantation mouse and rabbit embryos, and cell strains derived from them. In: The Ciba Foundation Symposium on the Preimplantation Stages of Pregnancy. Ed. by G. E. W. Wolstenholme and M. O'Connor. Little, Brown, and Co., Boston.

Daniel, J. C., Jr. (1963) Cleavage of rabbit ova in protein-free medium Am. Zool., 3:526.

———— (1964) Early growth of rabbit trophoblast. Am. Naturalist, 98:85–98.

———— (1965) Studies on the growth of five day old rabbit blastocysts *in vitro*. J. Embryol. Exp. Morphol., 13:83–95.

Dwinnell, L. A. (1939) Physiological contraction of double hearts in rabbit embryos. Proc. Soc. Exper. Biol. and Med., 42:264–267.

Givelber, H. M., and J. A. DiPaolo (1968) Growth of explanted eight day hamster embryos in circulating medium. Nature, 220:1131–1132.

Glenister, T. W. (1961) Organ culture as a new method for studying the implantation of mammalian blastocysts. Proc. of the Royal Soc. B., 154:428–431.

Ham, R. G. (1962) Clonal growth of diploid Chinese hamster cells in a synthetic medium supplemented with purified protein fractions. Exp. Cell Res., 28:489–500.

———— (1963) An improved nutrient solution for diploid Chinese hamster and human cell lines. Exp. Cell Res., 29:515–526.

————— (1965) Clonal growth of mammalian cells in a chemically defined, synthetic medium. Proc. Nat. Acad. Sci., 53:288–293.

Huff, R. L. (1962) *In vitro* cultivation of rabbit blastocysts. Am. Zool., 2:416.

————— and K. B. Eik-Nes (1966) Metabolism *in vitro* of acetate and certain steroids by six-day-old rabbit blastocysts. J. Reprod. Fertility, 11:57–63.

Jolly, J., and C. Lieure (1938) Recherches sur la culture des oeufs des mammifers. Arch. Anat. Microscop., 34:307–374.

Lutwak-Mann, C., M. F. Hay, and C. E. Adams (1962) The effect of ovariectomy on rabbit blastocysts. J. Endocrinol., 24: 185–197.

New, D. A. T. (1967) Development of explanted rat embryos in circulating medium. J. Embryol. Exp. Morphol., 17:513–525.

————— and J. C. Daniel (1969) Cultivation of rat embryos explanted at 7½–8½ days of gestation. Nature, 223:515–516.

————— and K. F. Stein (1963) Cultivation of mouse embryos *in vitro*. Nature, 199:297–299.

————— and K. F. Stein (1964) Cultivation of post-implantation mouse and rat embryos on plasma clots. J. Embryol. Exp. Morphol., 12:101–111.

Nicholas, J. S., and D. Rudnick (1934) The development of rat embryos in tissue culture. Proc. Nat. Acad. Sci., 20:656–658.

————— (1938) Development of rat embryos of egg-cylinder to head-fold stages in plasma cultures. J. Exp. Zool., 78:205–232.

Ogawa, S., and D. T. Imagawa (1969) Development of pulsating embryos from rabbit blastocysts cultivated *in vitro*. Nature, 223:409–410.

Purshottam, N., and G. Pincus (1961) *In vitro* cultivation of mammalian eggs. Anat. Record, 140:51–56.

Turbow, M. M. (1965) Teratogenic effect of trypan blue on rat embryos cultivated *in vitro*. Nature, 206:637.

Waddington, C. H., and A. J. Waterman (1933) The development *in vitro* of young rabbit embryos. Am. J. Anat., 67:355–370.

Waterman, A. J. (1933) Development of young rabbit blastocysts in tissue culture in grafts. Am. J. Anat., 53:317–347.

21

Blastocyst Transplantation in the Rabbit

R. E. STAPLES

Worcester Foundation for Experimental Biology
Shrewsbury, Massachusetts 01545

Rabbit ova were first successfully transferred to foster mothers in 1890 by Heape; many investigators have since employed this technique to advantage (see review by Austin, 1961), but, the successful transfer of the partially or fully expanded blastocyst was not accomplished until relatively recently. Expanded blastocysts were first transferred to study zygote development after exposure to low temperatures (Chang, 1950a) in synchronous and asynchronous does (Chang, 1950b), and, later, in nonovulated does (Chang, 1951). Blastocyst transfers were also completed by Shah (1956) to determine the effect of ambient heat upon developing zygotes; by Chang (1964) to study the effects of antifertility agents on blastocyst implantation; and by Staples (1967a) to test viability and development following *in vitro* culture for up to 24 hours.

In general, the blastocyst transfer technique is used either to provide a quasi-normal environment in which to test zygote viability and development following containment in an experimental *in vivo* or *in vitro* environment, or to determine the ability of an experimental doe to maintain the viability of untreated zygotes without having the litter exposed to the experimental conditions before the transfer. This technique is a powerful tool for the study of zygote requirements and sensitivities, especially if the

The background for this article was acquired while the author was supported by the PHS (Grant No. HD 01785). The excellent technical assistance of Mary Jane Hepinstall is gratefully acknowledged.

findings are considered with those obtained from available morphological and chemical techniques.

TRANSFER PROCEDURE

Preparation of Donors and Recipients

Donor females are mated to a fertile buck, or are artifically ovulated with an intravenous injection of 25 I.U. HCG, or 0.5 mg LH (Day 0 of gestation) and are artificially inseminated with semen collected in an artificial vagina (Gregoire, Bratton, and Foote, 1958; Gibson, Staples, and Newberne, 1966). These procedures are reliable and yield a normal complement of blastocysts if the does used are neither pregnant nor pseudopregnant. Pregnant and pseudopregnant does do not make good donors, even if artificially ovulated and inseminated because sperm transport is altered and the development of pyometra is likely (Austin, 1949; Murphree, Black, Otto, and Casida, 1951); as a result many ova are not fertilized. Both mature and immature donors may be superovulated (Kennelly and Foote, 1965; Hafez, 1964) but the response is usually poor if the zygotes are not flushed out prior to Day 5; only a few of the zygotes expand and those that do expand do so in varying degrees.

Recipients are ovulated synchronously with the donors (or no more than one day later), by mating them with a vasectomized buck or by injecting them with HCG or LH. If the recipients ovulate either one day before or two days after the donors, the blastocyst transfer has little chance of success (Chang, 1950b). Does used as recipients should be mature, healthy, primiparous females that have been caged individually for at least 18 days (Staples, 1967b). If primiparous females are not available, mature virgin does that were ovulated 18 days or more previously will suffice; for standard Dutch-Belted females, the minimum recommended body weight is 1.75 kg (Staples and Holtkamp, 1966). New Zealand does should weigh more than 3.20 kg. Only young, mature females should serve as recipients because in older does the ability to ovulate healthy, viable ova outlasts the ability of the uterus to maintain a pregnancy (Adams, 1964).

Just before the transfer and before the donors are killed, the recipients are anesthetized; pentobarbital sodium (60 mg/ml) is usually used (about 0.75 ml/kg body weight), but a short-acting barbiturate may also be employed. The doe is restrained gently while a marginal ear vein is punctured with a 22- to 25-gauge hypodermic needle. It is convenient to have this hubless needle attached to a polyethylene catheter (PE 20), about 16 inches long (Hulka, Mohr, and Lieberman, 1966), which obviates removal of the needle from the vein during the transfer. A 5-ml syringe containing the anesthetic is attached to the polyethylene tubing by a disposable 25-gauge needle. One milliliter of the pentobarbital solution is injected rapidly, but thereafter, surgical depth is approached gradually through a series of injections of no more than 0.2 ml for Dutch-Belted recipients (0.4 ml for larger breeds), administered 2 to 3 minutes apart. Indications of sufficient depth of anesthesia vary but, in general, at surgical depth: the hind legs of the animal can be extended without resistance, pinching of the web between the toes of the hind feet causes no more than a minor response, the animal does not respond to pinching of the skin of the abdomen, the white of the eye is not visible, the eyelid barely twitches if the upper eyelashes are touched, breathing becomes largely intercostal, and, the most significant indicator of anesthesic depth, the animal demonstrates only slight movement of the hind legs upon incision into the peritoneal cavity. Make an initial injection of about 1 ml of anesthetic, empty the rabbit's bladder by digital pressure, fasten the rabbit on an operating board or table, shave the abdomen with electric clippers, and vacuum the loose hair. Apply to the shaved area a bactericidal solution consisting of two parts ethanol and three parts distilled water, to which sufficient benzalkonium chloride (Winthrop Labs, New York, N.Y.) has been added to give a final concentration of 1:/1000 (vol/vol). This solution is also used to clean the investigator's hands. A 20 minute soak will sterilize the instruments particularily if one-eighth of the water is replaced by acetone.

At this point the donor is killed either by cervical dislocation or by intravenous injection of air. Fifty percent aqueous ethanol is used to mat the abdominal fur. The peritoneal cavity is entered at the linea alba, one uterine horn and the accompanying ovary are removed, the corpora lutea are counted, and then the uterine horn is placed on a sterile paper towel so that any free blood will be absorbed when the adhering fat and vaginal wall are removed. A blunted 18-gauge hypodermic needle is inserted into the uterotubal junction and is held in place with a rubber-tipped hemostat. The cervix is slit longitudinally or cut off, and the caudal tip of the uterus is carefully cleansed of blood. The hemostat is either held by hand or fastened to a ring stand, to suspend the uterine horn for flushing: use 2 or 3 ml of flushing media for the Dutch-Belted rabbit; 5 to 8 ml are required to flush the uterine horns of larger breeds. The flushed blastocysts are collected in a watch glass which is then kept in a light-free container until the other uterine horn is removed and flushed. All the blastocysts are counted and their diameters are measured with a stereoscopic microscope fitted with an ocular micrometer. Blastocyst size and their correlation with age are presented in Table 21-1.

While the blastocysts are being collected from the donor, the abdomen of

TABLE 21-1
Blastocyst size in Dutch-Belted rabbits

Hour after LH Injection	No. of Observations	Blastocyst Size (mm ± SE)
120	113	1.25 ± 0.032
127	119	1.65 ± 0.036
137	62	2.50 ± 0.070
144	142	3.41 ± 0.047

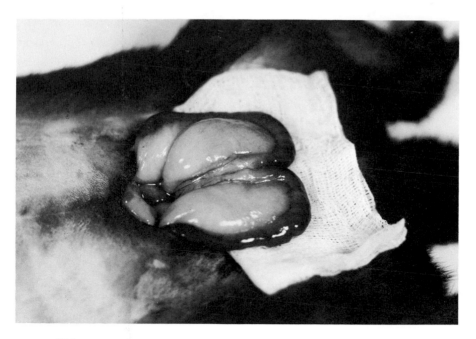

FIGURE 21-1

Uterine horns exposed immediately prior to transfer. The gauze is used for photographic purposes only; for routine transfer an autoclaved cloth equipped with a medial slit is used to cover all but the exposed uterus.

the anesthetized recipient is opened for about 2 inches along the linea alba between the posterior pair of nipples; an autoclaved cloth with a medial slit is used to cover the doe, then the uterus is exposed (see Fig. 21-1). Pseudopregnancy is determined by the presence of vascularized corpora lutea on the ovaries. A prospective recipient is not given blastocysts unless five or more corpora lutea are present.

A series of sterilized (autoclaved), fire-polished glass capillary tubes, about 4 inches long, are used to transfer the blastocysts. They are stored according to internal diameters (ranging from 0.3 to 5.0 mm depending upon blastocyst age) in gauzecapped glass vials. The tube selected, that size which will allow entry of the blastocysts without altering their shape, is attached to a 1 ml syringe by a rubber or plastic adapter. After 0.2 ml of sterile air is measured on the syringe, the blastocysts are sucked into the capillary tube with a minimal amount of fluid. No more than five are transferred to each uterine horn of the recipient if they are to be expected to survive to term.

TRANSFER METHODS

In one method of blastocyst transfer, a stab wound is made near the ovarian end of the uterine wall. The puncture is made with a pair of iridectomy scissors and the desired degree of blunt tearing is obtained by slowly opening the scissor blades. A piece of sterile gauze or Gelfoam

(Upjohn) is used to stanch any bleeding. The capillary tube containing the blastocysts is inserted into the wound (see Fig. 21-2a) toward the cervical end and is threaded to a depth such that the last blastocyst is about one-half inch down the cornu (see Fig. 21-2b). The blastocysts are deposited by gradual withdrawal of the capillary tube with simultaneous depression of the syringe plunger so that, ideally, the blastocysts do not move *in utero*. The plunger should reach the 0.15 ml level as the tip of the capillary tube emerges from the stab wound. To help prevent adhesions, a portion of the serous membrane of the uterus is freed to lie over a small stab wound; otherwise, one Lembert suture (triple-0 silk) through the serosa is used to cover the exposed endometrium and to prevent possible blastocyst expulsion by this route. The cornua are then replaced and the peritoneum is sutured with either continuous or interrupted sutures (Ethacon silk, size-0). An antibacterial agent (Furacin, Eaton Labs, N.Y.) is applied routinely and the skin is clamped with 16 mm wound-clamps. The recipient is given 125,000 I.U. of penicillin and 100 mg of streptomycin intramuscularly, and during recovery is put on its side on a heating pad and covered with a towel. The antibiotic doses are repeated daily for the next two days.

A second method of transfer used with very good results, particularly for the smaller blastocysts (e.g., Days 4 to $5\frac{1}{2}$), consists of transfer by the cervical route. The same surgical procedure is employed except that, if necessary, the bladder is brought out of the peritoneal incision, placed below the incision, and covered by saline-soaked gauze, and the ventral vaginal wall is incised just below the cervices, again avoiding major blood vessels (see Fig. 21-3a). The capillary pipette containing the blastocysts is threaded through the cervices as they are immobilized between thumb and finger (see Fig. 21-3b). The technique used for evacuating the blastocysts from the capillary tube is as described previously. The slit in the vaginal wall is not sutured. Very good results have been achieved following this route of transfer, and adhesions have not been a problem.

Although not considered an absolute necessity for the successful transplantation of blastocysts, the availability of a sterile area is a great convenience in the handling of mammalian zygotes. For this purpose we have devised the cabinet illustrated in Figure 21-4. Preliminary sterilization is achieved by 15-minute use of an overhead germicidal lamp. For the last five minutes of this period the germicidal lamps in the air duct are turned on. During use, adequate sterility is maintained through ultraviolet sterilization of the intake air together with the laminar air flow within the chamber; the air flow is interrupted only when zygotes are exposed. This chamber is used for the final sterilization of all equipment required for flushing, collecting, examination and photography, culturing or transference of zygotes. The media to be used are sterilized by millipore filtration, but even this procedure is done in the chamber. Blastocysts obtained are measured with an ocular micrometer that is inserted into the eyepiece of the dissecting microscope. They are subsequently maintained, until transferred, either in a light-free holding chamber (previously sterilized by ultraviolet light), or, in an incubator (see Fig. 21-5) if they are to be maintained at temperatures above room temperature (24°C). The media, watchglasses, and microscope

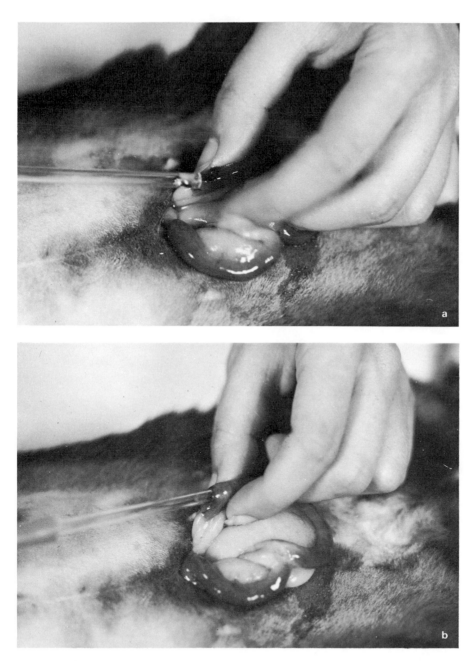

FIGURE 21-2

Blastocyst transfer. (a) *Illustrates uterine stab wound and glass tube that contains blastocysts.* (b) *Deposition of blastocysts.*

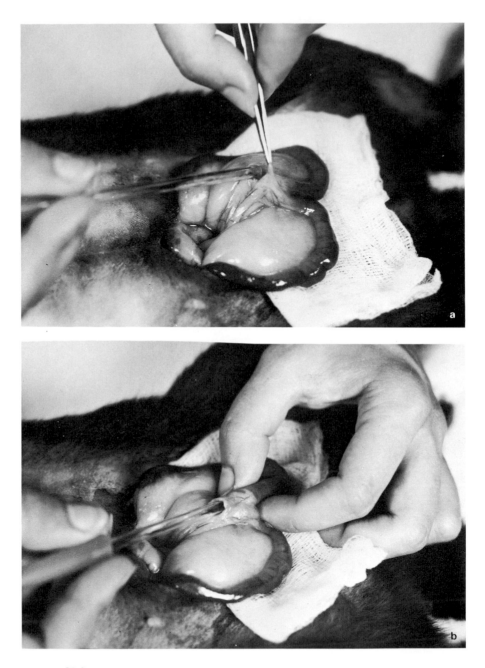

FIGURE 21-3

Cervical blastocyst transfer. (a) Slit made in vaginal wall. (b) Transfer tube in position for blastocyst deposition. Note corpora lutea on left ovary.

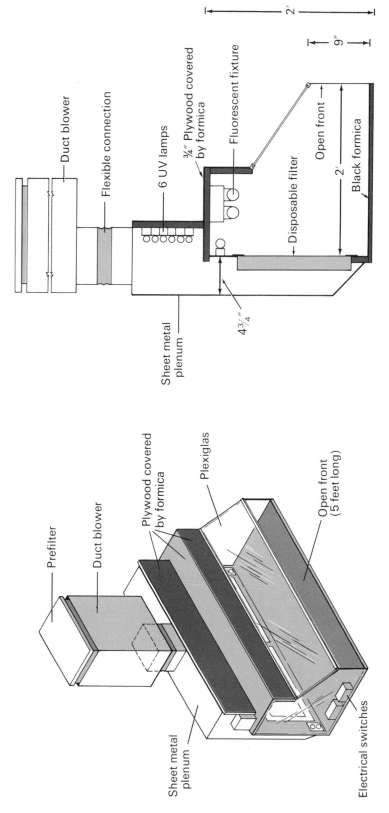

FIGURE 21-4

Sterilization cabinet. Air sucked through upper filter by fan in duct is blown over a bank of UV lamps and then through the filter in the cabinet. (The cabinet currently used has two such filters placed side by side; a Heppar filter replaces the bank of UV lamps.) A cloth covers the plexiglass during initial sterilization with the UV lamp situated in the upper portion of the cabinet. Gasses are mixed as desired in a proportioner, the mixture is humidified and sterilized in a large Erlenmeyer flask and is then distributed at desired flow rates by a four-tubed flowmeter. (Duct blower is Brundage Model 808–138 1/6 H.P., 489–746 R.P.M., 115 V, 1 phase, 400 C.F.M. at 1/4 S.P. at 683 R.P.M. Plenum of 20 Ga. galvanized sheet metal. Ultraviolet lamps are General Electric G15T8, 15 watts.)

FIGURE 21-5

Photograph of holding or culture chamber. The atmosphere mixture is humidified and sterilized by bubbling it through a candle filter that is immersed in water to which phenylmercuric nitrate has been added. A dial thermometer records atmospheric temperature within the chamber, and a thermocouple is used to regulate sample temperatures. A large Petri dish filled with sterile water for added humidity is contained within the chamber. For holding, the blastocysts are placed in watch glasses, as shown. For culturing, the chamber is loaded in the large sterilization cabinet, the doors are sealed with sterile masking tape and the entire sealed chamber is then carried to a Visidome incubator. The UV lamp shown is used if more cultures are added during a run—all glassware used for such experiments is opaque.

stages are held at any given temperature by being placed on copper plates, which are soldered to coiled copper tubing through which water is circulated from a constant temperature pump (Bronwill, Will Scientific, Inc.).

To keep the blastocysts in any given atmospheric mixture other than air, the gases to be used are either purchased premixed or are mixed in the desired proportions by volume via proportioners (Air Products and Chemicals, Inc., Allentown, Pa.). Each gas is filtered prior to mixing. The gaseous mixture is then bubbled through a column of water containing a nonvolatile poison (e.g., phenylmercuric nitrate) and is dispensed to desired locations by flow meters (Air Products and Chemicals, Inc.). For rapid equilibration, miniature bubbles are formed through use of Labpor filter candles (Daigger and Co., Chicago, Ill.). If the area to be gassed is at other than room temperature, the mixture is reequilibrated to the desired temperature to prevent hydration or dehydration of the media used. This gaseous mixture is dispensed through the sterilizing cabinet wall to the light-free hood which covers the warming plates, to the holding chamber, and to a large funnel suspended above the stage of the dissecting microscope. The rate of flow to each of these areas is regulated independently. Blastocysts are thus retained in the desired atmosphere from flushing to culture or transfer.

Photography of blastocysts or ova is accomplished without removing the zygotes from the sterile area by attachment of appropriate apparatus to one of the microscopes which project through the slanted plexiglass top of the chamber. An adequate seal is effected between the microscope and the plexiglass by use of a thin, fitted rubber sheet. Such a connection also allows simple removal of a microscope if it is not required for the current project.

CONSIDERATIONS

During the holding period, which varies depending upon the complexity of the experiment, partial collapse of some of the blastocysts may occur, particularly if physiological saline is used as the medium. If cultured, they round up again and continue to grow as well as do non-collapsed blastocysts. Viability is being determined by transfer.

A maximum of five blastocysts per horn are transferred to offer the best likelihood for survival. Crowding, which decreases likelihood for survival (Adams, 1960b; Hafez, 1964), is thought to have its lethal effect through hemodynamic factors (Hafez, 1965b). It has been reported that animals with five or more functional corpora lutea are more likely to maintain a litter to term (Schmidt and Bachnick, 1961) than animals with less than this number, but it has not yet been clearly determined whether a recipient with a total of five corpora lutea is as good a host for ten blastocysts as is one with ten or more corpora lutea. It is known, however, that up to the stage of blastocyst formation the number of corpora lutea present in the recipient is not directly related to blastocyst size (Adams, 1965).

The wide variability found in "between litter" and "within litter" blastocyst size for any given postpartum age necessitates careful grouping for studies involving the culture or transfer of blastocysts. This variability was not related to any of a number of factors studied by Beatty (1958); however, 50% to 75% of fetal size variability is considered to be due to maternal factors (Hafez, 1965a). Hafez (1962) found that within a litter, slow-cleaving ova do not survive or implant as well as do fast-cleaving ova, and it was suggested by Beatty (1958) and Hafez and Rajakoski (1964) that the small blastocysts within a litter may not be as likely to develop as the larger ones. Preliminary data collected in our laboratory confirm their suggestion.

On Day 12 all recipients are again laparotomized as described earlier, however, a short-acting barbiturate (e.g., Surital, Park-Davis) is more commonly used than pentobarbital sodium. Such laparotomy does not interfere with subsequent development (Adams, 1960a). The cornua are exposed, the number, size, and appearance of implantation sites are recorded, and any other findings of interest are recorded. Implantation sites in Dutch-Belted recipients that measure less than 14 mm between the crown of the site and the mesometrium are seldom found to contain live fetuses at term; in fact, these small sites are often not grossly evident on Day 27. In larger breeds surviving implantation sites measure at least 10 mm on Day 9 (Hafez, 1964) and 12 mm on Day 10 (Adams, 1960a).

The following quantitative information on native zygotes, collected from 13 artificially ovulated and inseminated Dutch-Belted does, is presented (See Table 21-2) to serve as a guide for certain reproductive parameters in this breed of rabbit. These control does, that weighed 2.11 ± 0.096 kg when injected with LH on Day 0, were laparotomized on Day 12, and autopsied on Day 27. On Day 27, one dead fetus was obtained and 13 resorption sites were grossly evident.

On Day 27 the recipients are killed routinely and their reproductive status is determined. Weights of the placentae and kits, whether dead or alive, are recorded, as is the sex of the offspring. (The body weight of

TABLE 21-2

Reproductive parameters of standard-sized Dutch-Belted rabbits[1]

	Corpora Lutea	Implants (Day 12)	Live Young (Day 27)
Total no.	105	99 (94.3%)	81 (77.1%)
Average no./doe ± SE	8.1 ± 0.45	7.6 ± 0.45	6.23 ± 0.588
Average size ± SE	—	20.1 ± 0.26 mm	23.96 ± 0.437 gm

[1] Data from thirteen does.

sixty-two fetuses taken from Dutch-Belted recipients on Day 27 was 29.5 ± 0.74 gm. The difference in average weight between these young and the native young presented in Table 21-2 is probably an expression of litter size.) Comparison is made between the Day 12 and Day 27 findings. Implantation sites disappearing before Day 27 represent embryo mortality shortly after implantation; atrophic placentae represent loss from soon after implantation to Day 17. Dead fetuses represent mortality later than Day 17 and are often associated with the period of embryo rotation (Adams, 1960b). In the rabbit, full growth of the maternal placenta is completed about Day 16. After Day 16 the fetal placenta becomes larger, but the maternal placenta becomes smaller. The maternal placenta is usually of normal size if embryo mortality occurs after Day 12 (Huggett and Hammond, 1952). Sometimes dead and partially macerated young are seen *in utero* on Day 27. Soft tissue changes are difficult, if not impossible, to evaluate; but, skeletal alterations are believed to precede fetal death.

RESULTS

The type of control data obtained in our laboratory after transferring expanded blastocysts by stab wound into one horn of untreated Dutch-Belted recipients is indicated in Table 21-3. Only 37% of the blasto-

TABLE 21–3

Fate of untreated expanded blastocysts (Day 5¼–5¾) upon transfer to one uterine horn of synchronous recipients[1]

No. of Blastocysts Transferred		No. of Recipients	
Implanting	Surviving to Day 27	With Implants	With Live Young on Day 27
200/305 (65.6%)	113/305 (37.0%)	64/75 (85.3%)	56/75 (74.7%)

[1] No more than five blastocysts per horn.

cysts transferred resulted in viable young at autopsy on Day 27, which is comparable to the results obtained by Chang (1950a,b; 1951). One-third of the blastocysts were not represented by implants on Day 12; almost one-third of those initiating implantation did not survive to term. Although this corresponds to the period of major embryonic loss in nonoperated does (Adams, 1960b), it is almost triple the normal loss. This phenomenon is not yet understood.

One possible explanation of embryonic loss is that the amount increases among small litters. Hafez (1961) indicated that when the number of implantations on Day 8 was less than four, the entire litter did not survive. This loss is of particular concern to those who transfer blastocysts, because if only five blastocysts are transferred per horn and only experimental zygotes are put into one side, it is possible that many recipients could be nonpregnant on Day 27, even though three of the five control blastocysts implanted normally. Although we have not duplicated exactly the conditions described by Hafez (1961), we do not consider it likely that the whole litter will degenerate in recipients possessing three or less implants; we believe this because: (1) in Hafez's article (1961), two recipients, #167 and #72, had only three implants each at laparotomy on Day 8; but, #167 gave birth to three kits, and #72 gave birth to two live and one stillborn kit; and (2) in our laboratory, seven of ten Dutch-Belted recipients of expanded blastocysts that had less than four implants on Day 12 carried live young to Day 27; on Day 12, seven of the 28 implants were obviously degenerating, but of the remaining 21 implants, 13 survived to Day 27 (62%).

It is possible, among recipients of transferred blastocysts, that the number of implants degenerating may be greater in litters of less than four than in larger litters. This was not indicated by preliminary data collected in our laboratory, or by data collected by Adams (1962) following the unilateral transfer of five, 60-hour zygotes to each of 12 recipients; however, until sufficient data is accumulated, it may be prudent to use as recipients only pregnant does that have had one oviduct occluded before insemination. Blastocysts could then be transferred into the empty horn and the native zygotes in the opposite horn would perhaps meet the minimum litter size requirements to prevent undue embryonic loss.

The main embryonic loss following transfer occurs between the time of transfer and Day 12; this corresponds to the period during which the

TABLE 21–4

Implant size (Day 12) in 29 Dutch-Belted recipients of expanded blastocysts

	Implants Yielding Live Young on Day 27	Implants Not Yielding Live Young on Day 27	Total
No. of implants	46	36	82
Average size (mm $\pm$ SE)	18.0 ± 0.43^1	12.9 ± 0.52^2	15.7 ± 0.44^1
Size distribution	$1 < 14$ mm[3]	$20 < 14$ mm[3]	

[1] Significantly smaller (P < .05) than native implants (Table 21–2).
[2] Significantly smaller (P < .05) than implants yielding live young on Day 27.
[3] On the basis of such measurements it is stated that in recipients of expanded zygotes, implants measuring < 14 mm seldom survive to Day 27.

developing embryo switches from the yolk sac to the chorioallantoic placenta for its nutrition (Adams, 1960b). The proportion of preimplantation loss that is due to ruptured blastocysts and blastocysts that escape by the cervical route after transfer is unknown; but from the measurement of implantation sites it is seen that even the Day 12 implants surviving to Day 27 (see Table 21-4) are smaller (P < 0.05) than corresponding native implants (see Table 21-2). Also, more than one-half of the implants that will not survive to Day 27 after transfer can be predicted at laparotomy on Day 12 (see Table 21-4).

EXAMINATION OF OFFSPRING

Under experimental conditions the Day 27 kits are routinely checked for gross, visceral, and skeletal malformations. To prepare rabbit fetuses for skeletal examination (Staples and Schnell, 1964), the viscera, pigmented skin, and fat pads are removed, and the specimens are air-dried for 12 hours. Maceration and staining are completed simultaneously in 1% aqueous KOH (about 300 ml/specimen) to which alizarin-red-S stain has been added (6 mg/liter). The solution will drip rather than string from the specimen upon lifting it from the container if maceration is complete; this takes up to two days. Following maceration the fetuses are rinsed in cold water, gently blotted or drained and immersed for 8 to 12 hours in a well-stirred mixture of two parts 70% aqueous ethanol, two parts glycerol, and one part benzyl alcohol. The cleared specimens are then transferred to a mixture of 70% ethanol and glycerol (equal parts) for inspecting and handling. If only a few specimens are to be cleared, processing is readily completed in 8-oz jars that are sealed with bakelite lids; after maceration has proceeded for one day, the old KOH solution is replaced with new. Compartmentalized plastic boxes are handy for processing a large number of specimens; such boxes can be filled from carboys and emptied by pouring, thus eliminating application and removal of many screw or snap lids. To save the 2:2:1 or 1:1 mixture for reuse, the plastic boxes are emptied into carboys either by pouring or by vacuum. If the fetuses cannot be processed beginning on the day they are obtained, they should be fixed in 70% aqueous ethanol, in the containers to be used for the clearing process; the peritoneum should be opened to facilitate rapid penetration of the alcohol. Formalin fixation is not used because it necessitates a more rigorous clearing procedure.

Finally, if allowed to carry the pregnancy to term, the recipients will deliver normally, but gestation will likely be one or two days longer than normally expected. The reason for this is unknown, but it is thought to be due to a slight delay in the rate of blastocyst development immediately following transfer. Subsequent maternal behavior, lactation, and growth of the young, is generally uneventful.

BIBLIOGRAPHY

Adams, C. E. (1960a) Studies on prenatal mortality in the rabbit, *Oryctolagus cuniculus:* the amount and distribution of loss before and after implantation. J. Endocrinol., 19:325–344.

—— (1960b) Prenatal mortality in the rabbit *Oryctolagus cuniculus.* J. Reprod. Fertility, 1:36–44.

—— (1962) Studies on prenatal mortality in the rabbit, *Oryctolagus cuniculus:* the effect of transferring varying numbers of eggs. J. Endocrinol., 24:471–490.

—— (1964) The influence of advanced maternal age on embryo survival in the rabbit. 5th Int. Congr. Anim. Reprod. Artif. Insem., 2:305–308.

—— (1965) Influence of number of corpora lutea on endometrial proliferation and embryo development in the rabbit. J. Endocrinol., 31:29–30.

Austin, C. R. (1949) Fertilization and the transport of gametes in the pseudopregnant rabbit. J. Endocrinol., 6:63–70.

—— (1961) The mammalian egg. Oxford: Blackwell Scientific Publ., Appendix no. 1., pp. 125–132.

Beatty, R. A. (1958) Variation in the number of corpora lutea and in the number and size of six-day blastocysts in rabbits subjected to superovulation treatment. J. Endocrinol., 17:248–260.

Chang, M. C. (1950a) Transplantation of rabbit blastocysts at late stage: probability of normal development and viability at low temperature. Science, 111:544–545.

—— (1950b) Development and fate of transferred rabbit ova or blastocyst in relation to the ovulation time of recipients. J. Exp. Zool., 114:197–225.

—— (1951) Maintenance of pregnancy in intact rabbits in the absence of corpora lutea. Endocrinology, 48:17–24.

—— (1964) Effects of certain antifertility agents on the development of rabbit ova. Fertility Sterility, 15:97–106.

Gibson, J. P., R. E. Staples, and J. W. Newberne (1966) Use of the rabbit for teratologic studies. Toxicol. Appl. Pharmacol., 9:398–407.

Gregoire, A. T., R. W. Bratton, and R. H. Foote (1958) Sperm output and fertility of rabbits ejaculated either once a week or once a day for forty-three weeks. J. Animal Sci., 17:243–248.

Hafez, E. S. E. (1961) An experimental study of ova reception and some related phenomena in the rabbit. J. Morphol., 108:327–345.

—— (1962) "Differential cleavage rate" in 2-day litter mate rabbit embryos. Proc. Soc. Exp. Biol. Med., 110:142–145.

—— (1964) Effects of over-crowding *in utero* on implantation and fetal development in the rabbit. J. Exp. Zool., 156:269–288.

—— (1965a) Maternal effects on implantation and related phenomena in the rabbit. Experientia, 21:1–11.

—— (1965b) Quantitative aspects of implantation, embryonic survival, and fetal development. Intern. J. Fertility, 10:235–251.

—— and E. Rajakoski (1964) Growth and survival of blastocysts in the domestic rabbit I. Effect of maternal factors. J. Reprod. Fertility, 7:229–240.

Heape, W. (1890) Preliminary note on the transplantation and growth of mammalian ova within a uterine foster-mother. Proc. Roy. Soc., B., 48:457.

Huggett, A. St.G., and J. Hammond (1952) Physiology of the placenta. In: Marshall's Physiology of Reproduction. Vol. II, Chap. 16. Ed. by A. S. Parkes. Longmans, Green, New York.

Hulka, J. F., K. Mohr, and M. W. Lieberman (1966) A simple catheter for intermittent intravenous anesthesia in animal surgery. Nature, 210:1389.

Kennelly, J. J., and R. H. Foote (1965) Superovulatory response of pre- and post-pubertal rabbits to commercially available gonadotrophins. J. Reprod. Fertility, 9:177–188.

Murphree, R. L., W. G. Black, G. Otto, and L. E. Casida (1951) Effect of site of insemination upon the fertility of gonadotrophin-treated rabbits of different reproductive stages. Endocrinology, 49:474–480.

Schmidt, K., and G. Bachnick (1961) Experiments to improve implantation of transplanted rabbit ova by means of progesterone and oestrone. Zuchthyg. Fort-PflStor. Besam. Haustiere, 5:332–338.

Shah, M. K. (1956) Reciprocal egg transplantations to study the embryo-uterine relationship in heat-induced failure of pregnancy in rabbits. Nature, 177:1134–1135.

Staples, R. E. (1967a) Development of 5-day rabbit blastocysts after culture at 37°C. J. Reprod. Fertility, 13:369–372.

—— (1967b) Behavioural induction of ovulation in the oestrous rabbit. J. Reprod. Fertility, 13:429–435.

—— and D. E. Holtkamp (1966) Influence of body weight upon corpus luteum formation and maintenance of pregnancy in the rabbit. J. Reprod. Fertility, 12:221–224.

—— and V. L. Schnell (1964) Refinements in rapid clearing technic in the KOH-Alizarin red-S method for fetal bone. Stain Technol., 39:62–63.

22

Methods for the Culture of Post-implantation Embryos of Rodents

D. A. T. NEW
Physiological Laboratory
Cambridge, England

The post-implantation embryos of rodents have been particularly convenient to study in culture. Nicholas and Rudnick (1934, 1938), and Jolly and Lieure (1938) obtained limited development of rat and guinea-pig embryos explanted with their membranes at stages between the primitive-streak and the first somites. The nutrient media were homologous sera or heparinised plasma and embryo extract. Preliminary studies were made by Nicholas (1938) on the development of rat embryos explanted into circulating medium. Grobstein (1950) reported little or no regular embryonic development of mouse egg cylinders of pre-streak to head-fold stages cultivated on plasma clots, but Smith (1964) was able to grow mouse embryos of four to eight somites on nutrient agar clots after tearing the yolk sac and amnion so that the embryos flattened out; these embryos developed to the sixteen-somite stage, but without a blood circulation. Givelber and DiPaolo (1968) have successfully grown hamster embryos of head-fold and early somite stages in circulating medium.

The problems of growing in culture whole rat and mouse embryos of stages from the primitive streak to fifty somites have been studied recently (New, 1966a, 1966b, 1967; New and Daniel, 1969; New and Coppola, 1970a, 1970b), and it is now possible to grow these embryos with considerably greater success than hitherto. This chapter will describe the methods of embryo culture which now appear most convenient and effective.

OBTAINING EMBRYOS

MATING

The duration of the estrous cycle in rats and mice varies among strains and among individuals, but it is commonly four days in rats and four to six days in mice. Estrus occurs once in each cycle, usually beginning during the night and lasting about twelve hours. Ovulation occurs while the animals are in estrus but both ova and sperm remain capable of fertilization for only a few hours. The chances of a successful mating of a particular female on a particular night are therefore, at best, about one-in-four for rats and one-in-five for mice. If necessary, the odds can be increased by prior injection of hormones (see Chap. 12 by Mintz).

Copulation is followed by the formation of a vaginal plug formed by a mixture of the secretions of the vesicular and coagulating glands of the male. This solid plug usually fills the vagina and persists for 18 to 24 hours, or occasionally longer. It can be seen easily by examination of the vulva, and provides a valuable indication that mating has occurred.

From about the tenth day of pregnancy in rats, and correspondingly earlier in mice, an experienced handler can determine to within a day the age of the embryos by palpating the pregnant females. For greater accuracy, or for timing the earlier stages, it is necessary to know within a few hours the time of fertilization. This can be done most simply by overnight caging of males and females that have previously been kept separate. On the following morning the females with vaginal plugs are removed and kept until the embryos have reached the required age. Good results can be obtained by allowing mice to mate overnight in small cages that have one male and two or three females per unit. Rats sometimes give better results if several males are placed in a cage with the females. The presence of a vaginal plug is proof of mating, but not of fertilization; however, more than half of the females with plugs are likely to be pregnant.

EXPLANTING THE EMBRYOS

Apparatus and materials required:

1. Dissecting microscope.
2. Dissecting scissors.
3. Dissecting forceps—2 pairs.
4. Watchmakers forceps—2 pairs.
5. Cataract knives—2.
6. Petri dishes, 10 cm in diameter.
7. Watch glass or cavity slides.
8. Pasteur pipettes with tip diameters varying between 2 and 4 mm.
9. Tyrode saline.

Sterilize before use

10. Provision for sterilizing the apparatus and explanting the embryos aseptically (for details see Paul, 1970, New, 1966a).

Rat embryos reach the primitive-streak and head-fold stages on Day 10 of pregnancy; by Day 11 they have developed five to fifteen parts of somites, and by Day 12 they have twenty to thirty pairs of somites. The corresponding stages in mice occur two days earlier. The embryos are completely surrounded by uterine decidual tissue that has proliferated to such an extent that it blocks the uterine lumen and produces a series of easily visible swellings in the uterus.

Aseptic precautions must be taken when explanting the embryos. The pregnant females are killed by breaking the neck; each uterine horn is transferred to a watch glass containing Tyrode saline; the uterine wall is then opened with forceps, care being taken not to squeeze or puncture the embryos, and the pear-shaped decidual swellings are dissected out. The most difficult part of the operation is the removal of the embryo and its membranes from the decidua (from this stage onward it is essential to work under a dissecting microscope). The best method for the younger embryos is to make a meridional incision with a cataract knife in the broad end of the decidual pear (see arrow in Fig. 22-1b), preferably by cutting along the groove representing the remains of the uterine lumen, and then to tear the decidua into two equal halves; the embryo and its membranes are usually intact and adhere to one of the halves (see Fig. 22-1c) from which they can easily be dissected. With rat embryos of more than 11 days gestation there is probably less chance of damaging the placentae if the decidua is removed with forceps.

Surrounding the embryo is the Reichert membrane with its adherent endoderm and trophoblast cells. This membrane must be torn open be-

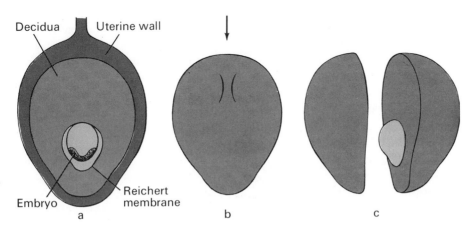

FIGURE 22-1

Explantation of mouse and rat embryos. (a) Section through uterus of eight-day pregnant mouse or ten-day pregnant rat. Decidua surrounds embryo and fills uterine lumen.
(b) Decidua after removal of uterine muscle layers. Decidua is next separated into two halves (from point indicated by arrow) and (c) embryo and membranes can then be easily dissected out.

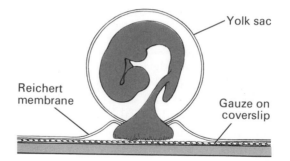

FIGURE 22-2

*Method of anchoring embryo in circulating medium.
The Reichert membrane with its adherent cells is
torn open and pressed down on to gauze
attached to a piece of glass coverslip or wire frame.*

cause it does not expand *in vitro* and it thus prevents normal expansion of the yolk sac. Fine watchmakers forceps are the best instruments for tearing the membrane, which is grasped just above the middle of the embryo, a place at which it usually stands furthest away from underlying tissues. If the embryos are to be grown in circulating medium the Reichert membrane can be used for anchoring them (see Fig. 22-2) ; if they are to be grown in static medium the membrane can be completely removed, but this is not essential and has no effect on embryonic development.

The explant now consists of a sphere of yolk sac with the embryo at one pole and the ectoplacental cone at the other (see Fig. 22-3b), with or without the opened Reichert membrane. It is transferred to the culture vessel and is then ready for incubation. With practice the explantation of each conceptus takes about 5 to 10 minutes. Delays of up to 3 hours between killing the rat or mouse and transferring the embryos to the incubator seem to have no effect on subsequent development; it is thus unnecessary to hurry the operation and thereby risk wounding the embryos. With reasonable care all the embryos in a uterus can be transferred undamaged to the culture vessels. Small injuries to the yolk sac usually heal and do not hinder embryonic development.

PETRI DISH CULTURES

Apparatus and materials required:

1. Incubator.
2. Sterilized Petri dish culture chambers (see Fig. 22-4).
3. Gas chambers and O_2/CO_2 supply.
4. Aseptically prepared rat serum.

A good nutrient for growing both rat and mouse embryos is rat serum. The serum may be obtained from male, pregnant female, or nonpregnant

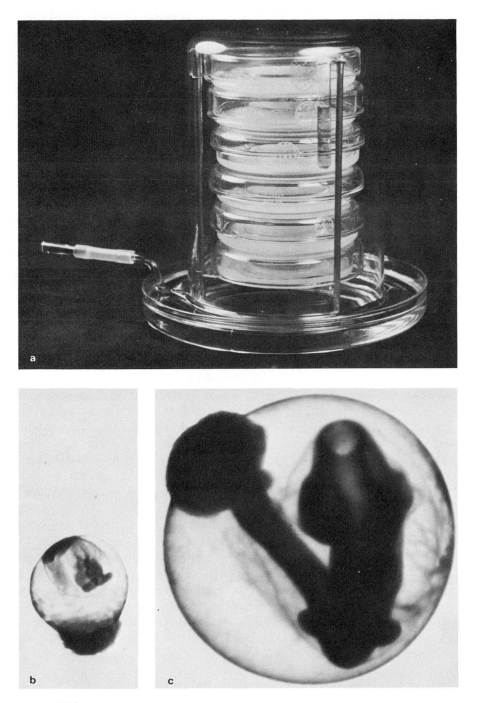

FIGURE 22-3

(a) *A cheap but effective gas chamber containing six Petri dish cultures, made by inverting an 800 ml beaker over a dish of liquid paraffin. The small tube inside the chamber contains bicarbonate and phenol red solution for determining the CO_2 concentration.* (b) *and* (c) *Embryo and membranes as explanted (7 somites = b) and after 40 hours in a Petri dish culture (27 somites = c). Limb buds and rudiments of eyes, ears and other organs have developed and the enlarged yolk sac is covered with a network of blood capillaries. The mass of tissue projecting from the spherical yolk sac is the allantoic placenta.*

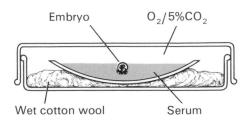

FIGURE 22-4

Petri dish culture. The explant floats in serum in a watch glass that is housed in a Petri dish and surrounded by wet cotton wool to maintain a constant humidity. These cultures are incubated in gas chambers containing the required O_2/CO_2 concentration.

female rats. Sera from other species are often unsatisfactory (New, 1966b) and chemically defined tissue culture media such as *199* and Waymouth's usually give poor results. Embryos of five to twelve somites will develop well on the surface of clots of plasma and embryo extract but they develop no better than in whole serum; furthermore the clots are more troublesome to prepare and more inconvenient for chemical analysis than whole serum.

The rat serum must be prepared aseptically. Blood is obtained from etherized adult rats by inserting a 10 ml hypodermic syringe with a #1 or #2 needle into the dorsal aorta at the junction of the iliac arteries. No anticoagulant is added but glass syringes must be siliconed to prevent premature clotting, and a fresh syringe needle is required for each animal. Each rat yields 6 to 12 ml of blood before being killed. The blood is transferred to centrifuge tubes where it is allowed to clot; the clot is then freed from the tube wall with a knife and left overnight to contract and release the serum. The following day the tubes are centrifuged (3000 rpm for 15 minutes) and the serum is decanted. If the serum is not needed at that time it can be stored, preferably frozen, in a refrigerator. The addition of 50 $\mu g/ml$ streptomycin diminishes the risk of infection without affecting the development of the embryos.

Each culture chamber (see Fig. 22-4) consists of a Petri dish containing a watch glass (4 cm diameter) surrounded by gauze wetted with 0.9% NaCl. Two or three embryos of the head-fold or early somite stages can be placed in each watch glass, but older embryos are best explanted one per watch glass because of their greater oxygen requirements. Adding 0.5 to 1.0 ml serum per embryo is enough; nothing is gained by using more or by transferring the embryos to fresh serum during the culture period. However, as will be discussed later, the younger embryos sink in the serum and the depth of the serum and the oxygen concentration of the gas phase must be adjusted to permit adequate respiration.

Older embryos grow much better in an atmosphere containing more than 60% O_2 than they do in air, and the cultures must be housed in a gas-tight container which can be supplied with extra oxygen. Containers manufactured by laboratory suppliers are satisfactory, but they are expensive and simple home-made equipment can be equally effective. Figure 22-3a shows a gas chamber made from an 800 ml beaker inverted over a large Petri dish; liquid paraffin in the dish provides a gas-tight seal, and gases are added through a glass tube passing under the spout of the beaker. Such a chamber will house six cultures in 7 cm Petri dishes, and will retain O_2 and CO_2 at any concentration for several days. Further details of this and of simple methods for controlling and for measuring O_2 and CO_2 are given in New (1966a). Desiccators are sometimes used as gas chambers, but they are likely to leak slowly and should be used only if provision is made for continual replacement of lost gas.

CULTURES IN CIRCULATING MEDIUM

Apparatus and materials required:

1. Hotbox, incubator or water bath (see Fig. 22-5a).
2. Circulators (see Fig. 22-3).
3. Embryo chambers (see Fig. 22-7).
4. Gauze coverslips.
5. Glass manipulating rods.
6. 10 ml syringes with 7 cm long, wide-bore needles.

} Sterilize before use

7. O_2/CO_2 supply.
8. Rat serum, prepared aseptically.
9. Pressure chamber (for embryos older than the 40-somite stage).

SERUM AS NUTRIENT MEDIUM

Whole rat serum, prepared in the same way as for the Petri-dish cultures, is a good nutrient medium. The serum may be obtained from either male or female rats (New, 1967), but dilution of the serum with chemically defined media such as *199* or Waymouth's gives inferior results. The embryos also grow well in human serum (Shepard *et al.,* 1969).

METHOD OF CIRCULATING THE SERUM

The circulator (drawn to scale in Fig. 22-6) was designed (New, 1967) to allow continuous observation of the embryo, to circulate the serum without the use of mechanical pumps, and to meet the requirements of sterility, cleanliness, and simplicity of construction. It consists essentially of a triangle of thick-walled glass tubing of 3 mm bore through which serum is continuously recirculated and oxygenated by a stream of O_2/CO_2 bubbles flowing up tube D. The bubbles are discharged into chamber B where serum drains from them and they collapse; if the rate of flow is fairly rapid the chamber may become filled with bubbles and a few may escape through tube A, but the loss of serum is usually negligible.

For steady circulation the level of serum in B must always be above the inlet from tube D. For a circulator using tubing of 3-mm-bore (designed as in Fig. 22-6) the minimum amount of serum required is about 5 ml and the maximum rate of flow is about 20 ml/min. The rate of flow can be estimated from the rate of bubble movement within tube D. If a stopcock

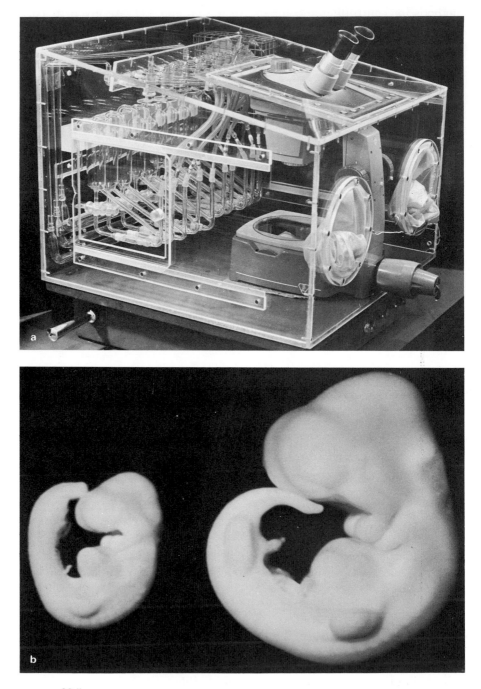

FIGURE 22-5

(a) *Hot box (base 45 × 45 cm) containing ten 'circulators' and a binocular dissecting microscope. The circulators are similar to that shown in fig. 22-6 but each incorporates a stopcock to regulate the flow of serum.* (b) *Two embryos taken from the same rat on the 12th day of pregnancy, the one on the left (27 somites) fixed immediately and the other grown to 40 somites in circulating serum.*

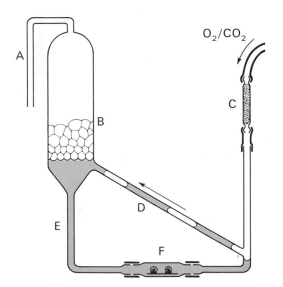

O$_2$/CO$_2$

FIGURE 22-6

A 'circulator' for growing embryos in flowing medium. The triangle of glass tube is filled with serum and circulation is maintained by O$_2$/CO$_2$ entering through the filter C and flowing as a stream of bubbles up tube D. The bubbles collapse in chamber B. The embryos are contained in the detachable chamber F. Two fifths of actual size.

is incorporated within the circulator at E, steady circulation can be obtained at any rate between 0 and 20 ml/min. Without a stopcock the minimum steady flow obtainable (by reducing the gas supply) is about 10 ml/min; at rates below this the flow becomes irregular.

If a flow faster than 20 ml/min is required the circulator must be made of larger-bore tubing. Thus a circulator of 5 mm bore gives flow rates up to 150 ml/min.

The gas is obtained from a cylinder of O$_2$/CO$_2$ or O$_2$/N$_2$/CO$_2$ mixed in the required proportions and equipped with a regulator valve to maintain the pressure at about 1 or 2 lb/sq inch. It is first humidified by bubbling through water, then passed through filter C which contains a fine powder (e.g., light kaolin, fine carborundum, or jeweller's rouge) packed tightly between two plugs of glass wool. The purpose of the filter is to remove any infective particles, and to control the flow of gas so that only a few milliliters per minute pass into the circulator forming a steady flow of bubbles in tube D.

Serum can be injected into or withdrawn from the circulator through the vertical side tube after detaching filter C. A syringe with a long wide-bore needle is better for this than Pasteur pipettes which are likely to break and leave lengths of fine glass tubing within the circulator. Good development is obtained using 2 or 3 ml of serum per embryo; little or no improvement is gained by using more than this or by adding fresh serum during the culture period.

Tube A, which is bent, is convenient for hooking the circulator onto a rack. Several circulators can thus be placed side by side in a small hotbox, incubator or water bath. If the same hotbox contains a low-power binocular microscope (see Fig. 22-5a) the embryos in each circulator can be examined in turn without cooling them or interrupting the circulation of the serum.

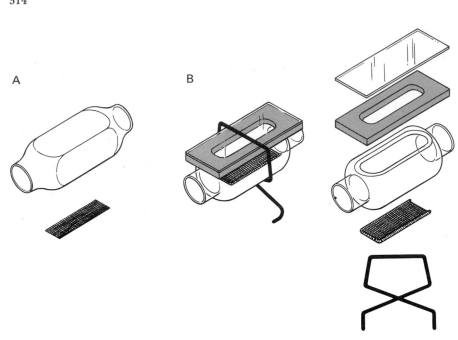

FIGURE 22-7

Two types of embryo chamber for use in circulators. (A) *Chamber square in cross section, open only at the ends, with gauze coverslip for attachment of embryos*
(B) *Chamber containing gauze mounted on wire frame, with detachable glass top and silicone rubber seal held in place with wire clip (parts shown assembled and separately).*

INSERTION OF THE EMBRYOS

The embryos are housed in the embryo chamber (see Fig. 22-6F) which is held in the circulator by short lengths of silicone tubing. Two types of embryo chamber have been proved satisfactory:

Chambers that open only at the ends (see Fig. 22-7A)

These are made from glass tubing that is square in cross-section; the embryos can thereby be observed through the flat sides without distortion. Tubing of internal dimensions 7 mm × 7 mm is suitable for all embryos explanted up to the thirty somite stage; 10 mm × 10 mm tubing is better for older embryos. The ends are rounded to facilitate attachment to the circulator.

To prevent the embryos from circulating with the serum they are anchored to a strip of (*e.g.* rayon) gauze 5 mm × 20 mm attached to a piece of glass coverslip of the same size. Good adhesion is obtained with gauze-to-glass and with embryo-to-gauze by the use of a thick solution of collagen. The solution is prepared (aseptically) by dissolving the tendons from rat tails in sterile 1:1000 acetic acid in water, and centrifuging. The gauze is soaked in the supernatant, placed wet on the coverslips, and allowed to dry overnight at 37°C.

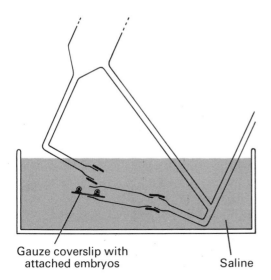

Gauze coverslip with
attached embryos

Saline

FIGURE 22-8

*Method of inserting embryos into
embryo chamber with lower part of
'circulator' immersed in saline.*

The gauze coverslips are immersed with the explanted embryos in a dish
of Tyrode saline. The embryos are then attached, two to each gauze
coverslip, by the Reichert membrane. The trophoblast layer on the outside
of the membrane is very sticky and if the membrane is torn open and
pressed against the gauze it adheres firmly (see Fig. 22-2). Glass rods that
are drawn out in a flame to give a rounded tip of 1 mm diameter or less are
useful for this.

The embryos are then transferred to the circulators, but to avoid damaging them this must be done without taking them from the saline. Each
circulator, with the embryo chamber attached at one end only, is supported
over the dish containing the embryos so that the lower part of the circulator
is below the level of the saline (see Fig. 22-8). A gauze coverslip with
attached embryos is raised gently from the floor of the dish and slid into the
embryo chamber. The circulator is then removed from the dish, care being
taken to keep it tilted enough to prevent the saline from spilling out of the
embryo chamber. Most of the remaining saline is displaced (and runs out of
the open end of the embryo chamber) by an injection of 2 or 3 ml of serum
into the side tube. The embryo chamber is then fixed in position and the
rest of the serum is added.

Chambers that have an open top

These chambers (see Fig. 22-7b) are made from glass tubing that is circular
in cross section. Tubing that has an internal diameter of 10 mm is suitable
for all embryos explanted at up to the forty-somite stage. An opening is
made at the top of the chamber and the edges are ground flat. The opening
is covered with a piece of optically flat glass, such as a microscope slide, and
leakage of serum is prevented by a seal cut from sheet silicone rubber. The

glass cover and silicone are held in place by a clip such as can be made from stainless steel wire about 0.5 mm in diameter.

The gauze can be attached to a stainless steel wire frame (as in Fig. 22-7b) which will permit the serum to flow under, as well as around and above, the explants. Alternatively, it may be attached to pieces of glass coverslip, as described above, or it may be attached directly to the floor of the embryo chamber.

Before inserting the embryos, the chamber (with the top open) is placed in the circulator and is partially filled with serum. The embryos are then injected by means of a pipette and are attached to the gauze inside the chamber.

This type of chamber has the advantages that it is not necessary to immerse the circulator in saline, and that the embryos are more accessible for subsequent operations. The disadvantages are that with more parts these chambers become difficult to make and to use, and it is technically more difficult to attach the embryos to the gauze inside the chamber than outside.

RESULTS

The youngest post-implantation embryos that can be grown in culture are egg-cylinder stages of seven to eight days gestation (rat), and the maximum stage of development that can be attained is about fifty-five somites. The final stage that can be expected varies with the age at explantation—it is not possible to grow the youngest embryos to fifty-five somites.

The growth and development of embryos explanted at the primitive streak stage or later closely resembles that of control embryos *in vivo*. They do not appear to decline gradually, but seem to attain a certain stage of development and then die rapidly; the first sign of this is usually the failure of the yolk sac blood circulation.

It is very important that the oxygen supply should be regulated according to the stage of development of the embryo. The older embryos grow best in rapidly circulating serum equilibrated with a gas phase of high oxygen concentration. On the other hand the young embryos can be killed easily by too much oxygen.

Embryos Explanted at the Egg-cylinder Stage
(Eighth or Ninth Day of Gestation in the Rat)

These embryos have been grown (New and Daniel, 1969) in circulating serum equilibrated with $20\%\,O_2/5\%\,CO_2$. For the first 36 hours of the culture period the serum was circulated at about 1 ml/min and then at about 10 ml/min. The serum was partially renewed after 48 hours.

The embryos developed to the ten- to twenty-somite stages with beating hearts. The yolk sac enlarged and formed a capillary network with conspicuous masses of erythrocytes, but there was usually a failure to establish a functioning blood circulation. A common abnormality of development was the formation of a double heart.

EMBRYOS EXPLANTED AT PRIMITIVE-STREAK TO HEAD-FOLD STAGES (10TH DAY OF GESTATION IN THE RAT)

In circulating serum primitive-streak embryos develop to 15 to 20 somites, and head-fold embryos to 20 to 25 somites with a functioning yolk-sac blood circulation. The serum should be circulated at five to 10 ml/min and equilibrated with $20\%\,O_2/5\%\,CO_2$. High oxygen concentrations are fatal.

The embryos can also be grown with a similar degree of success in Petri dish cultures if the depth of the serum above the embryos is adjusted for the O_2 concentration of the gas in the culture chamber. If $20\%\,O_2/5\%\,CO_2$ is used the serum should just cover the explants, but in $95\%\,O_2/5\%\,CO_2$ the embryos, at explantation, must be covered by a depth of 2 or 3 mm. (Explants of the primitive-streak and the head-fold stages do not float in the serum, and the ectoplacental cone anchors the conceptus to the floor of the watch glass.)

EMBRYOS EXPLANTED AT ABOUT 10 SOMITES (11TH DAY OF GESTATION IN THE RAT)

These embryos grow well to the 25- to 30-somite stage in Petri dish cultures incubated in $95\%\,O_2/5\%\,CO_2$; the explants float so respiration is unaffected by the depth of the serum. But the best results are obtained in *slowly* circulating serum, equilibrated with $95\%\,O_2/5\%\,CO_2$. About 50% of the embryos develop to the 30- to 35-somite stage and synthesize 0.3 to 0.5 mg of protein during the period in culture (see Figs. 22-9 and 22-10). The rate of serum flow should be about 1 ml/min (when the embryo chamber is 0.5 cm² in cross section); faster rates of flow are soon fatal.

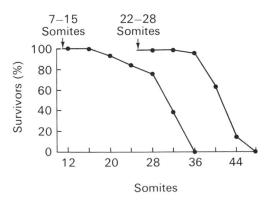

FIGURE 22-9

Percentage surviving to different stages of development as indicated by number of somites, of embryos explanted at 7 to 15 somites and 22 to 28 somites and grown in circulating homologous serum.

318

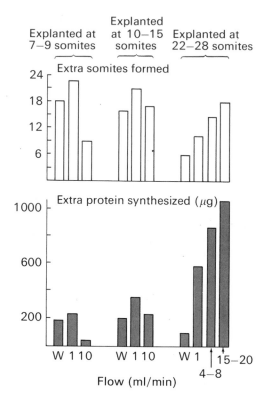

FIGURE 22-10

Development as indicated by extra somites formed and growth as indicated by extra protein synthesised, of embryos grown in watch glasses of serum (W) and in serum circulating at different rates. The serum in all these cultures was equilibrated with 95% O_2, 5% CO_2.

EMBRYOS EXPLANTED AT ABOUT 25 SOMITES
(12TH DAY OF GESTATION IN THE RAT)

Not much development can be obtained from embryos of this age in normal Petri dish cultures, although if the cultures are placed in hyperbaric O_2 at 15 lbs/sq inch the embryos may develop to 35 to 40 somites.

Much better results are obtained by growing the embryos in rapidly circulating serum equilibrated with 95% O_2/5% CO_2 (see Fig. 22-5). Under these conditions about 50% of the embryos develop to the forty to forty-six somite stage and synthesize about 1 to 1.5 protein (see Figs. 22-9 and 22-10). The optimum rate of serum flow appears to be about 10 to 20 ml/min; faster rates of flow have not yielded larger embryos.

EMBRYOS EXPLANTED AT ABOUT 40 SOMITES
(13TH DAY OF GESTATION IN THE RAT)

New and Coppola (1970b) were able to grow these embryos in circulators placed inside pressure chambers. Embryos explanted at 40 somites developed well to the fifty- to fifty-five-somite stage in circulating serum in hyperbaric O_2 at 15 lbs/sq inch. The final protein content of the embryos was about 2.5 to 3.0 mg, and they closely resembled control embryos *in vivo*. The yolk sacs, however, were harmed by the high oxygen concentration and failed to expand.

BIBLIOGRAPHY

Givelber, H. M., and J. A. DiPaolo (1968) Growth of explanted eight-day hamster embryos in circulating medium. Nature, 220:1131–1132.

Grobstein, C. (1950) Behaviour of the mouse embryonic shield in plasma clot culture. J. Exp. Zool., 115:297–314.

Jolly, J., and C. Lieure (1938) Recherches sur la culture des oeufs des mammifères. Arch Anat. Micr., 34:307–374.

New, D. A. T. (1966a) The culture of vertebrate embryos. Logos Press, London.

——— (1966b) Development of rat embryos cultured in blood sera. J. Reprod. Fertility, 12:509–524.

——— (1967) Development of explanted rat embryos in circulating medium. J. Embryol. Exp. Morphol., 17:513–525.

——— and P. T. Coppola (1970a) Effects of different oxygen concentrations on the development of rat embryos in culture. J. Reprod. Fertility, 21:109–118.

——— and P. T. Coppola (1970b) Development of explanted rat fetuses in hyperbaric oxygen. Teratology. 3:153–162.

——— and J. C. Daniel (1969) Cultivation of rat embryos explanted at 7.5 to 8.5 days of gestation. Nature, 223:515–516.

Nicholas, J. S. (1938) The development of rat embryos in a circulating medium. Anat. Record, 70:199–210.

——— and D. Rudnick (1934) The development of rat embryos in tissue culture. Proc. Nat. Acad. Sci., 20:656–658.

——— and D. Rudnick (1938) Development of rat embryos of egg cylinder to head-fold stages in plasma cultures. J. Exp. Zool., 78:205–232.

Paul, J. (1970) Cell and tissue culture. 4th ed. Livingstone, London.

Shepard, T. H., T. Tanimura, M. Robkin, and G. Bass (1969) *In vitro* study of rat embryos. I. Effects of decreased oxygen on embryonic heart rate. Teratology, 2:107–109.

Smith, L. J. (1964) The effects of transection and extirpation on axis formation and elongation in the young mouse embryo. J. Embryol. Exp. Morphol., 12:787–803.

23

Methods for Studying Ovoimplantation and Early Embryo-Placental Development *in Vitro*

T. W. GLENISTER
Professor of Embryology
University of London
Charing Cross Hospital Medical School
London, W.C.2. England

These techniques were designed to establish methods that would permit the investigation in extracorporeal environments of implantation processes and the factors controlling the differentiation, invasiveness, and functions of trophoblast. These techniques should provide conditions more readily controllable than those obtainable in an experimental animal.

It is generally accepted that the mutual relationship of trophoblast and endometrium during implantation and placental development is such that neither is solely the active agent. It is thus a balanced interaction of maternal systemic and uterine factors with embryonic factors that is responsible for the successful accomplishment of the developmental processes under consideration.

The study of embryo-endometrial relationships *in vitro,* both during and after implantation, offers obvious possibilities of eliminating maternal systemic factors and such uterine factors as the effects of the myometrium or the effects of circulation through endometrial blood vessels.

I have not been alone in recognizing the possibilities offered by the development of such *in vitro* methods for elucidating the question of what the requirements of normal ovoimplantation are—these possibilities have

These projects have been supported by the British Empire Cancer Campaign, the Medical Research Council and the Dan Mason Research Foundation. I am most indebted to R. H. Watts and to my research assistants Miss Janet Everett and Miss Barbara Cooling for helping me to develop these techniques.

also been discussed by Amoroso (1959), Meunier (1959), and Mayer (1960).

Having succeeded in procuring the attachment of rabbit blastocysts to endometrial explants maintained in organ culture (Glenister 1960, 1961a) it has been difficult subsequently to produce significant variations in the incidence of "ovoimplantation in organ culture" by alterations in the environment or in the physiological state of the endometrial explants (Glenister 1961b, 1962, 1963, 1965, 1967). Also, to date, it has not been possible to repeat the experiments in another species.

Despite these disappointments, organ culture combined with an ultra-structural comparative analysis of *in vivo* and *in vitro* material has been useful in the study of early embryo-endometrial relations, trophoblastic differentiation and function, and aspects of early mammalian development in culture.

RABBIT MATERIAL

Hysterectomy

The operation is performed using nembutal and/or ether anesthesia, and all aseptic precautions on a rabbit doe 6½ days after it has been observed to mate. If blood is required for the preparation of serum, it is obtained by cardiac puncture before the animal is killed. The uterine horns are placed in a sterile, covered glass receptacle and washed two or three times with sterile Tyrode solution to remove blood from the specimen.

Each horn is then transferred to a separate large sterile Petri dish. After this stage, all manipulations which require the lifting of lids or dishes are done under transparent perspex shields. Aseptic precautions are taken throughout and all solutions, glassware, and instruments, are prepared and sterilized by methods suitable for tissue culture work (Paul 1965).

Collection of Blastocysts

In turn, each uterine horn is trimmed free of mesometrium and fat using sharp scissors and then washed again with Tyrode solution to remove blood and detritus.

The uterine horns are opened under a dissecting microscope using a magnification of ×5. They are slit along their mesometrial border with fine curved scissors. The points of the scissors are guided along the groove between the placental folds of the endometrium to avoid puncturing the contained blastocysts.

The walls of the uterine horns are laid open carefully, revealing the blastocysts as shiny transparent spheres (3 or 4 mm in diameter). It is important that the blastocysts should not have expanded too much and not reached a stage of development when the zona pellucida is so stretched that it disintegrates when touched. When isolated at the correct stage of development the blastocysts can be lifted off the endometrium on the slightly parted blades of the curved scissors or with a small surgical spoon or scoop.

The blastocysts are then transferred to a watch glass containing equal parts of Tyrode Solution and rabbit serum, or tissue culture medium *199* (Glaxo Laboratories Ltd., Greenford, England) or Waymouth's Medium MB 752/1 (*Difco,* Baird and Tatlock, Ltd., London, England) .

Preparation of Endometrial Explants

Explants of endometrium are prepared by dissecting the endometrium from the myometrium. This is achieved most readily by laying flat a completely opened uterine horn in the Petri dish with the outer surface of the horn uppermost. The myometrium is then cut away in strips from the endometrium, by using fine iridectomy scissors. The resulting sheet of endometrium is cut into roughly rectangular explants measuring 10 to 15 mm by 5 to 7 mm. The explants are then thoroughly washed with Tyrode solution and stretched on strips of rayon (cellulose–acetate) fabric as in the Shaffer modification of the Fell technique for organ culture (Shaffer 1956) . The explants are next transferred to their definitive culture media.

Preparation of Culture Media

Solid media

Initially, the explants were incubated on clots of fowl plasma. These were prepared by delivering into watch glasses of about 4 cm diameter, fifteen drops of fowl plasma from a Pasteur pipette (of 1 mm bore at the tip) and adding five drops of chick embryo extract from a similar pipette. The mixture in the watch glass, stirred and spread into an even pool, coagulates into a firm clot within ten minutes. It was found however that such a medium liquefied too much for the requirements of these experiments.

A much more satisfactory solid medium consists of fifteen drops of a mixture of agar, chick embryo extract, egg albumen and Tyrode solution; clots of this mixture do not liquefy as do the clots of fowl plasma.

The medium is prepared in bulk as follows: 50 ml of Tyrode solution are poured into a 100 ml flask (previously calibrated with a mark indicating 50 ml) 0.3 g of agar are added and the solution is brought slowly to boil. It is then set aside and allowed to cool to about 60°C (a temperature at which it can be held comfortably in the palm of the hand) .

Meanwhile, an unfertilized hen's egg is washed with soap, water, and 70% alcohol; the shell is then broken and the egg is put into a deep Petri dish. Into another 100 ml flask, 10 ml of Tyrode solution are pipetted; to this are added 5 ml of albumen from the egg in the deep Petri dish. When the agar solution has cooled to 60°C the albumen/Tyrode mixture is poured into it and 1.5 ml of chick embryo extract are added. All the ingredients are mixed well and the medium is then ready for pipetting into watch glasses.

Culture Chambers

Culture chambers consist of Petri dishes 7 or 8 cm in diameter containing watch glasses of about 4 cm in diameter surrounded by cotton wool soaked in water or saline solution to maintain a humid atmosphere.

Fifteen drops of the nutrient medium, consisting of the above agar

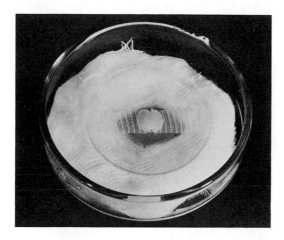

FIGURE 23-1

Photograph of a watch glass containing a clot of agar, embryo extract and albumen on which is laid an endometrial explant resting on a strip of rayon fabric. The whole is enclosed in a Petri dish containing moist cotton wool. (From: Embryo-endometrial relationships during nidation in organ culture, by T. W. Glenister. Journal of Obstetrics and Gynaecology of the British Commonwealth, October 1962.)

mixture, are placed in the watch glass with a Pasteur pipette while the medium is still warm. It is then allowed to solidify by cooling to room temperature.

COMBINED BLASTOCYST-ENDOMETRIAL EXPLANTS

A strip of rayon fabric bearing an endometrial explant is placed on the solid medium clot in one of the culture chambers described above. As blastocysts obtained from does $6\frac{1}{2}$ days postcoitum are invariably surrounded by a zona pellucida, and as rabbit blastocysts, unlike those of rodents such as mice, do not emerge spontaneously from the zona pellucida in culture, it is necessary to remove it prior to explantation in culture. The simplest way to do this is to tear off the outer mucin layer and the zona pellucida with steel watchmaker forceps. With practice, it is possible to do this without damaging the blastocyst, but only if the forceps have been carefully sharpened under a dissecting microscope so that the tips are very fine and meet exactly. The next step is to place about nine drops of liquid medium over the endometrial explant to form a supporting layer of fluid for the blastocyst. This liquid medium may be either serum, Waymouth medium, or the chemically defined *199* medium.

The blastocyst is eventually placed on the endometrial explant by means of an angled pipette with a bore sufficiently large to accommodate the blastocyst. The surface tension of the supporting layer of liquid medium is sufficient to hold the blastocyst on the endometrial surface.

The combined explants are then incubated at 36°C for up to a week, the explants being transferred to fresh medium after 48 or 72 hours. About three-fourths of the blastocysts so explanted "implant" successfully within 72 hours.

METHOD VARIATIONS

BLASTOCYST ATTACHMENT SURFACE

Instead of using endometrium from the doe that provided the blastocysts, endometrium may be used which has been obtained from other

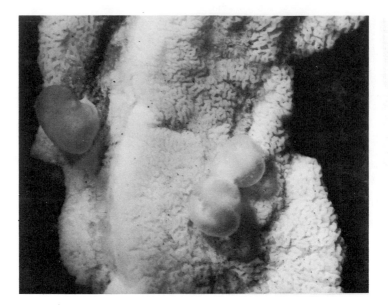

FIGURE 23-2
Photograph of three rabbit blastocysts attaching themselves to an explant of maternal endometrium 72 hours after explantation.

pregnant does or from adult virgin does. The proportion of blastocysts attaching themselves successfully to such "foreign" endometrium seems to be much the same as when maternal endometrium is used. If transitional epithelium-lined bladder wall is substituted for the endometrium, about half of the blastocysts so explanted attach themselves to the bladder lining (Glenister 1967).

Addition of Hormones

Two methods of studying the effects of added hormones have been devised:

1. The base of solid medium is made of agar with added albumen only, the embryo extract being omitted. Hormones are added to the solid medium and to the *199* medium which is used as the supporting fluid.

2. The chemically defined *199* medium is the only one used, the explants being supported on a stainless wire-mesh platform as described by Trowell (1959). (Hormones are added to the *199* medium in proportions to be described below.)

The culture chambers consist also of Petri dishes of 7 or 8 cm in diameter but instead of cotton wool and a watch glass they contain a shallow vitreosil basin (Baird and Tatlock, Ltd.) of about 4 cm in diameter. This small basin contains the liquid *199* medium and the wire-mesh platform measuring about 2 cm-square and 3 mm high stands in it. The platforms are made of 60-to-the-inch steel wire netting (supplied by The Expanded Metal

FIGURE 23-3

Photograph of a vitreosil basin containing a wire mesh platform on which is laid an endometrial explant. The basin is enclosed in a Petri dish containing moist gauze.

Company Ltd., 16 Caxton St., London, S.W.1). Medium is added to the basin till the surface of the liquid is level with the surface of the platform —on no account should it be submerged because it is important that only a thin film of fluid should cover the explants.

CONCENTRATION OF HORMONES

Steroid hormones (supplied by Organon Laboratories Ltd., Morden, Surrey, England), have been added to media in the proportion of 1 μg/ml of medium for estrogens and 2 μg/ml of medium for progesterone.

The concentrations used may justifiably be criticized as being unphysiological, the normal concentrations of these hormones in the blood being approximately one-tenth these amounts (Zander, 1954; Aitken, Preedy, Eton, and Short, 1958). Other workers, however, have found that similar high doses of estrogens and progesterone must be used in order to obtain a histologically recognizable response of the uterus *in vitro* (Gaillard, 1942; Kahn, 1958; Bimes, Planel, and David, 1961; Everett, 1962 and 1963).

Solutions of estrogens and progesterone, separately or together may be added to the culture medium. To obtain the desired concentrations the hormones are first dissolved in ethanol (10 mg progesterone in 5 ml alcohol; 10 mg estriol or 10 mg estrone in 10 ml alcohol), 0.05 ml of the alcoholic solution is added to 10 ml of Tyrode solution and three drops of this diluted solution are added to twelve drops of culture medium, giving approximately the final concentrations in the medium quoted earlier.

From the limited number of experiments performed it appears that once the blastocyst trophoblast has been removed from the zona pellucida it is capable of attaching itself to explants of endometrium irrespective of the endometrium's physiological state at the time of explantation and irrespective of the presence or absence of added female hormones in the culture medium. It should however be pointed out that, whereas *in vitro* implantation experiments yield a consistent attachment rate of 75% when biological media are used, the success rate falls close to 50% when a chemically defined synthetic medium is used.

The Gaseous Environment

In all the initial explantation experiments concerned with the study of ovoimplantation (Glenister, 1960, 1961a, 1961b), the cultures were incubated in air.

Subsequently the method was improved by incubating the explants in 95% oxygen and 5% carbon dioxide; this was suggested first by Trowell (1959). Using this gaseous environment, evidence was found which suggested that after a week in culture, not only do the endometrial constituents of the explants show far fewer signs of tissue damage than do comparable explants incubated in air, but that the environment favors blastocyst expansion and differentiation of syncytium from cellular trophoblast (Glenister, 1962, 1963, 1965).

Well developed embryonic structures, however, developed more rarely in blastocysts incubated in this gaseous environment than in those incubated in air. It appears therefore that although a high concentration of oxygen is favorable to the differentiation of syncytium and essential to endometrial survival in organ culture, it is deleterious to embryonic tissues other than trophoblast (Glenister 1967).

Recent experiments performed in our laboratory tend to confirm the significance, pointed out by New (1966), of the fact that the media used by several different workers for successful expansion of the blastocyst in culture had been equilibrated with 3 to 5% carbon dioxide. It is obvious from

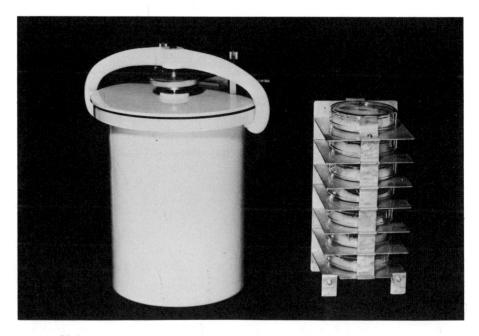

FIGURE 23-4

Photograph of an aluminum carrier for Petri dishes whose lids are slightly raised by wire clips. The carrier is made specially to fit inside the McIntosh and Fildes jar which is shown on the left of the picture.

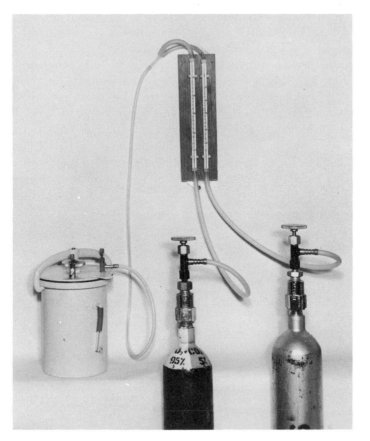

FIGURE 23-5

Photograph of the equipment used for gassing the McIntosh and Fildes jar with varied concentrations of oxygen, nitrogen, and carbon dioxide.

the above that it is still necessary to determine which proportions of nitrogen, oxygen and carbon dioxide most favor the development of embryonic structures, blastocyst expansion, trophoblastic differentiation and endometrial survival in organ culture.

To incubate explants in atmospheres other than air, the culture chambers are incubated in a McIntosh and Fildes "anaerobic" jar (obtained from Baird and Tatlock Ltd., London) containing the required gaseous mixture.

The Petri dishes containing the explants have their lids raised by wire clips (Grover 1961) and are stacked on the platforms of a specially designed aluminium carrier fitted inside the jar. The lowest platform of the carrier should be about an inch above the bottom of the jar so that some water may be placed in it. The inlet valve of the jar is connected to a glass tube which is long enough to deliver the incoming gases below the surface of the water contained in the bottom of the jar.

The appropriate gaseous mixture is passed through the jar at a rate of 200 ml/minute for 15 to 30 minutes. The gases are supplied in cylinders by the

British Oxygen Co.; they contain either 95% oxygen and 5% CO_2, or 95% air and 5% CO_2. The desired oxygen concentration is obtained by varying the proportions of gas delivered from each cylinder.

A simple method for providing the required gaseous concentrations was evolved by Dr. T. Fainstat at the Strangeways Laboratory, Cambridge. Each of the two gas cylinders is linked by a controllable reducing valve to a free float glass flowmeter, calibrated for a flow range of 20 to 200 ml/minute (supplied by Rotameter Manufacturing Co. Ltd., Croydon, Surrey, England). For convenience, the flowmeters are mounted in parallel on a wall, and the tubes leading from them are linked by a Y-shaped connection, so that the gases emenating from both flowmeters are mixed before reaching the inlet of the anaerobic jar.

The following table indicates the various percentage oxygen concentrations obtainable by varying the proportions supplied from the two gas cylinders.

TABLE 23-1

Flow rate from two individual gas cylinders, at a total of 200 ml/minute at room temperature.

(1) 95% O_2 + 5% CO_2 (ml/min)	(2) 95% Air + 5% CO_2 (ml/min)	% of O_2 Delivered
150	50	76
140	60	72.5
130	70	69
120	80	65
115	85	63
110	90	61
105	95	60
100	100	58
90	110	54
80	120	50
70	130	46
60	140	42.5
50	150	39
40	160	35
30	170	31

Check on oxygen concentration

A sample of a particular gaseous mixture is collected from the outlet of the McIntosh and Fildes jar, after the mixture has been allowed to flow through the jar for 15 minutes. Samples are collected for flow rates designed to produce oxygen concentrations of 31, 39, 50, 60, and 72.5%. A calibration line is then formed of oxygen percentage values recorded with an oxygen electrode against electrode readings established with 95% oxygen, air (20% oxygen), and nitrogen (0% oxygen).

The experimental readings have been found to be accurate to within 2% of the predicted concentration.

It has been confirmed that the oxygen concentration within the anaerobic jar has been maintained at its original level after three days in an incubator at 36°C, and it is now routine to regas the jars after three days of incubation.

CULTURING POSTIMPLANTATION EMBRYOS WITH DEVELOPING PLACENTAE

All the above methods can be used to culture implanted primitive streak and early somite embryos with their developing chorioallantoic placentae and related endometrium.

The constituents of the explant are isolated from uterine horns according to the following procedure. The horn is immersed in Tyrode solution and with fine curved scissors the antimesometrial half of uterine wall and related membranes of the cystic conceptus are cut away. The embryonic disc, developing placenta, and mesometrial placental folds of endometrium are thus exposed. Using either fine curved scissors or iridectomy scissors, a thin sliver of mesometrial endometrium bearing the developing placenta and embryonic disc is dissected off the mesometrial uterine wall. After trimming and washing, the composite explant is stretched carefully on a rectangular piece of rayon fabric, the endometrium being in contact with the fabric and the embryo lying on the superficial aspect of the endometrium with its endodermal aspect uppermost. The explant can then be incubated either on solid or on liquid medium.

Although the endometrial and placental components survive and develop satisfactorily in such cultures, the thickness of the explant is such that, although the embryonic tissues may develop well for the first 48 hours, the superficial half of the embryo soon degenerates. Also, as the developing embryonic tissues lie unsupported on the endometrium and placenta, mechanical factors interfere with the development of the embryo and its membranes.

It is possible to overcome some of these mechanical factors by culturing the explants on the undersurface of the vault of arches made of flexible polyurethane foam material (Flexocel from the Baxenden Chemical Company Ltd., Accrington, East Lancs., England).

The arches are cut from rectangular blocks of the foam material measuring 15 to 20 mm long, 10 mm high and 5 to 7 mm wide. The pillars of the arch are made 5 to 7 mm thick, leaving an interval of approximately 5 mm between them. Their inner aspects are cut to a height of approximately 7 mm, and a lateral incision 2 mm deep is made at the junction of the inner aspect of the pillars with the vault.

The arch is soaked in liquid culture medium and placed upside down in a shallow vitreosil basin (approximately 4 cm in diameter). The explant is

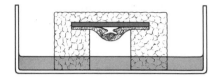

FIGURE 23-6

Diagram showing how a combined explant of endometrium developing placenta and early embryo is mounted for culture on a polyurethane foam arch.

then stretched onto the vault of the arch and the edges of the explant are slotted into the incisions at the junctions of the pillars with the vault. Once the explant has been mounted in this way, the arch is turned back so that the foot of each pillar is standing in the liquid medium contained in the basin and the graft fixed to the undersurface of the vault. This permits the foam to soak up medium and enables the explant to be bathed by a thin film of medium. It is important to ensure that the level of fluid is kept well below the level of the explant fixed to the vault; in this situation the tissues have free access to air or other gaseous mixtures inside the arch.

With this type of culture the embryo hangs from its developing placenta enabling various developmental processes, such as amnion formation, to take place without mechanical interference; embryonic development, in general, is thus facilitated and enhanced.

More recently the polyurethane foam material used for the arches has been supplanted by fine-pore cellulose sponge (American Sponge and Chamois Co. Inc., 4700 34th St., Long Island, N.Y. 11101). This material has the great advantage that it can be sectioned readily for histological examination should the explant become attached firmly to the sponge arch by the invasion of its interstices by migrating cells from the explant (Paul, 1965).

METHODS FOR OTHER SPECIES

As stated earlier it has not been possible as yet to reproduce ovoimplantation in organ culture by using material from another species. It may however be useful to give a brief account of the methods attempted.

Although Biggers, Gwatkin, and Brinster (1962) have succeeded in procuring the development of mouse ova to advanced blastocyst stages in organ cultures of fallopian tubes on a chemically defined medium, I have not been able to obtain ovoimplantation when whole uterine horns containing preimplantation blastocysts have been incubated as organ cultures using the methods described above. This has been attempted with mouse, rat, guinea pig and hamster material.

COLLECTION OF HAMSTER BLASTOCYSTS

Hysterectomy under nembutal or ether anesthesia is performed about three-and-a-half days after the hamster female has copulated. The uterine horns are separated from one another, each horn has a ligature tied around one end and each horn is placed in a separate large watch glass (approximately 6 cm diameter). The lumen of each horn is then distended gently by injecting Tyrode solution or other physiological solution by means of a Pasteur pipette inserted through the open end of the horn. The fluid in the horn is allowed to flow back into the watch glass, the ligature is cut, and the horn is then flushed with more solution. In this way it is not at all unusual to recover twelve blastocysts from each pregnant female.

The blastocysts are surrounded by a zona pellucida when recovered and, as the blastocysts would have divested themselves of their zona within a few hours had they remained *in vivo,* several have been explanted without prior removal of the zona in the hope that they would emerge from it in culture. Other blastocysts had their zona pellucida removed with pronase (Mintz, 1962, 1964).

The modification of the method used by Dr. Gwatkin is convenient because it prevents the denuded blastocysts from becoming stuck to the glassware. The pronase is made up as a 0.25% solution in phosphate buffered saline of pH 7.2 and which contains 1% polyvinyl pyrrolidone. The blastocysts are incubated in this solution for five minutes and then washed three or four times in culture medium.

Handling of Hamster Blastocysts

These blastocysts are considerably smaller than those of the rabbit and cannot therefore be handled in the same way. They have a diameter of 0.1 or 0.2 mm and thus considerable care must be exercised when transferring from one receptacle to another, or when explanting uterine tissue.

A convenient way of effecting such transfers has been found to be a small pipette with an internal bore of approximately 0.5 mm and which possesses a glass bulb of a few cubic millimeters near its tip. When such a pipette is joined to a length of thin rubber tubing (rather than being fitted with a rubber teat) it is possible to effect sufficiently small movements of fluid so as to handle and transfer the blastocysts with precision.

Preparation of Hamster Uterine and Endometrial Explants

Because of the very small size of the blastocysts it is no use explanting them on strips of endometrium as had been done for rabbit material. Surface tension disperses them over the surface of the endometrial explant and they usually disappear within the culture dish.

Efforts have therefore been made to devise means of keeping the blastocysts in contact with endometrium to give them a chance of implanting. Using rings of uterine horn lying on a hard medium and pipetting the blastocysts into the lumen has failed. Also unsuccessful was the placing of stainless steel rings or small sheets of agar with punch holes onto endometrial strips (in this way wells were formed over the endometrium into which blastocysts were pipetted). The only method which has kept the blastocysts in contact with endometrium has been the one that makes use of polyurethane foam blocks.

A boat-shaped cradle is cut out to a depth of about 5 mm in a rectangular block 2 cm by 1 cm and about 1 cm high. Endometrial strips are made by the method described for the rabbit. These strips are used to line the cradle of foam material which is soaked in liquid culture medium and placed in a shallow vitreosil basin (as used in the polyurethane foam arch method).

The blastocysts are pipetted into the cradle and incubated. Although none have so far been implanted, blastocysts have been observed to remain in position after 48 hours and the endometrium thrives in this type of culture.

HISTOLOGICAL PREPARATION

Apparent attachment of a blastocyst to an endometrial explant when examined under a dissecting microscope is insufficient evidence of implantation, even if it adheres to the endometrium after frequent washing. Frequently blastocysts appear to have become attached to the endometrium, but on histological examination they are found to be only adherent to it by means of cellular débris. It is therefore essential to establish by histological means that the trophoblast has fused with and penetrated maternal tissue.

To achieve this and the combined explants are fixed carefully in Bouin's fluid and embedded in paraffin wax. All specimens are sectioned serially, mounted, and stained with routine histological stains.

BIBLIOGRAPHY

Aitken, E. H., J. R. K. Preedy, B. Eton, and R. V. Short (1958) Oestrogen and progesterone levels in foetal and maternal plasma at parturition. Lancet, 2:1096–1099.

Amoroso, E. C. (1959) The attachment cone of the guinea-pig blastocyst as observed under time-lapse phase-contrast cinematography. In: Implantation of Ova. Ed. by P. Eckstein. pp. 50–54. Mem. Soc. Endocrin., no. 6. Cambridge University Press, Cambridge.

Bimes, C., H. Planel, and J. F. David (1961) Action des Hormones génitales sur la souche tumorale HeLa et diverses catégories cellulaires normales cultivees *in vitro*. C. R. Soc. Biol. (Paris), 155:138–141.

Biggers, J. D., R. B. L. Gwatkin, and R. L. Brinster (1962) Development of mouse embryos in organ cultures of Fallopian tubes on a chemically defined medium. Nature, 194:747–749.

Everett, J. (1962) The influence of oestriol and progesterone on the endometrium of the guinea-pig *in vitro*. J. Endocrinol. 24:491–496.

———— (1963) Action of Oestrogens and Progesterone on Uterine Epithelial Mitosis in Organ Culture. Nature, 198:896.

Gaillard, P. J. (1942) Hormones regulating growth and differentiation in embryonic explants. Hermann, Paris.

Glenister, T. W. (1960) Experimental nidation of blastocysts in organ culture. Bull. Soc. Roy. Belg. Gynecol. Obstet., 30:635–640.

———— (1961a) Organ culture as a new method for studying the implantation of mammalian blastocysts. Proc. Roy. Soc., series B., 154:428–441.

———— (1961b) Observations on the behaviour in organ culture of rabbit trophoblast from implanting blastocysts and early placentae. J. Anat., 95:474–484.

———— (1962) Embryo-endometrial relationships during nidation in organ culture. J. Obstet. Gynaecol. Brit. Commonwealth, 69:809–814.

———— (1963) Observations on mammalian blastocysts implanting in organ culture. In: Delayed Implantation. Ed. by A. C. Enders. University of Chicago Press, Chicago, pp. 171–182.

———— (1965) The behaviour of trophoblast when blastocysts effect nidation in organ culture. In: The Early Conceptus: Normal and Abnormal. p. 24. Ed. by W. W. Park. Livingstone, Edinburgh and London.

———— (1967) Organ Culture and its combination with electron microscopy in the study of nidation processes. In: Fertility and Sterility. Ed. by B. Westin and N. Wiqvist. pp. 385–394: Excerpta Medica Foundation.

Grover, J. W. (1961) The enzymatic dissociation and reproducible reaggregation *in vitro* of 11-day embryonic chick lung. Devel. Biol., 3:555–568.

Kahn, R. H. (1958) Epithelial differentiation *in vitro*. Anat. Record, 130:321–322.

Mayer, G. (1960) Morphologie et physiologie comparées de l'ovo-implantation—résultats et problèmes. In: Les fonctions de nidation utérine et leurs troubles. Ed. by G. Masson. Masson, Paris, pp. 1–32.

Meunier, J. M. (1959) Culture *in vitro* de fragments d'endomètre de lapin. C. R. Acad. Sci. (Paris), 248:304–307.

Mintz, B. (1962) Experimental study of the developing mammalian egg. Removal of the zona pellucida. Science, 138:594–595.

———— (1964) Formation of genetically mosaic mouse embryos, and early development of "lethal (t^{12}/t^{12})-normal" mosaics. J. Exp. Zool., 157:273–292.

New, D. A. T. (1966) The culture of vertebrate embryos. Academic Press; Logos Press, London.

Paul, J. (1965) Cell and tissue culture. 3d ed. Livingstone, Edinburgh and London.

Shaffer, B. M. (1956) Culture of organs from embryonic chick on cellulose-acetate fabric. Exp. Cell. Res., 11:244–248.

Trowell, O. A. (1959) The culture of mature organs in a synthetic medium. Exp. Cell Res., 16:118–147.

Zander, J. (1954) Progesterone in human blood and tissues. Nature (London), 174:406–407.

24

Methods for Studying Changes
in Capillary Permeability
of the Rat Endometrium

A. PSYCHOYOS
*Institut National de la Santé
et de la Recherche Médicale
Hôpital de Bicêtre
94 Bicêtre, France*

The endometrium of the rat, properly sensitized by the ovarian hormones, responds with a decidual reaction to the blastocyst stimulus and to a variety of mechanical, electrical, or chemical ones, as well as to systemically offered inducers. Whatever the inducing stimulus may be, an increase in the permeability of the local capillaries is the *sine qua non* condition for decidualisation and it always precedes and accompanies the decidual metamorphosis of the endometrial cells.

This vascular modification can be easily revealed by the intravenous injection of macromolecular dyes (m.w. of about 960.000). The most often used of these dyes are Geigy Blue, Pontamine Blue, or Evans Blue, all of which are able to bind with serum albumin so that when they are injected intravenously they cannot leave the circulation system except at sites at which the capillary permeability is abnormally increased. Thus the endometrial regions, at which the stimuli have occured, accumulate the dye and become macroscopically distinct; decidual transformation of the subepithelial cells of these regions usually occurs at least 12 hours later. A serum albumin labeled by fluoresceine may be used instead of the above dyes, or, in the case of quantitative estimation, serum albumin containing a radioactive tracer may be employed.

In the study of the decidualisation process and the very early ovo-endometrial interactions, these techniques dispense with the need for serial section of the whole uterus in order to reveal the site of the embryo-maternal contact. During normal pregnancy in the rat, the uterine sites at which the blastocysts are present are colored on the afternoon of Day L4[1] (Psychoyos, 1960). In the mouse this test becomes positive late on Day L3 (Orsini and McLaren, 1967) and in the hamster at three days and 10 to 12 hours after ovulation (Orsini, 1963). To our knowledge, information for other species is missing.

THE BLUE DYES TEST

METHOD

Geigy, Pontamine, or Evans Blue is diluted at 0.5-1% in saline. The solution can be kept at room temperature without being altered, but should be shaken lightly before use. A half ml of this solution is injected intravenously for each 100 g of body weight. A cutaneous lesion, made immediately after injection by pinching the paw of the animal with forceps, may serve as an indicator for the optimal time of observation. The coloration of the lesioned area shows the sharpest contrast with the rest of the skin about 15 minutes after injection and it is then that the uterus of the animals should be examined.

EXAMINATION *in Vivo*

When laparotomy is performed and the uterus is examined *in situ* for blue spots, care must be taken not to touch the uterine horns directly as a nonspecific accumulation of the dye may be induced. However, even so, the blue spots resulting from the deciduogenic stimulus can be easily distinguished since they are limited to the endometrium and can be seen through the transparent muscular layer. Examination of frozen sections shows that the coloration, which is dependent on the increase in permeability produced by the blastocyst, is, in the early stages, limited to the stroma underlying the antimesometrial epithelium which is in contact with the blastocyst. When the blue test is applied at a later stage, the colored regions are given a spherical shape by the uniform spreading in the stroma of this vascular modification. The size of the spots and the intensity of the color are therefore dependent upon the time at which the ovo-endometrial interactions have been initiated. It may happen that such differences are found among spots of the same horn, as the blastocysts of the lower part of the uterus usually implant first. A pale coloration of all uterine spots may be due however to an insufficient dose of colorant, or to a too short interval between the injection of the colorant and the time of observation. It is therefore useful to compare the degree of dye accumulation in the uterus with the one of the control lesion performed in the skin. In any case, whenever difficulties arise in distinguishing the uterine blue spots, a simple way to remove the difficulty is to press the horn slightly between two slides.

---

[1] In this text L1 is the first day of appearance of leucocytic vaginal smear (the day following the discovery of sperm).

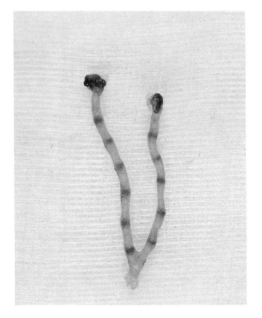

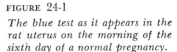

FIGURE 24-1

The blue test as it appears in the rat uterus on the morning of the sixth day of a normal pregnancy.

POSTMORTEM EXAMINATION

As mentioned previously, the sharpest contrast of the colored spots is obtained when the uterus is examined about 15 minutes after the injection of the dye. Sacrifice should be done at this time by decapitation. A complete exsanguination of the uterus by perfusion of saline through the abdominal aorta is recommended for special morphological studies. This material is fixed in AFA (30 ml of 95% alcohol, 10 ml commercial formalin, 10 ml glacial acetic acid, and 50 ml water). Excellent results are obtained by the benzyl-benzoate clearing technique and observation under oblique illumination (Orsini, 1963). This method consists of fixation of the uteri in AFA and then slow dehydration through ascending alcohols. Hydrogen peroxide is added to the 70 and 80% alcohol, and from absolute alcohol the tracts are passed through absolute alcohol-benzol to benzol, and thereafter to benzyl-benzoate in which they are studied and may be stored. The colorant is retained throughout the clearing process and the preparation obtained is extremely elegant.

THE FLUORESCEINE TEST

Possibilities similar to the use of macromolecular dyes are offered by the injection of a serum-albumin labelled by fluoresceine—50 μg of this protein diluted in 0.5 ml of saline are injected intravenously per 100 g of body weight. The optimal time for observation is, in this instance, 30 to 40 minutes after injection. The fluorescent uterine spots can be studied under a UV scanner that has long wave (3660 Å) illumination. This technique offers the advantage that, in parallel with the study of the capillary permeability modifications, macroscopical and microscopical study of the periuterine lymphatic vessels can be made (Psychoyos, unpublished).

QUANTITATIVE METHOD

The techniques described above offer a qualitative appraisal of the degree of capillary permeability. A quantitative estimation of this degree can be obtained by the use of a serum-albumin labelled with I^{131} injected together with a suspension of red cells labeled with Cr^{51}. Because of the different half-life period of these two elements, this combination allows the activity due to the serum-albumin I^{131} contained in the uterine blood to be calculated and thus the quantity of the extravascular protein which reflects the degree of permeability can be determined (Psychoyos, 1961; Bitton *et al.*, 1965).

A solution of human serum-albumin I^{131} of a specific activity of 0.58 mC/mg is injected intravenously at the dose of 5 μC per animal and is immediately followed by the intravenous injection of the red-cell suspension. Labelling of the cells is obtained by adding 400μC of Cr^{51} to 10 ml of heparinized rat blood; the blood is then incubated at 37°C for one hour and the red cells are separated by centrifugation and washed three times in saline; 5 ml of the last suspension of 1:3 of red cells in saline, is injected per animal. Thirty minutes after the double injection, the animals are decapitated. Two successive counts of six-day intervals are made so that the radioactivity due to I^{131} and the Cr^{51} contained in the uterus and in 0.1 ml of blood, is calculated with the formulae:

$$I = \frac{K^2X - Y}{K^2 - K^1} \quad \text{and} \quad Cr = \frac{Y - K^1X}{K^2 - K^1}$$

where I = the radioactivity due to I^{131} content of the uterus or the blood, Cr = the radioactivity due to Cr^{51} content, K^1 = the coefficient of the radioactivity decrease of the I^{131} for the time interval between the two counts, K^2 = this coefficient for Cr^{51}, X = the first count and Y = the second one.

The blood volume of the uterus (BV) is then determined according to

$$BV = \frac{Cr}{Crb}$$

where Crb = the activity due to the Cr^{51} content of 0.1 ml of blood from the same animal. Radioactivity due to the I^{131} content of the uterine blood (Iub) is found by multiplying BV to Ib, where Ib = the radioactivity due to I^{131} content of 0.1 ml of the blood. The radioactivity due to the I^{131} content in the extravascular space of the uterus (It) is then obtained by the formula: It = I-Iub and the corresponding weight of the extravascular serum-albumin I^{131} calculated according to $\frac{It}{\rho}$, where ρ = the specific activity of the injected protein.

BIBLIOGRAPHY

Bitton, V., G. Vassent, and A. Psychoyos (1965) Reponse vasculaire de l'uterus au traumatisme, au cours de la pseudogestation chez la Ratte. C.R.Ac.Sci., Paris, 261:3474–77.

Orsini, M. W. (1963) Morphological Evidence on the Intrauterine Career of the Ovum. In: Delayed Implantation. Ed. by A. C. Enders. Univ. of Chicago Press, Chicago.

—— and A. McLaren, (1967) Loss of the zona pellucida in mice, and the effect of tubal ligation and ovariectomy. J. Reprod. Fertility, 13:485–99.

Psychoyos, A. (1960) Nouvelle contribution à l'étude de la nidation de l'oeuf chez la Ratte. C.R.Ac.Sci., Paris, 251:3073–75.

—— (1961) Permeabilité capillaire et decidualisation utérine. C.R.Ac.Sci., Paris, 252:1515–17.

25

Observing Inside
the Living Uterus

BENT G. BÖVING
Department of Embryology
Carnegie Institution of Washington
115 West University Parkway
Baltimore, Maryland 21210

The developmental process associating the conceptus and uterus has customarily been studied by methods that begin either by separating the conceptus and the uterus and thereby destroying the association or by killing the components and so destroying the process. Studying the process of association while it is still in progress might be more to the point, however. In general, one could look for interactions of anatomical components and study their reciprocal mechanical and chemical behavior; in particular, one might hope to see a mechanical interaction causing the consistent and species-specific orientation of the embryonic pole of the conceptus with respect to the mesometrial-antimesometrial axis of the uterus. To chemically explore this or any other anatomically selective interaction, the reacting parts must not only be distinguishable but also accessible.

The dark interior of a uterus with a healthy circulation cannot be illuminated and studied without breaching the uterus. Breaching however alters the uterine mechanics and chemistry, particularly the circulation, which has not only its usual supportive and homeostatic roles but is also the specific target of placental development (Böving, 1962, 1967). Accordingly, the technical objective is to achieve maximum visibility and access with minimum mechanical and chemical disturbance of the conceptus *in utero*.

Rabbits are the best subjects for these studies. They mate anytime in season, and implantation occurs seven days after mating. Since the blastocyst is unusually large (5 mm), it is easily seen and manipulated, and is sturdy thanks to an encasement by lemmas that persist throughout the

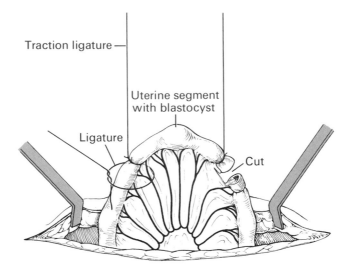

FIGURE 25-1

Swelling and translucency of the rabbit uterus mark the location of a blastocyst within. For observation, the uterus is doubly ligated and cut about 15 mm on either side of it, preferably in a region free of mesometrial vessels. The outer ligature goes deeper into the mesometrium to include mesometrial vessels near the region to be cut; the inner ligature encompasses only uterus. The cut first divides the uterus and then extends slightly at a right angle to trim mesometrium from the first few millimeters of the uterine segment.

muscular phase of attachment and into the subsequent adhesive phase (Böving, 1963). The uterus has two separate cylindrical horns, each about 6 mm in diameter and 200 mm long. The mesometrial blood vessels to it are so situated that segments of uterus can be tied off without unduly curbing circulation (see Fig. 25-1), and the uterine wall is thin.

DESIGN AND CONSTRUCTION OF APPARATUS

A plane window should be set perpendicular to the axis of cylindrical uteri in order to maximize the viewing diameter and to prevent any distortion of the observed part of the uterus. Two such windows, one at each end of a segment of uterus, provide a straight optical path for illuminating the blastocyst within and seeing it against a plain background (see Fig. 25-4). The ends of the uterine segment rest in holes in opposite sides of a plastic box. The tapered stopper-shaped windows, when pushed into the uterine lumen, circumferentially compress the uterine wall against the confines of the hole, anchor the uterine segment, and provide hemostasis. The stopper windows also retain the uterine contents. Access to the contents is gained by thin plastic tubing through a hole in a stopper window (see Fig. 25-5) or by a hypodermic needle directed first through a rubber stopper in the top of the plastic box and then through the uterine

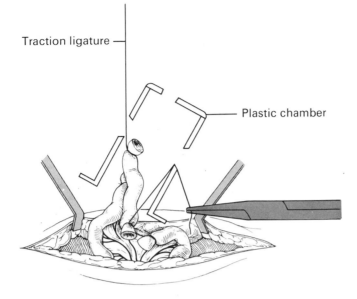

FIGURE 25-2

The uterine segment is pulled into position by traction on the ligatures at its ends, supplemented by fine forceps applied only near the cut. The first end through a mounting hole may be anchored by clamping a hemostat on the suture before maneuvering the second end into place.

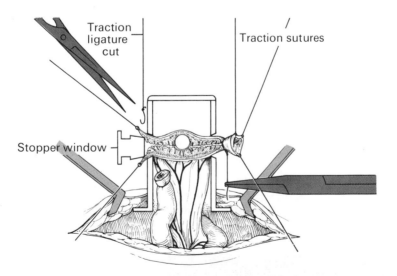

FIGURE 25-3

After each end of the uterine segment rests in a hole in the plastic chamber, twists and tensions are relieved, and the stopper windows are inserted.

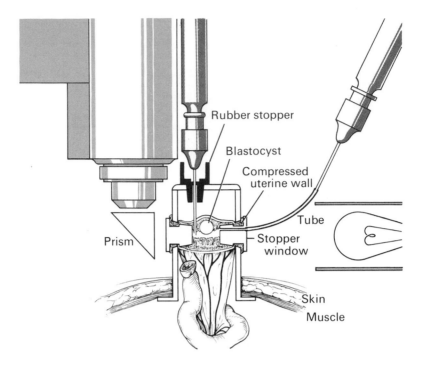

FIGURE 25-4

The chamber is sutured into place as a ventral hernia. A rubber stopper is placed in the roof. Through it fluid is injected and air aspirated until no peritoneal surface remains exposed to air. If endometrial folds obscure the view, a little fluid may be injected into the uterine lumen. The groove at the large end of the stopper windows keeps them from being squeezed out by the uterus. The prism permits the animal to lie on its back, which makes open drop ether easier to administer, and keeps any leaking fluid or blood away from the stopper windows.

wall. The box, open to the peritoneal cavity, curbs drying and cooling of the exteriorized uterine segment, maintains optical alignment of stopper windows and uterus, and permits clamping so that body movements do not jiggle the structures being observed, manipulated, or photographed.

The box (see Fig. 25-6) and stopper windows are of transparent methyl methacrylate polymer (Lucite, Plexiglas, or Perspex). The mounting holes in the box should be taper-reamed to be just large enough to admit a

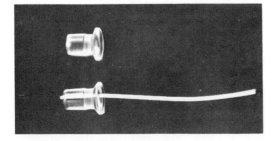

FIGURE 25-5

Stopper windows are turned from a methyl methacrylate polymer such as Lucite, Plexiglas, or Perspex. There is a slight taper, a shallow groove at the large end, and a rim just outside that. The plane ends are polished. A polyethylene tube may be placed through a stopper window.

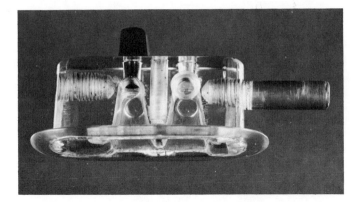

FIGURE 25-6

A double model chamber for low magnification is shaped like a high hat and has holes in opposite sides to hold the ends of the two uterine segments and the stopper windows, one of which is shown in the right hole. A vaccine vial type of rubber stopper goes through the top of the chamber and is seen through the left hole. To prevent respiratory and other body motions from moving the field in view, the short rod protruding at the right is fastened with a clamp.

stopper window, and the edges of the holes should be polished to avoid cutting the uterus. The stopper windows taper from a diameter of about that of the uterus to 1 mm larger, then have a groove and a lip. The groove provides snap-fastener action that prevents the uterus from extruding the stopper window. The lip facilitates handling and keeps the cut edge of the uterus outside the field of view. Both ends of the stopper windows are plane and polished.

ALTERNATIVE APPARATUS DESIGNS

Various advantages may be traded off in design. Doubling (see Fig. 25-6) lets a second uterine segment be used as a control or reserve, but takes longer to set up. An outside window-to-window dimension of 15 to 25 mm has proved best for the low magnifications used to study uterine muscular grasp of the blastocyst, adhesion, and lemma shedding. An outside dimension of 10 or 12 mm adapts to the shorter working distance of phase and other microscopy at about ×100 magnification (see Figs. 25-7 and 25-8) .

METHOD

Preparation includes clipping, intravenous sodium pentobarbital anesthesia supplemented by open-drop ether if necessary, and midventral incision. Choose a segment of uterus that is about 30 mm long and has a swelling and translucency from a blastocyst near the midpoint. Several mesometrial vessels should be present near the middle, but at the ends of

FIGURE 25-7

A chamber for high magnification must accommodate to a short working distance by using shorter stopper windows and a shorter distance between the sides of the box. Since there is no room for a 90° prism, the chamber is used on its side when the optical system is vertical. In the upper side of the chamber, barely visible between the objective and the adjusting screw, is a hole to permit adding fluid around the uterine segment and pushing it to adjust the field of view. Near the open base of the chamber are holes to suture the chamber to the abdominal muscles.

the segment there should be a gap between such vessels (see Fig. 25-1). Place double ligatures at both ends. The outer ones should go far enough into the mesometrium to catch all visible vessels that go to the region between the pair of ligatures. The inner ligatures include only myometrium, and should be cut long to provide for traction. After waiting about five minutes to be sure the ligated segment retains good color, cut the uterus between each pair of ligatures. There should be little or no bleeding, and no loss of contents from the uterus. Pull the traction ties through the bottom of the plastic box and out through the mounting holes in the sides, (see Fig. 25-2). Then pull the ends of the uterine segment up into the box,

FIGURE 25-8

A chamber for high magnification uses short stoppers with a plateau, the polished surface of which has been filed to about the size and shape of the field registered by the camera frame. The screws push the chamber walls and stopper windows apart; after installing the uterine segment the screws are backed off to let the stoppers approach each other.

taking care to apply instruments only near the cut edges, which later will be trimmed. Place three traction sutures near each cut edge. Pull the traction sutures apart to open the ends of the uterine segment, and start the stopper windows into the uterine lumen; push them farther in as the ligature around the uterus is cut and removed (see Fig. 25-3). After the stoppers have been pushed in as far as they will go, trim away the overhanging cut edge of the uterus. Close the abdominal incision so that the open base of the chamber is internal to the abdominal muscles and the rest of the chamber projects like an umbilical hernia (see Fig. 25-4). Through the rubber stopper in the top of the chamber, inject physiological salt solution around the uterine segment, and aspirate air bubbles. Adjust the clamp, light, and optics, and attempt observation. If endometrial folds are in the way, as they usually are, they may occasionally be pushed aside by a needle inserted through both the rubber stopper and the wall of the uterus. If they cannot be thus pushed aside, inject fluid through the needle into the uterine lumen, distending the uterine segment no more than is necessary for visibility. Alternatively, thin polyethylene tubing through a hole in a stopper window can be used for injecting fluid into the uterine lumen or sampling its contents.

DISCUSSION

Chemical as well as mechanical consequences of the injected fluid deserve attention. Tyrode's solution has usually been employed, except when 0.85% NaCl solution was chosen to avoid buffering when estimating pH with indicators. Admittedly, neither "physiological" fluid is any more natural than the various media used for blastocyst culture; the normal uterine fluid, what little there is of it, is a heterogeneous mixture of mucus and cell debris as well as of electrolytes. However, since the chamber-mounted specimen, unlike cultured specimens, is connected to a living animal by a functioning circulatory system, one may expect regulation to approximate the natural state within minutes, if one may judge from pH regulation.

Leucocytes were present in abnormal numbers in fluid that was aspirated from the uterine lumen 24 hours after mounting with sterile but not antibiotic technique. Accordingly, the chamber is recommended only for acute experiments or for observations of a few hours' duration.

Transparent chambers for *in vivo* observation of other structures are described by Algire (1952), Bourne (1967), Clark (1952), and Zintel (1936).

BIBLIOGRAPHY

Algire, G. H. (1952) Transparent chamber technique. In: Laboratory Technique in Biology and Medicine. 3d ed. Ed. by E. V. Cowdry. Williams and Wilkins, Baltimore, pp. 354–356.

Bourne, G. H. (1967) *In Vivo* Techniques in Histology. Williams and Wilkins, Baltimore.

Böving, B. G. (1952) Internal observation of rabbit uterus. Science, 116:211–214.

———— (1962) Anatomical analysis of rabbit trophoblast invasion. Contrib. Embryol., 37:33–55.

———— (1963) Implantation mechanisms. In: Mechanisms Concerned with Conception. Ed. by C. G. Hartman., Pergamon, New York, pp. 321–396.

———— (1967) Chemo-mechanics of implantation. In: Comparative Aspects of Reproductive Failure. Ed. by K. Benirschke. Springer-Verlag, New York, pp. 142–153.

Clark, E. R. (1952) Transparent chamber technique. In: Laboratory Technique in Biology and Medicine. 3d ed. Ed. by E. V. Cowdry. Williams and Wilkins, Baltimore, pp. 351–354.

Zintel, H. A. (1936) A new transparent chamber for exteriorizing a loop of intestine and its mesentery. Anat. Record, 66:437–447.

26

Study of the Exposed Guinea Pig Fetus

ANDREW M. NEMETH
Department of Anatomy
University of Pennsylvania
Philadelphia, Pennsylvania 19104

A guinea pig fetus can be made accessible for direct study by delivering it into a warm saline bath through incisions in the mother's abdominal wall and uterus. The bath approximates intrauterine conditions for the fetus. The placenta remains attached to the uterus and, if the fetus, the umbilical cord and the lower parts of the mother are kept entirely submerged in the bath, the fetus will usually survive in good condition for several hours. The circulatory, respiratory and nervous systems of the mammalian fetus, as well as placental transport, have been studied in such preparations (Cohnstein and Zuntz, 1884; Huggett, 1926; Barcroft, 1946; Kennedy and Clark, 1941; Flexner, Tyler, and Gallant, 1950; Shepherd and Whelan, 1951; Hershfield and Nemeth, 1968).

REPRODUCTIVE CYCLE

Many features of the guinea pig reproductive cycle lend themselves to the study of the life-history and physiology of the fetus. With a gestation period of approximately 68 days, the fetus on Day 35 of gestation weighs 4 to 6 grams and at birth 70 to 100 grams (Draper, 1920; Ibsen, 1928). This size makes experimentation convenient, but not cumbersome. The species is fecund and a young, mature female can be bred repeatedly for at least one year (Eckstein and Zuckerman, 1956). A litter usually consists of 2 to 4 fetuses; a litter of 12 was once observed in our colony.

It is possible to determine accurately the day of conception of a guinea pig litter (Scott, 1937). In our laboratory, however, gestational age is usually estimated before an experiment by palpating the abdomen of the mother and judging fetal size; after an experiment, measurements of crown-to-rump length and weight can be used to estimate fetal age by reference to growth-tables that have been prepared by Draper (1920) and Ibsen (1928). In our experience these tables are satisfactory if the litter size is no more than five; growth during intrauterine life is rapid enough to make weight and length remarkably accurate parameters by which to order different litters in the life-history of the fetus.

MORPHOLOGY

Day 35 of gestation is a significant point in the development of the guinea pig fetus. By that time the external form of the fetus approaches that of the newborn (Harman and Prickett, 1932, 1933; Scott, 1937; Hard, 1946; Mossman, 1937; Wilson, 1928) and, thenceforth, development is largely a matter of growth and maturation in cytological detail and in function. Almost all biochemical and physiological studies have dealt with this latter period.

The later part of guinea pig gestation is represented by the drawing (cross-section) in Figure 26-1. It is important to know that the fetus is covered by two sacs: the outer vitelline membrane or vitelline placenta (Amoroso, 1961) which is vascularized, and the inner amniotic membrane which is not; these membranes are attached to the top of the disc-shaped allantoic placenta (Enders, 1965).

The vitelline vessels in the umbilical cord, as in their course in the amniotic membrane, consist of an artery and a vein; these are branches of the fetal mesenteric vessels. As the vitelline artery enters the vitelline membrane, it immediately gives rise to 4 or 5 major branches, each branch of which supplies a sector of the membrane; veins accompany the arterial branches.

The allantoic vessels in the umbilical cord consist of two arteries and one vein. The arteries arise from the iliac arteries of the fetus, the vein drains into the fetal liver sinusoids and ductus venosus. Though the vitelline membrane is attached to the allantoic placenta, there do not appear to be any connections between the blood vessels of the two structures.

To remove the guinea pig fetus from the uterus, the vitelline and amniotic membranes must be cut; this stops vitelline circulation and loses amniotic fluid. But, placental (allantoic placenta) circulation of the fetus is not interrupted and appears able to support normal fetal function, at least for a time, even though the fetus is out of the uterus. The criteria which define normal functioning will be discussed after the procedure for exposing the fetus and keeping the placental circulation intact is described in detail.

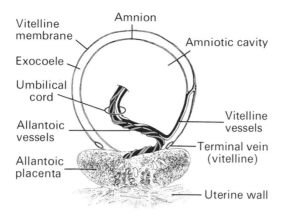

FIGURE 26-1

Membranes and placenta of the guinea pig fetus during the latter part of gestation. (Cross-section with fetus omitted). The allantoic vessels in the umbilical cord consist of two arteries and a vein. The vitelline vessels in their transit in the cord and also in the amniotic membrane consist of one artery and one vein; the vitelline artery gives off 4 or 5 major branches immediately upon entering the vitelline membrane; each branch supplies a sector of the membrane; veins accompany the arterial branches. The placenta is attached to the mesometrial side of the uterus. There are usually 1 or 2 fetuses in each horn of the uterus.

SURGICAL EXPOSURE

There are many ways to induce anesthesia in small laboratory animals. For the pregnant guinea pig weighing between 400 and 1000 grams, we have used the following procedure: inject 0.2 ml sodium pentobarbital (50 mg/ml) intraperitoneally, wait 15 to 30 minutes and then administer diethylether with an inhalation cone until the animal shows little or no response when abdominal skin is pinched. To give the ether, hold the animal on a table in a standing position at first; care must be taken that the animal inhales the ether fumes with enough air. (In our experience with guinea pigs, pentobarbital alone often causes respiratory arrest before satisfactory surgical anesthesia is achieved. On the other hand, ether without sedation frequently causes the guinea pig to regurgitate the contents of its stomach, aspirate and die of asphyxia. Fasting the animal overnight avoids this problem to some extent, but fasting is not an acceptable preparation for some experiments.)

After the pregnant guinea pig is anesthetized, remove the hair from the abdominal skin with hair-clippers and place it on its back onto a board and tie it in place by its feet. Put the board into a saline bath kept at 35° to 37°C so that the animal's hind quarters are submerged. The head-end of the board should be 20° to 30° above the horizontal plane. Satisfactory anesthesia can be maintained by occasional administration of ether.

The guinea pig has a bicornate uterus and usually there will be one or two fetuses in a horn. The horns are suspended from the dorsal abdominal wall by mesenteries.

Make a paramedian abdominal incision through the skin approximately 2 to 4 cm from the midline, then cut through the muscle to the peritoneal membrane with a knife. It is usually not necessary to tie any blood vessels to prevent bleeding. The incision should be long enough to deliver a fetus, but not so long as to make it difficult to keep the abdominal contents in their places.

Cut the peritoneal membrane with sharp, blunt-nosed scissors. Bring a horn of the uterus to the incision, pushing aside the intestines. A few small gauze pads soaked in saline can be packed into the abdominal cavity to keep the intestines in place and out of the way. The antimesometrial side of the uterine horn will be at the point of incision; the placenta is always implanted on the mesometrial side making it easy to avoid disturbing it. Make a short longitudinal incision in the uterus with scissors. Cut first through the vitelline membrane, then through the amniotic membrane; slide the pregnant guinea pig further into the saline so that the abdominal incision is submerged. Deliver the fetus into the saline bath taking care not to pull or handle the umbilical cord, and keeping the fetus, cord, and incision submerged at all times. The incision can be partly closed with clips after the fetus has been delivered. One may raise the board on the side opposite the incision to submerge the incision deeper into the bath. A fetus in this preparation will survive for several hours exhibiting spontaneous movements. The umbilical (allantoic) arteries can be seen to pulsate. The blood in the umbilical (allantoic) vein will remain bright red; the blood in the arteries, dark purple. Blood flow ceases in the vitelline artery and vein of the umbilical cord when the vitelline membrane is cut. The fetus and its placental circulation are now accessible for direct study.

FETAL STUDIES

Many different kinds of experiments can be carried out with the exposed guinea pig fetus. A number of studies will be described to give an idea of the possibilities and to indicate some of the ways in which the preparation can be modified for different purposes.

ELECTROENCEPHALOGRAM

Flexner and co-workers (1950) found spontaneous electrical activity on the surface of the frontal cortex of the fetus on Day 45 of gestation. The appearance of electrical activity at this stage was correlated with the growth of nerve cell processes, the exclusion of sodium ions and the concentration of potassium ions within the cells, and changes in the activities of various enzymes, including a rapid increase in acetylcholinesterase activity.

To measure the electroencephalogram, the fetus must be free of anesthetic during the recording. This was accomplished as follows: first, the pregnant guinea pig was anesthetized with diethylether, then the spinal cord was transected at the midthoracic level, and finally the animal was allowed to recover. The maternal skin was then insensitive and laparotomy could be performed on the anesthetic-free mother. The fetal head was positioned in the saline bath so that the top of the skull was just above the surface. The tissues covering the frontal cortex were removed and recording electrodes were placed on the frontal cortex. Electrical activity was absent in fetuses younger than 45 days; in fetuses older than 45 days, it was regularly observed. Clamping the umbilical cord stopped this electrical activity. Thus it was concluded that the electrical activity in the frontal cortex depends on a well-oxygenated blood supply, good circulation, and on efficient placental gas exchange. The presence of good spontaneous electrical activity on the frontal cortex of the exposed guinea pig fetus after Day 45 of gestation certainly indicates that the circulation and oxygen supply of the fetus and mother in this preparation are normal, or at least adequate to support many complex physiological processes.

Placental Transfer of Free Fatty Acids

Hershfield and Nemeth (1968) measured the rates of transfer (in equivalents/min) of free linoleic and palmitic acid from maternal to fetal plasma across the allantoic placenta using radioisotopically labelled fatty acids. The approach was based on the assumption that at any moment the specific radioactivity of a substance crossing the placenta from maternal to fetal plasma equals the specific radioactivity of the substance in maternal plasma. This idea is expressed in the formula

$$mFA\text{-}C^{14}/mFA = tFA\text{-}C^{14}/tFA$$

where mFA represents the concentration ($m\mu$eq/ml) of a free fatty acid in maternal plasma, $mFA\text{-}C^{14}$ represents the radioactivity (cpm/ml) of the fatty acid, and tFA and $tFA\text{-}C^{14}$ represent the amount and radioactivity of the free fatty acid that is transferred to fetal plasma. To calculate tFA over any period of time, $mFA\text{-}C^{14}/mFA$ in the formula must represent the average specific radioactivity of the fatty acid and $tFA\text{-}C^{14}$, the total radioactivity transferred from maternal to fetal plasma during the time interval.

In addition to exposing the guinea pig fetus and delivering it into a saline bath, a maternal jugular vein and carotid artery were exposed. The artery was cannulated. C^{14}-palmitic and C^{14}-linoleic acids were injected together into the jugular vein. Maternal blood samples were taken from the carotid artery at intervals of 30 seconds, 2, 4, and 8 minutes after injection to estimate $mFA\text{-}C^{14}/mFA$ at these times. To measure $tFA\text{-}C^{14}$ all the fetal blood flowing from the placenta in the umbilical (allantoic) vein was collected for 1 minute after the start of the injection of radioisotopes into the mother. This was done in the following manner: a 20-gauge hypodermic needle on a heparin-coated 5 ml syringe was placed in the umbilical (allantoic) vein of the exposed fetus so that the tip pointed towards the placenta and was almost within the placenta. When suction was exerted by gently pulling the plunger of the syringe, the vein collapsed around the needle except at the tip where the walls of the vein were fixed by the placenta; after suction was begun, the umbilical vein remained collapsed and blood emerging from the placenta could not flow around the needle to the fetus. The flow rate of fetal blood through the placenta into the umbilical vein remained within the normal range during the one-minute collection period.

Decay curves of the specific radioactivities of free linoleic and palmitic acid in maternal plasma were constructed from the values measured at 30 seconds, 2, 4, and 8 minutes after injection; zero-time values were obtained by extrapolation. From these curves the average specific radioactivity of a fatty acid over the 0 to 1 minute interval was estimated graphically; rate of transfer (tFA) for each acid was then calculated. It was found that during the latter part of gestation the rate of transfer of linoleic acid from maternal to fetal plasma was approximately twice that of the transfer of palmitic acid, although the concentration of free palmitic acid in maternal plasma, on the average, was 1.5 times greater than the concentration of linoleic acid. Fasting the mother before the experiment did not alter these ratios but did increase the concentrations of free fatty acids in maternal plasma, as well as their rates of transfer from maternal to fetal plasma. The measured rate of linoleic acid transfer from mother to fetus was sufficient to account for the accumulation of this essential fatty acid in the guinea pig fetus during normal growth and development.

The rate of transfer of a substance from fetal to maternal plasma is more difficult to measure than the maternal-to-fetal way because the total venous drainage of the uterus at a placental site is carried by several vessels and, thus, cannot be collected easily. Schenker, Dauber, and Schmid (1964) circumvented this difficulty in a special case when measuring bilirubin transport from fetus to mother. Bilirubin is concentrated by the adult liver and excreted into the bile duct. The maternal and fetal bile ducts were cannulated; radioactive bilirubin was injected into the fetal circulation. The bile fluids were then analyzed for the radioactive bilirubin, and in this way, fetal to maternal plasma transport could be estimated.

Closure of the Ductus Arteriosus

The exteriorized fetus, as the fetus *in utero* (Wells, 1947; Jost, 1947; Brambell, 1958; Leissring and Anderson, 1961), will tolerate a considerable amount of surgery just as long as placental gas exchange remains adequate. Kennedy and Clark (1941) were able to observe the closing of the ductus arteriosus in the exposed guinea pig fetus near term by removing the anterior chest wall of the fetus under the saline. The ductus remained open for several hours if the fetus was left undisturbed, but it closed following rhythmic inflation of the fetal lungs through a tracheal cannula. On cessation of inflation the ductus reopened. This sequence could be repeated several times in the same fetus.

It should be emphasized that the results of experiments with the guinea pig cannot be projected to other species, regardless of how closely related they may be to the guinea pig. Further experimentation is necessary. For example, some developmental events such as the appearance of alkaline phosphatase activity in the intestinal epithelium (Moog, 1951, 1960) or the appearance of electrical activity in the frontal cortex, occur several days before birth in the guinea pig, but several days after birth in the rat and mouse. Other changes, such as an increase in liver glucose-6-phosphatase activity, have been correlated with birth in all the species studied thus far (Nemeth, 1954; Weber, 1955; Dawkins, 1966). Liver tryptophan pyrrolase activity, on the other hand, develops at birth in the guinea pig and rabbit but only 12 to 16 days after birth in the rat (Nemeth and Nachmias, 1958; Nemeth, 1959; Auerbach and Waisman, 1959). It appears that the same developmental event may be initiated very differently in different species.

BIBLIOGRAPHY

Amoroso, E. C. (1961) Histology of the placenta. Brit. Med. Bull., 17:81–84.

Auerbach, V. H., and H. A. Waisman (1959) Tryptophan pyrrolase, histidase and transaminase activity in the liver of the developing rat. J. Biol. Chem., 234:304–306.

Barcroft, J. (1946) Researches on prenatal life. Oxford: Blackwell.

Brambell, F. W. R. (1958) The passive immunity of the young mammal. Biol. Rev., 33:488–531.

Cohnstein, J., and N. Zuntz (1884) Untersuchungen über das blut, den kreislauf und die athmung beim säugethier-fötus. Pflug. Arch. ges. Physiol., 34:173–233.

Dawkins, M. J. R. (1966) Biochemical aspects of developing function in newborn mammalian liver. Brit. Med. Bull., 22:27–33.

Draper, R. L. (1920) The prenatal growth of the guinea pig. Anat. Record, 18:369–392.

Eckstein, P., and S. Zuckerman (1956) The oestrous cycle in the mammalia. In: Marshall's Physiology of Reproduction, 3d ed., vol. I, pp. 280–282. Ed. by A. S. Parkes. Longmans, Green, London.

Enders, A. C. (1965) A comparative study of the trophoblast in several hemochorial placentas. Am. J. Anat., 116:29–68.

Flexner, L. B., D. B. Tyler, and L. J. Gallant (1950) Onset of electrical activity in developing cerebral cortex of fetal guinea pig. J. Neurophysiol., 13:427–430.

Hard, W. L. (1946) A histochemical and quantitative study of phosphatase in the placenta and fetal membranes of the guinea pig. Am. J. Anat., 78:47–77.

Harman, M. T., and M. Prickett (1932) The development of the external form of the guinea pig between the ages of eleven and twenty days of gestation. Am. J. Anat., 49:351–378.

——— (1933) The development of the external form of the guinea pig between the ages of 21 days and 35 days of gestation. J. Morphol., 54:493–519.

Hershfield, M. S., and A. M. Nemeth (1968) Placental transport of free palmitic and linoleic acids. J. Lipid Res., 9:460–468.

Huggett, A. St.G. (1926) Factors influencing the fetal respiratory center. J. Physiol., 61:30–31.

Ibsen, H. L. (1928) Prenatal growth in guinea pigs, with special reference to environmental factors affecting weight at birth. J. Exp. Zool., 51:51–94.

Jost, A. (1947) Experiences de decapitation de l'embryon de lapin. C. R. Acad. Sci. (Paris), 225:322–324.

Kennedy, J. A., and S. L. Clark (1941) Observations on the ductus arteriosus of the guinea pig in relation to its method of closure. Anat. Record, 79:349–372.

Leissring, J. C., and J. W. Anderson (1961) The transfer of serum proteins from mother to young in the guinea pig. I. Prenatal rates and routes. Am. J. Anat., 109:149–155.

Moog, F. (1951) The functional differentiation of the small intestine. II. The Differentiation of alkaline phosphomonoesterase in the duodenum of the mouse. J. Exp. Zool., 118:187–207.

Moog, F., and E. Ortiz (1960) The functional differentiation of the small intestine. VII. The duodenum of the fetal guinea pig. J. Embryol. Exp. Morphol., 8:182–194.

Mossman, H. W. (1937) Comparative morphogenesis of the fetal membranes and accessory uterine structures. Carnegie Contr. to Embryology, 26:129–246.

Nemeth, A. M. (1954) Glucose-6-phosphatase in the liver of the fetal guinea pig. J. Biol. Chem., 208:773–776.

——— (1959) Mechanisms controlling changes in tryptophan pyrrolase activity in developing mammalian liver. J. Biol. Chem., 234:2921–2924.

——— and V. T. Nachmias (1958) Changes in tryptophan pyrrolase activity in developing liver. Science, 128:1085–1086.

Schenker, S., W. H. Dauber, and R. Schmid (1964) Bilirubin metabolism in the fetus. J. Clin. Invest., 43:32–39.

Scott, J. P. (1937) The embryology of the guinea pig. Am. J. Anat., 60:397–432.

Shepherd, J. T., and R. F. Whelan (1951) The blood flow in the umbilical cord of the fetal guinea pig. J. Physiol., 115:150–157.

Weber, G., and A. Cantarow (1955) Glucose-6-phosphatase activity in regenerating fetal and newborn rat liver. Cancer Res., 15:679–684.

Wells, L. J. (1947) Progress of studies designed to determine whether the fetal hypophysis produces hormones that influence development. Anat. Record, 97:409.

Wilson, J. T. (1928) On the question of the interpretation of the structural features of the early blastocyst of the guinea pig. J. Anat., 62:346–358.

27

Fluorescent Antibody Methods

LAUREL E. GLASS
Department of Anatomy
University of California Medical School
San Francisco, California 94122

For most animals, the antibody, with its specificity of formation and reaction, provides a primary protection against the invasion of foreign organisms and molecules. For the developmental biologist, antibody specificity can be, in addition, a significant tool for the detection and analysis of molecular patterns and syntheses during differentiation. Embryologists have used immunological techniques to examine the relationships between chemical and morphological differentiation, to answer such questions as: Do organ-specific molecules appear before, simultaneously with, or after characteristic morphological changes in a tissue? What is the distribution in the embryo of organ-specific molecules? Does their distribution coincide with the cellular limits from which an organ will arise, or is the antigen broader (or more limited) in distribution? Can causal relations be deduced between differentiation and particular antigens and their interactions?

Antibody *specificity* means that the serum antibody molecules, generally globulins, react only with the substance that produces them (the homologous antigen) or with substances showing a very close chemical relationship to the homologous antigen (cross-reaction with heterologous antigens). The majority of antigenic substances are protein in nature, although complex carbohydrates, and probably other molecular types, stimulate antibody production in some species.

Against purified antigens, antibodies with three general types of specificity may be developed. (1) The antibody may show *species specificity;* it will react (though to different degrees) with antigens from different tissues of

The research for this article was supported in part by AEC Contract No. AT (11-1)-34, Project Agreement No. 53. Sincere appreciation is expressed to Miss Evelyn Gabriels, an excellent technician, and to David Akers for photographic assistance.

the same species, but will show limited or no cross-reaction with antigens from other species. For example, an antibody developed for the rat kidney may show cross-reactivity with the rat brain, the rat spleen, and other rat organs, but will not react with guinea pig, mouse, or human organ antigens. (2) The antibody may show *organ specificity;* it will react with antigen from a particular organ even though the antigen is derived from a different species. For example, an antibody for rat kidney may cross-react with the antigen from mouse kidney, chicken kidney, human kidney, etc., though not with antigen from rat brain, rat heart, or mouse or chicken brain. (3) The antibody may react *only* with its homologous antigen; by appropriate treatment, cross-reacting antibodies can be removed from an antiserum, usually by precipitating them out by addition of heterologous antigen. In such a case, for example, an antiserum specific to rat kidney will react *only* with rat kidney, and will not cross-react with other rat organ antigens nor with kidney antigen from other species. This limited reactivity means that antibodies can be used to detect particular antigens in mixtures that can not be resolved otherwise.

FLUORESCENT ANTIBODY TECHNIQUES

LABELED SCIENTIFIC REAGENTS

The fluorescent antibody and other immunohistological techniques are special applications of immunological procedures. They take advantage of antigen-antibody specificities to ask not only "whether" and "when" but also to demonstrate "where" specific macromolecules are appearing during development.

The search for chemically labeled reagents that can be used to detect and localize antigens or antibodies in tissues was guided by the following criteria: (1) the antibody molecules must retain specificity during and after the necessary chemical manipulation; (2) there must be a stable chemical linkage between the antibody and its label; (3) it must be possible to separate the labeled antibody from unconjugated tracer material; (4) optical rather than analytic or radiographic methods are preferable for localization, since it should be possible to determine not only the organ, but also the cells within the organ, that contain the antigen in question (Coons *et al.,* 1942). Fluorescein isocyanate and isothiocyanate as antibody labels largely satisfy these criteria.

FLUORESCEIN-CONJUGATION PROCEDURES

The use of fluorescent tags for an antibody is relatively recent. Aromatic isocyanates were first conjugated to protein through carbamido linkage by Hopkins and Wormall (1933), who concluded that the most likely reaction site was the ε-amino group of lysine. The general conditions for the conjugation of isocyanates of the higher aromatic hydrocarbons to protein were worked out by Creech and Jones (1941a,b) and used by them in the conjugation of fluorescein isocyanate to antibody solutions (Coons *et al.*, 1942).

The Coons group first employed anthracene as the fluorescent agent and coupled it with an antipneumococcus III rabbit serum (1941). However, since the normal blue fluorescence of mammalian connective tissue was enhanced by formalin fixation, the blue-fluorescing antibodies were difficult to distinguish. Consequently, these investigators switched to coupling with fluorescein, which had a bright yellow-green fluorescence (1942). Their studies were interrupted by World War II, and it was not until 1950 that a second paper on this method was published. It detailed an improved technique for conjugation of fluorescein isocyanate with the protein antibody (Coons and Kaplan, 1950).

Since fluorescein isocyanate is very unstable, it must conjugate with antibody protein virtually as soon as it is generated. So long as isocyanate was the primary label available for fluorescent antibody work, the method was confined to only a few laboratories, since phosgene gas is required to generate the fluorescent label. In 1957, Goldman and Carver discovered that measured amounts of fluorescein isocyanate could be adsorbed to small pieces of filter paper and would remain stable in a dessicator chamber for several months. In 1960, Riggs *et al.* developed a method that made fluorescein conjugation accessible to most investigators. Their method uses the stable compound, fluorescein isothiocyanate, which is now available commercially (Cherry *et al.*, 1960, list suppliers). Although fluorescein is the most frequently used fluorochrome for labeling antibodies other fluorescent compounds, e.g., lissamine rhodamine B (RB 200; Chadwick *et al.*, 1958) and 1-dimethylaminonapthalene-5-sulphonic acid (DANS; Clayton, 1954; Clayton and Feldman, 1955) have been used to good advantage.

FLUORESCENCE MICROSCOPY

The most important advantage of fluorescence microscopy is its great sensitivity. High contrast is given by extremely low concentrations of fluorescing materials. The principal difficulty is low light intensity, since high contrast is of no use if the fluorescence intensity is too low to permit accurate detection or measurement. The problem in fluorescence microscopy is like that of distinguishing between 3 and 20, whereas in ordinary bright-field microscopy, the problem is one of distinguishing between 5000 and 5100. "It is like observing pin points in a sheet of black paper held up against a dim light, as against observing slightly cloudy spots on a clear glass surface against a bright light . . ." (Price and Schwartz, 1956; see also Mellors, 1959).

Despite its specificity and its consequent high potential for analytic studies of molecular differentiation during development, the fluorescent antibody method has been little used by embryologists, probably because of a variety of technical problems that may result in confusing data. Because these difficulties are so exasperating and because the method has so much potential usefulness, this paper will describe, step-by-step, the procedures needed to use the method. Steps which have caused us problems will be indicated, and small details we find convenient will be mentioned. Many different, satisfactory procedures are employed by other workers, and especially useful discussions of the method and rationale for fluorescent antibody procedures will be found in Coons (1956), Mellors (1959), Clayton (1960), and Nairn (1962).

MATERIALS AND METHODS

CAUTIONS TO THE NOVICE

Be hypercompulsive about "contamination"! Nonspecific fluorescence can be introduced easily, and the source of the contamination and the step at which the technique goes bad are sometimes very hard to find. Specifically:

1. Keep the instruments and glassware used in fluorescent antibody procedures separate from all others in the laboratory. Petris for moist chambers, rinsing beakers, staining dishes used for hydration of tissues, Millipore filters and syringes, and everything else used in the process, must be isolated. We use a separate cupboard and drawers to store the fluorescent antibody "equipment." New glassware is used to replace broken pieces. The dishes are washed separately and in a separate pan from other laboratory glassware. The laboratory bench where the antibody procedures are carried out is used for nothing else that conceivably could bring in nonspecific fluorescence or quenching.

2. Keep potentially contaminating materials entirely out of the laboratory room where the fluorescent staining is done; carry out conjugation procedures and standard histological staining in a different room. Remember, for example, that eosin is fluorescent, as is the picric acid in Bouin's, as are most standard mountants, and as are *many* other histological reagents in routine use. They must be kept out of the immunohistology area, and, if one is making a fluorescent run, off one's hands.

3. It is less important, but still useful, to remember that quenching of fluorescence is caused by heavy metal ions, e.g., Hg^{+2}. Therefore, keep compounds containing these materials out of the fluorescent antibody area.

TISSUE PREPARATION

We work primarily with mouse ovary, female reproductive tract, and preimplantation and early postimplantation embryos (Glass, 1961, 1963, 1966, 1968; Glass and McClure, 1965). The procedures to be described are appropriate with these materials; obviously, each experimental system must be checked carefully to be sure that particular methods preserve antigenicity and do not introduce serious artifact. (Figures 27-1a to 27-1f illustrate the technique with mouse preimplantation embryos.)

Tissues are removed from the living, anesthetized animal, and immediately placed in fixative. Generally, ether anesthesia is used. When another anesthetic is desired, pentobarbital is administered intraperitoneally, using the careful dose-weight table developed by Pilgrim and deOme (1955).

Fixation, embedding, and sectioning

(a) Frozen sections. Tissues may be placed alone or with a drop of agar or (rabbit) serum, or other medium, into one-half of an empty pharmacist's capsule; a bit of rabbit serum on the outside rim serves as a "glue" to hold the top in place. The capsule can be lowered or dropped into liquid nitrogen $(-195°C)$. Tissues can also be placed on aluminum-foil boats or in stainless-steel cassettes and dropped into the nitrogen. After fast-freezing, the capsules are removed from the nitrogen and stored in a freezer $(-20°C)$ until sectioned in a cryostat.

We have had good luck obtaining relatively thin cryostat sections, 4 to 10μ, using a rotary microtome and a Palmborg window (Coons et al., 1951). Seven microns is quite convenient. Sometimes we have managed 10 to 12 sections in a series, but usually one must be content to work with only a few sections in sequence.

Sections are taken from the knife onto a cold coverslip and, generally, are air-dried. In some laboratories, such sections are freeze-dried. (See Meryman, 1960; Nairn, 1962, pp. 104ff.).

(b) Chemical fixation and embedding in low-temperature paraffin. Routinely, we fix mouse tissues in Carnoy's II Fluid (absolute ethanol, 60 parts: chloroform, 30 parts: glacial acetic acid, 10 parts); this is a very rapid fixative and contains no fluorescent constituents. Another nonfluorescent fixative we have used with moderate success is a modified Tellesniczky's Fluid (formalin, 2 parts: 70% ethanol, 20 parts: glacial acetic acid, 1 part). Though less harsh than Carnoy's, Tellesniczky's fluid is also less rapid, and fixation times must be increased to between 8 and 12 hours, thereby increasing the chance that the desired antigens may be leached away.

Fixed mouse ovaries and oviducts are embedded according to the following schedule:

1. Carnoy's, 60 minutes at room temperature (90-minute maximum). N.B.: Short times are possible partly because large volumes of fluid are used, e.g., 30 ml of fixative for the paired ovaries and oviducts from one mouse.

2. Absolute ethanol, three changes, 15 minutes each.

3. Toluene, 50% absolute ethanol, 50%, 15 minutes.

4. Toluene, two changes, 10 minutes each; during second toluene, place in 54°C paraffin oven. (Xylol can be used also but is said to make tissues more brittle.)

5. Low temperature Tissuemat, melting point 52.5°C, three changes, 15 minutes each. N.B.: Our system is destroyed by temperatures of 55°C and above. Therefore, the Tissuemat (paraffin) is kept at the lowest possible temperature during embedding. We use an immunological water bath in which heavy-bottomed, one-ounce glass vials are placed.

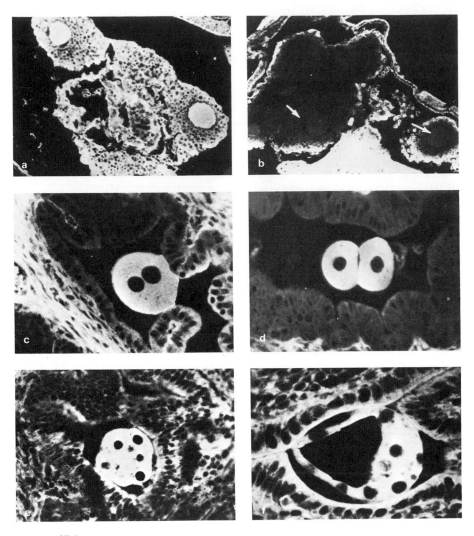

FIGURE 27-1

Mouse oviducts from uninjected or albumin-injected mice were fixed in Carnoy's, embedded in paraffin, and sectioned serially. Fluorescent antibody procedures were indirect. Oviductal sections were reacted first with nonfluorescent rabbit antiserum directed against mouse serum antigens or against bovine plasma albumin; then the sections were treated with fluorescein-labeled sheep antiserum against rabbit globulin (anti-R*). Fluorescence indicates the presence of molecules similar to or identical with mouse serum antigens or with the intravenously injected bovine plasma albumin. (From Glass, 1963).

(a) *Secondary oocytes just after ovulation in oviductal ampulla of unmated female. Serum-like antigens are present in the ooplasm, in the cytoplasm of adjacent follicle cells, and in the oviductal epithelium and lamina propria. ×58*

(b) *Fertilized ova (arrows) six or seven hours after mating. A formed, male pronucleus is peripheral in each zygote. None of the ova have emitted a second polar body, but the second meiotic spindle has turned and is perpendicular to the surface; chromosomes are at the metaphase plate. Serum-like antigens are not detectable in the ooplasm; a similar absence of detectable antigen is also characteristic for albumin-injected ova. ×58*

(c) *Pronuclear ootid from an albumin-injected mouse. Antigens like those of the intravenously injected foreign albumin are present in the cytoplasm. Reappearance of serum-like antigens in the ootid characteristically occurs after dispersal of the follicle cells;*

Three vials in a row contain the three changes of paraffin and dull-pointed watchmaker's forceps, warmed to 54°C in the paraffin oven, are used to transfer the tissues from one vial to the next. Water in the bath is kept about 54.5°C, and its level must be nearly at the top of the vials and well above the paraffin level (which is only 2 or 3 cm deep), otherwise, the paraffin surface does not stay melted. The cover of the water bath usually is kept on except when changing tissues; a swath of cheese cloth wrapped around the cover prevents water from condensing on the top and dripping into the paraffin.

6. Embed in fresh Tissuemat in disposable plastic embedding boats. We generally leave the boat in a paraffin oven and bring vial and tissue to it. Since one is working as close as possible to the paraffin melting point, embedding can be tedious.

The total time from the animal to the paraffin block is approximately three to three and one-half hours.

If the tissues cannot be run through the total process on the same day that they are fixed, they may be taken from Carnoy's to absolute to 95% ethanol (three changes, 15 minutes each, in each reagent) and then may be stored overnight in 70% ethanol. For embedding, the reverse sequence is followed, to absolute ethanol and to paraffin.

Routinely, tissues are sectioned serially on a rotary microtome at 5 microns; thicker sections (or tissue folds) can make the fluorescence data more difficult to read. Since the paraffin has a low melting point, sectioning is easier with a cold knife. During sectioning we fill a funnel with dry ice and support it on a ring stand so that the vapor is directed onto the paraffin block and knife. Sections are affixed to the slide with a very small amount of egg albumin.

(c) *Freeze-substitution.* An empirical "poor man's" freeze-substitution procedure is used sometimes. Tissues are fast-frozen in liquid nitrogen. (Fast-freezing is supposed to result in smaller ice crystals within and between the cells, and, thus, to minimize tissue distortion during freezing and thawing; see Meryman, 1960.) After freezing, the tissues are transferred to Carnoy's that has been precooled to −70°C by standing in a beaker placed in an acetone and dry ice slush. (Less frequent renewal of the dry ice is required if the container is left in a cold room.) After two to eight hours in the −70°C Carnoy's, the fixative and tissues are placed in a freezer at −20°C for 24 to 48 hours, and then in a cold room at 0° to 4°C for 24 to 48 hours. Finally, the tissues are brought to room temperature and embedded in low melting point paraffin.

---

since there are many confirming data, these observations are interpreted as indicating that serum antigens are transferred from the blood into oviductal eggs. ×116

(d) *Two-celled embryo in oviductal isthmus from albumin-injected mouse; albumin-like antigens are present in the blastomere cytoplasm, absent from the oviductal epithelium, and present in the lamina propria. ×116*

(e) *Morula in ovarian end of uterus from albumin-injected mouse; blastomere cytoplasm and cytoplasm of endometrial epithelium and stromal cells contain albumin-like antigen. ×116*

(f) *Unimplanted blastocyst from uterus of albumin-injected mouse. ×116*

This schedule was chosen pragmatically, not by systematic testing, and, obviously, will vary with the size and texture of the tissue. Systematic analyses are reported by Feder and Sidman (1958) and by Balfour (1961).

(d) Other fixation, embedding, and supporting procedures are discussed by Nairn (1962) and briefly but helpfully by Clayton (1960). Embedding in "Carbowax," a low molecular weight polyethylene glycol, also has been utilized (George and Walton, 1962).

(e) Rationale for use of various fixation and embedding procedures. For some easily destroyed antigen systems, frozen sectioning may be the most desirable and, perhaps, the required method for handling the tissue sections. However, frozen sections are much inferior morphologically to freeze-substituted or paraffin-embedded sections; distorted morphology makes discrimination between antigenic sites much less sensitive.

More serious is the fact that soluble antigens may be displaced significantly in frozen sections because of transient thawing at the knife edge during sectioning, and random diffusion during subsequent exposure of the unfixed section to a pool of fixative (*e.g.,* ethanol, methanol, formalin, etc.; see Coons, 1956) and/or antibody. Such displacement can be random; therefore, it may be detected as "nonspecific fluorescence," which is observed at different, inconsistent sites on several sections from the same tissue. Worse, translocation of soluble antigens in unfixed tissue may be nonrandom. A substrate-specific reprecipitation of the dissolved antigen can occur and, since it is dependent on the normal morphology and physiology of the test tissue (Gomori, 1952; Casselman, 1959), such antigen localization may be replicable and consistent. Nonetheless, the observed fluorescence will be artefactual, and will not represent the normal *in vivo* position of the antigen.

The theoretical probability that soluble antigens will be translocated in frozen sections, plus the poor morphology of such sections, have convinced us that the problem of dislocation artifact may be more critical than fixation artifact with our soluble antigen systems. Therefore, whole-tissue fixation and low-temperature paraffin embedding are the usual methods employed in our immunohistological studies. Unless the crucial antigen systems are destroyed thereby, consideration of appropriate chemical-fixation and paraffin-embedding procedures is recommended to other workers.

A caution is necessary. Experimental comparisons of fluorescent antibody localizations on sections from Carnoy-fixed, paraffin-embedded tissues and on frozen sections show very similar (or identical) tissue localizations for the specific test antigens. Fluorescence is slightly less intense on the fixed-embedded sections, presumably because of antigen denaturation by the fixation and embedding processes. Few critical studies have been done on the effect of different denaturing agents (fixatives in this discussion) on antigenic reactions. Heidelberger's group studied reactions between anti-egg albumin (anti-Ea) and egg albumin that had been denatured by acid-, alkali-, heat-, or water-methods (DnEa) and then characterized by ultracentrifuge, diffusion, and viscosity measurements. Antigen-antibody precipitation between anti-Ea and DnEa did occur. However, as compared with the homologous reaction, much larger quantities of heterologous DnEa than of homologous Ea were required to produce quantitatively comparable reactions with the anti-Ea sera. Apparently only a fraction of the Ea antibodies were precipitable by the DnEa (McPherson and Heidelberger, 1940, 1945a, 1945b; Heidelberger, 1956, pp. 77ff.). The structure of the denatured protein was not random and disordered, but had a definite and typical steric specificity, a portion of which resembled that of undenatured egg albumin, although other portions were dissimilar (Heidelberger, 1956; see also Kabat and Mayer, 1961, p. 450). Our observations that fluorescence is of slightly lower intensity with fixed-embedded than with frozen sections are consistent with Heidelberger's anti-Ea:DnEa data. Some leaching also may occur during preparation of the tissue.

Presumably, therefore, fixation and embedding may lead to underestimation rather than overestimation of particular antigens; that is, a false negative is more likely than a false positive result. Clearly, some research problems (as, for example, studies attempting to detect the earliest stage at which a particular embryonic antigen appears) will require the use of frozen sections for, at least, the critical stages. Obviously, the fixation and embedding method of choice must be determined and controlled differently for each research problem and for each antigen-antibody system.

There are a few other data on the response of antigens to fixatives. Several frog oocyte antigens, for example, produce characteristic Ouchterlony agar diffusion patterns after precipitation by agents such as trichloracetic acid, fat solvents, heat, and ammonium sulfate (Barber, 1958). The problem is discussed also by Clayton (1960), Balfour (1961), and Glass (op. cit.); nonetheless, much more data about these questions would be very helpful.

The general problem of the artefactual localization or elimination or modification of cell constituents, inherent in all cytochemical methods, is discussed extensively by Gomori (1952), Casselman (1959), and Pearse (1961). Persons using fluorescent antibody techniques should be thoroughly familiar with these observational and theoretical considerations.

Antibody preparation

Basic information on antigen preparation, immunization schedules, and antibody characterization is found in *Experimental Immunochemistry* (Kabat and Mayer, 1961). Two volumes of a multivolume work titled *Methods in Immunochemistry* (Williams and Chase, 1968, 1969) are now available and are very useful.

(a) Antigen and bleedings. Rabbit antisera are developed in adult male rabbits. Since we are working with mixtures of serum antigens, some of which are in low concentration, and since the complete adjuvant is reported to enhance antibody production to minor constituents in an antigen mixture, the Freund technique is the major one used (Freund and Bonanto, 1944; Kabat and Mayer, 1961, Appendix F).[1] Alternatively, purified alum-precipitated antigens have been introduced by serial intravenous injections

---

[1] We obtain Freund's complete adjuvant from Difco Laboratories, Detroit, Michigan.

via the marginal ear vein (Kabat and Mayer, 1961, Appendix F) or intra-muscularly (Cherry *et al.,* 1960, Appendix B).

Before any antigen is injected, the rabbit is bled to obtain normal rabbit serum (NRS), which is used in subsequent studies with antiserum from that rabbit in NRS controls for specificity in fluorescent antibody localizations (see Table 27-1, Reaction 3, on page 371).

The antigen-adjuvant emulsion is injected subscapularly with approximately half the volume of emulsion being placed deep to each scapula. Insertion of the 18- to 20-gauge needle through the (tough) skin at the dorsal midline is easier if, after swabbing with alcohol, a shallow 1- to 2-mm slit is made through the epidermis with the corner of a razor blade. The needle need not be removed in transferring from one scapula to the other but can be withdrawn from the muscle and a fold of dorsal skin lifted allowing one to slip the needle point over the backbone.

Test-bleeding by marginal ear vein is made four to six weeks after injection. If preliminary tests show that reinjection is necessary, rabbits receive an initial intraperitoneal injection of 10 mg antigen in saline; two or three days later, 50 mg antigen in Freund's is administered subscapularly. Two weeks after reinjection, rabbits are test-bled.

A total of 50 to 250 mg of antigen protein is administered to a rabbit during the course of an experiment.

Test bleedings are characterized roughly and quickly by antigen dilution interface microprecipitin tests (Boyd 1956); antisera detecting antigen at protein concentrations of $5\mu g$ per ml are adequate for our purposes.

If interface tests show that the titer is adequate, about 50 ml of antiserum is obtained from each rabbit by bleeding directly from the central ear artery or by cardiac puncture. The arterial method is modified from Hammerstrom (1963). The rabbit is placed in a restraining cage and one ear is shaved clean around the central artery. Arterial circulation is increased by applying a little xylol to the ear distal to the entry site and rubbing the ear vigorously. A Vacutainer needle is inserted into the distended artery and, without applying negative pressure, up to 60 ml of blood will flow freely into a collecting tube or flask. We use a thin walled disposable Vacutainer needle made for use with negative pressure vials. When the desired amount of blood has been collected, the thumb is placed on top of the needle in the artery and the needle is withdrawn while applying slight local pressure.

For cardiac puncture we use a 50 ml syringe and an 18 gauge needle. Entrance is through the anterior abdominal wall just to the left of the xiphoid process. The needle is inserted at a sharp angle upward through the diaphragm into the ventricle; the fibrous pericardium is fused to the thoracic surface of the diaphragm and helps hold the heart in a constant relationship to these landmarks. There are many routes for a cardiac puncture. If done skillfully, the method is rapid, safe, and comfortable for the rabbit, but a bungled cardiac puncture will be lethal. If possible, one should be taught how to do it by an experienced technician.

(b) Characterization of antisera. In addition to the interface precipi-tin tests, our antisera are characterized routinely by microimmuno-electrophoresis (Scheidegger, 1955; Crowle, 1961; Wieme, 1965); antisera are run against their homologous antigen and against appropriate heterologous antigens. Generally, our injection antigen is mixed, and immunoelectrophoresis allows an estimate of the minimum number of reactive antibodies in the antiserum. We use a 2% agar gel on microscope slides, *p*H 8.2 barbitol buffer, ionic strength 0.05. Runs are 90 minutes at 5 milliamps per slide. Antigen concentrations of 5 mg of protein per ml and whole undiluted antiserum are convenient in our system. Slides are stained for protein using Amido Black 10B; alternatively, enzyme studies can be performed. If a rough estimate of relative amounts of antigen and/or antibody is desired (or sometimes for other reasons), agar diffusion studies are carried out, usually in small Petri dishes. For routine use, microimmunoelectrophoresis has replaced agar diffusion tests in our work.

(c) Absorptions. Generally, absorptions are necessary to remove cross-reacting antibodies and delimit the immunological specificity of an antiserum; each step of the absorption procedure is followed by immunoelectrophoresis. Complete absorption with homologous antigen is used to prepare nonreactive antisera as one control for the immunological specificity of fluorescence observed in the fluorescent antibody tests.

The optimal-proportions method and a serum-dilution titer method are described in Kabat and Mayer (1961, p. 14 and pp. 69–70, respectively), and may be used for rough initial estimates of the amount of absorbing antigen required. With our system of mouse serum antigen and rabbit antiserum, complete absorption of most homologous and of all cross-reacting antigens is obtained at proportions of approximately one part of antigen protein to ten parts of whole antiserum protein.

In absorptions the antigen-antibody mixture is incubated one hour at 37°C. The precipitate is centrifuged off in the high-speed head of a standard refrigerated centrifuge, and the supernatant is checked by interface precipitin tests. If antibody reactivity is still present, more antigen is added, and the whole process is repeated. The final supernatant is checked by precipitin tests and by immunoelectrophoresis against both the homologous and the cross-reacting antigen.

Fractionation of antisera

If fractionation of the antiserum is desired, the cold ethanol method, a DEAE-Sephadex batch procedure, DEAE column fractionation, or ammonium sulfate precipitation may be utilized.

(a) Cold ethanol fractionation (Deutsch, 1952). The cold ethanol method can be a rapid, easy technique, and requires no elaborate apparatus except a refrigerated centrifuge capable of reaching −8°C. The explicit procedural details given by Deutsch for fractionation of rabbit serum are very helpful for the non-biochemist. We place the serum, a Teflon-coated magnetic stirring bar, and a thermometer that can read to −10°C into a small Erlenmeyer flask, and put the flask into a larger beaker containing cold acetone (0° to 4°C). The beaker is set on a magnetic stirrer; the serum is stirred slowly but constantly during all subsequent procedures. The total volume of cold, 50 per cent ethanol to be used is placed in a separatory funnel or burette suspended from a ring stand. (A serological pipette controlled by finger can be used also.) As the ethanol is added dropwise to the stirred serum, a gradual decrease in the temperature of the serum-alcohol mixture is desired, so that denaturation of serum molecules by the ethanol will be minimal. The temperature decrease can be obtained readily by placing small chips of dry ice into the acetone; small chips should be used, since one wants to keep the temperature near the freezing point of the mixture, not below it. It is easier to keep the temperature stable by conducting the procedure in a cold room, although a cold room is not necessary for successful fractionation. After centrifugation of the globulin precipitates, the alcohol-containing supernatant should be decanted; immediately after decanting, the desired volume of cold phosphate-buffered saline should be added. If the surface is allowed to dry, the precipitate is difficult to get back into solution. In our hands, this procedure takes about half a day from whole serum to dialysis.

(b) DEAE-Sephadex batch method (Baumstark *et al.,* 1964). This procedure uses the chloride form of DEAE-Sephadex at pH 6.5 and requires 20 grams of DEAE-Sephadex per 50 ml of serum. We have not used the method, but are told that it is rapid and convenient, requires no special apparatus, and gives a good yield of antibody globulin. Minor, useful procedural addenda are included below (Johnson and Lasky, personal communication).

All twenty grams of DEAE-Sephadex are prepared at one time; about half is used in each of the two steps of the batch fractionation. Fifty ml of serum is added to a 400-ml beaker containing 10 grams of washed, moist DEAE-Sephadex equilibrated to pH 6.5. After stirring in the cold for one hour, the mixture is transferred to a Buchner funnel and washed with four 25-ml portions of 0.01 phosphate buffer, pH 6.5, and then filtered to dryness; the DEAE-Sephadex is discarded. The filtrate is transferred to a one-liter flask with a second 10-gram portion of washed moist DEAE-Sephadex equilibrated to pH 6.5; the mixture is stirred one hour in the cold. The mixture is placed then on a Buchner funnel and washed with a series of ten 20 ml aliquots of buffer, totaling 200 ml. The filtrate, which contains the γ_2-globulin, is adjusted immediately to a pH of 7.5, using 1.0 M potassium monophosphate. The DEAE-Sephadex is discarded. All reagents should be cold.

This procedure gives an excellent yield of γ_2-globulin, although other serum constituents are lost; it takes less than a full day in total time, but best results are obtained if the washed DEAE-Sephadex is allowed to swell overnight before equilibration.

(c) DEAE column fractionation (Riggs *et al.,* 1960). DEAE cellulose ion exchanger is used for antiserum fractionation by many laboratories. A stepwise elution schedule uses increasing concentrations of phosphate buffered saline.

(d) Details of a simple $(NH_4)_2SO_4$ fractionation procedure are given by Cherry *et al.* (1960, Appendix C).

Conjugation of antibodies with fluorescein

Many direct conjugation procedures have been described. Useful and simple labelling techniques are described in Riggs *et al.* (1958) and in Chadwick and Fothergill (1962). Unfortunately, since the procedure is so simple technically, we have not obtained satisfactory, specific conjugates using isothiocyanate on Celite pellets (Rinderknecht, 1962).

Crystalline fluorescein isothiocyanate or rhodamine isothiocyanate can be obtained commercially; a list of commercial sources is given in Cherry *et al.* (1960, Appendix F), but these materials are not of equal quality, particularly in the amount and ease of removal of fluorescent, nonspecific contaminants. It is worth checking with a laboratory that is successfully using immunohistological techniques before ordering reagents; additionally, it is worth trying another product if results with the first are unsatisfactory. In our experience, fluorescein-labeled antisera have been more useful than rhodamine conjugates.

After conjugation, hydrolyzed fluorochrome and other unreacted fluorescent materials must be removed from solution, since they can cause serious problems by giving immunologically nonspecific staining of the tissues. Passage of the conjugated antiserum through a Sephadex column (G-25 or G-50) in the proportions, roughly, of 1 ml of antiserum per each 1-cm by 3-cm column of Sephadex is straightforward and effective; 0.01 M phosphate buffered saline is used as eluent (*p*H 7.1, Fothergill and Nairn, 1961; *p*H 7.4, George and Walton, 1961; *p*H 6.5, Rinderknecht, 1962). The conjugated antiserum comes off the column before the unreacted fluorescent material.

Alternatively, Cherry *et al.* (1960, Appendix D) suggest mixing the antiserum with an equal volume of Dowex 2-X4 (chloride form), 20- to 50-mesh, anion-exchange resin. After intermittent shaking for one hour in the cold, the conjugate is drained from the resin and is dialyzed against phosphate-buffered saline overnight. The time required for dialysis can be reduced if Dowex is added to the beaker containing the dialysis fluid (roughly one gram of Dowex to 2.5 ml of conjugate).

These procedures for removal of fluorescent contaminants are far more desirable than adsorption of the fluorescent conjugate with activated charcoal or with tissue powders, procedures that cause significant loss of volume (and, sometimes, of titer) by the antiserum; moreover, repeated treatments generally are required to obtain "clean" fluorescent antisera.

Reaction of antibody with tissue

Both "direct" and "indirect" methods have been used in immunohistological studies, each with good rationale. In the direct method, the specific antiserum being tested is conjugated with a fluorescent agent; then, the fluorescent antiserum is reacted directly with tissue or cells containing the homologous antigen (see Fig. 27-2). The single-step, direct method is preferred by many because it is faster and because specificity must be controlled for only one antiserum; the opportunity for nonspecific fluorescent contaminations is lessened thereby. In addition, there is less chance of losing tissue from the slide, because the total time of exposure to fluid is shorter.

The indirect fluorescent antibody procedure requires two steps: First, unconjugated, specific antiserum (developed in, e.g., rabbit) is reacted with the tissue containing homologous antigen. Second, antiserum prepared in

368

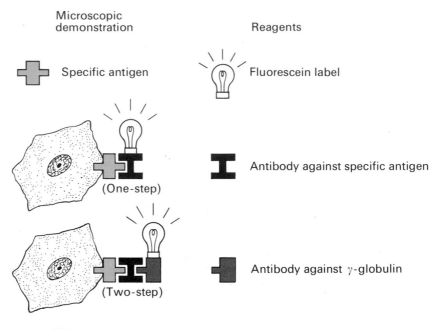

Microscopic
demonstration

Reagents

Specific antigen

Fluorescein label

(One-step)

Antibody against specific antigen

(Two-step)

Antibody against γ-globulin

FIGURE 27-2

Diagrammatic representation of "direct" (top) and "indirect" (bottom) fluorescent antibody procedures. (From: Analytic Cytology, 2nd edition, ed. by Robert C. Mellors. Copyright 1959, McGraw-Hill Book Company.)

another species (e.g., sheep or goat) and directed against gamma globulins of the first species (e.g., sheep antiserum against rabbit gamma globulin) is conjugated with fluorescein; the conjugated antiglobulin then is reacted with the tissue section. In our example, fluorescein-labeled anti-rabbit globulin serum reacts with rabbit globulin antibodies that are attached to specific tissue antigen. Thus, the reaction is two-step (see Figs. 27-2, 3). The indirect method is 4 to 12 times more sensitive than the direct method, presumably because of a kind of amplification of reactive sites by interposition of the unlabeled globulin between the tissue antigen and the fluorescent reagent (Holborow, 1964). If several antisera are to be used, and indirect method is a significant time-saver, because only a single, high titer, anti-globulin serum need be conjugated and characterized. Conjugation of a single antiserum is useful also because indeterminate protein denaturation occurs during conjugation. Moreover, the fluorescein-protein ratios may differ for different labeled antisera. It is concluded that the indirect method facilitates comparison between the differential reactions of several antisera; that is, the specificities of the experimental antisera differ but the fluorescent test reagent used to compare them is constant.

(a) Staining procedures. The indirect, two-step method, rigorously controlled for immunological specificity of the fluorescent localization, is our primary method. The procedure we use follows.

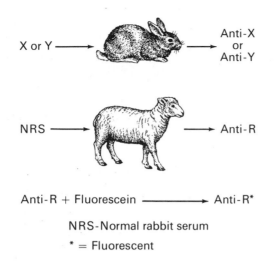

X or Y ⟶ 🐰 ⟶ Anti-X or Anti-Y

NRS ⟶ 🐑 ⟶ Anti-R

Anti-R + Fluorescein ⟶ Anti-R*

NRS-Normal rabbit serum

* = Fluorescent

Generalized Reaction

| Tissue section | + | Anti-X or Anti-Y | ⟶ | Tissue : X (No fluorescence) | Anti- X | + | Anti- R* | ⟶ | Tissue X | : | Anti- X | : | Anti- R* (Localized fluorescence) |

FIGURE 27-3

Diagrammatic representation of antisera preparation and reactions in "indirect" fluorescent antibody method. Species other than rabbit and sheep may be used as antibody producers.

1. Hydrate tissue to phosphate-buffered saline, pH 7.0 (exact pH is not crucial; pH 6.8 to 7.5 is all right); 5 minutes toluene; 2 minutes each for absolute, 95%, and 70% ethanol; 5 minutes in buffered saline.

2. "Step 1": Add drops of nonfluorescent rabbit antiserum to tissue sections. N.B.: Antiserum may be whole serum or globulins from ethanol or DEAE-Sephadex, or DEAE batch or $(NH_4)_2SO_4$ fractionation. The antiserum should be characterized by precipitin and immunoelectrophoresis tests against homologous and appropriate heterologous antigens, and should be absorbed as much as necessary to ensure specificity of reaction.

    Incubate slide in moist chamber, 5 to 15 minutes, 37°C. Wash slide in 800 to 1000 ml of phosphate-buffered saline stirred constantly, 20 minutes, room temperature.

3. "Step 2": Add drops of fluorescent antiserum (sheep anti-rabbit globulin serum). N.B.: Fluorescent antiserum should be run through a millipore filter each day just before use. Incubate in moist chamber, 5 minutes, 37°C. Wash slide as above in 800 to 1000 ml of stirred, phosphate-buffered saline, 20 minutes, room temperature. Mount in 50% glycerine is distilled water.

4. Never let tissue sections dry.

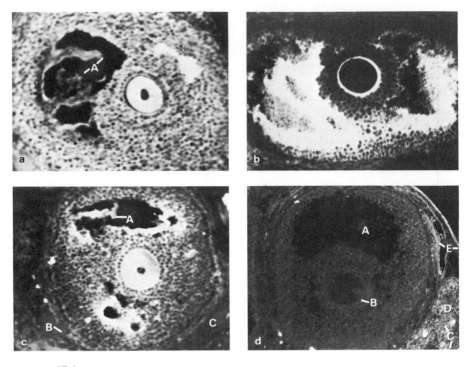

FIGURE 27-4

Mouse tertiary ovarian follicles reacted with rabbit sera of different specificities. Fluorescence indicates the presence of specific antigens against which the test antisera were directed; the intensity and location of fluorescence is consistently different with each treatment.

(a) Tertiary follicle from ovary of uninjected mouse; section treated with nonfluorescent antiserum against mouse serum antigens and then with fluorescent sheep anti-rabbit globulin serum (anti-R*). ×81. (From Glass, 1961).

(b) Tertiary follicle from ovary of mouse injected intravenously with bovine gamma globulin; section treated with nonfluorescent anti-bovine gamma globulin followed by anti-R*. Compare and contrast the sites of fluorescence in Figures a and c. ×72. (From Glass, 1966).

(c) Middle-sized tertiary follicle from ovary of mouse injected intravenously with bovine plasma albumin; section reacted with nonfluorescent anti-bovine albumin serum followed by anti-R*. Contrast intensity and location of fluorescence in the ooplasm, granulosa cells, and ovarian stroma with that observed in Figures a, b, and d. A, Follicular fluid; B, follicular thecae; C, yellow-orange autofluorescent particles. ×81. (From Glass, 1961).

(d) Tertiary follicle of similar size and from the same ovary as in (c). Treated with normal rabbit serum followed by anti-R*. Photographic exposure twice as long as for (c). Printed for optimum reproduction of follicle. A, Follicular fluid; B, oocyte; C, yellow-orange autofluorescent particles; D, old corpus luteum. ×81. (From Glass, 1961).

(b) Controls to ensure immunologically specific fluorescence. The adequacy of all fluorescent antibody data depends entirely on rigorous controls for the immunological specificity of the fluorescence observed. Appropriate controls are summarized in Table 27-1 and are illustrated partially in Figure 27-4. (See also discussions in Coons, 1956.)

In general, control studies must use various combinations of absorbed and unabsorbed antisera on adjacent serial sections from the same animal, and/or on sections from similarly and differently treated animals (Reactions 4, 5, 7, and 8 in Table 27-1). "Blocking" or inhibition reactions

TABLE 27-1

Necessary controls for immunological specificity in fluorescent antibody studies: indirect method[1]

	Reaction	Animal Treatment	Tissue Treatment		Microscopic Observations
			"Step 1" (Nonfluorescent Serum)	"Step 2" (Fluorescent Serum[2])	
"Baseline" controls	1	As exptl.	Hydrated and mounted	—	Autofluorescence
	2	As exptl.	Saline	Anti–R*	Slight fluorescence, immunologically nonspecific
	3	As exptl.	NRS[3]	Anti–R*	Slight fluorescence, immunologically nonspecific
"Antiserum" controls	4	X–injected[4]	Anti–X, absorbed with X	Anti–R*	Slight fluorescence, same intensity and locations as reaction 3
	5	Y–injected or Uninjected	Anti–X, unabsorbed[5]	Anti–R*	Slight fluorescence, same intensity and locations as reaction 3
	6	X–injected	Anti–X ("Blocking" reaction)	Anti–X*	Fluorescence more intense than in reactions 4 and 5; less intense than reactions 7 and 8. Localization of fluorescence as in reaction 7
"Differential" controls	7	X–injected	Anti–X	Anti–R*	Consistent, characteristic fluorescence at particular tissue sites; specific to anti-X and different from anti-Y
	8	Y–injected	Anti–Y	Anti–R*	Consistent, characteristic fluorescence at particular tissue sites; specific to anti-Y and different from anti-X

[1] If the direct method is used, the reagents listed in "Step 1" column are themselves fluorescein-labeled. Exceptions are reaction 2, in which fluorescein isothiocynate in saline is used, and the "blocking" reaction 6, which remain the same.
[2] Fluorescein-labeled antiserum against rabbit globulins is designated as Anti-R*.
[3] Normal rabbit serum; NRS is from the preinjection bleeding of each rabbit whose antiserum is used in fluorescent antibody tests.
[4] Intravenous injection via tail vein.
[5] No cross-reaction with Y in agar diffusion or immunoelectrophoresis.

provide an additional control when the direct method is used (Reaction 6 in Table 27-1). In this test, unlabeled specific antiserum is reacted with the section; then, a fluorescein-labeled aliquot of the same antiserum is added. Theoretically, the unlabeled antibodies have occupied all the available antigenic sites on the tissue, and the labeled antibodies have no attachment sites available. In practice, some of the unlabeled antibodies are replaced by some of the labeled antibodies during the second step of the reaction. Therefore, fluorescence on these "blocked" slides is less than on experimental slides but higher than on the other "antiserum" and "baseline" controls.

Nonspecific affinities may exist between rabbit serum and/or fluorescein and the tissue. To identify these, NRS controls are used routinely; if possible, the NRS is from the rabbit whose antiserum is used on the experimental slides (see Reaction 3 in Table 27-1 and Fig. 27-4d). Fluorescein in saline is also used as a control (see Reaction 2 in Table 27-1). Tissue autofluorescence under the conditions of the experiment is determined by hydrating the tissue and mounting it for observation (see Reaction 1 in Table 27-1).

In combination, these baseline and antiserum controls, plus the fact that antisera of different specificities are localized consistently at different tissue sites, provide adequate demonstration that particular fluorescent antibody localizations are immunologically specific. (See also Coons, 1956.)

(c) Variations on the basic immunohistological technique. In addition to those already described, other technical variations have been introduced that increase the flexibility and sensitivity of the fluorescent antibody method. Among these is complement staining; this is similar to the indirect method, except that complement, in addition to the unconjugated antiserum, is added to the tissue (or smear) in Step 1. The conjugated globulin used in Step 2 is directed against the species supplying the complement rather than against the species supplying the unconjugated antiserum (Cherry *et al.*, 1960; Nairn, 1962).

A double-labeling method, intriguing in its theoretical potential but difficult in its operation, uses two antisera of different specificities, each coupled to a different and contrasting fluorescent label, e.g., one conjugated with fluorescein (greenish yellow) and the other with rhodamine (orangish red); see Clayton (1954), Clayton and Feldman (1955).

When cells contain antibody, a two-step method may be used that reacts the tissue first with antigen and then with labeled antiserum specific to the antigen (Mellors, 1959). A combination of agar diffusion, enzyme assays and fluorescent antibody is utilized in the "Echo" technique (Nace and Suyama, 1961).

FLUORESCENCE MICROSCOPY

Light source

Fluorescence microscopy generally uses a mercury arc lamp of high intensity. The Osram HBO 200 mercury arc lamp is a convenient light source of small size and high intensity; shielded housing for the lamp is readily available and easy to handle. (Other light sources available are listed in

Cherry *et al.*, 1960, Appendix A). We find that a wooden barrier placed between the lamp housing and the microscope reduces stray light and decreases eye fatigue; passage for the neck of the lamp, through which light passes toward the microscope, is provided by a slot in the barrier.

Filters

Alternative filter systems are described by Cherry *et al.* (1960) and Nairn (1962). The primary filter, between the light source and the tissue, must transmit wavelengths that stimulate emission by the fluorochrome. Secondary filters are interposed between the fluor and the eye; these should cut out the inciting wavelengths but transmit the wavelengths at which the fluorescence was emitted. Our system utilizes 1- to 3-mm Schott BG 12 filters as primary, with OG 4 and OG 5 as secondary filters.

Optics

Recommended combinations of microscope oculars and objectives vary greatly from laboratory to laboratory. We use ×10, ×12.5, or ×16 oculars (Zeiss, Leitz) designed to give a wide, flat field; if possible, different oculars should be tried with one's own system before purchase. We use both achromatic and apochromatic objectives with as large a numerical aperture at each magnification as is available. We do not use fluorite objectives. Oil immersion magnifications are possible but not too satisfactory in our system because of the glycerine-mounted, paraffin-tacked cover slips. We use a bright field condenser.

Photography

A photographic as well as a written record of the observations is desirable, and is made easy by use of a trinocular microscope unit explicitly designed for fluorescence (and other) microscopy, in which objects of low light intensity are to be observed. In these trinocular heads, all the light passes to the binocular eyepieces *or* all the light passes to the monocular eyepiece; the beam is not split between them. A 35-mm camera back may be mounted on the monocular tube. The focusing device associated with the camera back must also be designed for fluorescence work, so that the light beam is not split but passes entirely to the eyepiece for focusing or entirely to the camera during exposure of the film. A photographic plateholder sometimes can be used, though light intensity generally is so low that focusing is very difficult.

For black and white pictures, we prefer Kodak High Contrast Copy Film to Kodak Tri-X. Color pictures can be obtained using Kodak High-Speed Ektachrome, Type B. With our material, exposures usually range from one to three minutes, although 10 to 12 minutes may sometimes be required. We have never used Anscochrome 500, but are told that it is very fast and that color reproduction is good.

Both Zeiss and Leitz have complete units for fluorescence microscopy

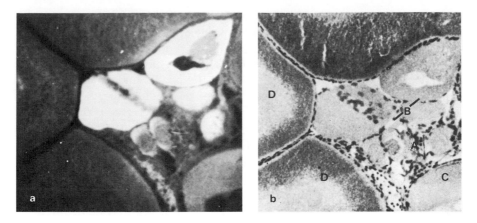

FIGURE 27-5

(a) *Ovarian section from frog ovary treated by the direct method with fluorescein-conjugated rabbit antiserum to adult female frog serum. Generalized, greenish yellow cytoplasmic fluorescence is present in the preyolk oocytes near the center of the picture; localized patches of fluorescence are present in the yolk-forming oocytes adjacent. Yellow-orange autofluorescent erythrocytes are present in the blood vessel passing longitudinally from the bottom of the picture. ×105.*

(b) *Same oocytes as in* (a), *stained with hematoxylin and eosin,* A, *Egg-nest oocytes;* B, *larger pre-yolk oocytes;* C, *oocyte early in yolk formation;* D, *oocyte later in yolk formation;* E, *blood-filled vessel. ×105. (From Glass, 1959).*

(which include a shielded light source with transformer and ignition; a trinocular microscope adaptable for fluorescence and light microscopy; recommended oculars, objectives and filters; camera back and focusing device). The Zeiss unit is particularly convenient for secondary filters, the Leitz unit for switching between fluorescence and light viewing. Other apparently excellent units are on the market, but we have no personal experience with them.

Histological staining

After fluorescence has been viewed and photographed, the cover slips can be removed from the slides by carefully cutting through the paraffin and then placing the slides (for a few minutes to an hour or so) upright in buffered saline. The cover slips slide gently off and the tissue can be stained with hematoxylin and eosin or other standard histological stains. If appropriate to the problem, light microscope photographs can be made of the identical stained sections on which particular fluorescent localizations were observed and photographed (see Figs. 27-5a, b). Alternatively, the sections can be viewed and photographed using phase-contrast optics.

CONCLUSIONS

The fluorescent antibody technique is a sensitive method for detecting the location of specific macromolecules in tissues. It requires careful characterization of the antisera so that they are immunologically specific, and rigorous controls for nonspecific fluorescence. When these criteria are met, this immunohistological procedure can provide unique, invaluable information about the behavior of specific large molecules during embryonic development. The method allows detection of their time of appearance and disappearance, of specific embryonic and adult antigens, of their apparent increase or decrease, and of their translocation during differentiation. The fluorescent antibody procedure allows correlation of the stages when tissue specific antigens appear in cells with the time when characteristic morphological differentiation is visible.

Enough such data, combined with data gathered by other methods, ultimately will help us comprehend the causal interactions between molecules and morphology during mammalian (and other animal) development. The results, actual and potential, are well worth the effort of establishing functional fluorescent antibody procedures.

BIBLIOGRAPHY

Balfour, B. M. (1961) Immunological studies on a freeze-substitution method of preparing tissue for fluorescent antibody staining. Immunology, 4:206–218.

Baumstark, J. S., R. J. Laffin, and W. A. Bardawil (1964) A preparative method for the separation of 7S gamma globulin from human serum. Arch. Biochem. Biophys., 108:514–522.

Boyd, W. C. (1956) Fundamentals of Immunology. Interscience, New York, 776 p.

Casselman, W. G. B. (1959) Histochemical technique. Methuen, London, 205 p.

Chadwick, C. S., and J. E. Fothergill (1962) Fluorochromes and their conjugation with proteins. In: Fluorescent Protein Tracing. Ed. by R. C. Nairn. Livingstone, Edinburgh, pp. 4–30.

Chadwick, C. S., M. G. McEntegart, and R. C. Nairn (1958) Fluorescent protein tracers: a trial of new fluorochromes and the development of an alternative to fluorescein. Immunology, 1:315–327.

Cherry, W. B., M. Goldman, and T. R. Carski (1960) Fluorescent antibody techniques in the diagnosis of communicable diseases. U. S. Dept. Health, Education and Welfare, Public Health Serv. Publ. no. 729.

Clayton, R. M. (1954) Localization of embryonic antigens by antisera labelled with fluorescent dyes. Nature, 174:1059.

——— (1960) Labelled antibodies in the study of differentiation. In: New approaches in cell biology. Ed. by P. M. B. Walker Academic Press, New York, pp. 67–88.

——— and M. Feldman (1955) Detection of antigens in the embryo by labelled antisera. Experientia, 11:29–31.

Coons, A. H. (1956) Histochemistry with labelled antibody. Int. Rev. Cytol., 5:1–23.

———, H. J. Creech, R. N. Jones, and E. Berliner (1942) The demonstration of pneumococcal antigen in tissues by the use of fluorescent antibody. J. Immunol., 45:159–170.

——— and M. H. Kaplan (1950) Localization of antigen in tissue cells. II: Improvements in a method for the detection of antigens by means of fluorescent antibodies. J. Exp. Med., 91:1–13.

————, E. H. Leduc, and M. H. Kaplan (1951) Localization of antigen in tissue cells. VI: The fate of injected foreign proteins in the mouse. J. Exp. Med., 93:173–188.

Creech, H. J., and R. N. Jones (1941a) The conjugation of horse serum albumin with isocyanates of certain polynuclear aromatic hydrocarbons. J. Am. Chem. Soc., 63:1661–1669.

———— (1941b) Conjugates synthesized from various proteins and the isocyanates of certain aromatic polynuclear hydrocarbons. J. Amer. Chem. Soc., 63:1670.

Crowle, A. J. (1961) Immunodiffusion. Academic Press, New York.

Deutsch, H. F. (1952) Separation of antibody-active proteins from various animal sera by ethanol fractionation techniques. Meth. Med. Res., 5:284–300.

Feder, N., and R. L. Sidman (1958) Methods and principles of fixation by freeze-substitution. J. Biophys. Biochem. Cytol., 4:593–602.

Fothergill, J. E., and R. C. Nairn (1961) Purification of fluorescent protein conjugates: Comparison of charcoal and Sephadex. Nature, 192:1073–1074.

Freund, J., and V. Bonanto (1944) The effect of paraffin-oil lanolin-like substances and killed tubercle bacilli on immunization with diptheric toxoid and *Bact. typhosum*. J. Immunol., 48:325–334.

George, W. H., and K. W. Walton (1961) Purification and concentration of dye-protein conjugates by gel filtration. Nature, 192:1188–1189.

———— (1962) Preservation of tissue sections in Coons' fluorescent antibody technique. Nature, 194:693.

Glass, L. E. (1959) Immunohistological localization of serum-like molecules in frog oocytes. J. Exp. Zool., 141:257–290.

———— (1961) Localization of autologous and heterologous serum antigens in the mouse ovary. Devel. Biol., 3:787–804.

———— (1963) Transfer of native and foreign serum antigens to oviducal mouse eggs. Am. Zool., 3:135–156.

———— (1966) Serum antigen transfer in the mouse ovary: Dissimilar localization of bovine albumin and globulin. Fertility Sterility, 17:226–233.

———— (1968) Immunocytological studies of the mouse oviduct. In: The mammalian oviduct. Ed. by E. S. E. Hafez and R. J. Blandau. Univ. of Chicago Press, Chicago, pp. 459–476.

———— and T. R. McClure (1965) Postnatal development of the mouse oviduct: Transfer of serum antigens to the tubal epithelium. In: Ciba Foundation Symposium on Preimplantation Stages of Pregnancy. Ed. by G. W. W. Wolstenholme and M. O'Connor. Churchill, London. pp. 294–321.

Goldman, M., and R. K. Carver (1957) Preserving fluorescein isocyanate for simplified preparation of antibody. Science, 126:839–840.

Gomori, G. 1952. Microscopic histochemistry: Principles and practice. Univ. of Chicago Press, Chicago.

Hammerstrom, R. A. (1963) New device for arterial bleeding of the rabbit. J. Lab. Clin. Med., 61:352–354.

Heidelberger, M. (1956) Lectures in immunochemistry. Academic Press, New York.

Holborow, E. J. (1964) Fluorescent antibody technique. In: Immunological methods. Ed. by J. F. Ackroyd. Blackwell, Oxford. pp. 155–174.

Hopkins, S. J., and A. Wormall (1933) Phenyl isocyanate protein compounds and their immunological properties. Biochem. J., 27:740–753.

Kabat, E. A., and M. M. Mayer (1961) Experimental immunochemistry. 3rd ed. Thomas, Springfield, Ill.

MacPherson, C. F. C., and M. Heidelberger (1940) Preparation and immunological properties of acid-denatured egg albumin. Proc. Soc. Exp. Biol. Med., 43:646–647.

————— (1945a) Denatured albumin. I. The preparation and purification of crystalline egg albumin denatured in various ways. J. Am. Chem. Soc., 67:574–578.

————— (1945b) Denatured albumin. III. Quantitative immunochemical studies on crystalline egg albumin. J. Am. Chem. Soc., 67:585–591.

Mellors, R. C. (1959) Fluorescent antibody method. In: Analytical cytology. Ed. by R. C. Mellors 2d ed. McGraw-Hill, New York, pp. 1–68.

Meryman, H. T., ed. (1960) Freezing and drying of biological materials. Ann. N.Y. Acad. Sci., 85:501–734.

Nace, G. W., and T. Suyama (1961) The echo technique: a procedure for the cellular localization of specific enzymes and other antigens utilizing preliminary agar diffusion with fluorescent antibody. J. Histochem. Cytochem., 9:596.

Nairn, R. C., ed. (1962) Fluorescent protein tracing. Livingstone, Edinburgh.

Pearse, A. G. E. 1961. Histochemistry, theoretical and applied. 2d ed. Churchill, London.

Pilgrim, H. I., and K. B. DeOme (1955) Intraperitoneal pentobarbital anesthesia in mice. Exp. Med. Surg., 13:401–403.

Price, G. R., and S. Schwartz (1956) Fluorescence microscopy. In: Physical techniques in biological research. Ed. by G. Oster and A. W. Pollister. III. Cells and tissues. Academic Press, New York, pp. 91–148.

Riggs, J. L., P. C. Loh, and W. C. Eveland (1960) A simple fractionation method for preparation of fluorescein-labeled gamma globulin. Proc. Soc. Exp. Biol. Med., 105:655–658.

Riggs, J. L., R. J. Seiwald, J. H. Burckhalter, C. M. Downs, and T. G. Metcalf (1958) Isothiocyanate compounds as fluorescent labeling agents for immune serum. Am. J. Pathol., 34:1081–1093.

Rinderknecht, H. (1962) Ultra-rapid fluorescent labeling of proteins. Nature, 193:167–168.

Scheidegger, J. J. (1955) Une micro-méthode de l'immuno-électrophorèse. Inter. Arch. Allergy Appl. Immunol., 7:103.

Wieme, R. J. (1965) Agar gel electrophoresis. Elsevier, Amsterdam. 425p.

Williams, C. A., and M. W. Chase, Eds. Methods in Immunochemistry. 4 vols. Academic Press, New York. Vol. 1, 1968; Vol. II, 1969; Vols. III and IV, in press.

28

The Study of Endocytosis
and Lysosome Function
in Embryotrophic Nutrition

F. BECK
Department of Anatomy
The London Hospital Medical College
Turner Street
London, E.1., England

J. B. LLOYD
Department of Biochemistry
University College
Cardiff, Wales

L. M. PARRY
Department of Anatomy
The London Hospital Medical College
Turner Street
London, E.1., England

Among the many functions performed by lysosomes in the economy of the cell, that of intracellular digestion of pinocytosed material is of particular relevance to embryotrophic nutrition. Observations by numerous authors have permitted the construction of a simplified diagram of the processes, which, together with the nomenclature proposed by de Duve and Wattiaux (1966), is presented in Figure 28-1. It should be stressed that lysosomes perform tasks in addition to intracellular digestion, but these are not directly relevant to the problem of embryotrophic nutrition and are therefore not shown.

Macromolecules taken into cells by endocytosis become segregated in phagosomes (heterophagosomes), which are formed by fusion of pinocytic or of phagocytic vacuoles. Eventually, probably by a variety of distinct methods, phagosomes become associated with lysosomal enzymes to form heterolysosomes or digestive vacuoles; within the latter, ingested material is hydrolyzed, and the products of digestion are able to diffuse into the general cytoplasm of the cell.

The authors are grateful to the Spastics Society and to Tenovus for financial support that made it possible to develop many of the techniques described. Mr. R. A. M. Williams helped us greatly with histochemical sections.

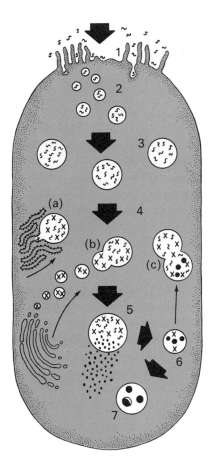

FIGURE 28-1

(1) *Mechanism of Intracellular Digestion. Absorptive cell surface often has a border of microvilli (see Fig. 28-6). Material (∽) is taken in by endocytosis (pinocytosis or phagocytosis) and comes to lie in small vesicles bounded by plasma membrane.*
(2) *These endocytic vesicles fuse to form larger bodies—heterophagosomes* (3) *Hetero-phagosomes (see Fig. 28-11) acquire hydrolytic enzymes (x) to become heterolysosomes* (4), *by* (a) *direct secretion of enzyme by granular endoplasmic reticulum (personal observations in rat yolk-sac epithelium (see Fig. 28-15) or* (b) *fusion with primary lysosomes budded from, or associated with, the Golgi apparatus (see Fig. 28-10) or* (c) *fusion with telolysosomes (see Fig. 28-9). Digestion takes place in heterolysosomes with the release of small molecular-weight products (∴) into the cytoplasm* (5). *Indigestible residues (●) persist in heterolysosomes, and these become telolysosomes* (6), *or, if enzymic activity has been lost, postlysosomes* (7). *(The word* lysosome *is used as a generic term and denotes any membrane-bound particle containing acid hydrolases. The term therefore embraces primary lysosomes, heterolysosomes, and telolysosomes.)*

HISTOCHEMICAL AND CYTOCHEMICAL DEMONSTRATION OF LYSOSOMAL ENZYMES

Many of the hydrolytic enzymes capable of breaking down embryotrophe within the cells of fetal membranes can be demonstrated by histochemical means. The most reliably demonstrable enzyme of this group is, without doubt, acid phosphatase (Gomori, 1952; Barka and Anderson, 1962), but workable methods are available for lysosomal aryl sulphatase

(Goldfischer, 1965), β-glucuronidase (Hayashi, Nakajima, and Fishman, 1964), certain nonspecific esterases (Shnitka and Seligman, 1961), N-acetyl-β-glucosaminidase (Hayashi, 1965), and cathepsin C (McDonald, Callahan, Ellis, and Smith, in press).

METHODS FOR ACID PHOSPHATASE

PRINCIPLES. In the azo dye methods, the commonly used substrates (sodium a-naphthyl phosphate or the phosphate ester of a substituted naphthol) are hydrolyzed enzymically. The insoluble and highly substantive naphthols so produced are simultaneously coupled with freshly prepared hexazotized pararosanilin.

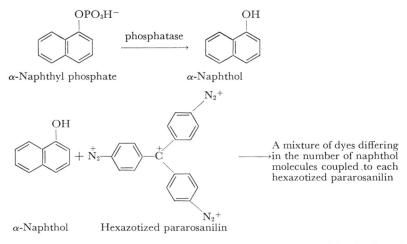

The Gomori (1952) method relies on the precipitation of lead phosphate from orthophosphate released at enzymically active sites by hydrolysis of the substrate β-glycerophosphate. Lead phosphate is electron dense, and the method is therefore suitable for electron microscopy. For light microscopy the colorless lead phosphate is converted into black lead sulphide with a solution of dilute ammonium sulphide.

METHOD. For light microscopy, small blocks of placental tissue or thin membranes, such as the visceral layer of the rodent yolk-sac, should be fixed overnight at 4°C in 4% formaldehyde containing 1% calcium chloride and adjusted to pH 7 with sodium hydroxide. Straus (1964a) recommends the addition of sucrose to a hypertonic concentration of 30%; we have found this modification to be of value in our laboratory. Tissues may then be stored in 30% sucrose in distilled water. Fixed tissue blocks should be quenched in isopentane at −70°C, and 7μ cryostat sections mounted on slides to be air-dried at room temperature for one or two hours. Thin membranes may be stained in bulk and subsequently embedded in paraffin prior to sectioning (Beck, Lloyd, and Griffiths, 1967a).

For electron microscopy, very small blocks of tissue, 15μ cryostat sections, or thin membranes, such as the visceral layer of the rodent yolk-sac, should be fixed for between one and twelve hours (depending upon the tissue) at 4°C in a glutaraldehyde-formalin mixture devised by Karnovsky (1965a). This is prepared by heating 2 g of paraformaldehyde in 25 ml of water to 65°C with constant stirring, and adding normal sodium hydroxide until the solution becomes clear (only two or three drops are required). The solution is cooled under running tap water; 5 ml of 50% glutaraldehyde (Sigma), 20 ml of $0.2M$ sodium cacodylate HCl buffer (4.28 g of sodium cacodylate per 100 ml) at pH 7.2, and 25 mg of calcium chloride are added, and the whole is thoroughly mixed. We have found that the addition of sucrose to make a 1% solution improves the morphological preservation of the tissue. Fixed tissues are washed for two or three days in $0.1M$ sodium cacodylate HCl buffer at pH 7.2, and may be stored for some weeks in this solution with the addition of sucrose to a concentration of 6.8%.

The simultaneous coupling azo dye method of Barka and Anderson (1962), using sodium α-naphthyl phosphate as substrate, has proved most satisfactory for the light microscopic demonstration of acid phosphatase in the visceral yolk-sac epithelium (see Fig. 28-2) and trophoblastic giant cells of the rat placenta and in the syncytiotrophoblast of the human and ferret placentae. A 0.4% substrate solution of sodium α-naphthyl phosphate made up in Michaelis veronal acetate stock (i.e., 19.428 g sodium acetate $3H_2O$ and 29.428 g of sodium barbitone in 1 liter of deionized water) is used; it is stable for some time at 4°C. The coupling agent is hexazotized pararosanalin, which is freshly prepared before use. A stock solution, made by adding 2 g of pararosanilin hydrochloride to 50 ml of $2N$ hydrochloric acid, heating gently, and filtering, may be kept, and hexazotized pararosanilin is prepared from this by adding it to an equal volume of 4% sodium nitrite in distilled water. Not all preparations of pararosanilin are pure enough to produce good results, and failure can often be turned into success by changing the batch of pararosanilin used.

The stain is prepared by adding 5 ml of the substrate solution to 13 ml of distilled water, followed by 1.6 ml of hexazotized pararosanilin. The pH is adjusted to 6.5 with normal sodium hydroxide (about 0.6 ml is required), and the whole mixture filtered before use. The staining time required will vary with the tissue, being about 20 minutes for bulk-stained preparations and frozen sections from rat placental membranes. Appropriate controls (heat inactivation at 90° to 100°C, incubation without substrate and for certain isoenzymes, preincubation with $10^{-2}M$ sodium fluoride) must be carried out (Beck and Lloyd, 1969). Bulk-stained membranes are dehydrated and embedded in paraffin prior to sectioning at 7μ. Frozen and paraffin sections can be lightly counterstained with haematoxylin or with chloroform-washed methyl green at pH 4, dehydrated rapidly, and mounted in DPX; alternatively, frozen sections can be mounted directly in aqueous media such as glycerine jelly.

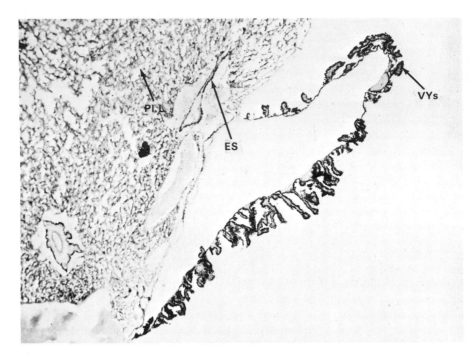

FIGURE 28-2

Rat visceral yolk-sac VYs and placental labyrinth (Pl.L) at 20½ days of pregnancy stained for acid phosphatase. A strongly positive reaction is given by the yolk-sac epithelium and a moderately strong one by the nearby endodermal sinus (ES). The placental labyrinth gives only a weak reaction for acid phosphatase. ×16. (Reproduced, with permission, from Beck and Lloyd, 1969.)

Gomori's method (Barka and Anderson, 1963) for acid phosphatase is the best available for the electron microscopic demonstration of the enzyme (see Fig. 28-3); it also has a place in light microscopy, not only because it can confirm results obtained by azo dye methods, but also because there is some evidence that the method may demonstrate enzymes which are exclusively lysosomal (Maggi, 1969), with substrate specificities differing slightly from naphthol AS-BI phosphatase (Rosenbaum and Rolon, 1962).

The incubating medium is prepared by adding 10 ml of 1.25% sodium β-glycerophosphate (adjusted to pH 5 with normal HCl) to 10 ml of distilled water and 10 ml of $0.1M$ tris maleate buffer at pH 5; 20 ml of 0.2% lead nitrate at pH 5 are then added to the buffered glycerophosphate drop by drop with constant stirring, and the final pH of the mixture checked with a pH meter. The incubating solution must be clear; a cloudy solution (often due to impurities in the buffer) must be discarded. Tris maleate buffer is made up from (A) $0.2M$ sodium hydroxide and (B) $0.2M$ tris acid maleate, which is 24.2 g tris (hydroxy-methyl) amino methane and 23.2 g of analytical grade anhydrous maleic acid *or* 19.6 g of maleic anhydride in a liter of distilled water. About 7 ml of A and 50 ml of B, made up to 100 ml with distilled water, give an $0.1M$ solution of pH 5.2.

Mounted, air-dried cryostat sections, very small blocks of tissue or thin fetal membranes, are stained in Gomori medium (30 minutes to 2 hours

FIGURE 28-3

Rat visceral yolk-sac epithelial cell at 13½ days of gestation. Glutaraldehyde-paraformaldehyde fix followed by acid phosphatase staining and Caulfield's osmic acid postfix. Fine sections stained with lead citrate. Apically located, phosphatase-negative heterophagosomes may be distinguished from more deeply placed phosphatase-positive hetero-lysosomes. ×8892.

are required for the visceral layer of the yolk-sac epithelium in the rat), and then washed for about 3 minutes in 0.1*M* cacodylate HCl buffer at *p*H 7.2. Glutaraldehyde-formalin-fixed material must be used for electron microscopy, but formalin-fixed tissue is adequate for examination by light microscopy (the post-staining sequence then consists of three quick rinses in distilled water). For electron microscopy, incubated tissues are post-fixed for two hours in Caulfield's (1957) osmic acid fixative. This is made up by adding 5 ml of 0.1*N* HCl to 5 ml of veronal acetate stock buffer (14.714 g sodium veronal and 9.714 g $CH_3COONa.3H_2O$, diluted to 500 ml); 12.5 ml of 2% osmic acid, which must be dissolved overnight, and 2.5 ml of distilled water are then added, the *p*H adjusted to 7.4 with 0.1*N* HCl, and 0.045 g of sucrose added per ml of fixative. A 20- to 30-minute wash with distilled water follows postfixation, and the tissues are then ready for embedding and fine sectioning by routine methods. Grids may be examined unstained or stained with lead citrate (Reynolds, 1963). When examination by light microscopy is required, Gomori-incubated tissue that was washed in distilled water (*vide supra*) is placed in 1% ammonium sulphide for 30 seconds, and then rapidly rinsed in water. Frozen sections may be counterstained and mounted directly in aqueous mountants, or dehydrated in a graded series of lead-saturated alcohols, cleared in lead-saturated xylol, and mounted in DPX. Thin membranes or small pieces of placental tissue stained in bulk are dehydrated in lead-saturated alcohols and xylol, and embedded in lead-saturated paraffin. Sections may be processed by normal histological techniques, using lead-saturated solutions. Methyl green or nuclear fast red are convenient counterstains for use with this technique.

Methods for Other Lysosomal Enzymes

Nonspecific esterases

PRINCIPLES. Although biochemical studies on liver homogenates subjected to differential centrifugation have consistently shown that nonspecific aliesterases sediment quantitatively in the microsomal fraction (Underhay, Holt, Beaufay, and de Duve, 1956; Carruthers, Woemley, Baumler, and Lilga, 1960; Markert and Hunter, 1959), several available histochemical techniques demonstrate a granular localization of the enzymes corresponding to the distribution of lysosomes in the cells concerned. Numerous explanations have been advanced to account for this discrepancy (Holt, 1963; Shibko and Tappel, 1964), but whatever the explanation, it seems clear that the selective histochemical demonstration of lysosomal esterase involves a sensitive cytoplasmic enzyme that is more inhibited than a resistant lysosomal isozyme. Within these limitations, it is therefore often possible to confirm the presence of lysosomes in the cells of fetal membranes by staining for nonspecific esterases. Three methods achieve this.

Holt (1958) has used halogen-substituted indoxyl esters (5-bromoindoxyl acetate, 5-bromo-4-chloroindoxyl acetate) as substrates yielding indoxyl compounds that are rapidly oxidized to derivatives of indigo by the incorporation of a ferro-ferricyanide redox buffer system in the reaction mixture.

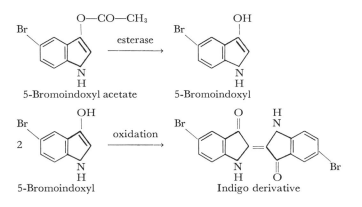

Shnitka and Seligman (1961) indicated that the redox system itself in-hibits the cytoplasmic demonstration of esterase more than it does the lysosomal, and have shown convincingly that the concentration of ferro-ferricyanide mixture in the substrate medium directly influences the locali-zation of the final reaction product. Another method in common use employs naphthol-AS acetate as substrate (Shnitka and Seligman, 1961) and relies on the use of organophosphate inhibitors to eliminate interfer-ence from cytoplasmic enzymes.

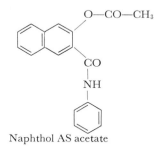

Naphthol AS acetate

Finally, Wachstein, Meisel, and Falcon (1961) used thiolacetic acid as a substrate that—with appropriate inhibitors—can be used to demon-strate esterases that appear to have a lysosomal distribution. The hydro-gen sulphide formed as a result of enzymic activity is precipitated as lead sulphide, and the reaction can therefore be used at the electron microscopic level.

$$CH_3—CO—SH \xrightarrow{\text{esterase}} CH_3—COO^- + H_2S$$
Thiolacetic acid

FIXATION. As described for acid phosphatase (*vide supra,* see p. 380), cold fixation overnight in calcium formol is generally satisfactory. Formalin fixation tends to inhibit cytoplasmic esterase more than lysosomal (Shnitka and Seligman, 1961) and the fixation time can therefore be varied in an attempt to display the latter to best advantage. Frost and Brandes (1967) report that fixation in formalin without sucrose resulted in the loss of cytoplasmic esterase, the latter being better preserved if the fixative is rendered isosmotic with sucrose; 16 to 24 hours has been found to be optimal for the demonstration of lysosome-like bodies in the visceral yolk-sac epithelium of the rat. Holt (1958) recommends that fixed tissues be stored by washing out the formalin in physiological saline, transferring to ice-cold $0.88M$ sucrose containing 1% gum acacia, infiltrating at 0° to 2°C prior to freezing in isopentane at −70°C, and cryostat sectioning at −20°C. For electron microscopy, very small pieces of thin fetal membrane or placental tissue are fixed in Karnovsky's (1965a) glutaraldehyde-formalin fix (see p. 381), and washed and stored in cacodylate buffer as described under the section for acid phosphatase.

HALOGEN-SUBSTITUTED INDOXYL ACETATE. The staining medium involves the use of three stock solutions: (1) $0.1M$ tris-hydrochloric acid buffer at pH

8.3; (2) 2M sodium chloride solution; (3) a mixture of equal volumes of 0.1M potassium ferricyanide and 0.1M potassium ferrocyanide stored at 4°C. First, 10 ml of tris buffer are mixed with 25 ml of sodium chloride solution, 5 ml of the ferro-ferricyanide redox system, and 10 ml of distilled water. The whole is then rapidly mixed with 5.8 mg of 5-bromoindoxyl acetate (George T. Gurr) in 0.5 ml of acetone, and cryostat sections or thin placental membranes are incubated at 37°C for ½ to 2 hours. Frozen sections may be counterstained with nuclear fast red, dehydrated, and mounted in DPX; bulk-stained material can be embedded in paraffin and subsequently sectioned and counterstained. The final medium described here contains 5mM of both ferro- and ferricyanides, and is suitable for the demonstration of lysosomes in the rat visceral yolk-sac epithelium; by decreasing the concentration of the redox system, cytoplasmic esterases are also stained. In some placental material, it is conceivable that higher concentrations would be required for the optimal demonstration of lysosomal activity.

THIOLACETIC ACID. Two basic solutions are prepared. The first is 0.15 ml of thiolacetic acid which is diluted to 5 ml with distilled water, and brought to pH 5.5 with 0.1N NaOH, 0.2M acetate buffer at pH 5.5 being added to make up 100 ml. The solution is made up fresh before use. The second is a 0.5% solution of lead nitrate in distilled water; this solution is stable. To prepare the final substrate, 1 ml of the lead nitrate solution is added drop by drop with constant stirring to 20 ml of the acetate-buffered thiolacetic acid solution; after standing for a few minutes, the whole is filtered and the clear filtrate used for staining.

The thiolacetic acid esterase technique stains both the azolesterases of motor end plates, as well as nonspecific ali-esterases. Although confusion of lysosomal enzymes with acetylcholinesterases and pseudocholinesterases is unlikely in placental material, it remains important to distinguish between cytoplasmic and lysosomal nonspecific esterases. In order to do this, an hour's preincubation of frozen sections or thin fetal membranes at 37°C in 10^{-5} M E600 (Diethyl-p-nitrophenylphosphate) in acetate buffer at pH 6 inhibits most of the cytoplasmic enzyme without apparently destroying lysosomal activity. Extreme caution should be observed when E600—a highly toxic compound—is being used.

After about an hour's incubation in substrate, sections are washed in distilled water; they are then mounted in glycerine jelly, or dehydrated and mounted in DPX. Bulk-stained tissues are dehydrated in a graded series of lead-saturated alcohols, embedded in paraffin, and subsequently sectioned at 7μ. Paraffin sections should be counterstained and dehydrated in lead-saturated solutions whenever possible. For electron microscopy, histochemically stained material is washed for about a minute in 0.1M cacodylate-HCl buffer and then postfixed for two hours in Caulfield's (1957) osmic acid fixative (see p. 383). Postfixed tissues are washed in distilled water for 20 to 30 minutes, dehydrated, and embedded in Araldite, as for conventional electron microscopy. Fine sections are prepared and should be examined unstained and stained with lead citrate (Reynolds, 1963).

Aryl sulphatase

PRINCIPLE. Aryl sulphatases A, B, and C may be distinguished in certain rat tissues (Roy 1960) ; of these, A and B are lysosomal with acid pH optima, whereas aryl sulphatase C, which is microsomal in origin, has a pH optimum of about 8. To some extent, varying the pH of the incubating medium enables a histochemical distinction between lysosomal and nonlysosomal sulphatases to be made. Goldfischer (1965) has described the most satisfactory method for the histochemical demonstration of these enzymes; he uses p-nitrocatechol sulphate as substrate and lead nitrate for the capture of liberated sulphate ions.

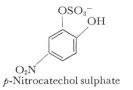

p-Nitrocatechol sulphate

Although Goldfischer's results with a variety of tissues, and our own experience with rat kidney and intestine, have generally been adequate for the demonstration of lysosomes, it must be admitted that rat placental tissue has given less encouraging results. Localization has been at the histological rather than at the cytochemical level.

METHOD. For light microscopy, fixation in cold hypertonic calcium formol (*vide supra,* p. 380) is recommended; overnight fixation is generally sufficient for thin fetal membranes, but larger pieces of placental material may require longer periods. Tissues may be stored in 30% sucrose in distilled water.

The incubating medium is prepared by dissolving 15 to 30 mg of p-nitrocatechol sulphate (sigma) in 5 ml of veronal acetate buffer (see p. 383) at pH 5.4. Next 0.16 ml of 24 percent lead nitrate is added, and the medium is adjusted to pH 5.5; a precipitate that may form at this point goes into solution when the pH is brought below 5.7. Air-dried cryostat sections are stained in the incubating medium at 37°C, for between 30 minutes and 2 hours, and then washed briefly in distilled water. They are transferred to a 1% ammonium sulphide solution for 30 seconds, again quickly rinsed in distilled water, counterstained in methyl green if desired, dehydrated and cleared in lead-saturated solutions, and mounted in D.P.X. Small blocks of tissue or thin membranes may be bulk-stained and then paraffin-embedded. Control sections should be heat-inactivated before incubation, and also incubated in substrate-free medium.

Goldfischer (1965) states that the method may be adapted for electron microscopy, by fixing thin membranes for 2 to 3 hours in 2% glutaraldehyde in $0.1M$ cacodylate buffer at pH 7.2 (Sabatini, Bensch, and Barnett, 1963) . Prolonged fixation is necessary for denser pieces of tissue. After staining in the above incubation medium, tissues are briefly rinsed in cacodylate-HCl buffer, *p*H 7.2, postfixed in Caulfield's osmic acid (Caulfield, 1957), dehydrated, and embedded in the routine way.

β-Glucuronidase

PRINCIPLE. Recently a much-improved method for the demonstration of this lysosomal and microsomal group of enzymes has become available (Hayashi, Nakajima, and Fishman, 1964). The substrate used in naphthol AS-BI-β-D-glucuronic acid (Calbiochem) which yields naphthol AS-BI; the latter is captured by means of hexazotized pararosanilin.

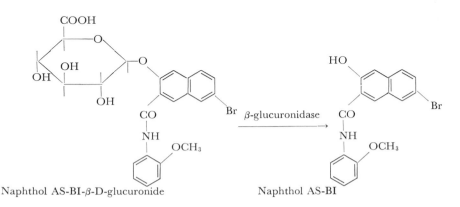

Naphthol AS-BI-β-D-glucuronide   Naphthol AS-BI

METHOD. Thin membranes or tissue slices are fixed in formol calcium (see p. 380) at 4°C for 24 hours, and may be stored in Holt's (1958) hypertonic gum sucrose (1% gum acacia in 0.88*M* sucrose).

For the incubation medium, a substrate stock solution is prepared by dissolving 28 mg of naphthol AS-BI-β-D-glucuronic acid in 1.2 ml of 0.05*M* sodium bicarbonate (0.42 grams of $NaHCO_3$ in 100 ml of distilled water), and making up to 100 ml with 0.2*M* acetate buffer at *p*H 5. This is stable for some weeks at room temperature. Next 0.3 ml of pararosanilin-HCl solution (see p. 381 for preparation) and 0.3 ml of a fresh solution of 4% sodium nitrite are mixed, and 10 ml of the substrate stock solution is added. The *p*H is adjusted to 5.2 with 1*N* NaOH, and the mixture made up to a final volume of 20 ml with distilled water. The resultant solution is filtered, giving a slightly yellow, clear incubation medium.

Air-dried cryostat sections are incubated for 30 minutes at 37°C, rinsed well in distilled water, counterstained in methyl green if necessary, dehydrated, cleared, and mounted.

Control sections are heat-inactivated before incubation, incubated in substrate-free media, or incubated in medium containing 1 to 3 m*M* saccharo-1,4-lactone.

N-acetyl-β-glucosaminidase

The method for demonstrating this enzyme is not included here since the authors have had no opportunity of testing it. The reader is referred to

Hayashi (1965) for details, but the technique has limitations; these are summarized by Beck and Lloyd (1969).

CYTOLOGICAL DEMONSTRATION OF HETEROPHAGOSOMES

Numerous methods for the demonstration of phagosomes are available, and are widely applicable to studies of the breakdown of embryotrophe by fetal membranes. Besides the characteristic morphological appearances of a phagocytic tissue, particularly at the level of the electron microscope, histologically or cytochemically identifiable macromolecules may be injected into pregnant animals (or placed directly into the uterine lumen at operation) and their subsequent fate traced in appropriate preparations of the fetal membranes. In this context the acid bisazo dyes typified by trypan blue are perhaps the most commonly used markers, but other compounds have recently been employed, each having advantages of its own. (For a fuller discussion, see Beck and Lloyd, 1969; Daems, Wisse, and Brederoo, 1969.) Thus fluorescent-labeled proteins have been used by Mayersbach (1958) to study protein uptake by rabbit fetal membranes, and Larsen and Davies (1962) have stained specific antigens within heterolysosomes by means of fluorescent-labeled antibodies. Autoradiography has been used by Anderson (1959) to follow injected radioalbumin in rat placental material and the non-ionic detergent Triton WR-1 1339 (Wattiaux, Wibo, and Baudhuin, 1963) may also be used as a marker for heterolysosomes (Schultz and Schultz, 1966). The last is very useful biochemically, but lacks precision histochemically. Krzyzowska-Gruca and Schiebler (1967) have injected ferritin into pregnant rats and subsequently localized it in the phagosomes of the visceral yolk-sac epithelium. *It should be stressed, however, that there are some anatomical situations at which embryotrophe is being digested (e.g., the uterine symplasma in the ferret) that are inaccessible to parenterally administered materials.*

Phagocytic Tissue Morphology

Initial observations should be made by light microscopy on paraffin sections. Bouin fixation for 24 hours, followed by storage in 70% alcohol, suffices for most fetal membranes, but very large pieces of placental material may be fixed for longer periods. An evaluation of the extent of embryotrophic nutrition must involve a study of portions of *all* the extraembryonic membranes present at the appropriate stage of pregnancy. Slides should be stained with haematoxylin and eosin, and with a suitable trichrome stain—e.g., the azan procedure developed by Heidenhain (1915). The epithelia of some fetal membranes may, by these simple methods, be shown to contain either recognizable cellular elements, e.g., maternal red-blood corpuscles, in the process of digestion (see Fig. 28-4), or the breakdown products of their digestive processes, such as intracellular hematin crystals. Intracellular vacuoles, especially where associated with a cuticular

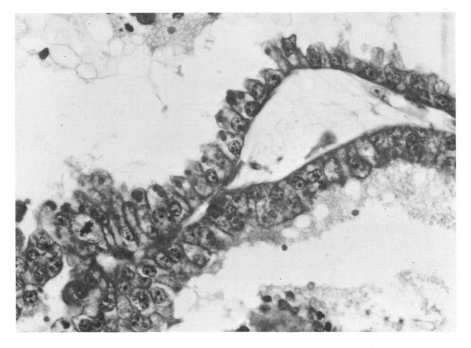

FIGURE 28-4

Ferret chorion in the region of the maternal haematoma normally found on the antimesometrial side of the uterus at 20 days gestation. Many of the trophoblastic cells contain maternal red blood corpuscles in the supranuclear vacuoles. Haematoxylin and chromotrope. ×500.

border, often indicate active pinocytosis (see Fig. 28-5) and trichrome stains may suggest that the vacuolar contents are protein in nature, with tinctorial properties similar to secretions that come into contact with the cells under consideration. It is of course important to remember that the presence of vacuolated cytoplasm can imply secretion and storage as well as absorption, and that the staining procedures described are in no way specific; nevertheless, these methods serve to focus attention on those fetal membranes that are *likely* to be concerned with absorption of embryotrophe.

More detailed morphological examination must be carried out with the electron microscope. For this purpose, good initial fixation of tissues is obtained with Karnovsky's (1965a) glutaraldehyde-formalin mixture, containing 1% sucrose at 4°C (*vide supra*, p. 381). The fixative should be dropped directly on the fetal membranes *in situ* immediately upon death; small pieces are then removed with fine instruments and fixed overnight. Alternatively, the rapid removal and immediate fixation of small pieces of membrane yields comparable results, but even a few minutes delay or storage in normal saline results in gross distortions of the morphological pattern. Fixed tissues are washed for 2 or 3 days in buffered cacodylate and postfixed in Caulfield's (1957) osmic acid fixative (*vide supra*, p. 383). It is often very useful to postfix for 30 minutes to an hour in Luft's (1956)

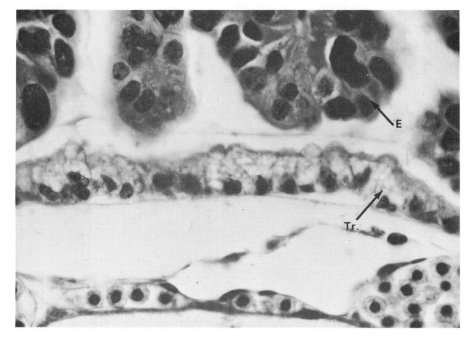

FIGURE 28-5

Ferret trophoblast (Tr) *lying adjacent to uterine epithelium* (E) , *which latter has been transformed into a symplasma, 20 days of gestation. The trophoblastic cells are highly vacuolated and have a well-marked cuticular border facing the uterine epithelium. Haematoxylin and chromotrope. ×620.*

permanganate fixative instead of osmic acid; permanganate fixation is of particular value in studying membranes, a special advantage in investigating intracellular digestive processes (see Fig. 28-6). Primary fixation with glutaraldehyde seems to provide sufficient rigidity to prevent the destructive effects of permanganate on delicate embryological tissue. Luft's (1956) fixative is prepared from a (refrigerated) stock solution of 1.2% potassium permanganate in distilled water, by mixing with an equal volume of veronal acetate buffer at pH 7.4 to 7.6. The typical appearances of an actively phagocytic fetal membrane are shown in Figures 28-7 through 28-12.

THE STUDY OF PHAGOSOMES BY INJECTION OF MARKER MOLECULES

Two representative methods are given, but many others are now available (*vide supra,* p. 389) .

Acid bisazo dyes

First used by Wislocki (1920, 1921) to demonstrate placental endocytosis, this method originated in the classical studies of Goldmann (1909, 1912) on vital staining of a variety of tissues in health and disease. The acid bisazo

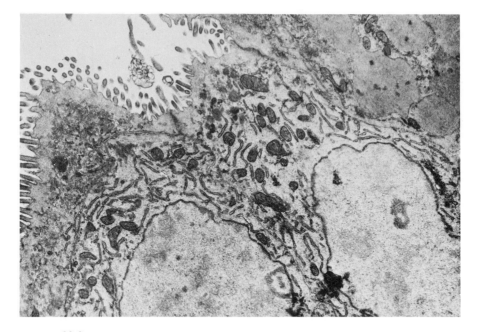

FIGURE 28-6

Apical portion of rat visceral yolk-sac epithelial cells fixed with glutaraldehyde-para-formaldehyde followed by permanganate postfix. Fine sections stained with lead citrate. A clear zone of cytoplasm appears to be situated between the microvilli at the cell sur-face and the subapical system of microtubules. ×9,500.

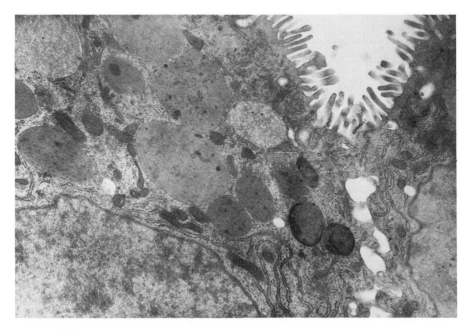

FIGURE 28-7

Supranuclear region of rat visceral yolk-sac epithelial cell at 15 days of gestation. Glutaraldehyde-paraformaldehyde fix, Caulfield's osmic acid postfix. Fine section stained with lead citrate (Reynolds, 1963). Micropinocytic vesicles are seen arising from the plasma membrane between the microvilli. A system of microtubules is apparent in the subapical regions of the cell. ×11,850. (Reproduced, with permission, from Lloyd et al. 1968.)

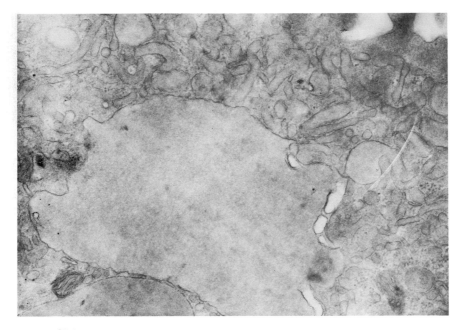

FIGURE 28-8

Subapical portion of rat visceral yolk-sac epithelial cell at 15½ days of gestation showing apparent fusion of microtubules and pinocytic vesicles to form heterophagosome. Glutaraldehyde-paraformaldehyde fix, Caulfield's osmic acid postfix. Fine sections stained with lead citrate. ×31,850. (Reproduced, with permission, from Lloyd et al., 1968.)

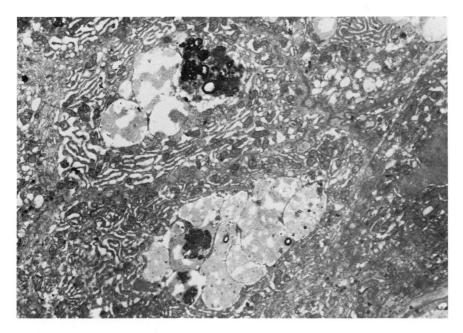

FIGURE 28-9

Transverse section through the supranuclear portion of the rat visceral yolk-sac epithelium. Glutaraldehyde fix, acid phosphatase stain, and Caulfield's osmic acid postfix. Fine sections stained with lead citrate. The central portion of each cell contains telolysosomes in close association and possibly fusing with heterophagosomes and heterolysosomes. ×6,850.

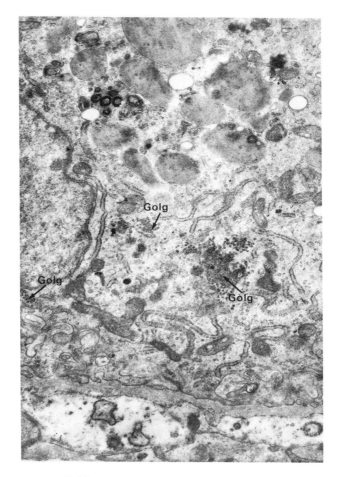

FIGURE 28-10

Rat visceral yolk-sac epithelium at 20½ days of gestation. Glutaraldehyde-paraformaldehyde fix followed by acid phosphatase staining and Caulfield's osmic acid postfix. Fine sections stained with lead citrate. A slight deposit of lead is seen in the cisternae of the granular endoplasmic reticulum. Phosphatase-positive Golgi regions (Golg) seem to be giving rise to primary lysosomes, which are passing into the cell cytoplasm. ×14,450.

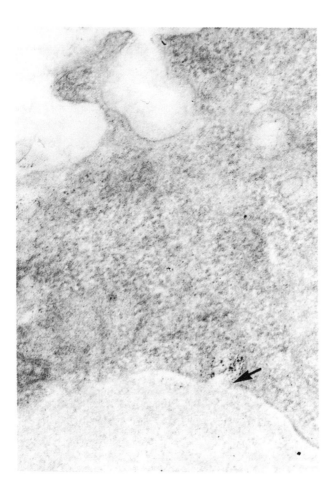

FIGURE 28-11

Apical part of rat visceral yolk-sac epithelium at 9½ days of gestation. Glutaraldehyde-paraformaldehyde fix followed by acid phosphatase staining and Caulfield's osmic acid postfix. Fine sections stained with lead citrate. A phosphatase-positive organelle—possibly a primary lysosome—appears to be fusing (at arrow) with the side of a large phosphatase-negative heterophagosome. (×76,750).

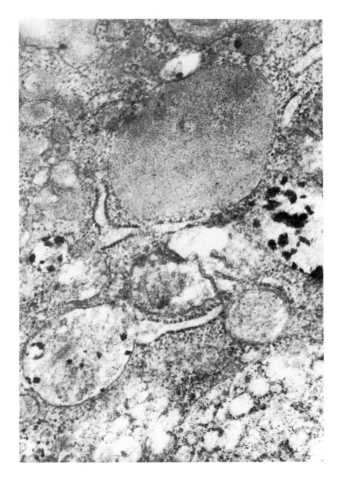

FIGURE 28-12

Apical part of rat visceral yolk-sac epithelium at 14½ days of gestation. Glutaraldehyde fix followed by acid phosphatase staining and Caulfield's osmic acid postfix. Fine section stained with lead citrate. A profile of granular endoplasmic reticulum is seen to open directly into a heterolysosome. ×44,450. (Reproduced, with permission, from Lloyd et al., 1968.)

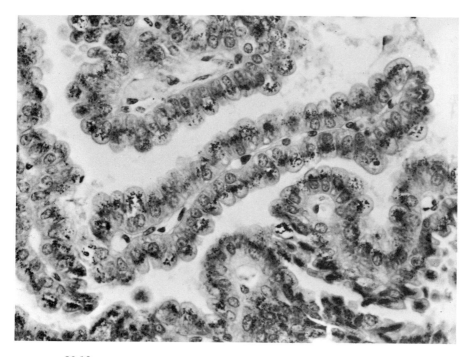

FIGURE 28-13

Visceral yolk-sac from 20-day pregnant rat injected subcutaneously with 75 mg/kg trypan blue 24 hours previously. The dye is clearly seen in granules (heterolysosomes) in the supranuclear part of the epithelial cells. Counterstained with nuclear-fast red. ×325. (Reproduced, with permission, from Lloyd et al., 1968.)

dye trypan blue binds strongly to albumin and is taken up by phagocytic tissue in combination with protein. It is easily obtainable commercially, but samples for injection should be filtered, dialyzed against running tap water for 24 hours in order to remove the salt and other impurities that are frequently present (Lloyd and Beck, 1963), and then evaporated to dryness on a rotary evaporator. More elaborate methods for the identification and purification of dye samples have been developed (see Beck and Lloyd, 1963; Lloyd and Beck, 1964; and p. 406), but these are not required for routine histological work, unless vital staining cannot be obtained with a given sample.

For the optimum demonstration of endocytosis between 50 and 75 mg per kg of body weight should be injected subcutaneously into pregnant animals 24 hours before death. A 1% aqueous solution should be used, and no more than 1 ml should be injected at any site. Tissues are prepared for light microscopy as described on p. 389, but storage in 70% alcohol should not be prolonged, since this tends to wash out some of the dye; by the same token, fixation in *alcoholic* Bouin's solution should be avoided. Tissues are counterstained in an 0.1% solution of nuclear-fast red in 5% potassium alum (Sams and Davies, 1967). The dye-albumin complex finds its way initially into heterophagosomes, which become faintly stained. Since the dye is not

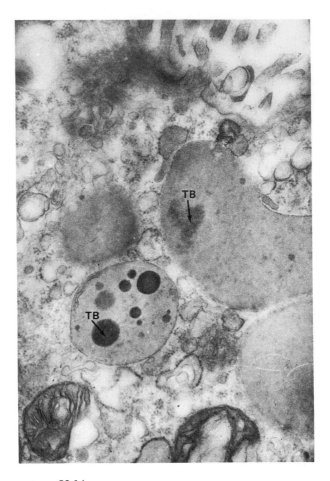

FIGURE 28-14

Apical portion of visceral yolk-sac epithelium from rat treated with 75 mg/kg trypan blue 8 hours previously. The dye (TB) is concentrated in a unit-membrane bound vesicle. In another, more apically situated vesicle, the dye is less concentrated. ×31,200. (Reproduced, with permission, from Lloyd et al., 1968.)

digested by lysosomal enzymes, it gradually accumulates in heterolysosomes and residual bodies, which therefore become stained intensely blue (see Fig. 28-13). Subcutaneous injection makes it possible to demonstrate the absorption of embryotrophe directly or indirectly from the maternal blood stream, but it is not applicable to a study of the digestion of the pabulum formed from endometrial tissues or from uterine milk if the dye is not secreted into the latter. Direct injection of trypan blue into the uterine lumen may therefore, in certain circumstances, yield further information regarding the phagocytic properties of fetal membranes. For this purpose the dialyzed dyestuff should be made neutral by adjusting to *p*H 7 with NaOH immediately before use. About 0.1 or 0.2 ml of a 1% solution may be injected into each horn of the pregnant rat uterus at 17 days of ges-

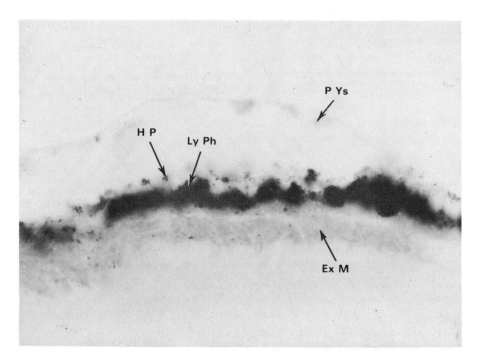

FIGURE 28-15

Visceral yolk-sac epithelium at 8½ days of gestation from a conceptus whose mother had been injected with horseradish peroxidase 6 hours previously. Double staining for peroxidase and acid phosphatase has been performed. The injected peroxidase is seen as small granules in the apical region of the cells (HP); in the deeper parts of the cells, fusion with lysosomes has occurred and digestive vacuoles or heterolysosomes have been formed (LyPh). No peroxidase is seen in either the extraembryonic mesoderm (ExM) or in the parietal yolk-sac epithelium (PYs). (Reproduced, with permission, from Beck et al., 1967a). ×985.

tation. Although the injected material diffuses throughout the uterine lumen, differences in concentration at various intrauterine sites cannot be excluded and, for biochemical assay, numerous small doses injected into portions of the uterus isolated by ligature may be attempted.

Trypan blue is visible electronmicroscopically (see Fig. 28-14) and can be demonstrated if the tissue is prepared by fixation in glutaraldehyde-formalin and postfixed in Caulfield's (1957) osmic acid (see p. 390 for details).

Horseradish peroxidase

PRINCIPLE. Straus (1964a,b,c) elegantly demonstrated that the plant enzyme horseradish peroxidase, when injected intravenously into rats, was taken up by the cells of the reticulo-endothelial system and by the proximal convoluted tubule cells of the kidney. The injected enzyme could be demonstrated within the heterophagosomes of these cells by appropriate histochemical procedures, and preparations to show its gradual association with acid phosphatase in heterolysosomes could be made. Beck *et al.*, (1967a) were able to apply these techniques to demonstrate the uptake of macro-molecules by rat fetal membranes, and to show that endocytosis is followed by the association of the ingested material with lysosomal acid phosphatase. The advantage of this method is that it can be carried out in conjunction with biochemical studies; the intracellular degradation of peroxidase can thus be followed quantitatively. The method is illustrated by reference to the visceral yolk-sac epithelium of the rat.

METHOD. About 1.5 ml of a 1% solution of horseradish peroxidase (Sigma type II) dissolved in normal saline or preferably in rat serum, is injected into the exposed jugular vein of a lightly ether-anesthetized pregnant rat. After varying periods of time (5 minutes to 48 hours) the abdomen is opened, and individual conceptuses removed. The fetal membranes are fixed in 4% formaldehyde containing 30% sucrose at 4°C. Time of fixation varies with the size of the specimen, but is generally between 18 and 24 hours. On the other hand a small egg cylinder at $8\frac{1}{2}$ days of gestation would require only $1\frac{1}{2}$ hours of fixation, since enzyme activity is completely abolished if tissues are overfixed. Fixed material is washed, and may be stored in 30% sucrose at 4°C for at least a week. Next, 8 to 15μ air-dried cryostat sections, thin fetal membranes, or very small pieces of placental material are stained for acid phosphatase by the azo dye method (see p. 380 for details) and again washed in 30% sucrose. Peroxidase staining is then performed on phosphatase-stained material (see Fig. 28-15), as well as on some unstained specimens (see Fig. 28-16). The principle underlying peroxidase staining is the following.

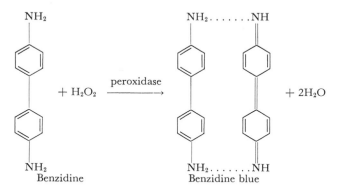

Benzidine   Benzidine blue

Specimens are treated at 0°C for between 5 and 90 seconds in a 70% alcoholic solution, containing 0.3% benzidine and between 0.015 and 0.03% hydrogen peroxide. The staining time depends on the development of a satisfactory blue color reaction, for which the temperature is very important. After staining, the specimen is washed in a large volume of ice-cooled 9% sodium nitroprusside in 70% alcohol, and then left in an ice-cooled stabilization bath of the same composition but also containing 0.01M acetate buffer at pH 5.0 for one hour. Frozen sections may then be counter-stained with nuclear fast red or methyl green and mounted in glycerine jelly, or dehydrated and mounted in D.P.X. Bulk-stained specimens can be *rap-*

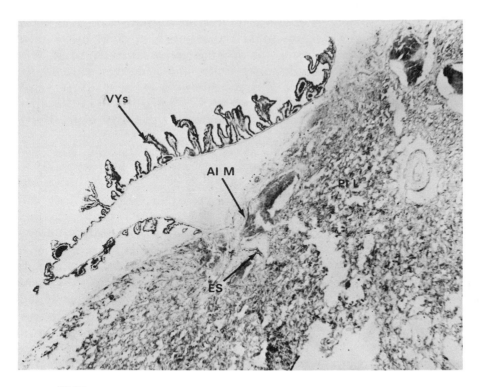

FIGURE 28-16

Peroxidase stained visceral yolk-sac (VYs) and placental labyrinth (PlL) at 20½ days of pregnancy from a rat injected 6 hours previously with the enzyme. There is a concentration of peroxidase in the yolk-sac epithelium but not in the underlying mesenchyme; no reaction is given by the allantoic mesoderm (AlM), the placental labyrinth, or the endodermal sinuses (ES). ×36.

idly dehydrated in graded alcohols and embedded in paraffin. The blue reaction product will withstand subsequent sectioning, deparaffinization, counterstaining in nuclear fast red or chloroform-washed methyl green, dehydration, and mounting in D.P.X.

Electron microscopic visualization

PRINCIPLE. Karnovsky (1965b) and Graham and Karnovsky (1966) have developed a method for the cytochemical localization of horseradish peroxidase at the ultrastructural level, in which 3,3′-diaminobenzidine is the oxidizable substrate yielding a reaction product that is noncrystalline, insoluble, and electron opaque after osmic postfixation.

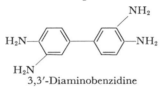

3,3′-Diaminobenzidine

METHOD. Pieces of placental material or thin membranes are fixed in formaldehyde-glutaraldehyde fixative (see p. 381 for details) for 2 to 5 hours. Frozen sections or thin membranes are then incubated at room temperature for 3 to 10 minutes in a solution made by dissolving 5 mg of 3,3′-diaminobenzidine tetrahydrochloride (Sigma) in 10 ml of 0.05M Tris-HCl buffer at pH 7.6. Enough hydrogen peroxide is added to make 0.01%. After incubation, material is washed in distilled water (three rapid changes) and postfixed for 90 minutes in Caulfield's osmic acid fixative. After Araldite embedding, fine sections are cut by the usual methods and may be stained with lead citrate.

BIOCHEMICAL METHODS FOR THE STUDY OF LYSOSOMES AND PHAGOSOMES

Modern biochemical techniques afford a powerful adjunct to the morphological methods described above. Their greatest contribution is their ability to furnish quantitative information, but their scope extends beyond the simple measurement of enzyme activities in isolated tissues, and includes techniques enabling the worker to investigate the constituents and functions of subcellular organelles. Although capable of great precision, biochemical methods have the disadvantage of requiring relatively large quantities of tissue. Furthermore, the material studied is frequently heterogeneous, and this sometimes has the effect of obscuring a histochemically precise localization of an enzyme or marker. For example, in biochemical studies on the rat conceptus, Beck *et al.* (1967a) found that the specific activity of acid phosphatase in the yolk-sac was only twice that in the chorioallantoic placenta, although histochemistry of yolk-sac had revealed a high concentration of the enzyme in the apical parts of epithelial cells. The biochemical assays were, of course, performed on homogenates of whole yolk-sac, which includes a considerable proportion of splanchnopleuric mesoderm and vitelline vessels, as well as phosphatase-rich epithelial cells.

Quantitative data on naturally occurring enzyme levels in tissues and on uptake of administered macromolecular tracers may be obtained by applying appropriate assay techniques following homogenization. For this purpose, complete tissue disruption is required and a variable-speed overhead-drive mixer is the instrument of choice. A number of commercially available mixers (e.g., Virtis, Gardiner, New York) incorporate a variety of sizes of blades and mixing vessels, thus allowing the homogenization of both small and large quantities of tissue. Since one intends *not* to preserve the integrity of the cytoplasmic organelles, the suspending medium need not be osmotically protected, water or a simple buffer solution being adequate. The incorporation of a detergent such as 0.1% Triton X-100 ensures full release of latent membrane-bound enzymes.

Lysosomal Enzymes

Of the growing number of enzymes considered on biochemical grounds to be lysosomal (see Barrett, 1969), most have been so characterized in only one or two tissues, and are therefore unsuitable as markers for lysosomes. In studies of fetal membranes, where enzyme assays are undertaken to give some measure of lysosome abundance, it is wise to choose enzymes widely accepted as having a major lysosomal component and preferably those whose intracellular location can be confirmed histochemically in the particular tissue concerned. Two such enzymes are acid phosphatase and β-glucuronidase, and the assays to be described have been chosen for their sensitivity, an important consideration when only small quantities of tissue are available. Other enzymes suitable as lysosomal markers are acid deoxynbonuclease and acid proteinase (de Duve *et al.*, 1955, and Barrett, 1967, give their most acceptable assays). Arylsulphatase is not recommended, because, although the assay with nitrocatechol sulphate is sensitive and measures principally the lysosomal arylsulphatases, complications due to nonlinear kinetics and enzyme activation by inorganic ions can lead to confusing results.

Enzymes contained within intact lysosomes display latency, only a fraction of the total activity being "free", i.e., accessible to substrate, and the following assays may be modified to measure the free and total activities in homogenates or cell fractions prepared in osmotically protected medium. For measurements of free activity, incubation periods should be short (10 minutes at most), and the incubation mixture should incorporate $0.25M$ sucrose. Total activity is also measured under these conditions, but in the presence of 0.1% Triton X-100; the sucrose concentration must be identical in both estimations, since sucrose partially inhibits some lysosomal enzymes, e.g., β-glucuronidase (Gianetto and de Duve, 1955).

Method for acid phosphatase (Modified from Torriani, 1961)

PRINCIPLE. After destruction of microsomal phosphatases by incubation at 37°C and pH 5 for 10 minutes, acid phosphatase activity is measured by the release of p-nitrophenol from p-nitrophenyl phosphate. The p-nitrophenol released is measured spectrophotometrically by its absorption at 420 mμ in alkaline solution. A high concentration of inorganic phosphate is added at the end of the incubation to inhibit any alkaline phosphatases.

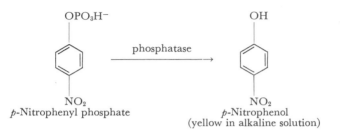

| p-Nitrophenyl phosphate | p-Nitrophenol (yellow in alkaline solution) |

REAGENTS

1. 0.2 M sodium acetate HCl buffer, pH 5.0.

2. 0.04 M sodium p-nitrophenyl phosphate, the solution to be prepared immediately before use.

3. 1 M tris—HCl buffer, pH 8.5, containing 0.4 M dipotassium phosphate.

PROCEDURE. To 1.2 ml of sodium acetate buffer, add 0.3 ml of homogenate and maintain at 37°C for 10 minutes. Add 0.5 ml of sodium p-nitrophenyl phosphate and incubate at 37°C for a suitable period. Add 2.0 ml of Tris buffer and measure the extinction at 420 mμ. From the value, subtract that for a control treated similarly but without incubation.

Method for β-glucuronidase (Gianetto and de Duve, 1955)

PRINCIPLE. Phenolphthalein-β-D-glucuronide is incubated with the enzyme at pH 5, the reaction being stopped by raising the pH. The liberated phenolphthalein is estimated spectrophotometrically in alkaline solution after removal of denatured protein.

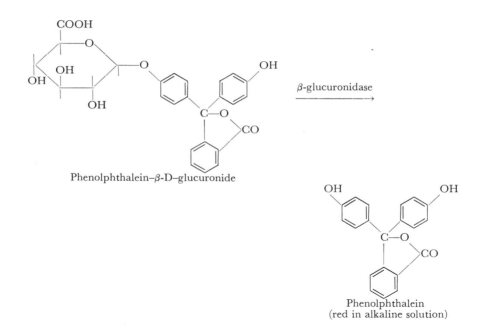

Phenolphthalein–β-D–glucuronide

Phenolphthalein
(red in alkaline solution)

REAGENTS

1. 0.1 M buffer, sodium acetate acetic acid, pH 5.0
2. 0.0025 M phenolphthalein–β-D–glucuronide in acetate buffer.
3. Alkaline solution containing glycine (0.133 M), sodium chloride (0.0067 M) and sodium carbonate (0.083 M), adjusted to pH 10.7 with concentrated NaOH.

PROCEDURE. To 1 ml of phenolphthalein–β-D–glucuronide, add 0.5 ml of buffer. Add 0.5 ml of enzyme, and incubate at 37°C for 10 minutes. Stop the reaction by adding 6 ml of alkaline solution, clarify by filtration or centrifugation at 1000 g for 10 minutes, and read the extinction at 550 mμ using glass cells. For controls, incubate the enzyme and the diluted substrate separately. Then to the diluted substrate, add first the alkaline solution and second the enzyme. A calibration curve of light absorption at 550 mμ against phenolphthalein concentration should be constructed. It is linear.

Method for protein (Lowry *et al.,* 1951)

Tissue concentrations (specific activities) of enzymes are commonly calculated with reference to protein content. The method of Lowry, Rosebrough, Farr, and Randall (1951) commands very wide acceptance.

REAGENTS
1. 2 per cent Na_2CO_3.
2. 1 per cent $CuSO_4.5H_2O$.
3. 2 per cent potassium sodium tartrate.
4. Folin and Ciocalteau's reagent, diluted with an equal volume of water.
5. Copper carbonate reagent (prepare this on the same day it is to be used) : to 0.5 ml of $CuSO_4$ solution, add 0.5 ml of potassium sodium tartrate solution. Then add 50 ml of Na_2CO_3 solution.

PROCEDURE. Digest 0.5 ml of homogenate (or a smaller volume diluted to 0.5 ml) with 0.5 ml of 1N NaOH for 30 minutes at room temperature to dissolve all protein. Add 5 ml of Alkaline Copper Reagent, mix well (e.g., with vortex mixer) and leave at room temperature for 10 minutes. Add 0.5 ml of diluted Folin and Ciocalteau reagent, mix rapidly and thoroughly, and leave at room temperature for 30 minutes. Read the optical density at 750 mμ (or 500 mμ if too strong at 750 mμ). A calibration curve using a pure protein such as crystalline bovine serum albumin should be prepared; it is not strictly linear at either wavelength.

PHAGOCYTOSIS

Quantitation of the uptake of macromolecules into certain fetal membranes has been achieved by some researchers. Lloyd *et al.* (1968) measured the quantity of trypan blue in whole homogenates and subcellular fractions of rat yolk-sac at varying times after maternal injection. Maternally administered horseradish peroxidase has also been used as a marker in similar studies, and it was possible to confirm biochemically the considerable uptake into the yolk-sac as compared with the chorioallantoic placenta (Beck *et al.,* 1967a). Fridhandler and Zipper (1964) incubated rat yolk-sac *in vitro* in a medium containing neonatal rat hemoglobin labeled with C^{14} and were able to detect both high and low molecular weight materials in the tissue. This was evidence both of uptake of hemoglobin and of a degree of proteolysis, presumably by intracellular enzymes. Uptake of Triton WR-1339 into rat yolk-sac has been demonstrated histologically but not yet biochemically. Although the detergent is not easily accessible to chemical assay, the molecule may be radioiodinated (Scanu and Oriente, 1961) and the product is apparently treated by cells exactly like the unlabeled material. Other markers that could probably be used to study phagocytosis in fetal membranes include carbon-14-labeled dextran (Bowers and de Duve, 1967), chromium-51-labeled erythrocytes (Bowers and de Duve, 1967), and radioactively labeled homologous or heterologous proteins (Mego and McQueen, 1965; Gordon and Jacques, 1966). As discussed briefly below (p. 414), recent work in our laboratories utilizes I^{125}-labeled albumin as a marker for endocytosis and digestion in the cultured rat yolk-sac.

As in the histological section above, the methods to be described use trypan blue and horseradish peroxidase.

Trypan blue

As for histological studies, animals should be injected subcutaneously at a dose of 25 to 75 mg per kg of body weight.

In quantitative experiments on dye uptake by fetal membranes, it is important to use a purified dye of known identity. Mislabeling of commercial samples of this group of dyes is not uncommon, and methods have been described (Lloyd and Beck, 1963, 1964) for their positive identification. Authentic samples of trypan blue containing no more than traces of colored impurities should be purified by the methods described below.

Partial purification of commercial trypan blue

PRINCIPLE. Commercial samples of trypan blue regularly contain small quantities of red and purple compounds of unknown identity. More seriously, they contain variable (often large) quantities of salt (Lloyd and Beck, 1963). The following procedure rids the dye of salt and of any insoluble matter, but not of the colored impurities; it also converts trypan blue into its free acid form. Methods have been described (Beck and Lloyd, 1963) for the preparation of chromatographically pure trypan blue, but these are not required if good commercial samples are used (*vide supra*).

PROCEDURE. Dissolve 2 g dye in 500 ml of water at 80°C and stir for 1 hour. Cool, place in a bag of dialysis tubing, and dialyze against running water for 24 hours. Filter through sintered glass and pass the filtrate through a column of cation-exchange resin (e.g., Dowex 50W) in the H$^+$ form. Evaporate the eluate (which should contain no sodium) to dryness at reduced pressure and dry at 50°C *in vacuo* in the presence of P$_2$O$_5$. The dye may be dissolved in water up to at least 1%, but fails to dissolve fully at 2%.

Estimations of trypan blue in tissue

PRINCIPLE. Dyes such as trypan blue are almost wholly protein-bound *in vivo* and published methods for their estimation in tissue (Judah and

Willoughby, 1962; Young, 1964) involve precipitation of the dye-protein complex, its dissociation with base, and estimation of the free dye spectro-photometrically. Since some loss of dye inevitably accompanies these methods, we have found it more convenient and accurate to measure dye in the presence of protein. Tissues or homogenates are digested in N NaOH, and the extinction of the solution measured at 560 mμ, the λ maximum for trypan blue in N NaOH. Digests of control tissues have a slight absorption at this wavelength, directly proportional to their protein content. The presence of tissue components does not alter the absorption spectrum of trypan blue in N NaOH. The method has been used successfully with whole rat yolk-sac and with homogenates and ultracentrifugal fractions of that tissue.

PROCEDURE. Homogenize the sample in N NaOH, and add N NaOH to a known volume. After 20 minutes, read the extinction at 560 mμ using glass cells. Subtract from the value that obtained from similar treatment of a dye-free sample of the same material and of the same protein content. A calibration curve relating light absorption in N NaOH to trypan blue concentration should be prepared, using the same dye sample as employed for the injections.

Dye concentrations may be calculated with reference to protein content. Trypan blue does not interfere with the determination of protein (see p. 405) if the extinctions are read at 750 mμ.

Horseradish peroxidase

A later section (p. 411) describes the doses, vehicles, and routes of administration suggested for investigations of peroxidase uptake by fetal membranes. Tissues to be examined biochemically should be homogenized in ice-cold water shortly before assay.

Assay of peroxidase (After Straus, 1962)

PRINCIPLE. The enzyme catalyzes the oxidation by H_2O_2 of N,N-dimethyl-p-phenylenediamine. A red pigment is produced, and its rate of formation is determined spectrophotometrically. A preincubation of the amine and the tissue extract is employed, as autoxidation of the diamine proceeds rapidly in the period immediately after these are mixed.

As is pointed out by Straus (1958), many tissues contain some innate peroxidase activity; it is therefore necessary to make determinations on tissue from untreated animals.

REAGENTS

1. 0.15% hydrogen peroxide solution
2. 0.1 M phosphate buffer, pH 6.2.
3. N,N-dimethyl-p-phenylenediamine dihydrochloride (Merck, Germany) solution, 0.6 mg/ml, to be prepared immediately before use.

PROCEDURE. To 2.4 ml of buffer add 0.2 ml of N,N-dimethyl-*p*-phenylenedi-amine and 0.2 ml of enzyme source. Maintain at 20°C for one hour. Start reaction by adding 0.2 ml of hydrogen peroxide (time zero), transfer to a 1-cm glass cuvette, and measure the extinction at 520 mμ every 15 seconds for 3 minutes, the control cuvette containing water. The enzyme activity is proportional to the rate of increase in optical density. Samples giving a ΔE_{520} per minute of more than 0.2 should be diluted and reassayed. A blank value that was obtained as above but by replacing the enzyme source with water should be subtracted; this should be in the region of 0.006 per minute.

Jacques (personal communication) has recently pointed out that peroxidase is strongly adsorbed to glass surfaces and recommends the incorporation of 0.01% Triton X-100 into the assay system.

Isolation of lysosomes and phagosomes

Detailed biochemical studies on the properties of lysosomes and phagosomes have been made possible by the development of methods of cell fractionation by differential centrifugation of homogenates. It was this technique that first pointed to the existence of lysosomes, even before they had been identified morphologically (de Duve *et al.*, 1955). It may be used in two distinct ways in the study of lysosomes. The first is to study the distribution pattern among the ultracentrifugal fractions of both lysosomal enzymes and of some phagocytosed marker in order to obtain evidence of its uptake into lysosomes. The second use is as a means of preparing more or less pure samples of lysosomes for investigations into such properties as their digestive capacity, the membrane-bound latency of their enzymes, or the susceptibility of their enzymes to inhibitors (Beck, Lloyd, and Griffiths, 1967b)

PRINCIPLE. Tissue is homogenized so as to cause minimal damage to organelles; this requires the use of isosmotic media and relatively gentle methods of tissue disruption. The suspension is then subjected to successive centrifugations at increasing speeds, each centrifugation yielding a pellet of precipitated material. Since every cytoplasmic particle has its own sedimentation coefficient, largely a reflexion of its size, but also depending on its shape and density, each type of organelle appears principally in one of these pellets. Each of the fractions isolated is then assayed for its content of marker lysosomal enzymes and of protein. De Duve *et al.* (1955) have described a graphical method for the presentation of data from such experiments, in which the fractions are represented on the ordinate scale by relative specific activity (ratio of percent of total enzyme activity to percent of total protein), and cumulatively on the abscissa by the percent of total protein content. Concentrations of phagocytosed markers may be similarly assayed and graphed. Figure 28-17 shows the results of an experiment carried out on the distribution of acid phosphatase and trypan blue in the subcellular fractions of liver from control and dye-injected rats. It illustrates that, although acid phosphatase is found in each fraction, the highest *concentra-*

tion is in the L (light mitochondrial) fraction, presumably because the majority of liver lysosomes sediment there. Recently developed techniques (Baudhuin, Evrard, and Berthet, 1967) for the reliable electron microscopic assessment of fractions make possible morphological confirmation of this latter inference. Figure 28-17 also shows that trypan blue is distributed similarly to acid phosphatase, suggesting an association of the dye with lysosomes, and that the presence of trypan blue in lysosomes does not greatly affect their sedimentation properties.

As far as we know, the only applications of differential centrifugation methods to studies of fetal membranes have been our own investigations on rat yolk-sac (Lloyd *et al.*, 1968), a recent study by Schultz (1969) on rat yolk-sac and chorio-allantoic placenta, and a publication by Contractor (1969). Our method was based closely on the now classic fractionation scheme devised for liver by Appelmans, Wattiaux, and de Duve (1955). Although this method will now be described, it must be emphasized strongly that the details of, for example, homogenization and suitable centrifugal forces which follow almost infinite variation, depend on the nature of the tissue and the aims of the investigation. It must not be asumed that a fractionation scheme devised for one tissue will yield similar results with another. This is particularly true of lysosomes, which exhibit great variability of size and stability in different cell types. The reader is referred to comprehensive reviews by de Duve (1965, 1967) for fuller discussion of these points.

Procedures for differential centrifugation of rat yolk-sac

Wistar rats, 15.5 days pregnant, were killed by a blow on the head and visceral yolk-sacs dissected free in ice-cold isotonic saline. The yolk-sacs were blotted dry and homogenized in ice-cold 0.25 M sucrose in an approximate volume of 1.3 ml per yolk-sac. A Teflon-on-glass Potter Elvejhem instrument (Tri-R Instruments, Jamaica, N.Y.) with a clearance between pestle and tube of 0.0013 cm was used, yolk-sacs being forced past the rotating pestle (2,500 rpm) four times in 30 seconds.

The resulting homogenate was separated into four particulate fractions and a final supernatant, these being designated N (nuclear), M (heavy mitochondrial), L (light mitochondrial), P (microsomal), and S (supernatant), following the usage of Gianetto and de Duve (1955). All operations were performed near 0°C, using ice-cold solutions and equipment. The first centrifugation was at 1000 g for 10 minutes (M.S.E. Mistral 6L, rotor No. 6875). The supernatant was carefully decanted, and the pellet suspended in a known volume of 0.25 M sucrose to constitute the N fraction. The supernatant was then centrifuged at 3300 g for 10 minutes (M.S.E. High Speed 18, rotor No. 69181) to yield a pellet (M fraction) and a supernatant. The L, P, and S fractions were prepared from this supernatant in an analagous manner, by successive centrifugations at 16,300 g for 20 minutes (M.S.E. High Speed 18, rotor No. 69181) and 100,000 g for 30 minutes (Beckman L2 preparative ultracentrifuge, rotor No. SW39L). Each pellet was resuspended in 0.25 M sucrose by one pass of the homogenizer.

410

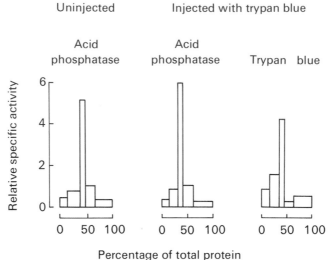

FIGURE 28-17

*Distribution of acid phosphatase and trypan blue in sub-
cellular fractions of rat liver. Male Wistar rats were injected
subcutaneously with trypan blue (100 mg/kg); after 16 hours
their livers were homogenized in 0.25 M sucrose, and N, M,
L, P, and S fractions prepared by successive centrifugations at
10,000 g, 33,000 g, 326,000 g, and 3,000,000 g. Each fraction was
analyzed for acid phosphatase (Gianetto and de Duve, 1955)
trypan blue (vide supra), and protein (Lowry et al.,
1951). Results are presented graphically in the order
of their isolation: N, M, L, P, and S, as described by de Duve
et al. (1955). (Reproduced, with permission, from Lloyd
et al., 1968.)*

DISCUSSION. In our own experiments (Lloyd *et al.*, 1968) carried out on
yolk-sacs from both trypan-blue-treated and control rats, the fractions isolated as above were subjected to ten cycles of freezing-and-thawing (to
rupture lysosomal membranes and release their contained enzymes) and
assayed for acid phosphatase, trypan blue, and protein. In control animals
the distribution of acid phosphatase differed from that in liver (see Fig.
28-17), in that the N and M fractions contained proportionately more of
the enzyme, although the highest specific activity still appeared in the L
fraction. This was not unexpected in view of the large size of many yolk-sac
lysosomes, as revealed by electron microscopy (see Figs. 28-3, 6, 7, 8, 9, 10,
12, 14). In rats which had received subcutaneous injections of trypan blue,
the acid phosphatase distribution did not differ appreciably from that in
control animals, but the dye was concentrated heavily in the N and M
fractions. Again this is consistent with the electron microscopic data, which
indicate both that there is a great diversity of lysosome size in the yolk-sac
and also that the large lysosomes (heterolysosomes) are those which contain
concentrated dye deposits.

These results emphasize the statements made earlier that a fractionation
scheme devised for one tissue cannot be expected to yield identical results

with another. In a tissue where lysosomes exist in a variety of sizes (such as the primary lysosomes, heterolysosomes, and telolysosomes in rat yolk-sac) it will be found that different populations sediment in different fractions upon centrifugation. A clear-cut distribution of acid hydrolases into a single ultracentrifugal fraction is therefore unlikely to be achieved in highly phagocytic tissues. On the other hand, it is often possible to isolate discrete classes of lysosome. For example, Maunsbach (1966), by means of dissection followed by homogenization and differential centrifugation at suitable speeds, was able to prepare purified fractions of the large autofluorescent granules (heterolysosomes) of rat kidney cortex.

EVIDENCE OF DIGESTIVE ABILITY

ESTIMATION OF PROTEIN DIGESTION BY SOME FETAL MEMBRANES *In Vivo*

A comprehensive study of embryotrophic nutrition must include not only the demonstration that a fetal membrane can take up macromolecules and expose them to intracellular hydrolytic enzymes, but also an attempt to prove that measurable intracellular digestion does in fact take place. This problem can be approached by tracing the disappearance of marker proteins from serially removed conceptuses in polyembryonic species.

PRINCIPLE. An easily measurable marker protein, such as horseradish peroxidase, is injected into experimental animals, and individual conceptuses are removed at timed intervals. Various fetal membranes are assayed to determine whether there has been uptake of the marker; in certain tissues there may have been. A sequential fall with time in the specific activity of the marker in such membranes is probably an indication of intracellular digestion, although the possibility of subsequent excretion cannot be excluded.

METHOD. Horseradish peroxidase (Sigma, type II) is made up to an 0.8% solution in rat serum (the latter prepared by centrifugation of clotted rat blood). At time 0, 12 mg per kg are injected into the exposed jugular vein of an 18.5-day pregnant rat, lightly anesthetized with ether; if time permits, the animal is allowed to gain consciousness. After a measured period the rat is reanaesthetized, the abdomen opened, and the uterus exposed. Beginning at the ovarian end of one horn, the thin uterine wall is cut antimesometrially with a fine pair of scissors, care being taken not to damage the visceral yolk-sac which clothes the underlying conceptus. A relatively small incision only is required, and uterine muscular tone helps to express the fetus with its fetal membranes intact. Two conceptuses are removed each time, the uterus is replaced without further manipulation, and the abdomen closed temporarily with Michel clips. The abdominal operation may be repeated at regular intervals over a period of 48 hours or more. The various fetal membranes from freshly removed conceptuses are separated in isotonic saline under a dissecting microscope and assayed for peroxidase by the method described on p. 407. In the rat, the visceral layer of the yolk-sac takes up large quantities of injected peroxidase, and the specific activity of the enzyme in yolk-sac homogenates can be shown to fall steadily with time.

Peroxidase activity in uninjected rat tissue is negligible compared with that after treatment (Beck *et al.,* 1967a), so that it is not necessary to remove any conceptuses before the initial injection; with some marker materials, however, it is conceivable that control conceptuses would have to be removed at time 0. The operative procedure described may in theory be performed at any stage of gestation, but in the rat it is found in practice that, before day 16 of gestation, the uterine manipulation required causes all the embryos to die when the first batch is removed. It is therefore sometimes necessary to inject many pregnant animals on a weight basis, killing them in batches of 2 or 3 at varying times after treatment, and assaying the membranes individually (to gain some measure of the scatter around the mean). It is then usually possible to construct a meaningful curve of peroxidase fall with time.

Estimation of Protein Digestion by Some Fetal Membranes *In Vitro*

PRINCIPLE. To overcome the limitations imposed by "open-ended" *in vivo* methods in clarifying the fate of ingested materials, *in vitro* experiments can be devised in which the fetal membranes responsible for embryotrophic nutrition are maintained in organ culture. Such closed systems are readily available for both histochemical study and biochemical analysis. Rat visceral yolk-sac can be cultured by an adaptation of the Trowell (1954) raft technique.

METHOD. The culture chamber (see Fig. 28-18) is a large dessicator containing moistened cotton wool (W), and having a gas inlet and outlet (G_1G_2). It is gassed every 24 hours with 5% CO_2 in air. Pieces of yolk-sac are cultured in 5 cm pyrex Petri dishes (D), in which (R), an easily constructed tantalum-wire gauze raft (Ethicon surgical tantalum gauze: Code TM-53), is placed on a filter-paper base. A square (P) of lens paper (Green's C105), washed four times in ether, and thoroughly rinsed in distilled water, is placed on the raft, and the whole sterilized by dry heat at 160°C for one hour. A measured volume of sterile culture medium—TC 199 (Burroughs Wellcome) containing antibiotics—is placed in the Petri dish to reach and moisten, but not to submerge, the lens paper. Up to ten Petri dishes may be accommodated in a single culture chamber; incubation is carried out at 37°C.

Rats, 17 days pregnant, are injected intraperitoneally with 25 mg/kg of horseradish peroxidase (Sigma type II) in saline. They are killed six hours later, and the conceptuses with their yolk-sacs are removed and manipulated in a sterile, balanced salt solution or in TC-199. Explants (E) are prepared

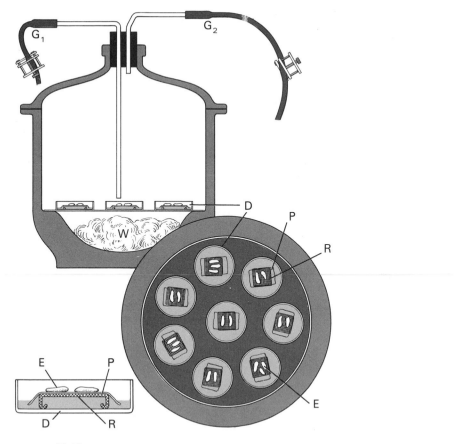

FIGURE 28-18
Diagram showing equipment required for culture of rat
visceral yolk-sac on a defined liquid medium. For annotations see text.

from uniform-sized pieces of the *villous* part of the yolk-sac and placed on the lens paper in the culture dishes. This is done under a dissecting microscope, care being taken to ensure that the villous surface of the yolk-sac lies uppermost on the rafts.

Ten minutes of incubation at 37°C are allowed for stabilization, and a number of the explants together with their corresponding media are then frozen and stored at −10°C for biochemical assay; this is taken as time zero. The remaining material is incubated for 24 hours, and small pieces of each explant are then fixed, the rest, together with the medium, being frozen and stored for subsequent assay. Explants may be weighed on a torsion balance, after harvesting; the difference between samples of material incubated for time zero and those grown for some hours thus provides additional confirmation for any fall in protein content after incubation as determined by the Lowry *et al.* (1951) method. Explants are homogenized in 2 ml of distilled water with a "Virtis" (Gardiner, New York) homogenizer, and both homogenates and media are assayed for peroxidase and total protein (for methods, see pp. 405, 407).

Between cultures the dessicator must be thoroughly washed, and the Petri dishes and tantalum rafts cleaned in Decon 75 (Medical Pharmaceutical Ltd., Shoreham-by-Sea, Sussex) for 24 hours, followed by thorough rinsing in several changes of distilled water.

Histological preparations of cultured material usually show good survival, and intracellular localization of horseradish peroxidase is clearly demonstrated histochemically. Biochemical results indicate about 45% net loss of peroxidase from the whole system after 24 hours of incubation. At the doses injected, very little peroxidase is detected in the medium and, since the enzyme is stable when incubated in TC199 at 37°C for 24 hours, it must have been degraded within the cells of the explant.

With higher intraperitoneal or intravenous doses of peroxidase, much enzyme can be detected in the culture medium after 24 hours of incubation. The significance of this observation is not clear; it may indicate that the explant is capable of exocytosis, but it may equally be the result of diffusion of adsorbed enzyme or a reflection of the small amount of cell death that is inevitable in such an organ-culture system.

Rats, 17 to 20 days pregnant, are used for culture because before day 17 there is insufficient villous yolk-sac to give meaningful readings on assay. The nonvillous part of the yolk-sac does not survive well in culture.

GENERAL COMMENTS

Continuing work in our laboratories has indicated that radioactively labelled proteins are in many respects even more useful than horseradish peroxidase as markers for protein degradation (Williams, Lloyd, and Beck, 1969 and in preparation). Their use overcomes the criticism that the disappearance of horseradish peroxidase might be due to its intracellular denaturation rather than to digestion. Furthermore, calf serum (which has some innate peroxidatic activity) can be added to the culture medium.

Other methods of organ culture are being developed which measure the rate of uptake of macromolecules added to the culture medium rather than injected into the pregnant animal. This allows the investigation of another parameter of embryotrophic nutrition, namely that of the power of endocytosis possessed by a fetal membrane.

Finally it must be admitted that many fetal membranes do not survive well in organ culture; histological and electron microscopic examination of cultured material must be carried out as a control (Beck, Parry, and Lloyd, in 1970).

BIBLIOGRAPHY

Anderson, J. W. (1959) The placental barrier to gammaglobulins in the rat. Anat. Record, 104:403–429.
Appelmans, F., R. Wattiaux, and C. de Duve (1955) Tissue fractionation studies.

V. The association of acid phosphatase with a special class of cytoplasmic granules in rat liver. Biochem. J., 59:438–445.

Barka, T., and P. J. Anderson (1962) Histochemical methods for acid phosphatase using hexazonium pararosanilin as coupler. J. Histochem. Cytochem., 10:741–753.

——— (1963) Textbook of histochemistry: Theory, practice and bibliography. Harper and Row, New York and London.

Barrett, A. J. (1967) Lysosomal acid proteinase of rabbit liver. Biochem. J., 104:601–608.

——— (1969) Properties of lysosomal enzymes. In: Lysosomes in Biology and Pathology, vol. II. Ed. by J. T. Dingle and H. B. Fell. North Holland, Amsterdam, pp. 245–312.

Baudhuin, P., P. Evrard, and J. Berthet (1967) Electron microscopic examination of subcellular fractions. I: The preparation of representative samples from suspensions of particles. J. Cell. Biol., 32:181–191.

Beck, F., and J. B. Lloyd (1963) The preparation and teratogenic properties of pure trypan blue and its common contaminants. J. Embryol. Exp. Morphol., 11:175–184.

——— (1969 Histochemistry and electromicroscopy of lysosomes. In: Lysosomes in Biology and Pathology, vol. II, Ed. by J. T. Dingle, and H. B. Fell. North Holland, Netherlands.

Beck, F., J. B. Lloyd, and A. Griffiths (1967a) A histochemical and biochemical study of some aspects of placental function in the rat using maternal injection of horseradish peroxidase. J. Anat., 101:461–478.

——— (1967b) Lysosomal enzyme inhibition by trypan blue: a theory of teratogenesis. Science (New York), 157:1180–1182.

Beck, F., L. M. Parry, and J. B. Lloyd (1970) In vitro studies on the degradation of horseradish peroxidase by rat yolk sac. Cytobiologie, 1:331–342.

Bowers, W. E., and C. de Duve (1967) Lysosomes in lymphoid tissue. III. Influence of various treatments of the animals on the distribution of acid hydrolases. J. Cell Biol., 32:349–364.

Carruthers, C., D. L. Woemley, A. Baumler, and K. Lilga (1960) The distribution of cytochrome oxidase, glucose-6-phosphatase and esterase in the fractions of liver cells prepared from glycerol homogenates. Arch. Biochem. Biophys., 87:266–272.

Caulfield, J. B. (1957) Effects of varying the vehicle for OsO_4 in tissue fixation. J. Biophys. Biochem. Cytol., 3:827–829.

Contractor, S. F. (1969) Lysosomes in the human placenta. Nature, 223:1275.

Daems, W. Th., E. Wisse, and P. Brederoo (1969) Electron microscopy of the vacuolar system. In: Lysosomes in Biology and Pathology, vol. I. Ed. by J. T. Dingle and H. B. Fell. North Holland, Amsterdam, pp. 64–112.

de Duve, C. (1965) The separation and characterization of subcellular particles. In: The Harvey Lectures, series 59, pp. 49–87. Academic Press, New York and London.

——— (1967) General Principles. In: Enzyme Cytology. Ed. by D. B. Roodyn. Academic Press, New York and London, pp. 1–26.

———, B. C. Pressman, R. Gianetto, R. Wattiaux, and F. Applemans (1955) Tissue fractionation studies. VI. Intracellular distribution patterns of enzymes in rat-liver tissue. Biochem. J., 60:604–617.

——— and R. Wattiaux (1966) Functions of lysosomes. Ann. Rev. Physiol., 28:435–492.

Fridhandler, L., and J. Zipper (1964) Studies in vitro of rat yolk-sac biosynthetic activities, respiration, and permeability to hemoglobin. Biochim. Biophys. Acta., 93:526–532.

Frost, J. L., and D. Brandes (1967) Nonspecific esterases in rat prostatic epithelial cells. J. Histochem. Cytochem., 15:589–595.

Gianetto, R., and C. de Duve (1955) Tissue fractionation studies. IV. Comparative study of the binding of acid phosphatase, β-glucuronidase, and cathepsin by rat-liver particles. Biochem. J., 59:433–445.

Goldfischer, S. (1965) The cytochemical demonstration of lysosomal aryl sulphatase activity by light and electron microscopy. J. Histochem. Cytochem., 13:520–523.

Goldmann, E. E. (1909) Die äussere und innere Sekretion des gesunden Organismus im Lichte der vitalen Färbung. Beitr. Klin. Chir., 64:192.

———— (1912) Die äussere und innere Sekretion des gesunden und des kranken Organismus. Beitr. Klin. Chir., 78:1.

Gomori, G. (1952) Microscopic histochemistry: Principles and practice. University of Chicago Press, Chicago.

Gordon, A. H., and P. Jacques (1966) Distribution of injected ^{131}I or ^{125}I labelled homologous plasma proteins among subcellular particles of rat liver. In: Labelled proteins in tracer studies. Ed. by L. Donato, G. Milhaud, and J. Sirchis, Euratom, pp. 127–132.

Graham, R. C., and M. J. Karnovsky (1966) The early stages of injected horseradish peroxidase in the proximal tubules of mouse kidney: Ultrastructural cytochemistry by a new method. J. Histochem. Cytochem., 14:291–301.

Hayashi, M. (1965) Histochemical demonstration of N-acetyl-beta-glucosaminidase employing naphthol AS-BI N-acetyl-beta-glucosaminide as substrate. J. Histochem. Cytochem., 13:355–360.

————, Y. Nakajima, and W. H. Fishman (1964) The cytologic demonstration of β-glucuronidase employing naphthol AS-BI glucuronide and hexazonium pararosanilin; a preliminary report. J. Histochem. Cytochem., 12:293–297.

Heidenhain, M. (1915) Über die Mallorysche Bindegewebsfärbung mit Karmin und Azokarmin als Vorfarben. Z. Wiss. Mikrosk., 32:361–372.

Holt, S. J. (1958) Indigogenic staining methods for esterases. In: General Cytochemical Methods, vol. 1. Ed. J. F. Danielli. Academic Press, New York.

———— (1963) Some observations on the occurrence and nature of esterases in lysosomes. In: Ciba Foundation Symposium on Lysosomes. Ed. by A. V. S. de Reuck and M. P. Cameron. Churchill, London.

Judah, J. D., and D. A. Willoughby (1962) A quantitative method for the study of capillary permeability: extraction and determination of trypan blue in tissues. J. Pathol. Bact., 83:567–572.

Karnovsky, M. J. (1965a) A formaldehyde-glutaraldehyde fixative of high osmolarity for use in electron microscopy. J. Cell Biol., 27:137A–138A.

———— (1965b) Vesicular transport of exogenous peroxidase across capillary endothelium into the T-system of muscle. J. Cell Biol., 27:49A–50A.

Krzyzowska-Gruca, St., and T. H. Schiebler (1967) Untersuchungen am Dottersackepithel der Ratte. Z. Zellforsch. Mikrosk. Anat., 79:157–171.

Larsen, J. F., and J. Davies (1962) The paraplacental chorion and accessory foetal membranes in the rabbit: Histology and electron microscopy. Anat. Record, 143:27–45.

Lloyd, J. B., and F. Beck (1963) An evaluation of acid disazo dyes by chloride determination and paper chromatography. Stain Technol., 38:165–171.

———— (1964) The identification of some acid disazo dyes by paper electrophoresis of their reduction products. Stain Technol., 39:7–12.

————, A. Griffiths, and L. M. Parry (1968) The mechanism of action of acid bisazo dyes. In: Biological Council Symposium, Interaction of Drugs and Subcellular Components in Animal Cells. Ed. by P. N. Campbell. Churchill, London, pp. 171–200.

Lowry, O. H., N. J. Rosebrough, A. L. Farr, and R. J. Randall (1951) Protein measurement with the Folin phenol reagent. J. Biol. Chem., 193:265–273.

Luft, J. H. (1956) Permanganate: A new fixative for electron microscopy. J. Biophys. Biochem. Cytol., 2:799–801.

McDonald, J. K., P. X. Callahan, S. Ellis, and R. E. Smith. Polypeptide degradation by dipeptidyl aminopeptidase I (cathepsin C). In: Tissue Proteinases. Ed. by A. J. Barrett and J. T. Dingle. North Holland, Amsterdam. (In press.)

Maggi, V. (1969) Lysosomal and non-lysosomal localization of acid hydrolases in animal cells. Biochem. J., 111:25–26.

Markert, C. L., and R. L. Hunter (1959) The distribution of esterases in mouse tissues. J. Histochem. Cytochem., 7:42–49.

Maunsbach, A. B. (1966) Isolation and purification of acid phosphatase-containing autofluorescent granules from homogenates of rat kidney cortex. J. Ultrastruct. Res., 16:13–34.

Mayersbach, H. (1958) Zur Frage des Proteinüberganges von der Mutter zum Foeten. I: Befunde am Ratten am Ende der Schwangerschaft. Z. Zellforsch. Mikrosk. Anat., 48:479–504.

Mego, J. L., and J. D. McQueen (1965) The uptake and degradation of injected labelled proteins by mouse-liver particles. Biochim. Biophys. Acta, 100:136–143.

Reynolds, E. S. (1963) The use of lead citrate at high pH as an electronopaque stain in electron microscopy. J. Cell Biol., 17:208–212.

Rosenbaum, R. M., and C. I. Rolon (1962) Species variability and the substrate specificity of intracellular phosphatases: a comparison of the lead salt and azo dye methods. Histochemie, 3:1–16.

Roy, A. B. (1960) The synthesis and hydrolysis of sulphate esters. Adv. Enzymol., 22:205–235.

Sabatini, D. D., K. Bensch, and R. J. Barnett (1963) The preservation of cellular ultrastructure and enzymatic activity by aldehyde fixation. J. Cell Biol., 17:19–58.

Sams, A., and F. M. R. Davies (1967) Commercial varieties of nuclear-fast red; their behaviour in staining after autoradiography. Stain Technol., 42:269–276.

Scanu, A., and P. Oriente (1961) Triton hyperlipemia in dogs. I. *In vitro* effects of the detergent on serum lipoproteins and chylomicrons. J. Exp. Med., 113:735–757.

Schultz, R. L. (1969) Effects of ovariectomy and hypervitaminosis-A on lysosomes of the rat conceptus. Teratology, 2:283–296.

———— and P. W. Schultz (1966) Morphological and histochemical changes in the rat conceptus following administration of a non-ionic detergent. Proc. Soc. Exp. Biol. Med., 122:874–877.

Shibko, S., and A. L. Tappel (1964) Distribution of esterases in rat liver. Arch. Biochem. Biophys., 106:259–266.

Shnitka, T. K., and A. M. Seligman (1961) Role of esteratic inhibition on localization of esterase and simultaneous cytochemical demonstration of inhibitor-sensitive and resistant enzyme species. J. Histochem. Cytochem., 9:504–527.

Straus, W. (1958) Colorimetric analysis with N,N-Dimethyl-p-phenylenediamine of the uptake of intravenously injected horseradish peroxidase by various tissues of the rat. J. Biophys. Biochem. Cytol., 4:541–550.

———— (1962) Colorimetric investigation of the uptake of an intravenously injected protein (horseradish peroxidase) by rat kidney and effect of competition by egg white. J. Cell Biol., 12:231–246.

———— (1964a) Factors affecting the state of injected horseradish peroxidase in animal tissues and procedures for the study of phagosomes and phago-lysosomes. J. Histochem. Cytochem., 12:470–480.

———— (1964b) Factors affecting the cytochemical reaction of peroxidase with benzidine and the stability of the blue reaction product. J. Histochem. Cytochem., 12:462–469.

———— (1964c) Cytochemical observations on the relationship between lysosomes and phagosomes in kidney and liver by combined staining for acid phosphatase and intravenously injected horseradish peroxidase. J. Cell Biol., 20:497–507.

Torriani, A. (1960) Influence of inorganic phosphate in the formation of phosphatases by *Escherichia coli*. Biochim. Biophys. Acta, 38:460–469.

Trowell, O. A. (1954) A modified technique for organ culture *in vitro*. Exp. Cell Res., 6:246–248.

Underhay, E., S. J. Holt, H. Beaufay, and C. de Duve (1956) Intracellular localization of esterase in rat liver. J. Biophys. Biochem. Cytol., 2:635–637.

Wachstein, M., E. Meisel, and C. Falcon (1961) Histochemistry of thiolacetic acid esterase: a comparison with nonspecific esterase with special regard to the effect of fixatives and inhibitors on intracellular localization. J. Histochem. Cytochem., 9:325–339.

Wattiaux, R., M. Wibo, and P. Baudhuin (1963) Influence of the injection of Triton WR-1339 on the properties of rat-liver lysosomes. In: Ciba Foundation Symposium on Lysosomes. Ed. by A. V. S. de Reuck and M. P. Cameron. Churchill, London, pp. 176–196.

Williams, K. E., J. B. Lloyd, and F. Beck (1959) The digestive capacity of rat yolk sac in organ culture. Biochem. J., 115:66.

Wislocki, G. B. (1920) Experimental studies on fetal absorption. I. The vitally stained fetus. II. The behavior of the fetal membranes and placenta of the cat toward colloidal dyes injected into the maternal blood stream. Contr. Embryol., 11:45–60.

———— (1921) Further experimental studies on fetal absorption. III. The behavior of the fetal membranes and placenta of the guinea pig toward trypan blue injected into the maternal blood stream. IV. The behavior of the placenta and fetal membranes of the rabbit toward trypan blue injected into the blood stream. Contr. Embryol., 13:89–101.

Young, D. A. B. (1964) A method for extraction of Evans Blue from plasma and tissues. Proc. Soc. Exp. Biol. Med., 116:220–222.

29

Methods for the Demonstration of Placental Circulation

ELIZABETH M. RAMSEY

Carnegie Institution of Washington
Department of Embryology
115 West University Parkway
Baltimore, Maryland 21210

The selection of an appropriate experimental animal is more important in the field of placental research than in many other areas of investigation. Basic biologic information can be gleaned from studies in any available placental animal, but morphological differences in placental type are so extensive (Amoroso, 1952; Wimsatt, 1962) that, if the data obtained are to be applied to a given species, the studies must be made in a related form.

Thus, if it is planned to set up an experimental model for the human being, the indiscriminate selection of any other species with a hemochorial placenta like that of man is not enough. The animal must also have a uterus unicornis. The uterus must have a thick muscular wall that is supplied by coiled endometrial arteries and these must undergo cyclic changes in response to hormonal stimulation, i.e., it must be a menstruating animal. This requirement rules out, among others, the guinea pig, even though it has a hemochorial placenta and the advantages of a short gestation period, ready availability, relatively low cost, and ease of handling. The choice must be a primate, and if it is a monkey, then it must be an Old World monkey because New World monkeys do not have a typical menstrual cycle. The higher primates, chimpanzees, orangutans, etc., are even closer to man than monkeys, but are hard to obtain and to handle and are very expensive. The smaller and quieter baboon (*Papio papio*) is currently gaining popularity at several of the primate centers, and the larger laboratories are experimenting with breeding and handling procedures. Cynomolgous (*Macaca irus*), the crab-eating monkey is being increasingly used in reproductive studies because it is more gentle than the rhesus, but its smallness is a disadvantage in some studies. Thus, the rhesus monkey (*Macaca mulatta*) remains the favorite, and the cumulative experience of many research colonies provides the valuable assistance of a large body of information on handling and care.

The foregoing extended analysis of the problems involved in the choice of an experimental model for the human is given as an illustration of the considerations which must be made when studies with any placental type are being planned.

Another factor which must be investigated before experimentation begins is the condition of the animal's health. Investigators are becoming increasingly aware of the importance of this. The ideal is to have a breeding colony of one's own, but that is not always possible. The alternative is to obtain animals from commercial sources. Unfortunately such animals frequently suffer from travel stress, malnutrition, or infection. All of these stresses may significantly affect circulation, especially during pregnancy. Purchased animals should therefore be carefully examined and given as long a period of acclimatization as is practical. Seriously debilitated or infected animals should be rigorously excluded because they do not produce reliable results. There is increasing appreciation of the need for baseline studies of both maternal and fetal blood to permit evaluation of experimental data on circulatory conditions. These should include the determination of blood pH, the degree of oxygenation, and the electrolyte concentration since these factors are influenced by the animal's health.

The effects of anesthesia, of the type and duration of surgical manipulation, of prolonged postural immobilization, of psychic factors in conscious animals, and of similar parameters of the experimental procedure must also be borne in mind. It may be emphasized that such effects should be considered not only in planning one's own experiments but also in evaluating the work of others. Recent work indicates the necessity of allowing a prolonged stabilization period after surgery and before making observations which will form valid "controls."

CIRCULATORY PATHWAYS

My experience has been limited to studies with monkeys and the specific information in the following descriptions is based largely upon that animal. All of the techniques, however, employ simple, general principles and procedures which can be adapted for use in other species. Indeed most of them are already in widespread use and some were originated in nonprimate animals and were secondarily applied to the monkey.

The earliest studies of placental vasculature consisted of painstaking gross dissections. More recently, emphasis has shifted to the intra- or postvital injection of colored material into the vascular channels. Such techniques, however, have the inherent disadvantage of establishing nonphysiological conditions and are likely to produce artifacts of a number of sorts. Some investigators have, therefore, chosen methods of study that do not place primary reliance upon injection, using it only as an accessory device. In the field of placental vasculature the most notable work of this sort has been done by Crawford (1955, 1962) whose study of the fetal placental circulation was carried out by a delicate method of progressive digestion of the tissue of the villi. Once the vascular tree is freed of surrounding tissue he injects with colored substances as desired. This method has limited applicability and requires great skill and experience. Most studies use the injection technique and to the many aspects of this basic procedure attention will now be directed.

INJECTION OF THE VASCULAR BED

Injection of the maternal placenta may be carried out via the aorta or a uterine artery in an anesthetized animal or, as is necessary in human studies, in specimens removed at operation or autopsy (Ramsey, 1949, 1954, 1956). The living animal is preferable, especially if its own blood pressure can be used as the propelling force. If that is not possible, normal blood pressure should be reproduced mechanically, with a manometer or recording unit interposed in the delivery system, and arrangements should be made for pulsatile flow (Rossman and Bartelmez, 1957; Panigel, 1962). The amount of injection material used should not produce an appreciable change in the animal's blood volume. In terminal experiments it is often convenient to make a small incision in the vena cava to prevent injection overloading and to permit free flow. Similarly, when injecting a fetus or the fetal placenta via an umbilical vessel, the drainage vessel may be opened.

If there is delay between the removal of surgical specimens and their injection, the vascular bed should be flushed to remove stagnant blood. An effective solution for the purpose is 1% sodium nitrite in isotonic saline solution with 1% histamine. Viscosity correction may be achieved by the addition of gum acacia or similar substances (Koenig *et al.*, 1945). The rinse is delivered intraarterially and blood pressure stability precautions should be observed as with the injection itself. The rinse should be run through the system until venous return is free and reasonably clear. Extensive clotting, of course, destroys the integrity of the final injection pattern. Air bubbles or other artifacts introduced during the injection have similar deleterious effects. Panigel (1962) recommends the use of a pump in specimens rendered resistant to gravity-flow by vascular spasm or the smallness of the vessels (young placentas of small animals). He illustrates some useful varieties of instrument. The pump may be used for introduction of the injection mass as well as the rinse. Volume precautions are not necessary with extirpated uteri when it is desired to inject the entire vascular bed. Large volumes are required in these circumstances, but it must be recognized that such injections demonstrate the total pattern rather than normal states of physiological patency, which may fluctuate from moment to moment during life (Martin *et al.*, 1964).

Thought should be given to the amount of extrauterine pelvic vasculature to be injected. If, in the species selected, such channels have never been adequately mapped, it is best to inject and analyze them in a preliminary study because arguments by analogy from species to species are not necessarily valid. Injection material may be wasted or lost if unsuspected anastomoses exist, for example, and resultant filling defects in the placenta will be frustrating at least and misleading at worst.

A similar caution with respect to mapping the vasculature of the uterine wall hardly seems necessary in the light of the intimate connection between these channels and the placenta, but it may be helpful to point out the necessity of studying veins, as well as arteries, even though they tend to be less dramatically involved in placentation.

When injecting the vascular bed of the uterus, whether it be pregnant or not, the investigator is confronted with a special complication; namely, the effect of myometrial contractions. Uterine activity patterns vary so extensively from animal to animal that they must be investigated separately for each. Furthermore, there are important differences in uterine and placental pattern and flow in the states of uterine relaxation and contraction (Corner *et al.,* 1963; Scoggin *et al.,* 1963). Results of injection should be analyzed with this variation in mind, and studies of circulatory dynamics should include continuous monitoring of intrauterine pressure.

Attention so far has been concentrated upon the maternal component of the placenta. Access to the fetus *in vivo* is the great problem. Fetuses of some animals, such as sheep and goats (Huggett, 1955), may be delivered with the membranes intact and may be maintained outside the maternal body for prolonged periods of observation. In such preparations injection may be made readily through the umbilical vessels or into a vessel of the fetal body. In other animals in which such a procedure is not possible—or for which suitable techniques have not yet been developed—hysterotomy can be performed without delivery of the fetus and an umbilical vessel can be injected *in situ.* Alternatively, a fetal limb may be delivered through the hysterotomy incision and injection made into the fetal femoral artery (Martin *et al.,* 1966). If the amnion is incised, a purse-string suture is placed around the opening to prevent escape of amniotic fluid and to preserve normal intrauterine pressure relations. Lost amniotic fluid can be replaced, in an acute experiment, with an equal amount of sterile physiologic saline solution. Umbilical or fetal femoral injection may be made at parturition or cesarean section by open and uncomplicated insertion of a needle or catheter, but pressure relations are no longer those prevailing during pregnancy. As with maternal injections, pressure must be carefully regulated in the fetal injections to insure the filling of all physiologically patent channels without overdistention. The fetus' own pressure will be the standard used.

In animals with a bidiscoid placenta (such as the rhesus monkey) fetal placental injection may also be made, *in vivo,* into one of the interplacental vessels connecting the two discs (Ramsey *et al.,* 1967). These vessels, which course within the chorionic plate, may be localized by transillumination and access to them is gained by extraamniotic hysterotomy. The fetal femoral route provides the more thorough and physiologic injection since the fetus' own blood pressure propels the injection mass, but the interplacental vessel route has the advantage that the injection material is not diluted to the same extent by fetal blood. The latter consideration is of particular importance when the injection mass is a radiopaque medium which is to be visualized by radioangiography (see below).

If both maternal and fetal injections are carried out at the same time, the full picture of placental vasculature is obtained as well as the relationship of the two components to one another (Ramsey *et al.*, 1967).

INJECTION MATERIALS

A wide assortment of substances has been used, with varying degrees of success, to demonstrate vascular pattern. No attempt will be made to compile an inclusive list of these agents, rather they will be considered by classes with major emphasis upon the substances that are most commonly used because they have been found most satisfactory for specific purposes. In selecting agents for the demonstration of placental vasculature, their ability to evoke constriction of the remarkably sensitive vascular bed must be considered. Such action is difficult to avoid and most undesirable as it seriously interferes with representative filling.

Fluid substances

The earliest injection materials employed in modern vascular studies were soluble dyes such as methylene, trypan, and Prussian blue. Although useful in studies of transparent capillary beds, like that in the mesentery, these solutions are unsatisfactory for demonstration of the vasculature of a parenchymatous organ like the uterus. They may be used successfully, however, to demonstrate the finer vasculature of the chorionic villi (Bøe, 1953). In general, deeply colored, particulate substances such as India ink and stained starch granules are preferable because they do not perfuse across the vessel wall. Ordinary commercial artists' India ink (Higgins) diluted 1:3 or 1:4 with isotonic saline and a few drops of concentrated ammonia is usually employed (Ramsey, 1949). The camphor that is used in India ink to maintain suspension of the lamp black makes the ink highly toxic, its intravascular injection is therefore not compatible with life. A nonirritating, nontoxic colloidal solution of mercuric sulphide (Hille) was employed by Daron (1936) who wished to make *in vivo* studies of the functional vascular bed. A concentration of 8% produced excellent results.

Suspensions of ordinary cornstarch can be prepared with granules of various sizes. Bartelmez (Rossman and Bartelmez, 1957), wishing to prevent the injection mass from entering the capillaries of the endometrium, incorporated starch granules averaging 10μ in diameter in a carmine gelatin medium (see the following section). He obtained full injection through the precapillary arterioles, and brought the terminals of the arterial tree clearly into view.

An unusual injection material was used in one of the monkey studies at the Carnegie Institution; namely, the animal's own blood colored with picric acid (Ramsey, 1958). For unrelated purposes it was desired to fix the living pregnant uterus *in situ*. To achieve this the monkey was anesthetized, laparotomy was performed and copious amounts of Bouin's fluid were poured over and around the uterus. After 3 to 7 minutes, all uterine ligaments and their contained blood vessels were clamped and the specimen was removed. In the brief period of persisting circulation before the ligaments were clamped, the interaction between the blood in the myometrial vessels and the penetrating fixative changed the color of the blood enough to make possible a very fine autoinjection.

Substances which solidify

John Hunter's historic observation of "curling arteries" in the human endometrium was made on a uterus in which the maternal arteries had been injected with liquid wax following a technique perfected in the mid-seventeenth century (Cole, 1921; Corner, 1963). Wax or gelatin are often used because they solidify and do not run out of the cut vessels when the uterus is sectioned. Latex and similar plastics that are soft enough for sectioning with an ordinary microtome fit into this category as well. If these injection masses are colored, they produce very useful specimens. Double injections of arteries and veins or of maternal and fetal circulations may be made by using contrasting colors so that the separate circuits are quickly identifiable.

More recently the substances-which-solidify technique has been carried a step further by using substances that form resistant casts that are not subject to the action of tissue-digesting agents. It is important that such substances have a low viscosity in their fluid state since they should flow freely into the vascular channels, including the capillaries. When they solidify they must not expand or contract significantly nor must they develop sufficient weight to press upon and occlude channels previously patent or to pull open physiologically constricted ones. Discrepancies among the results obtained by different investigators may result from variations in their media with respect to these criteria.

Various substances of this type have been employed: for example, Spanner (1935) used celluloid, Arts (1961) and Freese (1966) used Plastoid, Panigel (1962) used *Rhodopas,* Martin (1970) used Batson's acrylic (Batson, 1955). The chemical composition of the substances and their manufacturers may be found in the articles cited.

The basic physiological principles that must be respected in making injections with fluid media apply here also: introduce the substance *in vivo* whenever possible; do so under controlled physiological pressure and into an unobstructed bed with consideration for the general condition of the animal and careful recognition of the state of uterine contractility. As with the nonrigid masses differential coloring of the injection substance can produce very instructive specimens. Panigel's meticulous injections of individual fetal cotyledons with Rhodopas illustrate this possibility dramatically (Panigel, 1962).

Other injection materials

Direct observation of uteroplacental and fetoplacental blood channels and the passage of blood through them has been achieved by radioangiography. Borell *et al.* (1958 and 1965a, b), Nelson *et al.* (1961), Freese (1966), and Wigglesworth (1967, 1969) made their observations in human patients. Donner *et al.* (1963), Richart *et al.* (1964), and Freese *et al.* (1966) studied rhesus monkeys. Göthlin and Carter (1969) used rabbits.

Routes of injection vary according to the type of placenta under investigation. In the hemochorial placenta of larger animals, both afferent and efferent maternal blood streams cannot be demonstrated radiologically by femoral artery injection alone because the entering contrast material is diluted by the blood in the placental pool. To demonstrate venous drainage, the medium must be injected directly into the intervillous space (Ramsey *et al.*, 1966). Similar exceptions to simple generalizations will be found with other placental animals and compensations must be made.

The radiopaque materials chiefly used have been: Thorotrast by Reynolds (personal communication) in sheep, Renografin by Freese *et al.* (1966) in humans and monkeys, Urografin by Borell *et al.* (1958, 1965a, b) in humans. At the Carnegie Institution Hypaque, Renovist, Angioconray, and Conray *400* have been used in monkeys. The last, because of its lesser viscosity, has been most satisfactory (Ramsey *et al.*, 1966). (Chemical names and sources of supply of these agents are noted in the original publications by these authors.) All of these substances provide satisfactory visualization of the vascular channels, but it has been the goal of all investigators to find a medium that will also be nontoxic and have a viscosity as close as possible to that of the subject's blood. There can be little doubt that all of the substances mentioned have some deleterious effect upon both mother and fetus even when dosage is carefully controlled. In the Carnegie studies two injections of any of their media within a period of two or three hours were innocuous if the individual doses did not exceed 3 ml to the term fetus and 15 ml to the mother. Circulatory impairment was often manifested if subsequent injections were made on the same day. The local vascular effect of the various media is a further source of concern although specific evidence is still lacking.

Wigglesworth (1967) combines still radiography with injection of barium gelatin solution (Micropaque in 3% gelatin) for demonstration of the maternal placental vasculature in delivered human placentas.

Using a technique which has been successfully employed in scanning the lungs and other organs, Longo (1968; Power *et al.*, 1966) has injected macroaggregates of serum albumin, labeled with radioactive iodine, into the maternal and fetal circulations of pregnant sheep. By using different isotopes for the two circulations and making counts after the macrospheres have lodged within or just before the capillary bed (diameter of spheres 15 to 16μ), the pathways of blood flow can be traced by scanning, well-counting, or autoradiography of microscopic preparations.

VISUALIZATION

The methods of visualizing the injected vascular bed form a chapter in themselves. The simple old system of serially sectioning and staining is still basic, and it is wise in any comprehensive study to inject some of the specimens in such a way that this can be done (soft injection mass). Identification of channels is contingent, in the last analysis, upon demonstration of the histology of their walls. The value of cross checking data by applying several techniques has been noted earlier and is nowhere more important than here. If some of the study involves digesting soft tissue to study a cast, it is necessary to use control serial sections on comparable specimens.

Prior to sectioning (or dissection, if the injection mass is one that becomes too firm to cut), removal and fixation of the specimen must be performed by appropriate means. No matter how promptly and carefully these operations are performed some uterine contraction is stimulated and alterations occur in the vascular bed. Even the application of a fixing fluid *in situ* has this effect. Small uteri may be frozen rapidly *in situ* with less contraction initiated, but the procedure is of limited applicability (Gersch, 1948). The best one can do is to proceed quickly and carefully and realize that all specimens have the same artifact. Consequently, if the technique is kept constant the results are at least comparable, even if the results cannot be said to reflect living conditions. Prompt and simultaneous ligation of afferent and efferent blood channels traps the blood in the uterus and placenta, at least preventing the artifact of drainage.

Rapid, widespread excision of the injected specimen should be followed by its suspension in at least three times its own volume of fixing solution. The specimen should not touch the sides or bottom of the vessel. For routine use, 10% formalin is a reliable fixative. The histological stains to be applied to the microscopic sections will determine the choice of an alternative solution. When the specimen has hardened slightly and can be manipulated without trauma (a matter of a few hours, generally) amniotic fluid should be withdrawn by hypodermic needle and syringe, and replaced with an equal volume of concentrated formol. The specimen should be returned to formalin for several days before it is opened. The position of the fetus should be shifted at intervals so that the formol surrounds it fully. The preparation of blocks for sectioning can be carried out in accordance with the usual histological routines.

If the fetus is to be perfused, or if the fetal placenta is to be injected, it should be done as promptly as possible by incising the uterus without preliminary fixation.

Preparation of corrosion casts is achieved, following complete hardening of the injection mass, by digestion of soft tissues with a cytolytic enzyme such as trypsin (2% trypsin in 1% sodium bicarbonate) or with a strong alkali such as potassium hydroxide. The akali works faster, is equally reliable, and is less obnoxious in laboratories that are not equipped with a chemical hood to carry away the odors of tissue digestion. Both enzyme and alkali act more rapidly at a temperature of 37° to 50°C.

To present the findings of a study of vascular anatomy and the deductions drawn from them requires more than isolated photomicrographs of individual sections; reconstructions are necessary. The original Born (1883) method of constructing wax models is unsatisfactory when delicate interlacing structures like blood vessels are to be shown. The intricate scaffold of wires required to support the narrow wax strips is a serious obstacle and not worth the time and effort involved. A better, simpler method employs transparent plastic sheets upon which cross sections of vessels in successive serial sections are traced from images projected by either a camera lucida or a projecting microscope. The recommended types of plastic sheets, pens, and inks are described by Harris and Ramsey (1966) as well as the methods of orientation and superposition needed to achieve accurate three-dimensional magnification. Such models, if not too thick (number of sheets), may be demonstrated by stereophotography or, if unsuitable for that method, may be drawn.

Rigid plastic casts may be freed of residual tags of soft tissue by gentle washing and the casts of adventitious vessels can be removed with forceps, a fine dental drill, or a jeweler's saw. Like the plastic models, these casts can be photographed or drawn for publication or lantern slide presentation.

Visualization of channels injected with a radiopaque medium requires radioangiography in the hands of a skilled radiologist who has the necessary, highly sophisticated equipment (Donner *et al.,* 1963). The technical radiologic details are outside the province of this presentation; such a study of placental circulation would not be undertaken unless this essential assistance had been assured at the onset. However, some evaluation of the results of the two leading techniques is in order. Rapid serial X rays taken in two planes simultaneously by means of an automatic film-changing machine such as the Schoenander Elena provide clear definition of the uteroplacental and fetoplacental channels. The three-dimensional orientation of these channels can be ascertained by comparing anteroposterior and lateral views. The flat plates are particularly valuable for tracing these channels as they can be superimposed, shuffled, and otherwise employed as tools. Motion picture records of the progress of the injection mass as it enters the placental channels, on the other hand, provide a clearer concept of circulatory dynamics. If it is feasible to arrange for spot (still) films to be taken at intervals during the course of the cineradiography, the advantages of both methods become available.

CIRCULATORY DYNAMICS

Two of the morphological techniques already described shed light upon the mechanisms of uterine and umbilical flow which control placental circulation.

In the radioangiographic studies the time when the contrast medium appears in the various segments of the circuit can be used as a fair approximation of circulation time (Borell, 1965a, b), and counts and measurements of the number of channels filled (Borell, *ibid.;* Ramsey *et al.,* 1963; Martin *et al.,* 1964) give an indication of the state of patency of the vessels. Such data, based upon observation of X-ray photographs, do not have the accuracy of readings obtained by the use of flow meters and similar physiologic tools. Indeed, the limitations of the method render exclusive reliance upon quantitative statements based on radiography very unwise, although as qualitative indicators of the processes involved radiographic studies have great practical usefulness.

Longo's recent injections of radioactive albumin agglutinates have shown the irregular distribution of uteroplacental circulation and may be expected to produce further information about shunts, arteriovenous anastomoses, and a range of similar aspects of flow in ungulates and other animals.

BIBLIOGRAPHY

To avoid making this chapter read like a scientific cook book, details of materials and methods have been kept to a minimum. It is recognized, however, that investigators who may wish to employ any of the techniques cited will need such information and will find the accounts of their predecessors' experiences to be valuable. Bibliographic material has been assembled to facilitate reference to original sources and to collateral publications supplying helpful modifications or evaluations of given procedures. The starred items are particularly rich in practical suggestions and advice.

Amoroso, E. C. (1952) Placentation. In: Physiology of Reproduction. Ed. by Marshall 3rd ed. II, ch. 15. Longmans, Green, London.

Arts, N. F. Th. (1961) Investigations on the vascular system of the placenta. I. General introduction and the fetal vascular system. II. The maternal vascular system. Am. J. Obstet. Gynecol., 82:147–166.

Batson, O. V. (1955) Corrosion specimens prepared with a new material. Anat. Record, 121:425.

Bøe, F. (1953) Studies on the vascularization of the human placenta. Acta Obstet. et Gynecol. Scandinavica, 32: suppl. 5, pp. 1–92.

Borell, U., I. Fernström, and A. Westman (1958) Eine arteriographische Studie des Plazentarkreislaufs. Geburtsh. u. Frauenh., 18:1–9.

Borell, U., I. Fernström, L. Ohlson, and N. Wiqvist (1965a) Influence of uterine contractions on the uteroplacental blood flow at term. Am. J. Obstet. Gynecol., 93:44–57.

——— (1965b) Arteriographic study of the blood flow through the uterus and placenta at mid-pregnancy. Acta obstet. et gynecol. Scandinavica, 44:22–31.

Born, G. (1883) Die Plattenmodellirmethode. Arch. f. Mikr. Anat., 22:584–599.

Cole, F. J. (1921) The History of Anatomical Injections. In: Studies in the History and Method of Science, vol. 2. Ed. by Charles J. Singer. Oxford University Press, Cambridge.

Corner, G. W., Sr. (1963) Exploring the placental maze. Am. J. Obstet. Gynecol., 86:408–418.

———, E. M. Ramsey, and H. Stran (1963) Patterns of myometrial activity in the rhesus monkey in pregnancy. Am. J. Obstet. Gynecol., 85:179–185.

Crawford, J. M. (1962) Vascular anatomy of the human placenta. Am. J. Obstet. Gynecol., 84:1543–1567.

——— and A. Fraser (1955) The foetal placental circulation. A technique for its demonstration. J. Obstet. and Gynaec. Brit. Empire, 62:896–898.

Daron, G. H. (1936) The arterial pattern of the tunica mucosa of the uterus in *Macacus rhesus* Am. J. Anat., 58:349–419.

*Donner, M. W., E. M. Ramsey, and G. W. Corner, Jr. (1963) Maternal circulation in the placenta of the rhesus monkey; a radioangiographic study. Am. J. Roentgenol. Radium therapy Nucl. Med., 90:638–649.

Freese, U. E. (1966) The fetal-maternal circulation of the placenta. I. Histomorphologic, plastoid injection and x-ray cinematographic studies on human placentas. Am. J. Obstet. Gynecol., 94:354–360.

———, V. Ranninger, and H. Kaplin (1966) The fetal-maternal circulation of the placenta. II. An x-ray cinematographic study of pregnant rhesus monkeys. Am. J. Obstet. Gynecol., 94:361–366.

Gersch, I. (1932) The Altman technique for fixation by drying and freezing. Anat. Record, 53:309–337.

Göthlin, Jan and Anthony M. Carter (1969) Pelvic angiography in the female rabbit. Invest. Radiol., vol. 4, pp. 45–49.

Harris, J. W. S., and E. M. Ramsey (1966) The morphology of human uteroplacental vasculature. Carnegie Contrib. Embryol., 38:43–58.

Huggett, A. St. G., (1955) Growth, pregnancy and carbohydrate metabolism. Am. J. Obstet. Gynecol., 69:1103–1126.

Koenig, H., R. A. Groat, and W. F. Windle (1945) A physiological approach to perfusion-fixation of tissues with formalin. Stain Technol., 20:13–22.

Longo, L. D. (1968) In: Fetal homeostasis. Vol. 3. Ed. by Ralph M. Wynn. Proceedings of the Third Conference. New York Academy of Sciences. Interdisciplinary Communications Program, New York, pp. 108–128.

Martin, C. B., Jr., H. S., McGaughey, Jr., I. H. Kaiser, M. W. Donner, and E. M. Ramsey (1964) Intermittent functioning of the uteroplacental arteries. Am. J. Obstet. Gynecol., 90:819–823.

Martin, C. B., Jr., and E. M. Ramsey (1970) Gross anatomy of the placenta of rhesus monkeys. Obstet. Gynecol., 36:167–177.

Martin, C. B., Jr., E. M. Ramsey, and M. W. Donner (1966) The fetal placental circulation in rhesus monkeys demonstrated by radioangiography. Am. J. Obstet. Gynecol., 95:943–947.

Nelson, J. H., Jr., R. L. Bernstein, J. W. Huston, N. A. Garcia, and C. Gartenlaub. (1961) Percutaneous retrograde femoral arteriography in obstetrics and gynecology. Obstet. Gynecol. Survey, 16:1–19.

*Panigel, M. (1962) Placental perfusion experiments. Am. J. Obstet. Gynecol., 84:1664–1683.

Power, G. G., L. D. Longo, H. N. Wagner, D. E. Kuhl, and R. E. Forster (1966) Distribution of blood flow to the maternal and fetal portions of the sheep placenta using macroaggregates. Proceedings of the Fifty-Eighth Annual Meeting, J. Clin. Invest., 45:1058.

*Ramsey, E. M. (1949) The vascular pattern of the endometrium of the pregnant rhesus monkey (*Macaca mulatta*). Carnegie Inst. Wash., Contrib. Embryol., 33:113–147.

——— (1954) Venous drainage of the placenta of the rhesus monkey (*Macaca mulatta*). Carnegie Inst. Wash., Contrib. Embryol., 35:151–173.

——— (1956) Circulation in the maternal placenta of the rhesus monkey and man, with observations on the marginal lakes. Am. J. Anat., 98:159–190.

———— (1958) In: Conference on Oxygen Supply to the Human Fetus, Proceedings, Macy Foundation—C. I. O. M. S., Thomas, Springfield, Ill., p. 67–79.

————, G. W. Corner, Jr., and M. W. Donner (1963) Serial and cineradiographic visualization of maternal circulation in the primate (hemochorial) placenta. Am. J. Obstet. Gynecol., 86:213–225.

————, C. B. Martin, Jr., and M. W. Donner (1967) Fetal and maternal placental circulations. Am. J. Obstet. Gynecol., 98:419–423.

————, C. B. Martin, Jr., H. S. McGaughey, Jr., I. H. Kaiser, and M. W. Donner (1966) Venous drainage of the placenta in rhesus monkeys: radiographic studies. Am. J. Obstet. Gynecol., 95:948–955.

Reynolds, S. R. M. (1967) Personal communication.

Richart, R. M., G. B. Doyle, and G. C. Ramsay (1964) Visualization of the entire maternal placental circulation in the rhesus monkey. Am. J. Obstet. and Gynecol., 90:335–339.

*Rossman, I., and G. W. Bartelmez (1957) The injection of the blood vascular system of the uterus. Anat. Record, 128:223–231.

Scoggin, W. A., H. S. McGaughey, Jr., W. L. Johnson, and W. N. Thornton, Jr. (1963) Uterine contractility in primates: a comparative study. Surg. Forum, 14:384–385.

Spanner, R. (1935) Mütterlicher und kindlicher Kreislauf der menschlichen Placenta und seine Strombahnen. Zeitschr. f. Anat. u. Entwicklungsgesch., 105:163–242.

Wigglesworth, J. S. (1967) Vascular organization of the human placenta. Nature, 216:1120–1121.

Wigglesworth, J. S. (1969) Vascular anatomy of the human placenta and its significance for placental pathology. J. Obstet. Gynaec. Brit. Cwlth., vol. 76, pp. 979–989.

Wimsatt, W. A. (1962) Some aspects of the comparative anatomy of the mammalian placenta. Am. J. Obstet. Gynecol., 84:1568–1594.

30

Experimental Approaches to Placental Permeability

RICHARD L. SCHULTZ
Department of Human Biology
University of Colorado School of Dentistry
Denver, Colorado 80220

PHYLLIS W. SCHULTZ
Department of Biology
University of Colorado Denver Center
Denver, Colorado 80220

The usual approach to the study of placental permeability has been to administer material (s) to either the mother or the fetus, and then to determine, by qualitative or quantitative means, whether any of the material is on the other side of the placenta. Such studies have provided considerable information on molecular-size restriction during placental transport, species specificity of many compounds, active transport versus diffusion of many compounds, variation in gestational age as a factor in transport, and variation in the amount of transport as a function of the morphology of the placenta. Many articles have presented extensive bibliographies on the placental transport of ions (Flexner and Gellhorn, 1942; Comar, 1956; McCance and Widdowson, 1961; Metcalf, 1965; Zipkin and Babeaux, 1965), glucose (Widdas, 1952), salts and monomers (Widdas, 1961), lipids and carbohydrates (Huggett, 1954; Robertson and Sprecher, 1968), proteins (Ratner *et al.,* 1927; Brambell, 1954, 1958, 1966, 1969; Hemmings and Brambell, 1961), labeled compounds (Sternberg, 1962), drugs (Apgar and Papper, 1952; Dentkos, 1966; Moya, 1963; Scanlon, 1964; Villee, 1965), and other types of materials (Hagerman and Villee, 1960).

Most studies have not considered the multiple variables that exist simultaneously or independently during transfer of a compound across the pla-

The authors are deeply indebted to Dr. E. Marshall Johnson of the University of California for providing a manuscript prior to publication, to Dr. Lynn Larkin of the University of California for assistance in the operative techniques, and to Dr. Pierre Jacques of the University of Louvain for personal comments on interpretation of biochemical data.

cental membranes. In a recent evaluation, Shapiro *et al.* (1967) have presented a mathematical model that includes many possible variables in the analysis of placental transfer that need to be investigated and the various assumptions that require verification. Some of the variables considered in this model are: total area of the exchange membrane; the nature of the maternal and fetal capillaries (thickness, height, length, number, and blood-flow rate) ; decay for concurrent and countercurrent flow between the fetal and maternal capillary systems; the concentration and the dissociation function of an arbitrary substance in the maternal and fetal blood vessels at any given time; and the fetal and maternal shunting coefficient.

The majority of studies have given little indication of the cytological aspects of movement of materials through the cells and tissues of the placenta. A few ultrastructural studies have been concerned with the pathway of materials through the yolk sac of the rodent placenta, but none of these studies has indicated that the materials gain access to the fetus (Luse, 1957; Lambson, 1966; Schultz *et al.,* 1966; Carpenter and Fern, 1969). There have been no fractionation studies on placental or fetal tissues to indicate the role of the various organelles of these tissues in transport of materials from one circulation to the other.

No attempt will be made here to describe all the varied approaches to the questions of placental transport. Instead, we have chosen to present selected current methodology that may be of help in elucidating some of the mechanisms of transport.

FINE STRUCTURAL STUDIES OF PLACENTAL TRANSPORT

In this approach, the pregnant animal is administered an electron-dense or -opaque material (e.g., dextran, colloidal gold, ferritin), enzymes that can be observed by cytochemistry, or a radioactive compound that can be localized by electron-microscope autoradiography.

Information on the rate and path of the distribution of a compound can be obtained by removal of conceptuses from a single pregnant animal (rat) at selected time intervals (Schultz *et al.,* 1966). The animal is anesthetized and a midventral abdominal incision is made. The mesentery between the pregnant uterus and the uterine blood vessels is teased open with a hemostat, and a doubled piece of 000 surgical silk is passed through the opening. The doubled silk is cut so that two pieces lie in the opening in the mesentery. Each piece is tied tightly around one side of the conceptus to be removed (see Fig. 30-1). The uterus is cut through on the conceptual side of the silk, and the whole conceptus can be taken out without interfering with the uterine blood supply to the pregnant uterus. This operation can be repeated as many as four times on a single pregnant rat before the mother begins to be adversely affected by the repeated surgical trauma. No increase in resorption or decrease in fetal weight has been noted (Larkin and Schultz, 1968). Serial operations can be begun as early as the eighth day of pregnancy in the rat.

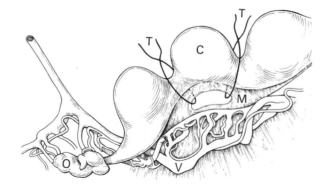

FIGURE 30-1

Diagram of the pregnant rat uterus, indicating the surgical approach to the removal of a single conceptus. The silk threads (T) lying in a hole in the mesentery (M) are tied on either side of a single conceptus (C) without interfering with the uterine blood vessels (V) or the blood supply to the ovary (O).

If only fetal samples are desired, the uterus can be opened on the antimesometrial side in such a manner that the embryo emerges from its membranes and can be removed without tying off the umbilical circulation. The cut surface of the uterus is then closed with a single tie of surgical silk, and other samples can be removed at a later time. In all cases, the abdominal wall of the mother is closed with silk and the skin reattached by skin clips.

The fetus, yolk sac, and chorioallantoic placenta are dissected free, cut into pieces 1 mm square, and prepared for electron microscopy. For morphological observations, the samples are fixed in buffered glutaraldehyde (3 to 6.25%), pH 7.4, dehydrated in methanol, infiltrated with a propylene oxide-Epon mixture, and embedded in Epon.

For enzyme cytochemistry (i.e., acid phosphatase), the best results in our laboratory are obtained by fixing the samples for 12 hours in either cold Holt's (Holt and Hicks, 1961) or Pease's (1962) fixative, storing overnight in cold, buffered sucrose solution, pH 7.3, rinsing in cold 7.5% sucrose, and reacting for acid phosphatase activity with Daems' modification of Gomori substrate, pH 5 (Daems, 1962) for 20 minutes at 37°C. Alkaline phosphatase or ATPase can be determined by modification of the pH or the substrate. The tissues are washed in 0.05M acetate buffer containing 4% formaldehyde and 7.5% sucrose, pH 5, postfixed in 1% osmium tetroxide, pH 7.4, and embedded as described above.

Horseradish peroxidase has been used extensively in *in vivo* and *in vitro* studies of pinocytosis. This enzyme can be injected intravenously into the pregnant animal, serial samples taken from the uterus at various time intervals, and the enzyme visualized by the method described in detail by Graham and Karnovsky (1966). Tritiated compounds can be injected intravenously, serial samples taken from the pregnant uterus, and the samples prepared for high-resolution autoradiography by the method of Caro and Tubergen (1962).

BIOCHEMICAL STUDIES OF CELL ORGANELLES

One biochemical approach to the study of placental transport involves the administration of a marker compound, the isolation of various cell organelles of the conceptual tissues by fractionation, the identification of the organelle (usually by enzyme activity), and the association of the injected compound with the organelle fraction. Biochemical investigations of young rat placental membranes can be done only on samples composed of tissues from many conceptuses. Two pooled samples from a single rat can be obtained by tying off the uterus and blood supply of one horn, removing the entire horn of the uterus and leaving the other horn for another pooled sample to be taken at a later time.

Samples of fetus, yolk sac, and chorioallantoic placenta are obtained by dissection as shown in Figure 30-2. Each of these pooled samples are homogenized in 0.25M sucrose with Dounce homogenizers. Care must be taken to limit the number of strokes with the pestle, since excessive homogenization appears to break the organelles. The homogenates are then subjected to differential centrifugation to obtain nuclear (N), heavy mitochondrial (M), light mitochondrial (L), microsomal (P), and supernatant (S) fractions (de Duve *et al.,* 1955).

The whole fractionation is performed near 0°C, using ice-cold solutions and equipment. To obtain the nuclear fraction, the crude homogenate is centrifuged for 10 minutes at 1700 rpm in an International refrigerated centrifuge, and, with the yolk sac and placental samples, the supernatant is decanted from the pellet. With the normal fetal sample, the pellet is gelatinous and the supernatant is removed by means of a pipette with a bent tip. The pellets are rehomogenized in 0.25M sucrose and recentrifuged. The final pellet is resuspended in sucrose as the N fraction. The combined supernatants and washings are thoroughly mixed; this is the "cytoplasmic extract" (E).

The cytoplasmic extract is centrifuged in a Spinco preparative centrifuge at 12,500 rpm for 3 minutes, 2 seconds (33,000 g minimum). The supernatant is removed by means of a pipette with a bent tip, care being taken to avoid touching the loosely packed material above the pellet. The pellets are washed by resuspension with a glass rod, those from each tissue being combined, and recentrifuged at the same speed to obtain the final M fraction. This and the remaining fractions are resuspended in a final volume of 0.25M sucrose by homogenizing the pellet in the centrifuge tube with the pestle of a large Dounce homogenizer.

The combined supernatants from the previous centrifugations are centrifuged at 25,000 rpm for 6 minutes, 42 seconds (250,000 g minimum), the supernatants are decanted, the pellets are resuspended and combined, and the fraction is recentrifuged at the same speed to obtain the L fraction.

The combined supernatants are centrifuged at 40,000 rpm for 30 minutes

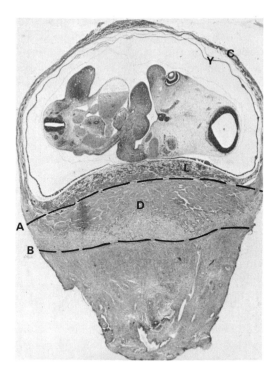

FIGURE 30-2

Section of 12-day rat conceptus showing the visceral yolk sac (Y), the chorioallantoic placenta (L), the deciduae capsularis (C) and basalis (D). Line A indicates the gross dissection separating the chorioallantoic membrane from the underlying decidua basalis. Line B indicates the division between decidua basalis and underlying myometrium. Papanicolaou stain. ×12. (After Schultz and Schultz, 1966.)

(3,000,000 *g* minimum), and the supernatant decanted to obtain the P fraction. Generally this fraction is not washed, in order to keep from diluting out the supernatant excessively.

The organelle distribution in each fraction can be ascertained by the use of assays for the following enzyme activities or chemical content: DNA, for nuclei; succinoxidase and/or cytochrome oxidase, for mitochondria; acid hydrolases, for lysosomes; catalase, uricase, and D-amino acid oxidases, for microbodies or peroxisomes; glucose-6-phosphatase and/or alkaline phosphatase, for microsomes (see de Duve, 1964).

The presence or absence of the compound administered to the mother or to the fetus can also be measured in the fractions listed above, by checking for, e.g., radioactivity of a labeled compound, enzyme activity of an injected enzyme, or chemical properties of the injected material. An enzyme, invertase, may be a valuable tool for this type of study, since it passes through the rat placenta to the fetus at relatively early stages of gestation (Schultz, 1966). When this enzyme is used, the homogenization and centrifugation must be done in 0.25*M* mannitol.

In a time study, one might anticipate that the material injected, particularly larger molecules, would first appear in the microsomal fraction of a particular placental membrane, indicating the formation of pinocytotic vesicles, then in the placental lysosomal fraction, and finally disappear from the cell of the placenta and appear in fetal tissue. This sequence has been observed biochemically in the liver (P. Jacques, personal communication) and relates to observations that have been made by light-microscope histochemistry of kidney cells (Straus, 1964).

Another method for correlating the cell organelle with the uptake of injected material is to subject the homogenate, the cytoplasmic extract (homogenate minus the nuclear fraction), or the "large granule fraction" (M + L) to density-gradient centrifugation (de Duve, 1964). Fractions can be obtained by various methods and analyzed for the various cell organelles and the injected compound. The theory and applications of gradient centrifugation have been described in detail by de Duve *et al.* (1959).

GAS EXCHANGE ACROSS THE PLACENTA

Gas exchange across the placenta has been investigated by the *in vivo* measurement of differences in gas tensions between maternal and fetal blood and by *in vivo* studies on the placental diffusion of carbon monoxide (Huggett, 1927; Barron, 1952; Young, 1952; Curtis *et al.*, 1955; Faber and Hart, 1966; Longo *et al.*, 1967; Bartels *et al.*, 1967).

Recently methods have been devised that allow for the short term *in vitro* survival of the postimplantation rat embryo (New and Stein, 1964; New, 1966). A modification of this method, described below, has made possible the direct measurement of the respiration of rat embryos (Netzloff *et al.*, 1968).

The culture medium is a 3:1:1 mixture of bovine serum, chicken embryo ultrafiltrate, and phosphate-buffered Ringer's solution. The implantation sites from pregnant rats are separated from one another in a Petri dish of culture media at 38°C. The uterine wall is incised at the antimesometrial border to the depth of the parietal layer of the yolk sac. The parietal yolk sac is excised, and the umbilical vessels external to the circumference of the visceral yolk sac are ligated with 7-0 silk thread, the ligature being located at the avascular region of the yolk sac. The yolk sac and its contents are left intact, and the chorioallantoic placenta is removed without loss of blood from the ligated vessels (see Fig. 30-3).

Using the free end of the silk tie around the umbilical vessels as a handle, the yolk-sac preparation is washed free of erythrocytes and other particulate matter in warm, phosphate-buffered Ringer-glucose solution. The preparations are placed in warm culture media in a Warburg vessel, and the manometric determination of oxygen uptake is begun following a 15-minute equilibration period.

Although embryonic viability begins to decrease rapidly following the removal of the embryo from the uterus, manometric determinations are possible for 75 minutes after removal. The data are considered acceptable only if both heart beat and yolk-sac perfusion are observed at the end of the experiment. This type of preparation has also been used to measure uptake of sodium, sulfate, and calcium ions by the embryo and yolk sac (Kernis and Johnson, 1968). The preparations are incubated in a medium containing radioactively labeled ions. The proximal yolk sacs and embryos are separated, dried, weighed, and counted for radioactivity. The results are interpreted as cpm per preparation and cpm per mg of dry weight.

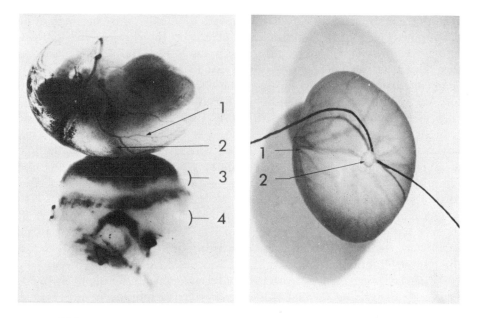

FIGURE 30-3

Isolated fetus and yolk sac preparation from a 14-day pregnant rat. Designated are (1) vitelline vessels coursing deep to the visceral yolk sac epithelium, (2) shadow of chorioallantoic vessels passing between the fetus and chorioallantoic placenta, (3) chorioallantoic placenta, and (4) reflected myometrium. (Courtesy of E. M. Johnson.)

IN VITRO EXPERIMENTS WITH ISOLATED PLACENTAL MEMBRANES

Several experimental approaches have involved the *in vitro* measurement of the physiological characteristics of the various placental membranes. The differentiation of the rat's yolk sac in organ culture has been reported by Sorokin and Padykula (1964). In this study, visceral yolk sacs were removed aseptically at various stages of pregnancy and cut into flat membranes. The samples were then transferred into culture flasks containing solid media. Culture media contained chicken or human cord serum (20 to 40%), glucose (0.1 to 0.3), and penicillin (100 units per ml), made up in Gey's balanced salt solution, pH 7.4, and solidified by the addition of 1.5% agar. One yolk-sac membrane, either whole or subdivided, was grown in each flask, and the medium was changed every second or third day by removing the fragments and placing them in a fresh flask. The specimens were removed at various time intervals (up to two weeks) and prepared for histological, histochemical, and ultrastructural study.

Prolonged absorptive capacity of the rat visceral yolk sac after explantation to the chick chorioallantoic membrane was studied by Ferm and Beaudoin (1960). The membranes at various ages of gestation were removed and cut into small pieces, 0.5 cm square, in warm, sterile chick-Ringer's solution. Windows were cut in the shell of fertile chicken eggs that had been incubated at 37°C for eight days, and under sterile conditions the egg-shell membrane was cut and removed to expose the chorioallantoic membrane. Pieces of rat yolk sac were placed on the chorioallantoic membrane over the site of a branching blood vessel. The grafts were placed so that the villous or maternal side of the tissue was uppermost. The windows were closed with cellophane tape and the eggs were returned to the incubator. After eight days of incubation of the graft, the absorptive activity of the yolk sac was determined by treating the chick chorioallantoic membrane with dyes. The dyes (1% trypan blue, 1% Niagara blue 2B, or 0.15% bovine azo protein) were either dropped directly onto the chorioallantoic membrane or injected into the chick yolk sac. The uptake of the dye by the rat yolk-sac graft was measured by gross and histological observations after fixation in 10% formalin.

The movement of ions across the isolated chorioallantoic membrane of the pig has been measured by Crawford and McCance (1960). A tissue sample approximately 3 cm square, consisting of all the uterine layers and the opposed chorioallantoic membrane, was obtained from the uterus of a 40- to 50-day pregnant pig by excision directly opposite the broad ligament in the area overlying the head of a fetus. The membrane in this area is almost free from visible blood vessels. The endometrium was teased away from the chorioallantoic membrane after pinning the tissue, fetal side up, to a block of paraffin immersed in warm bathing medium. The chorioallantoic membranes were inspected for any sign of discontinuity, tied to the ends of 4-cm lengths of thin-walled polyethylene tubing of 0.15 cm internal diameter in such a way as to form a sac, and suspended in chambers containing bathing medium. The amount of fluid in the membrane sac was adjusted so that the hydrostatic pressure inside the sac was equal to that outside in the bathing medium. Before and at the end of a given time period, the polyethylene tubing and the sac with its contents were rapidly weighed in air, and the difference between these two weights taken as a measure of the water that had passed across the epithelium. The movement of ions, sodium and potassium, across the membrane was measured by modifying their concentration in the bathing medium or within the membranous sac and measuring their concentration within the sac with a flame photometer.

Wild (1965) has reported on the *in vitro* passage of protein across the amnion, yolk sac, and paraplacental chorion of the rabbit. Perspex dialysis cells were constructed, each consisting of two chambers connecting by a circular window, 2.5 cm in diameter, in the facing ends. The chambers were held together by four bolts passing through the end pieces. Membranes were dissected from the conceptus of the 20- to 28-day pregnant rabbit and their orientation noted. After washing in 0.9% saline, the membranes were mounted between two gaskets that fitted over the facing ends of the chambers and had windows of the same diameter as those in the chambers. Evans blue (1%) was added to maternal serum diluted with oxygenated Krebs bicarbonate saline solution, and this fluid was then introduced into the appropriate chamber through an inoculum hole. The chambers were filled alternately; after adding a volume of serum to one chamber, an equal volume of oxygenated Krebs bicarbonate saline solution was added to the other, thus avoiding excessive distension of the membrane. Defects in the membrane were detectable visually at this stage by passage of the dye across the membrane. The cells were incubated at 37°C for three hours, after which the dialysates were concentrated by ultrafiltration and their protein content measured by the method of Lowry *et al.* (1951).

BIBLIOGRAPHY

Apgar, V., and E. M. Papper (1952) Transmission of drugs across the placenta. Curr. Res. Anesth. Anal., 31:309–320.

Barron, D. H. (1952) Some aspects of the transfer of oxygen across the syndesmochorial placenta of the sheep. Yale J. Biol. Med., 24:169–190.

Bartels, H., D. El Yassin, and W. Reinhardt (1967) Comparative studies of placental gas exchange in guinea pigs, rabbits, and goats. Resp. Physiol., 2:149–162.

Brambell, F. W. R. (1954) Transport of proteins across the fetal membranes. Symp. Quant. Biol. (Cold Spring Harbor), 19:71–81.

———— (1958) The passive immunity of the young mammal. Biol. Rev., 33:488–531.

———— (1966) The transmission of immunity from mother to young and the catabolism of immunoglobulins. Lancet, 2:1087–1093.

———— (1969) The transmission of immune globulins from the mother to the foetal and newborn young. Proc. Nutr. Soc., 28:35–41.

Caro, L. G., and R. P. van Tubergen (1962) High-resolution autoradiography. I. Methods. J. Cell Biol., 15:173–188.

Carpenter, S. J., and V. H. Ferm (1969) Uptake and storage of thorotrast by the rodent yolk sac placenta. Am. J. Anat., 125:429–439.

Comar, C. L. (1956) Radiocalcium studies in pregnancy. Ann. N.Y. Acad. Sci., 64:281–298.

Crawford, J. D., and R. A. McCance (1960) Sodium transport by the chorioallantoic membrane of the pig. J. Physiol., 151:458–471.

Curtis, G. W., E. J. Algeri, A. J. McBay, and R. Ford (1955) The transplacental diffusion of carbon monoxide. Arch. Path., 59:677–689.

Daems, W. T. (1962) Mouse liver lysosomes and storage: a morphological and histochemical study. Ph.D. thesis, University of Leiden.

de Duve, C. (1964) Principles of tissue fractionation. J. Theoret. Biol., 6:33–59.

————, J. Berthet, and H. Beaufay (1959) Gradient centrifugation of cell particles: Theory and applications. Progr. Biophys. Biophys. Chem., 9:325–369.

————, B. D. Pressman, R. Gianetto, R. Wattiaux, and F. Appelmans (1955) Tissue-fractionation studies. VI. Intracellular distribution patterns of enzymes in rat-liver tissue. Biochem. J., 60:604–617.

Dentkos, M. C. (1966) Passage of drugs across the placenta. Am. J. Hosp. Pharm., 23:139–144.

Faber, J. J., and I. M. Hart (1966) The rabbit placenta as an organ of diffusional exchange. Circ. Res., 19:816–833.

Ferm, V. H., and A. R. Beaudoin (1960) Absorptive phenomena in the explanted yolk-sac placenta of the rat. Anat. Record, 137:87–91.

Flexner, L. B., and A. Gellhorn (1942) The comparative physiology of placental transfer. Amer. J. Obstet. Gynecol., 43:965–974.

Graham, R. C., and M. J. Karnovsky (1966) The early stages of absorption of injected horseradish peroxidase in the proximal tubules of mouse kidney: Ultrastructural cytochemistry by a new technique. J. Histochem. Cytochem., 14:291–302.

Hagerman, D. D., and C. A. Villee (1960) Transport functions of the placenta. Physiol. Rev., 40:313–327.

Hemmings, W. A., and F. W. R. Brambell (1961) Protein transfer across the foetal membranes. Brit. Med. Bull., 17:96–101.

Holt, S. J., and R. M. Hicks (1961) The localization of acid phosphatase in rat liver cells as revealed by combined cytochemical staining and electron microscopy. J. Biophys. Biochem. Cytol., 11:47–66.

Huggett, A. St.G. (1927) Foetal blood-gas tensions and gas transfusion through the placenta of the goat. J. Physiol., 62:373–384.

———— (1954) The transport of lipins and carbohydrates across the placenta. Symp. Quant. Biol. (Cold Spring Harbor), 19:82–92.

Kernis, M. M., and E. M. Johnson (1968) Effects of diazo dyes on ion uptake by yolk sacs *in vitro*. Anat. Record, 160:375.

Lambson, R. O. (1966) An electron microscopic visualization of transport across rat visceral yolk sac. Am. J. Anat., 118:21–32.

Larkin, L. H., and R. L. Schultz (1968) Histochemical and autoradiographic studies of the formation of the metrial gland in the pregnant rat. Am. J. Anat. 122:607–620.

Longo, L. D., G. G. Power, and R. F. Forster, 1967 Respiratory function of the placenta as determined with carbon monoxide in sheep and dogs. J. Clin. Invest., 46:812–828.

Lowry, O. H., N. J. Rosebrough, A. L. Farr, and R. J. Randall (1951) Protein measurement with the Folin phenol reagent. J. Biol. Chem., 193:265–275.

Luse, S. A. (1957) The morphological manifestations of uptake of materials by the yolk sac of the pregnant rabbit. In: Gestation. Ed. by C. A. Villee. Josiah Macy, Jr., Foundation, Madison, New Jersey, pp. 115–141.

McCance, R. A., and E. M. Widdowson (1961) Mineral metabolsim of the foetus and new-born. Brit. Med. Bull., 17:132–136.

Metcalfe, J. (1965) Placental gas transfer. Anesthesiology, 26:460–464.

Moya, F. (1963) Considerations in maternal and placental physiology. Anesth. Analg., 42:661–664.

Netzloff, M. L., K. P. Chepenik, E. M. Johnson, and S. Kaplan (1968) Respiration of rat embryos in culture. Life Sci. 7:401–405.

New, D. A. T. (1966) Development of rat embryos cultured in blood sera. J. Reprod. Fertility, 12:509–524.

New, D. A. T., and K. F. Stein (1964) Cultivation of postimplantation mouse and rat embryos on plasma clots. J. Embryol. Exp. Morphol., 12:101–111.

Pease, D. C. (1962) Buffered formaldehyde as a killing agent and primary fixative for electron microscopy. Anat. Record, 142:342.

Ratner, B., H. C. Jackson, and H. L. Gruehl (1927) Transmission of protein hypersensitiveness from mother to offspring. I. Critique of palcental permeability. J. Immunol., 14:249–265.

Robertson, A. F., and H. Sprecher (1968) A review of human placental lipid metabolism and transport. Acta Paediat. Scand., 57:3–18.

Scanlon, R. T. (1964) Placental transmission of drugs. Clin. Proc. Child. Hosp., 20:116–119.

Schultz, P. W., J. F. Reger, and R. L. Schultz (1966) Effects of triton WR-1339 on the rat yolk-sac placenta. Am. J. Anat., 119:199–234.

Schultz, R. L. (1966) Placental transport of an active enzyme, invertase. Proc. Soc. Exp. Biol. Med., 122:1060–1062.

Schultz, R. L., and P. W. Schultz, (1966) Lysosomal enzyme changes in the rat conceptus following ovariectomy and the injection of a nonionic detergent. Life Sci., 5:1735–1742.

Shapiro, N. Z., T. Kirschbaum, and N. S. Assali (1967) Mental exercises in placental transfer. Am. J. Obstet. Gynecol., 97:130–137.

Sorokin, S. P., and H. A. Padykula (1964) Differentiation of the rat's yolk sac in organ culture. Am. J. Anat., 114:457–477.

Sternberg, J. (1962) Placental transfers: modern methods of study. Am. J. Obstet. Gynecol., 84:1731–1748.

Straus, W. (1964) Occurrence of phagosomes and phagolysosomes in different segments of the nephron in relation to the reabsorption, transport, digestion, and extrusion of intravenously injected horseradish peroxidase. J. Cell Biol., 21:295–308.

Villee, C. A. (1965) Placental transfer of drugs. Ann. N.Y. Acad. Sci., 123:237–244.

Widdas, W. F. (1952) Inability of diffusion to account for placental glucose transfer in the sheep and consideration of the kinetics of a possible carrier transfer. J. Physiol., 118:23–39.

——— (1961) Transport mechanisms in the foetus. Brit. Med. Bull., 17:107–111.

Wild, A. E. (1965) Protein composition of the rabbit foetal fluids. Proc. Roy. Soc., Series B, 163:90–115.

Young, I. M. (1952) CO_2 tension across the placental barrier and acid-base relationship between fetus and mother in the rabbit. Am. J. Physiol., 170:434–441.

Zipkin, I., and W. L. Babeaux (1965) Maternal transfer of fluoride. J. Oral Ther. Pharmacol., 1:652–665.

31

Mammary Gland Culture

EVELYN M. RIVERA
Department of Zoology
Michigan State University
East Lansing, Michigan 48823

The mammary gland offers a number of experimental advantages for the study of development. The events associated with its structural and functional specialization depend, to varying degrees, on hormones, nervous influences, genetic background, nutrition, and a host of other factors. Its structural modifications are distinct and characteristic, enabling the use of morphological guidelines in the assessment of the developmental process. The secretion of milk by the functional gland is vitally linked to the synthesis and elaboration of cell-specific products, the appearance of which provides valuable indices of differentiation. Tumors of the mammary gland have stimulated many facets of cancer research. The occurrence of a precancerous lesion in mammary-tumor-bearing mice offers a unique operational advantage, because it allows separation of the change from normal to cancerous into two stages: normal to precancerous and precancerous to cancerous. These several aspects of mammary gland development thus provide a promising potential for analysis of the differentiative process and the factors controlling its normal and abnormal manifestations.

In recent years, cell- and organ-culture approaches have become increas-

This chapter was written during tenure of a Research Grant and Research Career Development Award from the U.S. Public Health Service. I am greatly indebted to friends and colleagues for their generous cooperation in providing illustrative material and for their helpful suggestions. Their contributions are acknowledged in the legends to the pertinent figures and in the text as personal communications.

ingly useful in mammary developmental studies. The technical steps are concerned with isolation of the mammary gland, either entirely or in part, and its maintenance in an environment in which the experimental conditions can be controlled more precisely than by animal experimentation alone. Experiments have been designed to approach problems concerning the maintenance, growth, and function of mammary cultures, with particular attention paid to the acquisition of differentiated characteristics in the presence of specific hormones. Some of the earlier data have been discussed in a number of reviews (Dieterlen-Lièvre, 1964; Fell, 1964; Lasnitzki, 1965; Waymouth, 1966). Recent efforts, stimulated by developments in our conceptual framework of approach, have emphasized the need for more detailed inquiry regarding the nature of hormone-dependent differentiation (Topper, 1968; Turkington, 1968c; Jones, 1969; Larson, 1969; Patton, 1969).

The purpose of this paper is to summarize culture procedures that have been applied to the mammary gland and to provide pertinent references to analytical techniques used in the assessment of mammary development in culture. Both monolayer cell culture and organ culture methods will be described. The techniques and references provided herein are intended primarily to serve as starting points for investigators beginning mammary culture studies, and are therefore subject to further modification in accordance with experimental objectives.

MAMMARY MORPHOGENESIS *IN VIVO*

As a preliminary to discussion of culture procedures, it is useful to review briefly the morphological changes occurring during development *in vivo*. Several developmental stages of the mouse mammary gland, in which normal and abnormal structures are well illustrated, are summarized below. More detailed descriptions are given for the mouse by Nandi (1959) and for several mammalian species by Turner (1952).

EMBRYONIC RUDIMENTS

The five pairs of mammary anlage appear in both sexes at about Day 12 of embryonic life, as epidermal thickenings on the thoracic and inguinal body surface (Hardy, 1950; Raynaud and Raynaud, 1956; Raynaud, 1961). These thickenings, or mammary buds, become rounded by Day 14, at which time sex differences become apparent. In male embryos, the rudiments become surrounded by a marked condensation of mesenchyme, sometimes lose connection with the epidermis, and regress either before birth (Raynaud, 1961; Kratochwil, personal communication) or shortly thereafter (Moretti, personal communication). Between Days 15 and 18, the nipple rudiment forms in females by invagination of the epidermis at the circumference of a circle, although eversion of the nipple does not occur until six weeks after birth (Hardy, 1950). The primary mammary duct (or sprout) is formed at about Day 16 of gestation in the

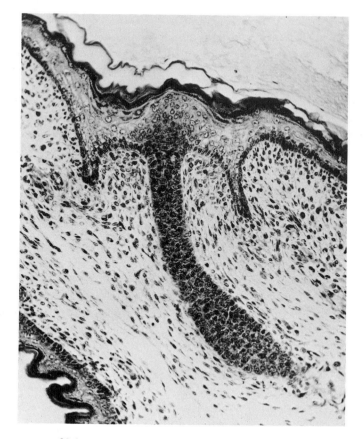

FIGURE 31-1

Section of mammary rudiment from an 18-day-old female mouse embryo. ×179. (From A. Raynaud and J. Raynaud, 1956, Ann, Inst. Pasteur, 90:187–219.)

subjacent mesenchyme, by tubular invagination of the epithelium from the circle formed by the nipple rudiment (see Fig. 31-1). Near term, secondary sprouts develop at the distal portions of the primary duct.

Duct-end bud development (prepubertal)

Development of the duct system is most rapid during the first week of postembryonic life, after which ramification and growth continue, but at a slower rate. In prepubertal mice, the gland consists of a branched duct system, the terminal portions of which are club-shaped end buds (see Fig. 31-2a).

Alveolar-ductual development (adult virgin)

After the onset of estrus, the gland continues to increase in size and extent of ductal branching (see Fig. 31-2b). Club-shaped end buds gradually disappear with the emergence of finer ducts. Small, isolated clusters of

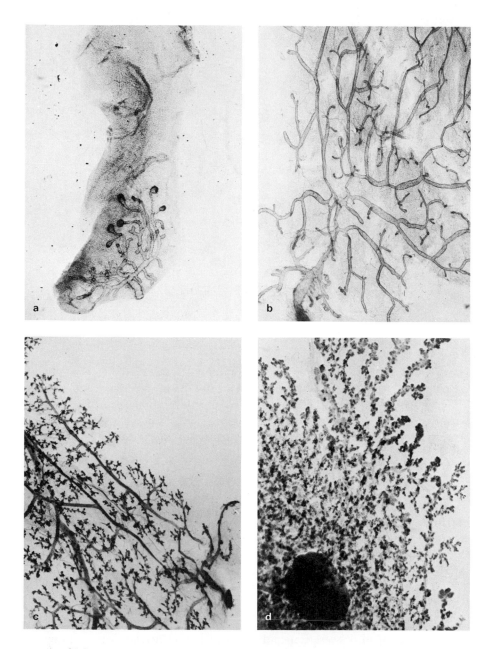

FIGURE 31-2

(a) *Whole mount of mammary gland from a 4-week-old BALB/c mouse. Note end buds and incompletely filled fat pad. ×9. (From Rivera, unpublished photograph)*
(b) *Whole mount of mammary gland from a 16-week-old virgin BALB/cfC3H mouse. Note absence of alveoli. ×8. (From Dr. R. Ichinose, University of California)*
(c) *Whole mount of mammary gland from an 18-week-old virgin C3H mouse, showing alveolar development. ×7. (From L. Young, University of California)*
(d) *Whole mount of mammary gland from an 18-day-pregnant BALB/cfC3H mouse, showing lobulo-alveolar proliferation. ×8. (From Dr. R. Ichinose, University of California.)*

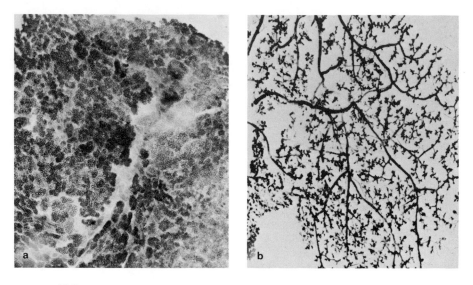

FIGURE 31-3

(a) *Whole amount of mammary gland from a lactating BALB/cfC3H mouse, one day post-partum. ×8. (From Dr. R. Ichinose, University of California)*
(b) *Whole mount of mammary gland from a C3H mouse, two weeks postweaning. ×8. (Rivera, unpublished photograph.)*

alveoli develop at the terminal portions of the ducts (see Fig. 31-2c). Strain differences are found in the extent of alveolar development in the glands of virgin females (Richardson and Hummel, 1959), but complete lobulo-alveolar development occurs only during pregnancy and lactation.

Lobulo-alveolar development (pregnancy)

During pregnancy the alveoli proliferate and become organized into alveolar lobules (see Fig. 31-2d). From the third trimester of pregnancy (about 14 days), the alveolar cells gradually acquire secretory characteristics: intracellular lipid vacuolation, secretion in the alveolar lumina, and distention of alveoli due to accumulation of luminal fluid ("pre-milk").

Lactational development (post-partum)

The process of milk secretion and removal is attained at parturition with the onset of suckling. Proliferation continues at least through midlactation (Munford, 1964). The glands are thicker than in pregnancy and assume a milky-white appearance owing to distention by milk (see Fig. 31-3a).

Involution (postlactation)

Following the cessation of suckling, the mammary gland regresses structurally and functionally, and eventually resumes the normal alveolar-ductal

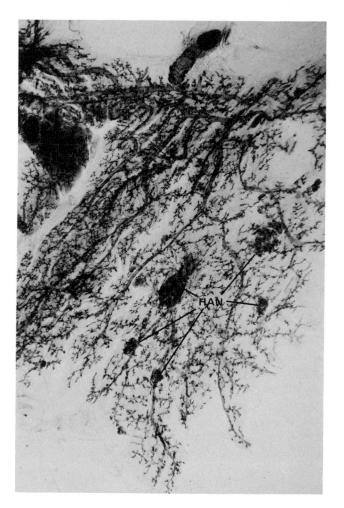

FIGURE 31-4

Whole mount of mammary gland from nonpregnant, multiparous C3H mouse, 11 months old. Several hyperplastic alveolar nodules (HAN) can be seen. ×8. (From L. Young, University of California.)

("resting") condition until the next pregnancy (see Fig. 31-3b). The rates of regression appear to differ among different mouse strains (Fekete, 1938).

Preneoplastic development

Hyperplastic alveolar nodules, which are morphologically identical to normal alveolar lobules, are found with high frequency in mouse strains having a high incidence of mammary tumors, for example, C3H, DBA, and RIII. These preneoplastic lesions are distinguished from normal lobules by their occurrence and persistence in nonpregnant, nonlactating female mice (see

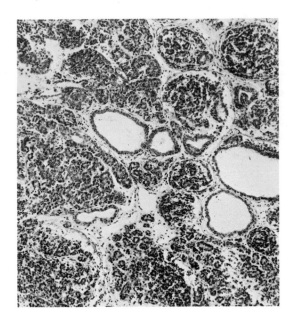

FIGURE 31-5

Section of mammary adenocarcinoma from C3H mouse, showing basic acinar structure and several cysts. ×89. (From Rivera et al., 1963, J. Nat. Cancer Inst. 31:671–678.)

Fig. 31-4) . The precancerous nature of hyperplastic alveolar nodules is now well established (DeOme *et al.,* 1962) .

Neoplastic development

The mammary tumors found in mice are typically adenocarcinomas (Dunn, 1959) , with well-differentiated acinar structure and small amounts of fibrous stroma (see Fig. 31-5) . They grow as well-circumscribed masses of tissue within a thin, fibrous capsule. In strain C3H mice, the incidence of tumors is 99 to 100% in eight-month-old virgins and in seven-month-old breeding females (Heston *et al.,* 1964) . Virus particles are associated with the development of both hyperplastic alveolar nodules and mammary tumors (DeOme, 1963; DeOme and Nandi, 1966) .

MAMMARY GLAND CULTURE

Monolayer cell cultures and organ cultures have both been used in studies of the mammary gland. (The term *monolayer* refers in this paper to the outgrowth of cells from the primary dispersed tissue cells.) In addition to hormone-target cell studies, monolayer cultures provide useful approaches to problems concerning the stability of mammary cells after dissociation, continued proliferation, and prolonged cultivation. The work of

Larson (1967; 1969) and colleagues (see below) on the functional decline of mammary cell cultures is of particular interest in this regard. The objectives of organ culture, on the other hand, are to inhibit outgrowth of cells from the explant and to maintain embryonic or differentiated cells as groups of normally associated components resembling as closely as possible their histological and physiological prototypes in the animal. Growth, development, and function may thus be analyzed in discrete tissues and presumptive organs in isolation from systemic influences. Each procedure has its advantages and limitations, and the approach selected is determined by the experimental objectives of the investigator.

MONOLAYER CULTURE

Mammary glands of mice (Lasfargues, 1957a, 1957b; Nikiforova, 1960; Daniel and DeOme, 1965), rats (Schingoethe *et al.,* 1967), cows (Ebner *et al.,* 1961a, 1961b; Twarog and Larson, 1962, 1964), and goats (Blanco *et al.,* 1967) have been successfully cultivated as monolayers. The basic procedures are described below.

Preparation of cells

After aseptic removal from the animal, the glands (or portions thereof) are minced with fine scissors and Bard-Parker blades, or are sliced first with a tissue slicer (Stadie and Riggs, 1944) and then minced in sterile balanced salt solution (BSS) with or without antibiotics. The tissue fragments are then dissociated by digestion with collagenase. Daniel and DeOme (1965) found that 0.05 to 0.1 g% of collagenase in BSS and magnetic stirring for 90 minutes at room temperature were adequate conditions for dissociation of prelactating mouse mammary tissue. Ebner *et al.* (1961a) used 0.2 to 0.3 mg/ml of collagenase in 4 ml of BSS and mechanical agitation for two hours at 37°C for dispersal of prelactating and lactating bovine mammary tissue. Better dispersion was obtained by replacing the collagenase with fresh solution after the first hour. Since the degree of dissociation (and possibly cell damage) varies with the time of exposure to the enzyme, the intensity of agitation, and the temperature of digestion, preliminary tests are recommended for mammary tissues from other species.

Following digestion, the cell suspension is centrifuged at moderate speed, the supernatant discarded, and the pellet of cells resuspended in nutrient solution. If further dissociation is indicated, a capillary pipette may be used for repeated cell withdrawal and expulsion until microscopic examination shows the desired extent of dispersal. The cells are then washed by centrifugation in several changes of BSS or nutrient medium and inoculated into appropriate culture vessels.

Aliquots of cell suspension are inoculated directly into the culture vessel or diluted to obtain a known ratio of packed cells per unit volume of medium (0.08 ml of packed cells/100 ml of medium; Daniel and DeOme, 1965) or a known number of cells per inoculum (200,000 cells in 0.5 ml of inoculum; Twarog and Larson, 1962). Cultivation is carried out at 37°C with medium changes every 48 hours (Daniel and DeOme, 1965) or twice a week (Ebner *et al.,* 1961a).

Bovine mammary cells have been serially propagated for periods up to 30 months (Ebner *et al.,* 1961a). Passaging is now (Larson, personal communication) carried out with the aid of a divalent ion-free solution (DVIF), versene, and trypsin. The cells are removed from the glass surface by a two-step procedure, as follows. After pouring off the medium, add DVIF solution (composition as described by Paul, 1965, but without versene) to the culture vessel and allow to stand for 5 to 10 minutes. Pour off, and add a solution containing trypsin *and* versene (0.5 ml of rehydrated Bacto-trypsin/100 ml of DVIF solution containing 0.2 g/liter of Na-versenate). Incubate for 10 minutes at 37°C, pour off, and wash cells by centrifugation in several changes of culture medium. Approximately 200,000 cells are used to inoculate subcultures.

Culture media

Table 31-1 shows the composition of the medium used for monolayer mammary cultures derived from several mammalian species.

Serum appears to be a strict requirement for the growth and establishment of cells grown as monolayers. Cells fail to adhere to the substrate in its absence, and on the few occasions when they do, they fail to proliferate (Daniel, personal communication). Serum, 10 to 20%, is normally used to establish growth, but after several passages, it may be desirable to reduce the serum concentration to 5%. Apparently, protein in the medium can sometimes interfere with certain types of analytical procedures on cell cultures (Larson, personal communication). When hormonal studies are planned, it is advisable to obtain serum from animals hypophysectomized several weeks prior to removal of blood. Although such an operation may not be feasible in large animals, it has been noted that serum processed from cows in various reproductive states may differentially affect growth responses of cell cultures (Ebner *et al.,* 1961a).

Culture vessels

Cultures are usually maintained in glass vessels having flat surfaces on which the cells can settle. Examples of the types in use include T-flasks, Leighton tubes, prescription bottles, milk-dilution bottles, diphtheria bottles, and Carrel flasks. Maximow slides with double coverslips appear satisfactory for small-scale cultures (Lasfargues, 1957b).

ORGAN CULTURE

The mammary gland of the mouse, at several stages of its development, will be used to illustrate the basic organ-culture procedures. In addition to the mouse, organ cultures have been prepared of the mammary glands from the rat (Trowell, 1959; Barnawell, 1965; Heuson *et al.*. 1967; Dilley and Nandi, 1968; Heuson and Legros, 1968; Turkington and Riddle,

TABLE 31-1

Culture media: mammary cell cultures

Species	Composition of Basal Medium	Study
Mouse	Medium *199*, 90%; newborn or fetal calf serum, 10%; insulin, 10 μg/ml; penicillin, 50 units/ml; streptomycin, 50 units/ml	Daniel and DeOme (1965)
	Human placental serum, 1 part; Simm's saline solution, 2 parts	Lasfargues (1957a, 1957b)
	Heparinated chicken plasma; Hank's saline solution; chick embryo extract; human serum	Nikiforova (1960)
Cow, goat	Medium *199*, 40–45%; Eagle Hela, 40–45%; bovine serum, 10–20%; penicillin, 100 units/ml; streptomycin, 100 mgs/ml; nystatin, 50 units/ml	Ebner *et al.* (1961a) Twarog and Larson (1962) Blanco *et al.* (1967)
Cow, rat	Eagle MEM, 95%; bovine serum, 5%; penicillin, 100 units/ml; streptomycin, 100 mgs/ml; nystatin, 50 units/ml	Schingoethe *et al.* (1967)

1969; Mishkinsky *et al.*, 1967), guinea pig (Gerritsen, 1960; Barnawell, 1965), rabbit (David and Propper, 1964; Barnawell, 1965; Propper and Gomot, 1967; Propper, 1968, 1969), hamster, dog (Barnawell, 1965, 1967), and human (Rovin, 1962; Barker *et al.*, 1964; Dickson, 1966; Whitescarver *et al.*, 1968).

Preparation of tissues

EMBRYONIC RUDIMENTS. The two horns of the uterus are removed from pregnant mice and placed into sterile Petri dishes for dissection of the embryos. Although mouse mammary rudiments are not visible with the dissecting microscope before Day 12 of gestation, ventral thoracic and inguinal skin containing the presumptive organs can be excised and explanted. Several investigators have reported mammary differentiation in skin cultures of this type from the mouse (Hardy, 1950; Balinsky, 1950; Lasfargues and Murray, 1959; Kratochwil, 1969), rat (Ceriani and Bern, 1968) and from the rabbit (David and Propper, 1964).

Alternatively, embryos may be selected when the mammary buds become

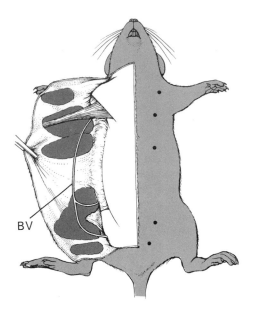

FIGURE 31-6

Diagram of ventral aspect of pre-pubertal mouse, showing positions of mammary fat pads (shadowed area) and nipples (black dots). Note thin muscle overlying second thoracic gland and blood vessel (BV).

apparent after Day 12 or Day 13 of embryonic life. In some embryos the epidermal buds may no longer be visible after the 17th day, so that identification of the rudiment from the exterior of the embryo becomes difficult (Moretti, personal communication). This difficulty may be circumvented by reflecting the skin from the body to expose the small primary duct, which is visible in the embryonic mesenchyme from about Day 18.

Kratochwil (1969) dissects mammary rudiments in a solution of Tyrode's and horse serum (1:1, v/v) under an atmosphere of 5% CO_2 in air. These conditions minimize the stickiness of tissues that occurs in alkaline solution. To separate the epithelial and mesenchymal components, the rudiments are incubated for 3 minutes at room temperature in Ca- and Mg-free Tyrode's solution containing a 2.25% trypsin (Difco 1:250) and 0.75% pancreatin (Difco N.F.). Collagenase (Worthington), 0.05% in Tyrode's solution, is also effective. After incubation, the rudiments are flushed through a capillary pipette to separate epithelium from mesenchyme.

WHOLE GLANDS (PREPUBERTAL). Mammary glands have been explanted *in toto* (parenchyma plus fat pad) from mice up to about 6 weeks of age (Prop, 1961, 1966; Koziorowska, 1962; Chapekar and Ranadive, 1963; Rivera, 1964a; Prop and Hendrix, 1965; Ichinose and Nandi, 1964, 1966; Gadkari *et al.*, 1968).

Female mice, usually 3 to 6 weeks old, are killed by cervical dislocation, washed in 70% ethanol, and pinned by the feet to a cork board or similar working surface. The glands are exposed by making a midventral incision along the length of the body, and loosening and pinning the skin to the cork board. Similarly, a middorsal incision may be made, the skin loosened and pinned down to the cork board as above, and the animal cut away from the skin and discarded (Prop, 1961). The positions of the five pairs of mammary glands in the mouse are shown in the diagram in Figure 31-6.

The second and third thoracic glands have usually been selected for culture because of their relative flatness and ease of dissection. Note that the second thoracic gland is partially obscured by a thin strip of muscle (see Fig. 31-6) which should be dissected before attempting to remove the mammary gland. The fourth glands (inguinal), although equally accessible, possess fat pads somewhat thicker than those of the thoracic glands. In 3- to 4-week-old mice, gland thickness is not a special problem, but those explanted from progressively older animals show less development in the thicker portions of the fat pad (Prop, 1966).

Using a dissecting microscope (at least 10- to 15-fold magnification), most of the fat pad is removed by cutting along its edges with iridectomy scissors (Aloe Surgical Supply) or a sharp scalpel. If care is taken to avoid cutting through the distal portions of the mammary ducts, the only portion of the parenchyma that need be cut is the primary duct at a point where it emerges from the undersurface of the nipple. Since the mammary parenchyma is sometimes difficult to discern in the fat pads of albino mice, visualization may be improved by painting the outside of the skin with gentian violet, 0.5 to 1.0%, or some similar dye. Caution should be taken to prevent the dye from coming in contact with the under-surface of the skin. After removal from the animal, the glands are spread out as flat as possible on stainless steel grids (Falcon Plastics) with or without lens paper and transferred to pre-prepared culture vessels. Because of the variation in morphology and size between glands, one of a pair is usually taken for cultivation and the contralateral one removed to serve as the initial control for morphological or biochemical analysis.

Ichinose and Nandi (1964, 1966) found that subcutaneous injections of 3- or 4-week-old mice with estradiol and progesterone (1 μg + 1 mg daily for nine days) increased the lobulo-alveolar growth response to hormones in organ culture. Ovarian hormone pretreatment also accelerated the onset of hormone-dependent casein synthesis in cultures of immature glands (Voytovich and Topper, 1967).

Isolated terminal end buds (Elias, 1961, 1962a), primary mammary ducts (Rivera, 1963), and minced fragments of the entire prepubertal gland (Lasfargues, 1960) have also been cultivated as organ cultures.

ADULT TISSUE EXPLANTS. (1) Normal tissues. Ductal-alveolar and lobulo-alveolar mammary tissues are obtained from adult nonpregnant, pregnant, and lactating mice. These developmental stages may also be prepared by appropriate hormone treatment of virgin females (Lyons et al., 1958; Nandi, 1959; Barnawell, 1965).

Owing to size limitations, mammary glands from adult mice are cultivated as tissue fragments, approximately 1.0 mm thick and 1 or 2 mm in diameter (Elias, 1957, 1959; Lasfargues, 1962; Rivera, 1964b; Juergens et al., 1965; Moretti and Abraham, 1966; Nicoll et al., 1966). The animals are killed and prepared as described in connection with whole gland cultures. Explants can be removed individually with iridectomy scissors from the edges of the glands. Alternatively, large portions of the gland may be removed and sliced into thin sections with razor blades or a tissue slicer

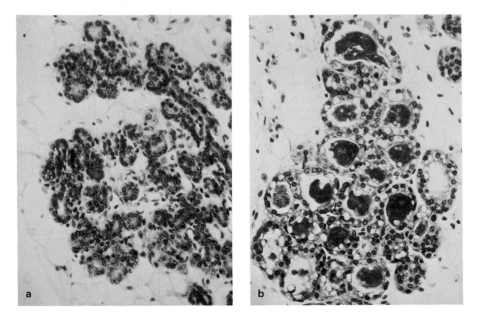

FIGURE 31-7

Sections of mammary gland from midpregnant C3H mouse showing (a) *nonsecretory appearance before culture, and* (b) *secretory response after five days culture in* medium 199 *supplemented with insulin, corticosterone, and prolactin.* ×208 *(From Rivera, unpublished photographs)*

(Barnawell, 1965; Barker *et al.*, 1964; Mishkinsky *et al.*, 1967) . The sections may then be cut into thin strips and finally into cubes or blocks of the desired dimensions. This latter procedure has the advantage of being faster than individual dissection of explants, especially when handling tissue from animals with thicker mammary glands than rodents. Figure 31-7 shows sections of midpregnant mammary-gland explants at the time of explantation and after several days in culture in the presence of hormones.

(2) Hyperplastic Alveolar Nodules. Nodules for explantation are se-lected from nonpregnant, nonlactating mice, in which they are readily identifiable as local areas of hyperplasia in otherwise resting glands (see Fig. 31-4) . In pigmented strains of mice, they appear as yellowish spots on gross examination. They are dissected individually with iridectomy scissors (Elias and Rivera, 1959; Ingraham, 1968) .

(3) Tumors. Mammary adenocarcinomas grow as highly vascularized nodular masses of tissue, the central portions becoming necrotic and liqui-fied with progressive increase in size. Small tumors (about 1 cm in diameter or less) are usually taken for culture to avoid explantation of necrotic areas. They are selected by skin palpation, removed from the animal, and placed into a sterile dish. The fibrous capsule is trimmed away with fine scissors and the tumor cut into small blocks with razor blades or a tissue slicer. Unlike normal mammary tissues, spontaneous mammary tumors from mice can be maintained in hormone-free synthetic medium (Elias and Rivera, 1959; Elias, 1962b; Moretti and DeOme, 1962; Rivera *et al.*, 1963) , and will grow and develop as tumors upon subsequent transplantation into host

mice. That hormones can influence mouse mammary tumor growth and ultrastructure has also been demonstrated (Rivera *et al.,* 1963; Turkington and Hilf, 1968; Turkington and Ward, 1969a; Michelson and Hagenau, 1969; Welsch and Rivera, 1970).

Wolff and collaborators (Wolff and Wolff, 1961; Wolff and Sigot, 1961; Wolff, 1964) have developed a rather unusual organ-culture approach to the study of tumors, wherein tumors of mammalian origin are cultured in xenoplastic combination with avian embryonic organs. A variety of tumors, including transplantable mouse mammary adenocarcinoma, fail to survive when cultivated alone, despite the inclusion of embryo extract in the standard medium. When fragments of tumor and embryonic organs (avian mesonephros, in particular) are mixed together and cultivated *en bloc,* the tumors proliferate and grow at the expense of the embryonic tissue. These cultures may be maintained for several weeks by transferring tumor explants to fresh embryonic tissue every 3 to 5 days. Interposing a vitelline membrane between tumor and mesonephros in culture showed that the former tissue subsisted on products of the mesonephros that were elaborated through the membrane (Wolff and Wolff, 1961).

A fibrin foam matrix procedure has also been used for passaging a transplantable mouse mammary tumor (CE 1460) alternately *in vitro* and *in vivo,* with essentially no change in the morphology and functional characteristics of the tumor (Kalus *et al.,* 1968).

Culture media

Chemically defined synthetic media have served as basal media to which hormones and other substances have been added. Medium 199, Trowell T8, Waymouth MB 752/1, and NCTC 109 are among those commonly used. Their compositions are given in the Handbook of Cell and Organ Culture (Merchant *et al.,* 1964). With the exception of Trowell T8, which contains 50 μg/ml of insulin, these media are protein-free and hormone-free.

Partially defined or natural media have also been useful. Balanced salt solutions may be supplemented with lactalbumin hydrolyzate and human male serum (Prop, 1961; Koziorowska, 1962; Chapekar and Ranadive, 1963) or with chick embryo extract (Wolff and Sigot, 1961; David and Propper, 1964). Clots composed of chick or cock plasma and chick-embryo extract provide a rich nutrient medium (Hardy, 1950; Rovin, 1962). A nonnutritive agar gel substrate (1g% agar in Gey's salt solution) is used in place of a clot by Wolff and collaborators (Wolff and Haffen, 1952), with nutrient medium consisting of equal parts of Tyrode's solution and chick-embryo extract.

Preparation of hormones

Protein and polypeptide hormones are usually readily soluble in neutral aqueous solutions. However, growth hormone (Rivera *et al.,* 1967), and sometimes prolactin, should be dissolved first in a minimum amount of dilute NaOH (*p*H 9 to 10), after which enough basal culture medium (e.g., Medium 199) is added to yield the desired stock concentration. Stock solutions are then sterilized by passage through millipore filters (0.45μ).

Insulin is dissolved first in a small amount of dilute HCl (0.005 N) and then prepared as described for pituitary hormones. Storage of protein hormone solutions for long periods of time should be avoided, since such solutions may prove unstable. It has also been shown that insulin, for example, adsorbs strongly to the surface of glass containers (Hill, 1959; Wiseman and Baltz, 1961).

Crystalline preparations of nonesterified steroid hormones, which are relatively insoluble in aqueous solution, are dissolved in absolute ethanol to yield the desired stock concentration. Sterilization is not required. The amount of alcohol solution added to the final medium should be kept as low as possible, about 0.5% or less.

Culture vessels

Most organ-culture assemblies for the mammary gland are based on the Petri dish-watch glass technique of Fell and Robison (1929) or its modification for liquid media by Chen (1954). Airtight assemblies of the Trowell-type (Trowell, 1959) or of the type devised by Moretti and Abraham (1966) have also been adapted for mammary studies, the latter specifically to enable measurements of CO_2 production. Grobstein culture vessels (Grobstein, 1956), widely used for tissue interaction studies, have recently been applied to mammary epithelium and mesenchyme interactions in culture (Kratochwil, 1969). The hanging-drop procedure has been modified to allow cultivation of mammary explants in small amounts of liquid media (Cooper *et al.*, 1967). Because of their labor-saving advantages and availability in bulk amounts, disposable, presterilized plastic dishes have, by and large, been the vessels commonly preferred. Falcon Plastics offers "organ-culture dishes" with built-in center well and ring of absorbent paper, with stainless-steel grids packaged separately. Other plastic disposable containers have also proved satisfactory, e.g., microdiffusion dishes (Stockdale and Topper, 1966), Petri dishes with two or more subdivisions (Barnawell, 1965), or standard size Petri dishes enclosing two or three small Petri dishes.

To maintain explants at the surface of liquid media, it is customary to employ a raft support of stainless-steel mesh, nylon, dacron, rayon acetate, or lens paper. Lens paper can be placed on top of stainless-steel grids, or, as is required for most of the polyester fabrics, treated with silicone to allow them to float without support. Siliconized lens paper may be prepared as follows. Cut lens paper into small squares (about 20×20 mm) and place them into a glass Petri dish. Add ether and set aside for several hours or overnight. After removing ether, rinse papers with several changes of distilled water, and immerse in a 10^3 aqueous dilution of silicone (Siliclad, Clay-Adams). Decant silicone solution, rinse with several changes of distilled water, and dry in a warm oven. (The preceding steps can all be carried out in a single dish.) With forceps, separate the dried papers and transfer them into clean Petri dishes. Sterilize in a dry heat oven at 120° to 130°C for four hours. The temperature and duration of sterilization are selected to avoid charring of the papers. It is useful to note that where the adipose tissue component is prominent, as in virgin and midpregnant mammary gland, the explants can float at the surface of liquid media

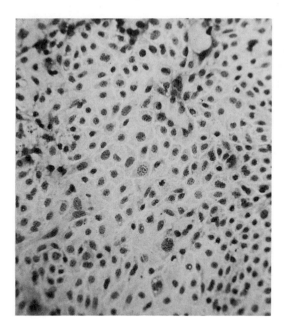

FIGURE 31-8

Monolayer culture of mouse mammary gland, showing representative area of epithelium; six-day culture. ×193. (From Dr. C. Daniel, University of California.)

without rafts (Mayne *et al.,* 1966; Mishkinsky *et al.,* 1967; El–Darwish and Rivera, 1970).

If the culture vessels are airtight, they can be gassed individually (Trowell, 1959; Moretti and Abraham, 1966). Vessels with loose-fitting tops, such as Petri dishes, are stacked into dessicators, large glass jars, or plastic boxes with tight-fitting lids. These boxes can be constructed from a variety of plastics, preferably heavy-duty polycarbonate. Where large-scale cultures are planned, the boxes may be connected in series with tubing (preferably silicone) to allow gassing from a single source. A mixture of 95% O_2 and 5% CO_2 is bubbled through distilled water in a gas-washing bottle into the glass or plastic container. When all the air has been displaced and the pH of the medium near neutral, the container may be sealed off. If continuous gas flow is desired, the container should be provided with a second opening for gas outlet. The humid atmosphere, which prevents drying of explants at the gas-medium interphase, is supplied by adding distilled water to the ring of absorbent paper in the Fell-type organ culture dish or by introducing dishes of distilled water into the container.

ANALYSIS OF DEVELOPMENT IN CULTURE

MORPHOLOGICAL METHODS

Identification of cell types in monolayer culture

The main cell types, epithelium and fibroblast, may be distinguished by the conventional criteria of morphology and behavior. Epithelium tends to form sheets of closely adherent polygonal cells (see Fig. 31-8), with free or ruffled membranes appearing at the edges of the colony. Fibroblasts are fusiform or stellate in shape with little tendency to form sheets; they move

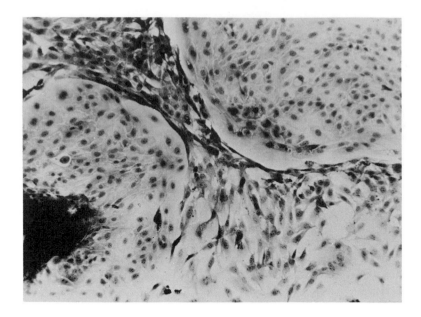

FIGURE 31-9

Monolayer culture of bovine mammary gland, showing epithelial areas separated by ridges of fibroblasts; six-day culture. ×91. (From Dr. B. L. Larson, from Ebner et al., 1961, Exp. Cell Res., 23:373–385.)

along the substrate as individual cells, whereas epithelial cells tend to move collectively in sheets. After cultures have been maintained for a week or more without passaging, fibroblasts tend to accumulate and pile up in ridges between sheets of epithelium (see Fig. 31-9). However, the two cell types do not appear to intermix.

A useful *in vivo* test for mammary epithelium would be to determine whether the cultivated cells are capable of generating mammary-gland outgrowths in gland-free fat pads (Daniel and DeOme, 1965). This method would demonstrate more conclusively the epithelial orgin of the cultivated cells. In this connection, Daniel (personal communication) noted recently that, although first-passage cultures of mouse mammary cells were entirely fibroblastic in appearance, these cells gave rise to mammary outgrowths when implanted into gland-free fat pads. The technique of clearing fat pads of mammary parenchyma is described by DeOme *et al.* (1959) and entails surgical removal of the mammary glands before they penetrate extensively into the fat pads. The remainder of the fat stroma is thus left in place and available for transplantation studies.

Phenotypic markers, such as β-lactoglobulin, are also useful in distinguishing cell types, especially during the early phases of cultivation (Twarog and Larson, 1962).

Fixation and staining

The fixation and staining of cultured cells and tissues follow, in general, the basic steps applied in standard histological technique, the choice of fixative and stain being determined by what one wishes to demonstrate. Several useful procedures are outlined by Merchant *et al.* (1964) and Paul (1965).

Monolayer cultures do not require embedding or sectioning, and may therefore be fixed and stained according to standard blood-smear procedures. A collodion stripping technique (Reissig *et al.,* 1956) is useful for removing monolayer cultures from large glass surfaces for subsequent fixation and staining. Cultures embedded in plasma clots should be washed thoroughly with a balanced salt solution before fixation to avoid artefacts caused by the proteinaceous medium.

Organ cultures are processed according to standard procedures for embedding in paraffin, sectioning, and mounting. Small explants, such as 1- to 2-mm alveolar lobules, may be prestained with 0.5% eosin in 70% ethanol to render them more readily visible during subsequent processing through the alcohols and paraffin. Barnawell (personal communication) has devised a tissue-processing container in which small explants may be placed for fixation, dehydration, and paraffin-infiltration in an autotechnicon. The container consists of a rectangular piece of stainless-steel mesh (20×50 mm), folded once, and enclosing a ring of polyvinyl tubing (10 mm in inside diameter). The explants are placed within the polyvinyl ring, and stainless-steel paper clips are used to hold together the two folds of the mesh. Explants may also be wrapped in lens paper and placed in standard-size autotechnicon containers.

Preparation of whole mounts

The procedure given below is modified from that in use at the Cancer Research Genetics Laboratory, University of California, Berkeley, in which iron hematoxylin stain is used. Ehrlich's hematoxylin, which is commercially available, is recommended here. Other stains, such as chromalum-gallocyanin (Prop, 1961) may be substituted for hematoxylin.

FIXATION. 15% formalin, overnight. To avoid folding, which tends to obscure the mammary parenchyma, spread the gland as flat as possible on a piece of lens paper. Place the paper with adhered gland into the fixative. The paper may be removed after fixation.

DEFATTING. Acetone, 24 hours; 100%, 95%, and 70% ethanol, one hour in each alcohol.

STAINING. Ehrlich's hematoxylin, 30 minutes. Glands will appear dark blue. Destain and differentiate in 1% HCl in 70% ethanol. Blue the glands in running tap water for 30 to 40 minutes.

CLEARING. 50%, 70%, 95%, and 100% ethanol, 5 minutes in each alcohol; toluene, two changes, 5 minutes each; store and examine in methyl salicylate.

After examination and photography of whole gland mounts, portions may be excised and prepared for histological sectioning, if desired.

Autoradiography

For the study of DNA synthesis by autoradiography (Stockdale *et al.*, 1966; Lockwood *et al.*, 1967a; Heuson *et al.*, 1967), explants are exposed to between 0.5 and 1.25 μc/ml of H^3-thymidine (specific activity, 6.0 to 7.2 c/mmole) in 2 ml of medium for 24 to 48 hours, fixed, sectioned at 5μ, and dipped in liquid photographic emulsion (Kodak, NTB 2). After a development time of 10 to 14 days, the sections are stained through the emulsion with hematoxylin (and eosin, if desired).

Preparation for electron microscopy

Explants are fixed for one hour in cold (4°C) 1% osmium tetroxide, buffered with acetate-veronal (*p*H 7.4) containing 0.045 g/ml sucrose. The tissues are then dehydrated in a cold, graded series of ethanols, and embedded in Epon. Sections are cut at 50 to 100 mμ, mounted on Formvar-coated copper specimen grids, stained with lead hydroxide, and examined in an electron microscope (Wellings *et al.*, 1966). Recently, explants have been fixed in cold 5% glutaraldehyde, post-fixed in osmium tetroxide, and embedded in Araldite (Michelson and Hagenau, 1969) or Maraglas (Mills and Topper, 1970).

BIOCHEMICAL METHODS

Nucleic acids

Various procedures are available for the preparation of tissue samples for subsequent determination of the nucleic acids in the tissue residue. Modifications of Schmidt and Thannhauser's procedure (1945) are commonly used, and involve extraction of the tissue with cold acid and lipid solvents, followed by incubation of the residue with alkali. RNA is degraded to acid-soluble ribomononucleotides, leaving DNA in acid-insoluble form. Acidification of the alkali digest results in an acid-soluble fraction containing RNA degradation products and a precipitate containing DNA. DNA and RNA are then determined in the separate fractions by one or more procedures. Table 31-2 lists procedures that have been applied to cultures of mammary tissue in some of the published work.

Slater (1961) and Denamur (1965) have discussed the difficulties presented by mammary tissues with respect to nucleic acid determinations. In a survey of several methods (diphenylamine, nitrophenylhydrazine, cysteine-sulfuric acid, and UV absorption for DNA; orcinol, bromphenylhydrazine, phosphorus, and UV absorption for RNA), Denamur found considerable variation in the results, particularly during the functional phases of the mammary gland. He therefore recommends alkaline digestion of defatted-dehydrated mammary extracts for 18 to 24 hours, followed by quantitative separation of the RNA 2', 3' mononucleotides by Dowex 1 column chromatography. The alkali-insoluble pellet containing DNA is digested with 1N

perchloric acid for one hour at 100°C, and the purine bases separated chromatographically on Amberlite resin. By these procedures, nucleic acid determinations may be made in as little as 5 mg of tissue residue (Denamur, personal communication).

Milk proteins

The terms used to designate the proteins of milk have often caused confusion: β-lactoglobulin, for example, which is synthesized in the mammary gland, refers to a discrete protein in the lactalbumin fraction, whereas the

TABLE 31-2

Procedures for nucleic acid determinations in mammary cultures

	Method	Study
DNA	Diphenylamine reaction	Heuson *et al.* (1967)
	UV absorption	Heuson *et al.* (1965)
	Incorporation of C^{14}– or H^3–thymidine	Stockdale and Topper (1966); Turkington and Topper (1966); Heuson *et al.* (1967) Turkington (1968a)
	Phosphorus	Heuson *et al.* (1965)
RNA	Orcinol reaction	Mayne *et al.* (1966)
	UV absorption	Mayne *et al.* (1966)
	Incorporation of H^3–uridine	Stockdale *et al.* (1966); Turkington (1968b)
	Incorporation of C^{14}–adenine (after prior separation of DNA)	Mayne *et al.* (1966)

lactoglobulin fraction refers to the immune proteins that are preformed in the blood. To clarify some of the confusion in terminology, the classic and contemporary nomenclature for the proteins of bovine milk are shown in Table 31-3.

CASEIN. (1) β-Casein. The synthesis of β-casein in monolayer cultures has been followed by measuring the rates of incorporation of H^3- or C^{14}-labeled essential amino acids in the casein fraction isolated by urea precipitation (Schingoethe *et al.*, 1967).

(2) Total casein. Total biosynthetic casein has been determined in mammary-organ cultures following cultivation with P_i^{32} or C^{14}-labeled amino acids (*Chlorella* hydrolysate). Casein is isolated by precipitation with carrier casein in the presence of rennin and calcium ions (Juergens *et al.*, 1965) and urea-gel electrophoresis (Turkington and Topper, 1966; Turkington, 1968b). The radioactivity of the four major casein bands revealed by electrophoresis comprises about 30% of that in the total calcium-rennin precipitable material (Turkington *et al.*, 1967).

β-LACTOGLOBULIN AND α-LACTALBUMIN. (1) β-lactoglobulin has been assayed in monolayer cultures by the following immunological procedures: precipitin reaction (Larson *et al.*, 1954; Ebner *et al.*, 1961b), agar diffusion (Larson and Twarog, 1961), and fluorescent-antibody technique (Larson and Twarog, 1962). The latter method is particularly helpful in the localization and identification of secretory cells in a mixed cell population.

TABLE 31-3

Proteins of bovine skim milk

Classical Nomenclature	Contemporary Nomenclature	Approximate % of Skim Milk Fraction	Other Characteristics
Casein (pptd. by acid at pH 4.6)	α–casein	45–63	Consists of a mixture of proteins; formed in the mammary gland
	β–casein	19–28	Formed in mammary gland
	γ–casein	3–7	Preformed in blood
Lactalbumin (sol. in ½ sat. ammonium sulfate	β–lactoglobulin	7–12	Formed in mammary gland
	α–lactalbumin	2–5	Formed in mammary gland
	Serum albumin	0.7–1.3	Preformed in blood
Lactoglobulin (γ–globulins; insol. in ½ sat. ammonium sulfate)	Euglobulin	0.8–1.7	Preformed in blood
	Pseudoglobulin	0.6–1.4	Preformed in blood
Proteose-peptone		2–6	Incompletely defined

SOURCE: Abbreviated from Thompson *et al.* (1965).

(2) Chemical procedures for β-lactoglobulin and α-lactalbumin involve cultivation with H^3- or C^{14}-labeled amino acids for between 4 and 24 hours. After the addition of skim milk carrier, the labeled proteins are isolated from the cultures and/or medium by differential precipitation with ammonium sulfate and cellulose-acetate strip electrophoresis (Groves and Larson, 1965; Lockwood *et al.*, 1966; Schingoethe *et al.*, 1967; Turkington, 1968b). In view of species differences in milk constituents, some caution should be taken in identifying milk whey proteins. Observations indicate that α-lactalbumin is widely distributed among mammals, but that β-lactoglobulin may be confined to milk of ruminants (Brew and Campbell, 1967). Thus, the mouse whey protein identified as β-lactoglobulin by comparing its electrophoretic mobility with the bovine protein (Lockwood *et al.*, 1966; Turkington, 1968b) should be characterized further before its final identity can be established.

Nonmilk protein

Two approaches have been used to measure cell protein in the presence of milk. In the first, total protein is measured colorimetrically (Lowry, 1951) and then corrected for retained milk (Folley and Greenbaum, 1947) to obtain an estimate of cell protein (Ebner *et al.*, 1961b). It should be noted that the calculation of retained milk is based on the percentage of lactose in milk, which is assumed to be constant at all times. When applied to protein determinations, it is also assumed that the ratio "milk lactose: milk protein" is also constant.

The second approach has been applied to mammary-organ cultures, and involves determination of protein in epithelial cells separated from the stroma at the time of assay. Tissues are cultivated with labeled amino acids, dissociated with collagenase according to the method of Lasfargues (1957b), casein removed, and radioactivity measured in the ammonium-sulfate-precipitable material (Lockwood *et al.*, 1966).

Lactose

Lactose may be determined in the medium or in cultivated cells by paper electrophoresis (Robinson and Rathbun, 1958) or by descending paper chromatography (Jerym and Isherwood, 1949), following deproteinization, deionization, and lyophilization of the medium and/or cells (Ebner *et al.*, 1961b; Twarog and Larson, 1964). For incorporation studies, the cultures are given 24-hour pulses of either glucose- or galactose-C^{14}, and radioactivity measured in the separated lactose (Ebner *et al.*, 1961b; Twarog and Larson, 1964).

Glycogen

Glycogen content can be measured by the anthrone reaction (Twarog and Larson, 1964) in cell extracts prepared according to the method of Roe *et al.* (1961), or by measuring the incorporation of glucose-C^{14} into glycogen isolated by paper electrophoresis (Ebner *et al.*, 1961b).

Glucose

Reducing sugars in cultivated cells have been determined by the methods of Somogyi (1945) and Nelson (1944) in deproteinized cell filtrates (Twarog and Larson, 1964), using the milk correction factor of Folley and Greenbaum (1947).

The depletion of glucose from the medium can be used as a measure of glucose utilization by cultivated tissue (Moretti and DeOme, 1962; Moretti and Abraham, 1966; Heuson *et al.*, 1967; Heuson and Legros, 1968), wherein glucose is estimated by the Somogyi (1945) or glucose oxidase method (Glucostat, Worthington Biochemicals).

Lactic acid, fatty acid, and CO_2 production from glucose-C^{14} may be determined by using the culture vessel described by Moretti and Abraham (1966), in which CO_2 is collected by hyamine in a small side well. Explants are cultivated with glucose-C^{14}, 2×10^5 counts/min/ml, and assays of the labeled compounds made at 24-hour intervals (Moretti and Abraham, 1966).

Enzymes

Many of the enzyme studies have been carried out in *mammary monolayer cultures,* and the assay procedures for the following enzymes are given in the papers of Larson and colleagues (Ebner *et al.,* 1961b; Twarog and Larson, 1964; Brodbeck and Ebner, 1966; Blanco *et al.,* 1967):

> Acid phosphatase
> Alkaline phosphatase
> β-galactosidase
> Catalase
> DPN cytochrome c reductase
> Galactose-1-phosphate uridyl transferase
> Glucose-6-phosphate dehydrogenase
> Hexokinase
> Isocitric dehydrogenase
> Lactate dehydrogenase
> Lactoperoxidase
> Lactose synthetase
> UDP-gal-4-epimerase
> UDPG pyrophosphorylase

The following enzymes have been assayed in *mammary organ cultures:*

> ATP citrate lyase (Jones and Forsyth, 1969);
> DNA polymerase (Lockwood *et al.,* 1967b; Turkington and Ward, 1969a);
> Glucose-6-phosphate and 6-phosphogluconate dehydrogenase (Jones and Forsyth, 1969; Leader and Barry, 1969; Rivera, 1969; Rivera and Cummins, 1969);
> Lactate dehydrogenase (Jones and Forsyth, 1969);
> Lactose synthetase (Turkington *et al.,* 1968; Turkington and Hill, 1969; Palmiter, 1969);
> Malate dehydrogenase (Jones and Forsyth, 1969);
> Malic enzyme (Jones and Forsyth, 1969);
> RNA polymerase (Turkington and Ward, 1969b);
> UDPG pyrophosphorylase (Jones and Forsyth, 1969).

Mammary tumor virus

An assay for mouse mammary tumor virus has been developed by Cardiff *et al.* (1968a,b) using radioisotope labeling and immunological precipitation techniques.

BIBLIOGRAPHY

Balinsky, B. I. (1950) On the developmental processes in mammary glands and other epidermal structures. Trans. Roy. Soc., Edinburgh, 62:1–31.

Barker, B. E., H. Fanger, and P. Farnes (1964) Human mammary slices in organ culture. I. Method of culture and preliminary observations on the effect of insulin. Exp. Cell Res., 35:437–448.

Barnawell, E. B. (1965) A comparative study of the responses of mammary tissues from several species to hormones *in vitro*. J. Exp. Zool., 160:189–206.

——— (1967) Analysis of the direct action of prolactin and steroids on mammary tissue of the dog in organ culture. Endocrinology, 80:1083–1089.

Blanco, A., U. Rifé, and B. L. Larson (1967) Lactate dehydrogenase isoenzymes during dedifferentiation in cultures of mammary secretory cells. Nature, 214:1331–1333.

Brodbeck, U., and K. E. Ebner (1966) The subcellular distribution of the A and B proteins of lactose synthetase in bovine and rat mammary tissue. J. Biol. Chem., 241:5526–5532.

Brew, K., and P. N. Campbell (1967) The characterization of the whey proteins of guinea-pig milk. Biochem. J., 102:258–264.

Cardiff, R. B., P. B. Blair, and K. B. DeOme (1968a) *In vitro* cultivation of mouse mammary tumor virus: replication of MTV in tissue culture. Virology, 36:313–317.

Cardiff, R. D., P. B. Blair, and P. Nakayama (1968b) *In vitro* cultivation of mouse mammary tumor virus: detection of MTV production by radioisotope labeling and identification by immune precipitation. Proc. Nat. Acad. Sci., 59:895–902.

Ceriani, R. L., and H. A. Bern (1968) Hormonal stimulation of rat fetal mammary rudiment in organ culture. In: Proc. Third Internat. Congr. Endocrinol., Internat. Congr. Series, no. 157, Excerpta Medica, abstr. 414.

Chapekar, T. N., and K. J. Ranadive (1963) *In vitro* studies on mouse mammary gland response to hormonal treatment. Indian J. Exp. Biol., 1:167–171.

Chen, J. M. (1954) The cultivation in fluid medium of organized liver, pancreas, and other tissues of foetal rats. Exp. Cell Res., 7:518–529.

Cooper, R. A., V. L. Jentoft, and S. R. Wellings (1967) A dish for hanging-drop organ culture, with particular reference to endocrine tissues. Am. Zool., 7:201.

Daniel, C. W., and K. B. DeOme (1965) Growth of mouse mammary glands *in vivo* after monolayer culture. Science, 149:634–636.

David, D., and A. Propper (1964) Sur la culture organotypique de la glande mammaire embryonnaire du Lapin. C. R. Soc. Biol., 158:2315–2317.

Denamur, R. (1965) Les acides nucléiques et les nucléotides libres de la glande mammaire pendant la lactogénèse et la galactopoïèse. In: Proc. Second Internat. Congr. Endocrinol., Internat. Congr. Series, no. 83, Excerpta Medica, pp. 434–462.

DeOme, K. B. (1963) The role of the mammary tumor virus in mouse mammary noduligenesis and tumorigenesis. In: Viruses, Nucleic Acids, and Cancer, Seventeenth Ann. Symp. Fundam. Cancer Res., pp. 498–507.

———, L. J. Faulkin, H. A. Bern, and P. B. Blair (1959) Development of mammary tumors from hyperplastic alveolar nodules transplanted into gland-free mammay fat pads of female C3H mice. Cancer Res., 19:515–520.

——— and S. Nandi (1966) The mammary-tumor system in mice, a brief review. In: Viruses Inducing Cancer. Ed. by W. J. Burdette, pp. 127–137.

————, S. Nandi, H. A. Bern, P. B. Blair, and D. R. Pitelka (1962) The preneoplastic hyperplastic alveolar nodule as the morphologic precursor of mammary cancer in mice. In: Proceedings of the International Conference on the Morphologic Precursors of Cancer, pp. 349–368.

Dickson, J. A. (1966) Tissue culture approach to the treatment of cancer. Brit. Med. J., 5491:817–823.

Dieterlen-Lièvre, F. (1964) Action d'hormones et de substances inhibitrices sur des organes cultivés *in vitro*. In: Les Cultures Organotypiques. Ed. by J. André-Thomas, pp. 135–172.

Dilley, W. G., and S. Nandi (1968) Rat mammary gland differentiation *in vitro* in the absence of steroids. Science, 161:59–60.

Dunn, T. (1959) Morphology of mammary tumors in mice. In: The Physiopathology of Cancer. Second ed. Ed. by F. Homburger, pp. 34–84.

Ebner, K. E., C. R. Hoover, E. C. Hageman, and B. L. Larson (1961a) Cultivation and properties of bovine mammary cell cultures. Exp. Cell Res., 23:373–385.

Ebner, K. E., E. C. Hageman, and B. L. Larson (1961b) Functional biochemical changes in bovine mammary cell cultures. Exp. Cell Res., 25:555–570.

El-Darwish, I., and E. M. Rivera (1970) Temporal efforts of hormones on DNA synthesis in mouse mammary gland *in vitro*. J. Exp. Zool., 173:285–292.

Elias, J. J. (1957) Cultivation of adult mouse mammary gland in hormone-enriched synthetic medium. Science, 126:842–844.

———— (1959) Effect of insulin and cortisol on organ cultures of adult mouse mammary gland. Proc. Soc. Exp. Biol. Med., 101:500–502.

———— (1961) Stimulation of secretion by insulin in organ cultures of mouse mammary duct end-buds. Anat. Record, 139:224.

———— (1962a) Response of mouse mammary duct end-buds to insulin in organ culture. Exp. Cell Res., 27:601–604.

———— (1962b) Normal, preneoplastic, and neoplastic mouse mammary tissues in organ cultures. In: Biological Interaction in Normal and Neoplastic Growth. Ed. by M. J. Brennan and W. L. Simpson, pp. 355–370.

———— and E. M. Rivera (1959) Comparison of the responses of normal, precancerous, and neoplastic mouse mammary tissues to hormones *in vitro*. Cancer Res., 19:505–511.

Fekete, E. (1938) A comparative morphological study of the mammary gland in a high- and a low-tumor strain. Am. J. Pathol., 14:557–578.

Fell, H. B. (1964) The role of organ cultures in the study of vitamins and hormones. Vitamins and Hormones, 22:81–127.

———— and R. Robison (1929) The growth, development, and phosphatase activity of embryonic avian femora and limb buds cultivated *in vitro*. Biochem. J., 23:767–784.

Folley, S. J., and A. L. Greenbaum (1947) Changes in the arginase and alkaline phosphatase contents of the mammary gland and liver of the rat during pregnancy, lactation, and mammary involution. Biochem. J., 41:261–269.

Gadkari, S. V., T. N. Chapekar, and K. J. Ranadive (1968). Response of mouse mammary glands to hormone treatment *in vitro*. Indian J. Exp. Biol., 6:75–79.

Gerritsen, G. C. (1960) Hormonal requirements for guinea pig mammary tissue *in vitro*. Fed. Proc., 19:384.

Grobstein, C. (1956) Transfilter induction of tubules in mouse metanephrogenic mesenchyme. Exp. Cell Res., 10:424–440.

Groves, T. D. D., and B. L. Larson (1965) Preparation of specifically labeled milk proteins using bovine mammary cell culture. Biochim. Biophys. Acta, 104:462–469.

Hardy, M. (1950) The development *in vitro* of the mammary glands of the mouse. J. Anat., 84:388–393.

Heston, W. E., G. Vlahakis, and Y. Tsubura (1964) Strain DD, a new high mammary tumor strain, and comparison of DD with strain C3H. J. Nat. Cancer Inst., 32:237–251.

Heuson, J. C., and N. Legros (1968) Study of the growth-promoting effect of insulin in relation to carbohydrate metabolism in organ culture of rat mammary carcinoma. European J. Cancer, 4:1–7.

Heuson, J. C., N. Legros, and A. Coune (1965) Mésure des acides nucléiques et des protéines tissulaires en culture d'organe. Rev. franç. études clin. et biol., 10:661–663.

Heuson, J. C., A. Coune, and R. Heimann (1967) Cell proliferation induced by insulin in organ culture of rat mammary carcinoma. Exp. Cell Res., 45:351–360.

Hill, J. B. (1959) Adsorption of I^{131}-insulin to glass. Endocrinology, 65:515–517.

Ichinose, R. R., and S. Nandi (1964) Lobulo-alveolar differentiation in mouse mammary tissues *in vitro*. Science, 145:496–497.

―――― (1966) Influence of hormones on lobulo-alveolar differentiation of mouse mammary glands *in vitro*. J. Endocrinol., 35:331–340.

Ingraham, R. L. (1968) Variability of mouse mammary hyperplastic nodules maintained in organotypic culture. J. Nat. Cancer Inst., 40:1273–1285.

Jerym, M. A., and F. A. Isherwood (1949) Improved separation of sugars on the paper partition chromatogram. Biochem. J., 44:402–407.

Jones, E. A. (1969) Recent developments in the biochemistry of the mammary gland. J. Dairy Res., 36:145–167.

―――― and I. A. Forsyth (1969) Increases in the enzyme activity of mouse mammary explants induced by prolactin. J. Endocrinol., 43:41–42.

Juergens, W. G., F. E. Stockdale, Y. J. Topper, and J. J. Elias (1965) Hormone-dependent differentiation of mammary gland *in vitro*. Proc. Nat. Acad. Sci., 56:629–634.

Kalus, M., L. Delmonte, J. J. Ghidoni, and R. A. Liebelt (1968) Transplantation of mouse mammary carcinoma through matrix tissue cultures. Tex. Repts. Biol. Med., 26:517–524.

Koziorowska, J. (1962) The influence of ovarian hormones and insulin on the mouse mammary glands cultivated *in vitro*. Acta Med. Polona, 3:237–245.

Kratochwil, K. (1969) Organ specificity in mesenchymal induction demonstrated in the embryonic development of the mammary gland of the mouse. Dev. Biol., 20:46–71.

Larson, B. L. (1967) Biochemical dedifferentiation in the *in vitro* mammary secretory cell. In: Carcinogenesis: A Broad Critique, Twentieth Ann. Symp. Fundament. Cancer Research, pp. 607–618.

―――― (1969) Biosynthesis of milk. J. Dairy Sci., 52:737–747.

――――, R. S. Gray, and G. W. Salisbury (1954) The proteins of bovine seminal plasma. II: Ultracentrifugal and immunological studies and comparison with blood and milk serm. J. Biol. Chem., 211:43–52.

Lasfargues, E. Y. (1957a) Cultivation and behavior *in vitro* of the normal mammary epithelium of the adult mouse. Anat. Record, 127:117–125.

―――― (1957b) Cultivation and behavior *in vitro* of the normal mammary epithelium of the adult mouse. II: Observations on the secretory activity. Exp. Cell Res., 13:553–562.

―――― (1960). Action de l'oestradiol et de la progestérone sur des cultures de glandes mammaires de jeunes Souris. C. R. Soc. Biol., 154:1720–1722.

―――― (1962) Concerning the role of insulin in the differentiation and functional activity of mouse mammary tissues. Exp. Cell Res., 28:531–542.

―――― and M. R. Murray (1959) Hormonal influences on the differentiation and growth of embryonic mouse mammary glands in organ culture. Dev. Biol., 1:413–435.

Lasnitzki, I. (1965) The action of hormones on cell and organ cultures. In: Cells and Tissues in Culture. I. Ed. by E. N. Willmer, pp. 591–658.

Leader, D. P., and J. M. Barry (1969) Increase in activity of glucose-6-phosphate dehydrogenase in mouse mammary tissue cultured with insulin. Biochem. J., 113:175–182.

Lockwood, D. H., R. W. Turkington, and Y. J. Topper (1966) Hormone-dependent development of milk protein synthesis in mammary gland *in vitro*. Biochim. Biophys. Acta, 130:493–501.

Lockwood, D. H., F. E. Stockdale, and Y. J. Topper (1967a) Hormone-dependent differentiation of mammary gland: sequence of action of hormones in relation to cell cycle. Science, 156:945–946.

Lockwood, D. H., A. E. Voytovich, F. E. Stockdale, and Y. J. Topper (1967b) Insulin-dependent DNA polymerase and DNA synthesis in mammary epithelial cells *in vitro*. Proc. Nat. Acad. Sci., 58:658–664.

Lowry, O. H., N. J. Rosebrough, A. L. Farr, and R. J. Randall (1951) Protein measurement with the Folin phenol reagent. J. Biol. Chem., 193:265–275.

Lyons, W. R., C. H. Li, and R. E. Johnson (1958) The hormonal control of mammary growth and lactation. Rec. Progr. Horm. Res., 14:219–248.

Mayne, R., J. M. Barry, and E. M. Rivera (1966) Stimulation by insulin of the formation of ribonucleic acid and protein by mammary tissues *in vitro*. Biochem. J., 99:688–693.

Mayne, R., I. A. Forsyth, and J. M. Barry (1968) Stimulation by hormones of RNA and protein formation in organ cultures of the mammary glands of pregnant mice. J. Endocrinol., 41:247–253.

Merchant, D. J., R. H. Kahn, and W. H. Murphy (1964) Handbook of Cell and Organ Culture. 2d ed. Burgess, Minneapolis.

Michelson, S. F., and F. Haguenau (1969) Organ cultures of mouse mammary tumors in various hormonal environments: an ultrastructural study. J. Nat. Cancer Inst., 42:545–558.

Mills, E. S., and Y. J. Topper (1970) Some ultrastructural effects of insulin, hydrocortisone, and prolactin on mammary gland explants. J. Cell Biol., 44:310–328.

Mishkinsky, J., S. Dikstein, M. Ben-David, J. Azeroual, and F. G. Sulman (1967) A sensitive *in vitro* method for prolactin determination. Proc. Soc. Exp. Biol. Med., 125:360–363.

Moretti, R. L., and K. B. DeOme (1962) Effect of insulin on glucose uptake by normal and neoplastic mouse mammary tissues in organ culture. J. Nat. Cancer Inst., 29:321–329.

Moretti, R. L., and S. Abraham (1966) Effects of insulin on glucose metabolism by explants of mouse mammary gland maintained in organ culture. Biochim. Biophys. Acta, 124:280–288.

Munford, R. E. (1964) A review of anatomical and biochemical changes in the mammary gland with particular reference to quantitative methods of assessing mammary development. Dairy Sci. Abstr., 26:293–304.

Nandi, S. (1959) Hormonal control of mammogenesis and lactogenesis in the C3H/He Crgl mouse. Univ. Calif. Publ. Zool., 65:1–128.

Nelson, N. (1944) A photometric adaptation of the Somogyi method for the determination of glucose. J. Biol. Chem., 153:375–380.

Nicoll, C. S., H. A. Bern, and D. B. Brown (1966) Occurrence of mammotrophic activity (prolactin) in the vertebrate adenohypophysis. J. Endocrinol., 34:343–354.

Nikiforova, E. N. (1960) Growth characteristics in mammary tissue cultures from mice of a noncancer lineage (E) associated with the addition of androsterone and estradiol-17β. Bull. Eks. Biol. Med. (U.S.S.R.), 49:92–97.

Palmiter, R. D. (1969) Hormonal induction and regulation of lactose synthetase in mouse mammary gland. Biochem. J., 113:409–417.

Patton, S. (1969) Milk. Sci. Am., 221:58–68.

Paul, J. (1965) Cell and Tissue Culture, 3d ed. Williams & Wilkins, Baltimore.

Prop, F. J. A. (1961) Effects of hormones on mouse mammary glands in vitro: Analysis of the factors that cause lobulo-alveolar development. Pathol. Biol., 9:640–645.

―――― (1966) Effect of donor age on hormone reactivity of mouse mammary gland organ cultures. Exp. Cell Res., 42:386–387.

―――― and S. E. A. M. Hendrix (1965) Effect of insulin on mitotic rate in organ cultures of total mammary glands of the mouse. Exp. Cell Res., 40:277–281.

Propper, A. (1968) Relations épidermo-mésodermiques dans la différenciation de l'ébauche mammaire d'embryon de lapin. Ann. Embryol. Morphogen., 1:151–160.

―――― (1969) Compétence de l'épiderme embryonnaire d'Oiseau vis-à-vis l'inducteur mammaire mésenchymateux. C. R. Acad. Sci., 268:1423–1426.

―――― and L. Gomot (1967) Interactions tissulaires au cours de l'organogenèse de la glande mammaire de l'embryon de Lapin. C. R. Acad. Sci., 264:2573–2575.

Raynaud, A. (1961) Morphogenesis of the mammary gland. In: Milk: The Mammary Gland and its Secretion. I. Ed. by S. K. Kon and A. T. Cowie, pp. 3–46.

―――― and J. Raynaud (1956) La production expérimentale de malformations mammaires chez le foetus de Souris, par l'action des hormones sexuelles. Ann. Inst. Pasteur, 90:187–219.

Reissig, M., D. W. Howes, and J. L. Melnick (1956) Sequence of morphological changes in epithelial cell cultures infected with poliovirus. J. Exp. Med., 104:289–304.

Richardson, F. L., and K. P. Hummel (1959) Mammary tumors and mammary gland development in virgin mice of strains C3H, RIII, and their F1 hybrids. J. Nat. Cancer Inst., 23:91–107.

Rivera, E. M. (1963) Hormonal requirements for survival and growth of mouse primary mammary ducts in organ culture. Proc. Soc. Exp. Biol. Med., 114:735–738.

―――― (1964a) Maintenance and development of whole mammary glands of mice in organ culture. J. Endocrinol., 30:33–39.

―――― (1964b) Differential responsiveness to hormones of C3H and A mouse mammary tissues in organ culture. Endocrinology 74:853–864.

―――― (1969) Some observations on the activities of HGH and HPL in mouse mammary organ cultures. In: Lactogenesis: The Initiation of Milk Secretion at Parturition. Ed. by M. Reynolds and S. J. Folley.

―――― and E. P. Cummins (1969) Hormonal stimulation of pentose phosphate enzymes in organ cultures of mouse mammary gland. Am. Zool., 9:578.

――――, J. J. Elias, H. A. Bern, N. P. Napalkov, and D. R. Pitelka (1963) Toxic effects of steroid hormones in organ cultures of mouse mammary tumors, with a comment on the occurrence of viral inclusion bodies. J. Nat. Cancer Inst., 31:671–687.

――――, I. A. Forsyth, and S. J. Folley (1967) Lactogenic activity of mammalian growth hormones in vitro. Proc. Soc. Exp. Biol. Med., 124:859–865.

Robison, H. M. C., and J. C. Rathbun (1958) Electrophoresis of free sugars in blood. Science, 127:1501–1502.

Roe, J. H., J. M. Bailey, R. R. Gray, and J. N. Robison (1961) Complete removal of glycogen from tissues by extraction with cold trichloroacetic acid solution. J. Biol. Chem., 236:1244–1246.

Rovin, S. (1962) The influence of carbon dioxide on the cultivation of human neoplastic explants in vitro. Cancer Res., 22:384–387.

Schingoethe, D. J., E. C. Hageman, and B. L. Larson (1967) Essential amino acids for milk protein synthesis in the *in vitro* secretory cell and stimulation by elevated levels. Biochim. Biophys. Acta, 148:469–474.

Schmidt, G., and S. J. Thannhauser (1945) A method for the determination of desoxyribonucleic acid, ribonucleic acid, and phosphoproteins in animal tissues. J. Biol. Chem., 161:83–89.

Slater, T. F. (1961) Interference in the diphenylamine reaction procedure for estimation of desoxyribonucleic acid. Nature, 189:834–835.

Somogyi, M. (1945). A new reagent for the determination of sugars. J. Biol. Chem., 160:61–73.

Stadie, W. C., and B. C. Riggs (1944) Microtome for the preparation of tissue slices for metabolic studies of surviving tissues *in vitro*. J. Biol. Chem., 154:687–690.

Stockdale, F. E., and Y. J. Topper (1966) The role of DNA synthesis and mitosis in hormone-dependent differentiation. Proc. Nat. Acad. Sci., 56:1283–1289.

Stockdale, F. E., W. G. Juergens, and Y. J. Topper (1966) A histological and biochemical study of hormone-dependent differentiation of mammary gland tissue *in vitro*. Dev. Biol., 13:266–281.

Thompson, M. P., N. P. Tarassuk, R. Jenness, H. A. Lillevik, U. S. Ashworth, and D. Rose (1965) Nomenclature of the proteins of cow's milk. 2nd rev. J. Dairy Sci., 48:159–169.

Topper, Y. J. (1968) Multiple hormone interactions related to the growth and differentiation of mammary gland *in vitro*. Trans. N.Y. Acad. Sci., 30:869–874.

Trowell, O. A. (1959) The culture of mature organs in a synthetic medium. Exp. Cell Res., 16:118–147.

Turkington, R. W. (1968a) Hormone-induced synthesis of DNA by mammary gland *in vitro*. Endocrinology, 82:540–546.

——— (1968b) Induction of milk protein synthesis by placental lactogen and prolactin *in vitro*. Endocrinology, 82:575–583.

——— (1968c) Hormone-dependent differentiation of mammary gland *in vitro*. In: Current Topics in Developmental Biology. III. Ed. by A. A. Moscona and A. Monroy, pp. 199–218.

——— and Y. J. Topper (1966) Stimulation of casein synthesis and histological development of mammary gland by human placental lactogen *in vitro*. Endocrinology, 79:175–181.

——— and R. Hilf (1968) Hormonal dependence of DNA synthesis in mammary carcinoma cells *in vitro*. Science, 160:1457–1459.

——— and R. L. Hill (1969) Lactose synthetase: progesterone inhibition of the induction of α-lactalbumin. Science, 163:1458–1460.

——— and M. Riddle (1969) Acquired hormonal dependence of milk protein synthesis in mammary carcinoma cells. Endocrinology, 84:1213–1217.

——— and O. T. Ward (1969a) DNA polymerase and DNA synthesis in mammary carcinoma cells. Biochim. Biophys. Acta, 174:282–290.

——— and O. T. Ward (1969b) Hormonal stimulation of RNA polymerase in mammary gland *in vitro*. Biochim. Biophys. Acta, 174:291–301.

———, D. H. Lockwood, and Y. J. Topper (1967) The induction of milk protein synthesis in post-mitotic mammary epithelial cells exposed to prolactin. Biochim. Biophys. Acta, 148:475–480.

———, K. Brew, T. C. Banaman, and R. L. Hill (1968) The hormonal control of lactose synthetase in the developing mouse mammary gland. J. Biol. Chem., 243:3382–3387.

Turner, C. W. (1952) The Mammary Gland. Lucas, Columbia, Mo.

Twarog, J. M., and B. L. Larson (1962) Cellular identification of β-lactoglobulin synthesis in bovine mammary cell cultures. Exp. Cell Res., 28:350–359.

———— (1964) Induced enzymatic changes in lactose synthesis and associated pathways of bovine mammary cell cultures. Exp. Cell Res., 34:88–99.

Voytovich, A. E., and Y. J. Topper (1967) Hormone-dependent differentiation of immature mouse mammary gland *in vitro*. Science, 158:1326–1327.

Waymouth, C. (1966) Cell, tissue and organ culture. In: Biology of the Laboratory Mouse. Ed. by E. S. Green, pp. 493–510.

Wellings, S. R., R. A. Cooper, and E. M. Rivera (1966) Electron microscopy of induced secretion in organ cultures of mammary tissue of the C3H/Crgl mouse. J. Nat. Cancer Inst., 36:657–672.

Welsch, C. W., and E. M. Rivera (1970) The differential effects of prolactin and estrogen on growth of dimethylbenzanthracene (DMBA) —induced rat mammary tumors *in vitro*. In: Tenth International Cancer Congress, Houston, Texas.

Whitescarver, J., L. Recher, J. A. Sykes, and L. Briggs (1968) Problems involved in culturing human breast tissue. Tex. Repts. Biol. Med., 26:613–628.

Wiseman, R., and B. Baltz (1961) Prevention of insulin-I^{131} adsorption to glass. Endocrinology, 68:354–356.

Wolff, Em. (1964) Chimères d'organes et cultures organotypiques de tumeurs cancéreuses. In: Les Cultures Organotypiques. Ed. by J. André-Thomas, pp. 337–376.

Wolff, Et., and K. Haffen (1952) Sur une méthode de culture d'organes embryonnaires *in vitro*. Tex. Repts. Biol. Med., 10:463–472.

Wolff, Et., and M. F. Sigot (1961) Les divers aspects d'une tumeur mammaire de Souris, associé à différents organes embryonnaires de Poulet en culture *in vitro*. C. R. Soc. Biol., 155:960–962.

Wolff, Et., and Em. Wolff (1961) Le rôle de mésonéphros de l'embryon de Poulet dans la nutrition de cellules cancéreuses, II: Etude par la méthode de la membrane vitelline. J. Embryol. Exp. Morphol., 9:678–690.

32

Culture of Mammalian Embryonic Ovaries and Oviducts

RUTH E. RUMERY
Department of Biological Structure
University of Washington
Seattle, Washington 98105

ELIZABETH PHINNEY
Department of Biological Structure
University of Washington
Seattle, Washington 98105

RICHARD J. BLANDAU
Department of Biological Structure
University of Washington
Seattle, Washington 98105

The techniques described in this chapter are intended for the use of graduate students and other investigators who have not had previous experience in tissue-culture methods. All procedures will be described in sufficient detail so that the beginner should be able to avoid many of the pitfalls and frustrations experienced by those just starting to work in this field.

Certain requirements must be satisfied in order to make the information gained from studies of living tissues and organs have meaning and be of value to the investigator. The tissue-culture medium is of prime importance; it must be adequate not only for survival of the tissues, but also for their growth and development within normal limits. Since no one medium can possibly meet the needs of all tissues, the composition of the medium will depend on the kind of tissue to be cultured. One must protect the culture against contamination from microorganisms, by applying sterile techniques and by using antibiotics. When the tissues have become established in culture, the medium must be renewed at regular intervals, so that the metabolism of the tissues is disturbed as little as possible. One should bear in mind that the organs and tissues in the living animal body are

The techniques described in this chapter have been developed largely under grants from the National Institute of Health, United States Public Health Service. Expert photographic assistance was provided by Roy Hayashi.

constantly influenced by circulating fluids that vary as the need for them dictates, and when an attempt is made to imitate these conditions *in vitro,* both knowledge of the correct procedures and extreme care are required.

CLEANING AND STERILIZATION

The preparation of the glassware for culturing is of prime importance; not only do the cells and tissues come in direct contact with the glass surfaces, but all constituents of the culture media are stored in and dispensed from glass containers. Glass syringes and pipettes that are used in the culture procedures must be cleaned with care. Most household detergents cannot be used for cleaning tissue-culture glassware, because residual materials remain on the glass surface. These are toxic to the tissues and are almost impossible to remove regardless of the amount of rinsing or other precautionary measures. Only those detergents which are manufactured specifically for cleaning tissue-culture glassware should be used.

*Glass Petri dishes, flasks, culture tubes and steel plates
of the Rose chambers*

1. Soak in tap water overnight.
2. Place the glassware in a stainless-steel pail filled with tap water in which Micro-Solv has been dissolved, and boil for 30 minutes
3. Refill the pail with warm water and Micro Solv. Scrub each piece thoroughly and rinse well in running tap water.
4. Transfer to another stainless steel pail containing 0.1N HCl, and allow the glassware to remain in the acid bath for 30 minutes.
5. As each piece is removed from the acid bath, it should be rinsed thoroughly, seven times in running tap water and three times in distilled water. All glassware should be dried completely before it is prepared for sterilization.

The Petri dishes are assembled and stacked in stainless steel canisters. All culture tubes are sorted by size and packed, open-end down, in stainless steel canisters. The Rose chamber plates are paired (top and bottom plates) and placed in stainless-steel canisters. All these are sterilized by dry heat at 180°C for two hours.

The bottles and flasks, all of which have screw caps, must be sterilized in an autoclave. The mouth of each bottle or flask is covered with Patapar paper, over which the cap is screwed loosely. Another piece of Patapar paper is wrapped over the cap and neck of the bottle and secured with autoclave tape.

Cover glasses

1. Wear surgical gloves while handling and washing the cover glasses in order to avoid finger marks on the glass surfaces.

2. Put each cover glass individually into a beaker of tap water to which Micro-Solv has been added. Boil for 30 minutes.
3. Discard and make a fresh solution for washing the cover glasses.
4. Hold each cover glass by opposite edges while scrubbing it between two nylon toothbrushes.
5. Rinse each side of the cover glass seven times in running tap water and three times in distilled water. Cover glasses are not put in an acid bath.
6. Store cover glasses in a covered glass container filled with absolute ethanol.
7. In order to avoid the formation of lint, dry each cover glass with lens paper and put in a 100×20 mm Petri dish. Place a piece of #50 Whatman filter paper, cut to the same size, between each cover glass. Each Petri dish will hold 20 cover glasses.
8. Sterilize in an autoclave.

Pipettes

1. Soak and boil in the manner described for the cover glasses. Pipettes are not placed in an acid bath.
2. Rinse each pipette seven times in running tap water and three times in distilled water by applying suction to draw the water through the pipettes.
3. Rinse in 95% ethanol to remove the excess water.
4. Drain and dry thoroughly in a pipette rack.
5. Plug the mouthpiece of each pipette with bacteriological cotton and place in a stainless steel container with the graduated tip down.
6. Sterilize in an autoclave.

Syringes

1. Soak and boil as described for the cover glasses. Syringes are not placed in an acid bath.
2. Rinse each syringe seven times in running tap water and three times in distilled water.
3. Invert barrel to drain and dry thoroughly.
4. When ready to sterilize, each syringe is placed, unassembled, in a sterilizer bag. The bag is completely closed by folding the open end over twice and fastening it with staples. Sterilize in an autoclave.

Instruments

Clean all metal instruments by brushing and washing in Micro-Solv and hot water. Wipe with a lintless linen towel. All instruments used for tissue culture may be safely sterilized by placing them in a cold germicidal solution. We have found zephiran chloride to be very useful for this purpose. The solution is compounded as follows: zephiran chloride, 30 ml; absolute ethanol, 1920 ml; distilled water, 1890 ml. Line each stainless-steel, covered instrument pan with bleached cheesecloth, pour in enough solution to half-fill it, and put the instruments in place. Add 4 to 6 antirust tablets to each pan, using care that the tablets do not touch the instruments. Before use, remove each instrument from the zephiran chloride solution with sterile forceps, wash it in a jar of sterile distilled water to remove the excess zephiran chloride, and place it on a sterile linen towel in the culture chamber.

Micro-stoppers

The micro-stoppers are cut so that the small tapered end fits snugly into the hub of a #25 gauge hypodermic needle. The total length of the cut stopper is reduced to 25 mm.

1. Soak in distilled water overnight.
2. Boil in distilled water without detergent.
3. Rinse in 95% ethanol made up in distilled water.
4. Spread to dry in a flat stainless steel pan.
5. Place in 100×15 mm Petri dishes and autoclave.

Stoppers (black rubber) and screw caps

1. Boil new, black rubber stoppers in dilute NaOH for 15 minutes, and rinse for 10 to 15 minutes in running tap water.
2. Clean the stoppers and screw caps in the same way as the Petri dishes, but omit the acid rinse.
3. Dry by spreading in a stainless steel pan.
4. Place the stoppers in 200×25 mm glass tubes that are fitted with cotton gauze plugs, and autoclave. Take care that the small end of each stopper is directed downward when put into the tube.
5. The screw caps are sterilized with the bottles and flasks to which they are fitted, as described above.

Cotton towels

1. Boil for 30 minutes in tap water containing Micro-Solv.
2. Rinse well in two to three changes of warm tap water.
3. Air-dry.
4. Fold each towel into a 4.5-inch square. Wrap separately in Dennison wrap and seal with autoclave tape. Sterilize in the autoclave.

Gaskets for Rose chambers

Cut a sheet of silicone rubber, ⅛ inch thick and 24-inches square, into 2-inch × 2-inch squares. Punch a hole 1 inch in diameter out of the center

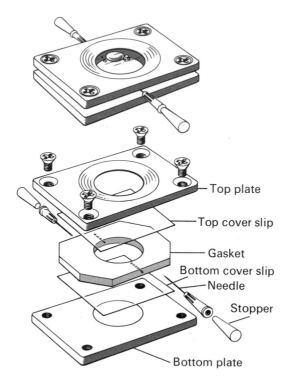

Top plate

Top cover slip

Gasket

Bottom cover slip

Needle

Stopper

Bottom plate

FIGURE 32-1

The various components of the modified Rose chamber shown individually and after the chamber is assembled.

of each square. The punch is made of stainless steel and sharpened to a fine, smooth bevel.

1. Boil in large beaker of tap water without detergent for 30 minutes to help remove oil and dirt.

2. Soak overnight in a beaker of distilled water.

3. The next morning, remove to a beaker of fresh distilled water and boil for 30 minutes without detergent. Handle gaskets with forceps.

4. Rinse by placing them in another beaker filled with distilled water; change distilled water two more times.

5. Air-dry them in a stainless-steel, flat pan. Handle only with forceps.

6. Siliconizing gaskets:

   a. Drop the gaskets into a large beaker containing the Siliclad mixture and let them soak for 20 minutes.

   b. Rinse in a beaker, filling three times with distilled water.

   c. Dry overnight by suspending the gaskets with a stainless steel hook attached to a wire.

   d. Insert two disposable (25 gauge, $\frac{5}{8}$ inch long) needles into each side of the gasket, as shown in Figure 32-1.

   e. Place the assembled gaskets in 150- × 20-mm Petri dishes, and autoclave.

MEDIA PREPARATION

The basic culture medium used in our laboratory is Eagle's Minimum Essential Medium (MEM) as modified by the suggestions of Pomerat and Steinberger. Gey's balanced salt solution serves as the base of Eagle's medium.

To avoid making up only one liter at a time, parts of the Gey's BSS, as 10X concentrated stock solutions, are prepared and autoclaved. These are made up as two separate solutions and will be labeled as solutions A and B. Separate solutions are prepared, since mixing prior to autoclaving results in precipitation.

Solution A consists of: NaCl, 80.0 g; KCl, 3.8 g; $Na_2HPO_4 \cdot 7H_2O$, 3.0 g; KH_2PO_4, 0.25 g. These salts are dissolved in 1,000 ml of glass triple-distilled water in the order given. A 2000-ml flask is convenient for continuous mixing.

Solution B consists of: $MgCl_2 \cdot 6H_2O$, 2.1 g; and $CaCl_2 \cdot 2H_2O$, 1.3 g. These salts are dissolved in 1000 ml of glass triple-distilled water in the order given. Again, a 2000-ml flask is convenient for continuous mixing.

These two solutions should be autoclaved separately. Screw-cap bottles are not as satisfactory for storing these solutions as serum bottles, which can be stoppered and sealed with aluminum caps. Solutions A and B are dispensed separately in 100-ml amounts into 125-ml serum bottles, stoppered with flange-type stoppers and capped with one-piece seal aluminum caps. The caps are crimped with a hand crimper. The stoppers and seals can only be purchased commercially in lots of 1000 or more, but smaller numbers can usually be obtained from a hospital pharmacy.

The refrigerated, sealed bottles are allowed to return to room temperature before being autoclaved at 15 lbs pressure for 60 minutes (slow exhaust). After autoclaving, the bottles are cooled before storing in the refrigerator. Solutions prepared in this manner may be stored for one year.

The glucose component for Gey's BSS is made up as a 10% solution and sterilized by filtering, because autoclaving may cause caramelization. Fifty g of glucose are dissolved in 500 ml of glass triple-distilled water. This solution is sterilized with Millipore filters and stored in the refrigerator.

The last solution needed to make Gey's BSS is sodium bicarbonate. This is purchased as a 10% solution and is kept refrigerated. Before use, the sodium bicarbonate solution should be brought to room temperature.

By preparing these concentrated stock solutions, the tissue culturist can prepare 10 liters of Gey's BSS before having to reweigh the ingredients. Thus, much time and effort are saved.

Once all the concentrated solutions have been prepared and sterilized, regular Gey's BSS and Eagle's medium can be prepared as follows: Remove the seals and stoppers, using a forceps; pour one bottle of 100 ml of 10X Solution A and one bottle of 100 ml of 10X Solution B into a 1000-ml sterile graduated cylinder. Then add 40 ml of the 10% glucose solution, and fill up to the 1000-ml mark with sterile glass triple-distilled water. Using a large funnel, pour this into a 2000-ml flat-bottomed boiling flask with a single 34/45 neck. To the Gey's BSS, now at 1X, add 2.5 ml of the 10% bicarbonate solution. If the bicarbonate solution is added before the Gey's BSS is diluted, a precipitate will form, and the entire solution will have to be discarded. If a precipitate should form in the 1X Gey's BSS even though the bicarbonate is added last, the bicarbonate solution may be contaminated.

If the Eagle's medium is to be stored for longer periods, not all the components are added to the Gey's BSS at once, since some are more unstable than others and should be added just before use. The components used in making the Eagle's medium are purchased in fairly large amounts, and in order to preserve uniformity in those that are frozen, they are thawed, dispensed into the amounts to be used, and refrozen. This eliminates frequent rethawing and refreezing.

The following components are added to the 1X Gey's BSS in a 2000-ml flat-bottomed boiling flask:

10.00 ml of MEM vitamin mixture (100X) ;

10.00 ml of MEM essential amino acids (50X) ;

10.00 ml of MEM nonessential amino acids (100X) ;

10.00 ml of sodium pyruvate 100mM (100X) ;

 1.25 ml of 1% phenol red solution;

 2.00 ml of 100,000 U per ml penicillin;

 0.50 ml of 200 mg per ml mycifradin sulfate.

The medium is dispensed and stored in 250-ml, screw-cap flasks, and must be refrigerated; it may be stored for up to four months.

Before using the Eagle's medium for culture, 50-ml amounts of the above incomplete Eagle's are dispensed into screw-cap Erlenmeyer flasks. To each 50 ml of this incomplete medium, add 1 ml of L-glutamine 200 mM (100X) and 5 ml of heat-activated horse serum. This completed Eagle's medium can be stored safely in the refrigerator for as long as three weeks.

When the Eagle's medium is to be used for culturing or feeding, the pH must be adjusted to a pH of 7.0 to 7.2 by the addition of a few drops of either a 10% bicarbonate solution or a 0.3N HCl solution. For the 0.3N HCl solution, add 2.79 ml of concentrated HCl to 100 ml of glass triple-distilled water, dispense into two 50-ml, screw-cap bottles, leaving caps slightly loose, and autoclave at 15 lbs pressure for 30 minutes with a slow exhaust.

In our laboratory, all prepared solutions are routinely checked for bacterial contamination. A few drops of each solution are placed in a general diagnostic medium, which is then incubated and checked periodically for bacterial growth.

An important tissue-culture technique used constantly to maintain sterility is "flaming." When a stopper, cap, or plug is removed from a piece of glassware, there is always the possibility of contamination. To lessen this possibility, the neck of every piece of glassware is rotated through the flame of a burner until hot. After using the glassware and before stoppering, capping, or plugging, the neck is again flamed. This technique *must* be used routinely when handling all sterile glassware.

MEDIA FILTRATION

Since the quantities of medium that need sterilization by filtration are small, a small-volume filter is used. The Millipore sterile, disposable 25-mm microsyringe filter with a pore size of 0.22μ is suitable. The filter comes assembled in a small plastic bag. All that needs to be added is a sterile, disposable, 18-gauge needle. Remove the bag, and place the unit on a sterile towel. Attach the needle, with its protective cover, to the plastic tip. Remove the bottle cap from the sterile bottle, exposing the inner, Patapar paper covering. Remove the protective cover on the needle and plunge the needle through the bottle's paper covering; the unit then comes to rest on the covered neck of the bottle. The medium to be run through the filter is drawn into a 5-ml or 10-ml syringe and delivered to the filter. During filtration, always apply positive pressure rather than negative pressure to the filter unit. Negative pressure is likely to pull the membrane filter loose from its supporting screen, thus allowing unsterile medium to pass through and contaminate the sterile filtered medium. When all the medium is filtered, remove and discard the filter unit. Remove the Patapar paper covering the neck of the bottle, flame the mouth of the bottle, and screw the cap on tightly.

MEMBRANE STRIP PREPARATION

Cellophane dialysis tubing, $\frac{3}{4}$ inch round diameter, $1\frac{3}{16}$ inch flat diameter, 24 Å pore size, is used. Use surgical gloves for handling the tubing to prevent contamination by natural oils. Cut the tubing into two-foot lengths for ease of handling. Slit the fold in the tubing on one side along the entire length. Because the walls of the tubing cannot be separated easily, it is advisable to attach a piece of scotch tape to each side and use these as levers to separate them. Cut the opened tube across its width into $\frac{1}{2}$-inch strips. This leaves the natural folds that occur when it is a flattened tube, which are useful in preventing excessive pressure on the tissue in the Rose chamber. Many of these strips can be cut at one time and stored in a dust-proof bag until ready to be sterilized. About 20 membranes can be sterilized at a time, for a minimum of 30 minutes in a 100- × 15-mm Petri dish containing approximately 50 ml of 70% ethanol (absolute ethanol and glass triple-distilled water). Remove the membranes one at a time with sterile cover-glass forceps, and rinse by submerging in 50 ml of sterile, glass triple-distilled water. When all membranes have been transferred to this first rinse, repeat the process in a second rinse of sterile, glass triple-distilled water, then rinse the strips twice in Hanks' BSS, transferring them one at a time with sterile forceps. Finally, transfer the strips to a dish of culture medium and store at least overnight in the refrigerator.

ROSE CHAMBER ASSEMBLY

When the tissue is minced and ready to be transferred to Rose chambers (see Figs. 32-1 and 32-2), place a sterile 100- × 15-mm Petri dish in the tissue-culture hood, and a 43- × 50-mm #1 cover glass in the dish. As skill increases, two or perhaps even three chambers may be set up at the same time. Place a small drop of medium, containing about 50 bits of tissue (sizes varying up to 1 mm) on each cover glass with a fine glass pipette (hand drawn). Withdraw the excess medium, leaving enough fluid to keep the tissue from drying. Since the *p*H of the medium becomes alkaline very rapidly when there is so little medium present, speed is essential in completing the preparation. Place the cover glass on the bottom plate of the Rose chamber, which is on a sterile towel, and cover it with the top of the Petri dish to decrease the chances of contamination. Recover the dialysis tubing strip, which is used to hold the tissue in place, from the conditioning

Rose chamber

Material :
Type 304 stainless steel sheet $\frac{1}{8}$″ thick

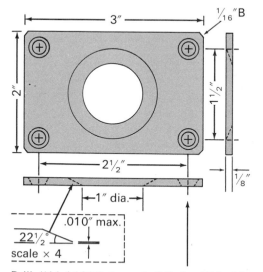

FIGURE 32-2

An engineer's drawing of measurements and materials of the top plate of the modified Rose chamber.

Drill #11 (.191″) through C.S. for #10–32 F. H. Phillips stainless steel screw × $\frac{1}{4}$″ long (4 plc.). 1 required as shown.

1 mating piece required. Drill #21 (.159″) through tap #10–32 (4 plc.).

medium, using two forceps. A cover-glass forceps is useful for separating the desired strip from the others in the dish, and a three- or four-toothed forceps is useful for holding one end securely. Hold the strip vertical, with one end touching some sterile filter paper, to drain off the excess medium; this is important to prevent the tissue from sliding to one side when the membrane is applied to the cover glass. As the membrane is lowered onto the cover glass, hold the ends higher than the center, forming a hammock. Bring the center of the membrane into contact with the tissue, then drop the ends quickly. Occasionally, air bubbles are caught under the strip; it is best to leave them. Move the cover glass about until most of the tissue lies in the center of the Rose-chamber opening. Place the gasket, with its needles in position, on the cover glass. Place another 43 × 50 mm #1 sterile cover glass on the gasket, and add the top of the Rose chamber. To complete the assemblage, insert the four screws into the screw holes and tighten carefully. When the screws are tightened just enough to hold the gasket in place, use a large serrated forceps to pull taut the protruding end of the dialysis strip. To avoid pulling the membrane through the chamber, it is necessary to hold the opposite side tightly. If the strip cannot be pulled taut because the screws have been tightened too much, loosen them slightly and try again. If there is uneven pressure on the cover glasses, they will crack. Therefore, each screw should be tightened gently, working diagonally, until the chamber is sealed. Attach a glass syringe, containing about 2 ml of culture medium, to one of the needles penetrating the gasket. Hold the chamber upright, and inject enough medium to almost fill the chamber. Leave a bubble of air to equalize the pressure as the temperature changes. Place a micro-stopper in each needle, giving the stopper a twist or two to make sure it is securely placed.

AGAR PREPARATION

As a base for the organ cultures, prepare strips of Bacto-agar, 2% in 0.7% NaCl, in the following manner. A reasonable quantity to prepare at one time is 500 ml. Weigh 10 g of Bacto-agar and place in a 2000-ml Erlenmeyer flask, using a powder funnel. Add 500 ml of glass triple-distilled water, making sure that the funnel is rinsed well. To dissolve the agar, it must be heated. As the water heats, swirl the flask to mix the agar. By the time the water reaches the boiling point, the agar will be dissolved. Permit the agar solution to cool as 3.5 g of NaCl are being weighed. Since NaCl impedes dissolution, add it after the agar is dissolved. Swirl the Erlenmeyer flask again to insure thorough dissolution. If at any time during the preparation the agar solution solidifies, it need only be reheated to liquefy it. For dispensing the completed agar-salt solution, all glassware used for measuring should be heated to insure against partial solidification if the solution comes in contact with the cold surfaces. The solution can be put into 25 × 150 mm screw-cap tubes or 50 or 100 ml screw-cap bottles. If dispensed into tubes, 15 to 20 ml are placed in each tube. If placed in bottles, a multiple of 15 to 20 ml should be measured out. The bottles should not be filled much over half full, or else the agar solution will boil over during autoclaving. Tighten the screw caps on the tubes or bottles, and then turn them back a generous quarter turn, before autoclaving them at 15 lbs pressure for 30 minutes with a slow exhaust. When the tubes or bottles are cool, tighten the screw caps and place the solidified agar solution in the refrigerator for storage.

ORGAN CULTURES

Remove 15 to 20 ml of the sterile Bacto-agar from the refrigerator and melt in a water bath or for one minute in the autoclave at 15 lbs pressure with a slow exhaust. Pour into a 100 × 15 mm Petri dish set on a flat surface to give a uniform agar depth of 2 to 3 mm. When it is cool, turn the dish over and draw lines on the bottom of the dish with a wax pencil, so that one may cut strips approximately 20 to 30 mm long and 5 to 7 mm wide. Add 15 to 20 ml of culture medium to the agar to condition it, at least one day before the agar strips are needed. The agar may be cut at any time after it is conditioned, using a sterile cataract or surgical knife.

To remove the agar strips, place a narrow, flat spatula under the strip and gently lift. Transfer the strip to a 60 × 15 mm plastic Petri dish to which 1 or 2 ml of culture medium have been added.

Place one or two strips in each Petri dish, depending on the number of whole organs to be cultured. Withdraw the fluid from the dish in order to straighten and position the agar strips. Pipette the organs with a fine pipette (hand-drawn) onto the strips, and remove any excess fluid around the organs by either pipetting or using the tip of a watchmaker's forceps as a capillary pipette. After the organs are positioned on the strips, pipette 2 ml of culture medium into the bottom, label the top of the dish, and place the dish on a tray in a humidified CO_2 incubator maintained at 37°C (see Fig. 32-3).

FEEDING THE CULTURES

The medium in the Rose chambers and the organ cultures must be renewed periodically, by withdrawing the used culture fluid and adding fresh medium.

For the organ cultures, use a 2- or 5-ml pipette to withdraw the metabolized medium and a 2-ml pipette to add the fresh medium. The same withdrawing pipette can be used for all the organ cultures, and the same feeding pipette can be used for all the feeding, provided that the sterile technique is maintained. Care must be taken during the feeding procedure not to drop medium onto the agar strips themselves, since this will dislodge the organs. The medium should be gently pipetted at the periphery of the Petri dish.

The Rose chamber feeding is somewhat more tedious. Syringes must be

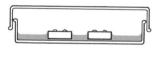

FIGURE 32-3

The method of positioning organs on agar strips in organ culture. The strips are placed in a 60 × 15 mm Petri dish as shown and the culture fluid added to the level indicated in the side view.

used, and the pressure within the chamber must not vary much from atmospheric pressure or the cover glasses will break. If one of the needles has been removed for microscopic observation and/or photography, it must be replaced by another sterile needle. Soak a 2-inch × 2-inch gauze sponge in 95% ethanol, and wipe the area of silicone rubber where the needle is to be inserted. Then insert a sterile, disposable, 25-gauge × ⅝-inch needle through the middle of the gasket. Turn the Rose chamber around, remove the micro-stopper in the other needle with large serrated forceps, and place it in a sterile Petri dish. If there are stoppers on both sides of the Rose chamber, because one needle has not been removed previously, take both stoppers out and place them in a sterile Petri dish. Attach a 5-ml glass syringe to one of the needles in the gasket, and apply a small amount of negative pressure to be certain that the needle is clear, as is shown by air entering the chamber. If the needle is not clear, remove it and insert a sterile needle as described above. Then withdraw the medium as completely as possible. Add fresh medium of pH 7.0 to 7.2, using a sterile glass syringe. Reinsert the stoppers, and give them a securing twist.

When working with the syringes, it is convenient to place them in tubes of 22-mm diameter that have been cut to 50-mm lengths. These are arranged in a slanted tube holder made especially for this purpose. By using these tubes, there is less possibility of contamination.

TISSUE PREPARATION

Mating and Timing

One of the most important aspects of the use of fetal tissues in culture experiments is the care of the animals and the timing of their mating. For the procedures described here, white mice of the Swiss-Webster strain are used. The cages for the mice are made of clear plastic, 11½ inches × 7¼ inches × 5¼ inches, with a stainless steel, perforated cover that fits loosely. Mouse chow in pellet form is offered *ad libitum* in a feeder suspended from one wall of the cage, and water is contained in a bottle with a stainless steel tube, which is fitted into a central hole in the cover. The bottom of each cage is covered by a 1.5-inch layer of medium fine shavings, which is changed every other day. The room is kept closed to avoid unnecessary noise, and temperature is maintained at 68°F.

For mating, two females are put in a cage with one male. Every morning at 8 o'clock the females are examined for the presence of a copulation plug. When a plug is seen, the female is designated as being in Day 0 of gestation. Females that mate on the same day are placed together in groups of not more than four animals to a cage, and each cage is labeled with the time and date of mating.

The female to be examined for a vaginal plug is picked up by firmly grasping the scapular skin with the left thumb and forefinger, then quickly turning the animal on its back, thus supporting its body with the same hand. The vaginal orifice can be examined easily by holding the mouse under a lamp. If the female has mated, the plug will frequently protrude slightly from the orifice. If, however, it is not visible immediately, the lips of the orifice can be opened, very gently, by using a clean pair of mosquito forceps. When the light is directed into the canal, the plug, if present, will be visible.

RECOVERY OF EMBRYONIC OVARIES AND OVIDUCTS

A 16-day pregnant mouse is killed by decapitation. Lay the mouse on her back on a paper towel, and pour 95% ethanol over the abdominal area to sterilize it. Use a pair of iris scissors and extra fine mouse-tooth forceps, soaked in 95% ethanol, to slit open the abdominal wall longitudinally. Fold the skin back to expose the fetuses. Use a new set of sterile scissors and forceps that have been stored in a sterile towel to remove the fetuses to a 60- × 15-mm Petri dish. Grasp each uterine cornu at the cervix, raise it and cut it loose from the connective tissue. Depending on the work to be done and the skill with which it can be done, up to four pregnant animals may be killed at one time.

Put the dishes containing the uteri under the tissue-culture hood before beginning the next procedure. Using iris scissors and an extra fine serrated forceps, open each cornu of the uterus along the antimesometrial border. Remove the fetuses and place them in a clean 60- × 15-mm Petri dish. Dissect the fetuses and recover the ovaries and oviducts under the binocular dissecting microscope. Use iridectomy scissors and watchmaker's forceps to remove the ovaries and oviducts from the fetuses. Lay a fetus on its left side and cut open the abdomen transversely to expose the closely associated kidney and gonad (see Figs. 32–4 and 32–5). Grasp the gonad near its supporting ligament with the watchmaker's forceps and cut it loose with the iridectomy scissors. Place the ovary and its capsule (see Fig. 32–6) in a 60- × 15-mm Petri dish containing 20 ml of cold culture medium. The medium should be cold to keep the tissues in better condition and to help maintain the proper *p*H. Turn the fetus over onto its right side and repeat

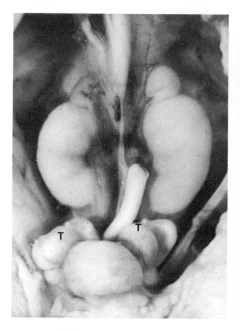

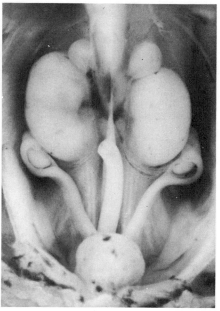

FIGURE 32-4

Abdominal cavity of a 16-day fetal male mouse showing the position of the gonads, (T-Testes). Fixed specimen. ×14.

FIGURE 32-5

Abdominal cavity of a 16-day fetal female mouse with the reproductive tract left in situ. Fixed specimen. ×14.

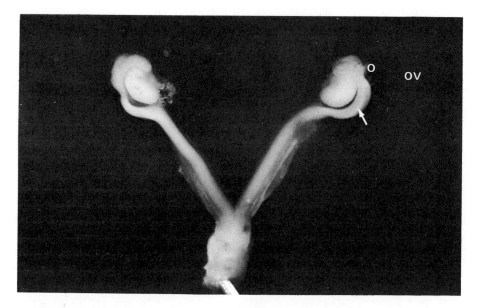

FIGURE 32-6

Reproductive tract of a 16-day fetal female mouse removed intact from the animal. The oviducts are simple tubular organs at this gestational age. Both ovaries and oviducts can be cut away easily by grasping the developing mesenteries. Fixed specimen, unstained. ×23. O, ovary; OV, oviducts; arrow, utero-tubal junctions.

the process. This procedure is repeated for each of the female fetuses. Transfer the ovaries and their capsules to another glass Petri dish containing fresh, cold culture medium. Using a clean pair of watchmaker's forceps and iridectomy scissors, dissect the ovaries from their capsules. Grasp the connective tissue of the capsule with the watchmaker's forceps, insert the iridectomy scissors alongside and cut the ovary away. Trim any extraneous connective tissue still adherent from the ovary. Remove the trimmed ovaries to another dish of fresh, cold medium.

If the ovarian tissue is to be cultured in Rose chambers, mince the tissue with two cataract knives into pieces no larger than 1 mm. Plastic Petri dishes should not be used for mincing, because the knives will cut into the dish and small pieces of plastic may adhere to the tissues. When the ovaries are to be explanted as organ cultures, pipette whole ovaries directly onto the prepared agar strips in the 60- × 15-mm plastic Petri dishes and position them.

To prepare the oviducts for tissue culture, remove the ovaries and attached oviducts (as described above) from the fetus and transfer them to a 60 × 15 mm Petri dish containing cold, fresh culture medium adjusted to the proper pH of 6.8 to 7.0. First, dissect the ovary away as explained in the previous section. Then hold the capsule with the watchmaker's forceps, and cut the tubular oviduct free with iridectomy scissors, cutting along the contour of the tube. Sever the oviduct at the utero-tubal junction. Remove all extraneous tissue from the dish.

At 16 days gestational age, the oviducts are simple tubular structures that are supported by the mesosalpinx within the capsule (see Fig. 32-6). There is no demarcation at the junction of the ampullar and the isthmic portions, but there is a noticeable increase in diameter at the junction of the oviduct and uterine cornu. The fimbriated end of the ampullar region is divided into two lips by a median cleft. This serves as an identification for the ampullar half when the oviduct is divided. Cut the oviducts in half, and remove each ampullar and isthmic segment to separate Petri dishes containing fresh, cold culture fluid, before mincing them into 1-mm pieces with two cataract knives. Transfer about 60 to 80 of these fragments from each segment with fine pipettes to separate cover glasses, which will become the bottom cover glasses of the Rose chambers. The chambers are assembled as described for the ovarian fragments.

If the oviducts are to be used for organ cultures, divide them into their respective ampullar and isthmic segments, but leave them whole. Transfer eight to ten of the whole oviduct segments from each region to separate agar strips with fine pipettes and place in position. Add 2 ml of fresh culture medium to each Petri dish, put the covers in place, and incubate the cultures at 37°C, in a humidified incubator with 5% CO_2 in air.

CULTURE GROWTH CHARACTERISTICS

Ovarian tissues

At 16 days gestational age, the ovaries have been seeded by oögonia, and early organization is in progress. The oögonia still have the capacity for considerable innate movement. Ovarian fragments of 1 mm or less begin to send out sheets of embryonic connective tissue within 48 hours after explanting. After five days in culture, the oögonia may be recognized readily, and some of them show definite signs of growth. Many of the oögonia disappear from the fragments, but others remain and grow to large sizes. Figure 32-7 is a phase-contrast photograph of a small segment of an ovarian explant. Numerous oöcytes are present. They vary greatly in size, and some of the larger ones are beginning to develop zonae pellucidae. Oöcytes of all sizes may survive for as long as 50 days in culture.

Oviductal tissues

The minced tissues, when placed below the dialysis membrane in the Rose chambers, become flattened against the surface of the cover glass. Fibroblasts grow out rapidly from the explanted tissues, forming cellular sheets that anchor the explants firmly to the glass (see Fig. 32-9). By the second day in culture, the luminal areas are well delineated, in that they are lined by large, round epithelial cells with prominent central nuclei. These areas vary in size and shape; some are quite irregular in appearance, others are long and narrow (see Figs. 32-10 and 32-11). Although cilia begin to appear on some of the epithelial cells on the fourth day in culture, they do not begin to beat before the fifth or the eighth day after explantation. Cilia continue to beat in ampullar explants for two or three weeks. In explants of the isthmic region, cilia beat for less than two weeks.

The onset of muscular contractions in cultures of oviductal tissue is variable. There is a tendency for muscle cells in ampullar explants to begin contracting somewhat earlier than those in isthmic tissues. However, when contractions in the latter segments have become established, they continue to contract for a longer period of time. The majority of the oviductal tissues in the Rose chambers survive for two or three weeks.

Ovaries

A larger number of oöcytes survive and grow in whole-organ-culture preparations than they do in the minced fragments cultured in the Rose chambers. Figure 32-8 is a 7μ section through a fixed and stained ovary that had

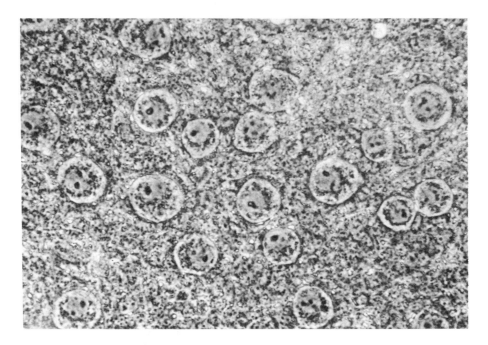

FIGURE 32-7

A small segment of a 16-day fetal mouse ovary that has been growing in a Rose chamber for 21 days. Oöcytes in various stages of growth are readily seen. Phase contrast, unstained, ×420.

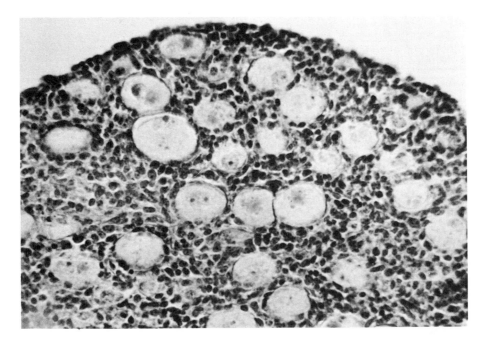

FIGURE 32-8

A 7μ section from a whole ovary of a 16-day fetal mouse that has been in organ culture for 21 days. Note the variation in size of the oöcytes and the developing zonae pellucidae, PAS stained. ×420.

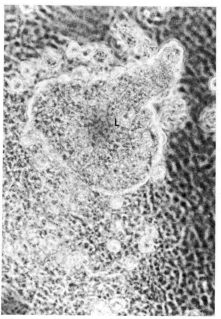

FIGURE 32-9

Oviductal tissue from a 16-day fetal mouse cultured in a Rose chamber for 48 hours. Note the early outgrowth of fibroblasts. Phase micrograph, unstained. ×154.

FIGURE 32-10

Ampullar tissue from an oviduct of a 16-day fetal mouse cultured in a Rose chamber for 14 days. The cells lining the triangular-shaped lumen (L) had beating cilia when the culture was photographed. Phase micrograph, unstained. ×370. (From "The Mammalian Oviduct," edited by E. S. E. Hafez and R. J. Blandau, University of Chicago Press, 1969.)

been in organ culture for 20 days. As may be seen, some eggs have survived and grown, becoming as much as 50 to 80μ in diameter. In organ culture, the oöcytes may survive for as long as 80 days.

Oviducts in organ culture

By the second or third day *in vitro*, the appearance of the explants begins to change. The most rapid alteration is in the ampullar explants. The lumina begin to dilate, and ridges develop along the walls. These ridges become folded and form fingerlike projections into the lumina. At the same time, the whole ampulla begins to curve, eventually forming a complete circle. The lumina may become so distended with fluid that vesicles are formed, which often are so large that they encompass the greater portion of the segment (see Fig. 32-12). Contractions are seen frequently in these vesiculated explants. Although the isthmic and ampullar explants develop similarly, the changes within the isthmus proceed at a slower rate. The walls of the isthmic segments become thicker, and vesiculation is less common.

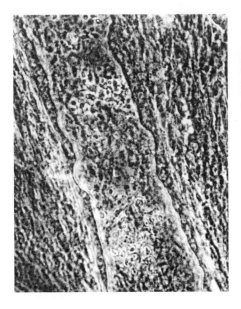

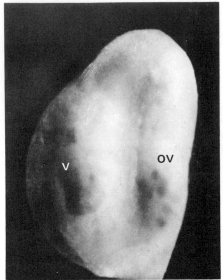

FIGURE 32-11

Another culture of ampullar tissue from a 16-day fetal mouse oviduct in culture for 16 days. The luminal area (L) contained cells with beating cilia and was bordered by contracting muscle when the culture was filmed. Phase micrograph, unstained. ×370. (From "The Mammalian Oviduct," edited by E. S. E. Hafez and R. J. Blandau, University of Chicago Press, 1969.)

FIGURE 32-12

A single large vesicle (V) formed within an ampullar segment of an oviduct (OV) in organ culture. The vesicle was lined by ciliated cells with actively beating cilia and the segment was contracting when photographed. Phase micrograph, unstained. ×49. (From "The Mammalian Oviduct," edited by E. S. E. Hafez and R. J. Blandau, University of Chicago Press, 1969.)

Cilia in both segments develop to the stage where they begin to beat by the sixth day after explantation. Beating cilia are not seen as commonly as in isthmic explants. The beginning of muscular contractions is observed in both segments by the sixth or seventh day after explantation, and they continue for two or three weeks. The majority of the organ cultures survive for three or four weeks.

SQUASH PREPARATIONS OF EXPLANTS

Some information may be gained by making squash preparations of the explants. Remove the explant from the agar strip with a fine pipette and drop into a small Petri dish containing the culture fluid. From here, transfer it with a pipette to the center of a vaseline ring on a microscope slide. Add a small drop of medium, and drop a round cover glass over the ringed area. When properly prepared, considerable cytologic detail may be observed under the phase microscope.

HISTOLOGICAL PROCEDURES

FIXATION AND STAINING OF TISSUES CULTURED IN ROSE CHAMBERS

1. Remove the cover glass with the culture carefully from the Rose chamber and put it in Hanks' balanced salt solution to separate the membrane from the culture and to wash the culture medium from the tissues. Fix the culture, without removing it from the cover glass, in a solution consisting of the following ingredients (as suggested by Professor W. S. Hsu) : 75 ml of distilled water; 15 ml of 40% formalin; 10 ml of glacial acetic acid; 1 g of picric acid; 1 g of chromic acid; 1 g of urea. The fixative is made up each time just prior to use.

2. Leave the culture in the fixative for one hour in the refrigerator, followed by one hour at room temperature.

3. Wash for two hours in running tap water.

4. Dehydrate gradually, beginning with 15% ethanol and progressing through 25%, 35%, 55% (15 minutes each), and 70% ethanol for overnight.

5. Stain with PAS and hematoxylin.

6. Dehydrate with absolute ethanol and xylene.

7. Mount in the usual manner with Permount. During the entire procedure, the culture remains on the cover glass.

FIXATION, EMBEDDING, AND STAINING OF OVARIES IN ORGAN CULTURES

1. Remove the ovaries from the agar strip with a fine pipette and wash in Hanks' balanced salt solution. Fix in a cold fixative (Hsu fixative described above), for 75 minutes in the refrigerator, and an additional 75 minutes at room temperature.

2. Wash for 2 hours in running tap water.

3. Dehydrate in 35% and 50% ethanol, and leave overnight in 70% ethanol.

4. Complete the dehydration by placing the tissue in several changes of 95% and absolute ethanol.

5. Use the double-embedding method of Peterfi. In this procedure, the tissue is placed first in a mixture of absolute ethanol, nitrocellulose, and methylbenzoate, then in the same mixture without ethanol.

6. Clear in several changes of benzene.

7. Embed in paraffin as usual.

8. Section serially at 7μ.

9. Stain with PAS and hematoxylin.

10. Dehydrate and clear in the usual manner.

11. Mount with Permount.

Fixation, Embedding, and Staining of Oviducts in Organ Cultures

1. Remove the segments from each area of the oviduct from agar strips with fine pipettes and wash in separate containers with Hanks' balanced salt solution for 10 minutes.

2. Process the tissues by the method published by Professor Luft, as follows. Put each segment in a chilled fixative, which consists of 1.66% osmic acid in 0.133 M phosphate buffer, with a 2.4% solution of glutaraldehyde added at the time of fixation. Fix for one hour in the refrigerator.

3. Pour off the fixative, add cold 35% ethanol, immediately withdraw it, and replace by fresh, cold 35% ethanol, in which the tissues remain 10 minutes in the refrigerator.

4. Dehydrate tissues in cold 50% ethanol, and then in 70% ethanol overnight in the refrigerator.

5. Complete the dehydration in 95% ethanol, absolute ethanol, and finally in three changes of acetone.

6. Infiltrate the tissues in a mixture of araldite, DDSA (dodecenyl succinic anhydride), DMP [2,4,6-tri (dimethylaminomethyl) phenol], and acetone over a period of 6 or 7 hours.

7. Embed in a mixture of araldite, DDSA, and DMP in Beem capsules.

8. Polymerize in an oven at 35°C, 45°C, and 60°C for 48 hours.

9. Cut blocks with tissue at 1.5μ with a glass knife on the ultramicrotome.

10. Sections are stained in Richardson's solution, which contains Mallory's Azur II and methylene blue with added sodium borate.

11. Mount cover glasses with Zeiss immersion oil and ring with Epoxy cement.

RECORDING THE OBSERVATIONS

The tissues growing in the Rose chambers may be observed and evaluated each day with the phase-contrast microscope. The microscope is enclosed in a plexiglass chamber in which the temperature is maintained at 37°C. The long-working-distance condenser and matching objectives must be used for making observations of the tissues in the Rose chambers. The most satisfactory light source for the phase microscope is a zirconium arc lamp cooled by a water jacket. A Zeiss green interference filter is used with the lamp, and polaroid filters are added to control the amount of illumination. Photographic records may be made using Kodak Plus-X film for still photography, and Gevaer-type negative film, #1.65, for cinematography.

APPENDIX: SOURCES FOR MATERIALS

Beckton-Dickinson and Co.
Rutherford, New Jersey 08307
  Glass hypodermic syringes, disposable hypodermic needles

Bellco Glass, Inc.
Vineland, New Jersey 08360
  V.I.P. "shortie" serological pipettes, rubber bulbs, and pipette cans for these special pipettes

A. J. Buck & Son., Inc.
1515 East North Avenue
Baltimore, Maryland 21213
  Steriroll wrapping paper, "Patapar"

Clay Adams, Inc.
141 East 25th Street
New York, New York
  Gold Seal cover glasses, fine dissecting instruments, Siliclad, cover glass forceps

Corning Glass Works
Corning, New York 14830
  Pyrex glassware, culture tubes and flasks

Dennison Mfg. Co.
Framingham, Massachusetts 01701
  Dennison wrap

Difco Laboratories
Detroit, Michigan 48201
  Bacto-agar, bicarbonate solution, phenol red solution

William Dixon, Inc.
32–42 East Kinny Street
Newark, New Jersey 07101
  Stainless-steel watchmaker's forceps (tweezer style, #5 Swiss)

Falcon Plastics
5500 West 83rd Street
Los Angeles, California 90045
  Disposable Petri dishes, disposable pipettes

Kontes Glass Co.
Vineland, New Jersey 08360
  2000-ml, flat-bottomed boiling flasks

Medical Instrument Shop
University of Washington School of Medicine
Seattle, Washington 98105
  Stainless-steel plates and screws for Rose chambers

Microbiological Associates, Inc.
4733 Bethesda Avenue
Bethesda, Maryland 20014
  Media, Micro-Solv, trays for Petri dishes, Hanks' BSS

Millipore Filter Corp.
Bedford, Massachusetts 01730
  Filters for sterilizing solutions

Minnesota Mining and Manufacturing Co.
St. Paul, Minnesota
  Autoclave tape, 3M #1222

Owens-Illinois Glass Co.
Toledo, Ohio 43601
  Kimax glassware, culture tubes, and flasks

Ronther-Reiss Corp.
Route #46, P. O. Box 35
Little Falls, New Jersey
  Silicone rubber sheets 24 × 24 × 1/8 inches, "M.D." color transparent

Will Ross, Inc.
4285 North Port Washington Rd.
Milwaukee, Wisconsin
  Sterilizer bags for syringes

Scientific Products
1210 Leon Place
Evanston, Illinois 60201
  Micro-stoppers

E. R. Squibb & Sons
New York, New York
  Penicillin

Sylvania Electric Products Co.
Salem, Massachusetts
  Zirconium arc lamps

T. C. Wheaton Co.
Millville, New Jersey 08332
  Serum bottles, sleeve-type stoppers, one-piece seal, hand crimper

The Upjohn Co.
Kalamazoo, Michigan
  Mycifradin sulfate

Union Carbide Corp.
6733 West 65th Street
Chicago, Illinois 60638
  Cellophane dialyzing tubing

Winthrop Laboratories
New York, New York
  Antirust tablets

BIBLIOGRAPHY

Blandau, R. J., E. Warrick, and R. E. Rumery (1965) *In vitro* cultivation of fetal mouse ovaries. Fertility Sterility, 16:705–715.

Hsu, W. S. Personal communication.

Luft, J. H. (1961) Improvements in Epoxy resin imbedding methods. J. Biophys. Biochem. Cytol. 9:409–414.

Moscona, A., O. A. Trowell, and E. N. Willmer (1965) Methods. In: Cells and Tissues in Culture, vol. I, pp. 19–98. Academic Press, New York.

Peterfi, T. (1964) Double Embedding. In: Laboratory Techniques in Biology and Medicine. 4th ed. V. M. Emmel and E. V. Cowdry, eds. Williams & Wilkins, Baltimore.

Pomerat, Charles M. Personal communication.

Price, D., and E. Ortiz (1965) The role of fetal androgen in sex differentiation in mammals. In: Organogenesis, pp. 629–652. Holt, Rinehart, and Winston, New York.

——— and J. J. P. Zaaijer (1969) Prenatal development of the oviduct *in vivo* and *in vitro*. International Symposium on The Mammalian Oviduct, pp. 29–46. University of Chicago Press, Chicago.

Richardson, K.C., L. Jarett, and E. H. Finke (1960) Embedding in Epoxy resins for ultrathin sectioning in electron microscopy. Stain Tech., 35:313–323.

Rumery, R. E. (1969) Fetal mouse oviducts in organ and tissue culture. International Symposium on the Mammalian Oviduct, pp. 445–457. University of Chicago Press, Chicago.

Steinberger, Anna. Personal communication.

33

Tissue Culture of Postnatal Male Mammalian Gonads

A. STEINBERGER and E. STEINBERGER

*Division of Endocrinology and Reproduction
Research Laboratories
Albert Einstein Medical Center
Philadelphia, Pennsylvania 19141*

Postnatal male mammalian gonads have been grown *in vitro* by a number of investigators (Champy, 1920; Champy and Morita, 1928; Esaki, 1928; Michailow, 1937; Martinovitch, 1937; Gaillard and Varossieau, 1938, 1940; Dux, 1940; Ferris and Plowright, 1958; Dulbecco and Vogt, 1954; Trowell, 1959; Lostroh, 1960; Jordan *et al.,* 1961; Hess *et al.,* 1963; Steinberger *et al.,* 1964 a,b,c; Steinberger and Steinberger, 1965, 1966 a,b, 1967 a,b; Kodani, 1962, 1966; Yao and Pace, 1964). Even though considerable success has been achieved in both cell and organ culture, growth of individual cell types and maintenance of the entire spermatogenic process *in vitro* continue to pose challenging problems.

Although many principles and techniques generally accepted for culture of various mammalian tissues are also applicable to culture of postnatal mammalian testis, the latter does require special considerations due to its many specific characteristics. The testis is a complex organ that in sexually mature animals is composed of many cell types with different properties. Some cell types undergo differentiation and mitotic divisions (spermatogonia, interstitial cells); others undergo miotic division (spermatocytes); yet others undergo differentiation without division (spermatids); and still others undergo no differentiation and no division (Sertoli cells).

Supported in part by United States Public Health Service Grant HD 00399.

Until becoming sexually mature, the testes undergo continuous morphologic and functional changes that are characteristic for each species and age
of the animal (Clermont and Perey, 1957). In seasonal breeders such as the
woodchuck, there are also changes related to the time of year. The microscopic appearance of seminiferous tubules varies depending on specific cell
associations (stages of spermatogenesis). To assess the extent of spermatogenic development, one must examine a large sample containing many
tubular cross sections. This is particularly important in the study of differentiation of germ cells in culture. Recently, much has been learned of the
dynamics of spermatogenesis in a number of mammalian species (Ortavant,
1954; Oakberg, 1956; Clermont et al., 1959; Heller and Clermont, 1963).
This knowledge and the improvements made in culture techniques and in
the composition of chemically defined media, make it possible to approach
the problems of gonadal cultures with better insight and understanding.
Methods for organ and cell culture of gonadal tissue will be described
under separate headings.

ORGAN CULTURE

Organ cultures of mammalian gonads have been used primarily
for the study of the extent of differentiation possible under artificial conditions, and also more recently for determination of factors which are necessary for normal spermatogenesis. Testes were obtained from animals of
various species and ages, and many different culture conditions were used
for their growth. Most early investigators used the classical plasma clot
method in which the plasma clot serves as substrate and source of nutrients.

Testes from newborn mice and rats were grown on clots composed of fowl
plasma and fowl embryo extract (Martinovitch, 1937). The germ cells
differentiated through the premiotic stages, but degenerated after the twentieth day. Gaillard and Varossieau (1938, 1940) grew fragments of testes
from sexually mature and immature rats on clots containing homologous
plasma. They reported formation of spermatids in testes of sexually immature rats after two or three days of cultivation, and better differentiation
when plasma from adult rather than younger animals was used. These
cultures, however, completely degenerated on the ninth or tenth day and
testes from adult animals became necrotic during the first day of cultivation. A solid medium containing agar instead of plasma was used for
cultures of mouse testes by Lostroh (1960). Within a short time the cultures
underwent degenerative changes.

Use of solid nutrient media has a distinct disadvantage: the tissue fragments must be transferred to fresh areas at frequent time intervals, and with
plasma clots there is the additional problem of clot liquefaction.

A method utilizing liquid synthetic medium for organ cultures was
introduced by Trowell (1959). According to this method, small tissue
fragments were grown at the liquid-gas interface, supported by stainless-
steel wire mesh grids, which were covered with lens paper or a thin sheet of
agar (2% in 0.7% NaCl). Each culture dish, containing a single grid, was

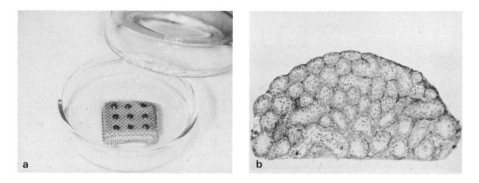

FIGURE 33-1

(a) *Culture dish containing nine explants for organ culture. Approx. ½ actual size.*
(b) *Cross section of a whole explant grown in culture for three weeks. ×40. (Note good maintenance of tubular structure and cellular elements.)*

enclosed in a metal device, which was incubated at 37°C and was perfused with 5% CO_2 in oxygen. Using this method, Trowell (1959) attempted to grow fragments or individual tubules of testes from adult rats. The explants, however, became necrotic after three to six days and cultures of individual tubules were completely unsuccessful. Later, Steinberger *et al.* (1964c) developed a modification of the above method, which permits long-term cultivation of testes from several mammalian species, including men (A. Steinberger and E. Steinberger, 1967; E. Steinberger, 1967). This method will be described in greater detail.

CULTURE TECHNIQUE

Stainless steel wire mesh grids are covered with a millipore filter or a thin sheet of agar, are placed inside Petri dishes, and sufficient medium is added to displace all the air under the grid. Care must be taken to avoid formation of air bubbles since they interfere with the proper diffusion of the medium. Thus prepared culture dishes can be kept in the incubator for several hours, until the tissue has been secured. It is essential to use fresh tissue because anoxia and autolysis occur very rapidly in the testis. Immediately after the donor animal is killed the intact gonads with their capsules (tunica albuginea) are removed through an abdominal incision and placed into a dish containing a small amount of culture media. Very small gonads can be cultured whole, with or without the capsule; otherwise, the capsule is removed and the gonads are cut into fragments of approximately one cubic millimeter with a pair of sharp blades. The explants are then arranged on top of the grids in the culture dishes (see Fig. 33-1a). In most laboratory animals, the capsule is loosely connected to the underlying tissues and can be removed easily. In some species (human and woodchuck), the capsule is attached more firmly and has to be dissected away taking utmost care in avoiding tissue damage due to excessive pressure, drying, or rise in *p*H. This point cannot be overemphasized, since it was one of the major causes of poor culture results in our laboratory and in others. Because germ cells are very easily displaced from their location, excessive pressure may completely alter the histologic organization within the seminiferous tubules. This is particularly undesirable if the cultures are to be used for the study of spermatogenic development. The cultures are incubated at 31°C in an atmosphere of 5% CO_2 and 95% air. The culture medium is maintained at pH 7.0 ± 0.2 and is changed every three or four days.

Fixation and Histology

At any time, one or several fragments can be removed from the grid for histologic examination by simply cutting the filter or agar around the fragments and lifting them together from below. This technique avoids direct handling of the tissue and thus the possibility of mechanical injury. The fragments are fixed in Bouin's solution, embedded in paraffin, and cut at 4 microns. The sections can be stained with various dyes, but for observation of the acrosome in spermatids, the periodic acid Schiff-hematoxylin stain is most useful. Microscopic appearance of a histological section from a cultured fragment is shown in Figure 33-1b. Regardless of standardized procedures used to set up the cultures, we found considerable variability in the microscopic appearance of the individual fragments. It is, therefore, essential in evaluating the results to examine several fragments from separate grids.

EFFECT OF INCUBATION TEMPERATURE

The incubation temperature of 37°C, which is commonly used for culture of mammalian tissues, is not suitable for culture of male gonads. The latter are normally maintained *in vivo* at temperatures between 32° and 35°C, and it is known that temperatures higher than normal scrotal temperatures are damaging to germinal elements (Steinberger and Dixon, 1959).

When the effect of various temperatures on rat testes in culture was investigated (Steinberger *et al.*, 1964c), the optimal range was found to be between 31° and 33°C. Incubation at 35°C, or higher, resulted in rapid degeneration of germinal cells. Therefore, the cultures should be incubated at temperatures comparable to or slightly lower than the normal scrotal temperature of the animal. The fact that an incubation temperature of 37°C was used by Trowell (1959) for cultures of rat testes may have been one of the causes for the rapid tissue degeneration.

EFFECT OF OXYGEN CONCENTRATION

The diffusion of oxygen through tissues can be enhanced by increasing the partial pressure of oxygen in the surrounding gas phase. However, the benefit of deeper oxygen penetration is often minimized by the well-known "oxygen poisoning" effect on many vital enzyme systems.

Thus, cultures of rat testes were maintained better in 5% CO_2 and 95% air than in atmosphere containing pure oxygen instead of air. Cultures grown in the higher oxygen concentration consistently had smaller necrotic centers, indicating better penetration of oxygen, but the tubular wall appeared more distorted, and the number of mitoses and germinal elements was greatly diminished. The germinal cells also survived for shorter periods. On the other hand, the presence of central necrosis had no effect on the seminiferous tubules located more peripherally.

Effect of Culture Media

The study of nutritional and hormonal factors needed for normal differentiation of gonads in culture necessitates the use of chemically defined media. The first attempt to grow adult rat testes in a synthetic medium for periods longer than a few days was not successful (Trowell, 1959). Whether failure of the tissue to survive was due to the composition of the culture medium or other factors is not clear.

The effectiveness of various chemically defined media and several complex supplements on testes of several mammalian species was investigated later by Steinberger *et al.* (1964a,c), and Steinberger and Steinberger (1966a; 1967). The tubular architecture, supporting cells, primitive type-A spermatogonia, and mitotic activity were maintained for several months in Eagle's Minimum Essential Medium (Eagle, 1959) supplemented with 1 mM of sodium pyruvate and 0.1 mM of the following amino acids: alanine, aspartic acid, glutamic acid, glycine, proline, and serine. In sexually mature testes the spermatocytes survived for three to four weeks and the spermatids survived for several days. Addition of 10% calf serum to the medium improved the over-all viability of cultures that were grown longer than four weeks; homologous serum had no advantage over the calf serum.

Progressive differentiation of germ cells was observed only when a combination of vitamins A, C, and E, or doubled concentration of glutamine, was present in the medium (Steinberger *et al.,* 1964a; Steinberger and Steinberger 1965, 1966a). The differentiation proceeded through all premeiotic stages (see Figs. 33-2a,b,c and 33-3a,b,c), but never resulted in completion of the meiotic division or the formation of spermatids. The pachytene spermatocytes, which formed in culture, degenerated after the fourth week of cultivation even though the tubular structure, supporting cells, and primitive spermatogonia remained well maintained (see Fig. 33-2e).

Effect of Hormones

A number of gonadotropic hormones have been tested for their ability to promote initiation or maintenance of spermatogenesis in culture. Variable results were reported. Gaillard and Varossieau (1940) observed a higher degree of differentiation in cultures of rat testes grown adjacent to explants of anterior pituitary; the enhanced differentiation was attributed to gonadotropic activity of the pituitary. Lostroh (1960) reported that addition of human chorionic gonadotropin to the medium ameliorated

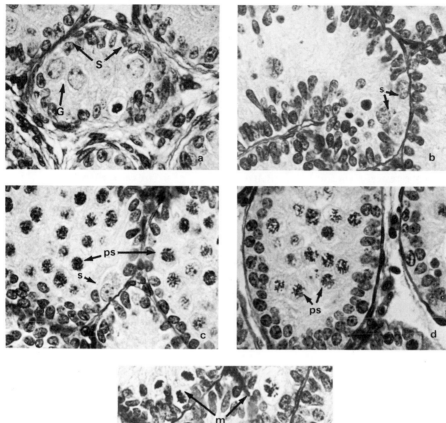

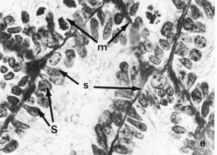

FIGURE 33-2

(a) *Section from testis of four-day-old rat. Seminiferous tubules contain only supporting cells (S) and gonocytes (G). More advanced germinal cells are not present.*
(b) *Testes from four-day-old rat after three weeks of cultivation in chemically defined medium, without vitamins A, C, and E, or additional glutamine. Primitive spermatogonia (s) are present but fail to differentiate into spermatocytes. (c) Similar culture, but grown in presence of vitamins A, C, and E, or doubled concentration of glutamine. Spermatocytes differentiated to pachytene stage (ps) of miotic division. Spermatogenic development similar to that occurring in vivo; compare with d following. (d) Section from testes of 24-day-old rat. Progression of spermatogenesis to late pachytene spermatocytes. (e) Testes from four-day-old rat following five weeks of cultivation in either medium. Spermatocytes are no longer present. Supporting cells (S) and spermatogonia (s) are well maintained; numerous mitotic figures (m). Magnification in all photomicrographs ×360. (a) through (d) from Exptl. Cell Res., reproduced with the permission of Academic Press.*

somewhat the degenerative changes that were occurring in cultures of mouse testis that were grown in the absence of the hormone. Other investigators did not observe enhancement of spermatogenesis when gonadotropic or other hormones were added to the culture medium (Steinberger *et al.,* 1964a; Steinberger and Steinberger, 1967).

Radioautographic Study of Spermatogenesis in Culture

For study of spermatogenic development in cultures of sexually mature testes, a method that utilizes tagging of germ cells with H³thymidine and radioautography was found to be very useful. It is well known that thymidine becomes incorporated into cellular DNA and cells that become labelled with H³thymidine retain the label in the absence of frequent cell divisions.

In germinal cells the synthesis of DNA is completed prior to the leptotene stage of the meiotic division. Thus the fate of labelled preleptotene cells in culture can be followed easily by utilizing radioautography methods. The experiment can be performed as follows: tritiated thymidine of high specific activity is injected into the animal by either subcutaneous or intraperitoneal route 0.8 to 1μCi per gram of body weight. The animal is killed three to four hours later, part of the testes is used to initiate cultures and a part is fixed for radioautography. The cultured fragments are also fixed for radioautography after various growth intervals. Presence of label in cells more advanced than those labelled prior to cultivation indicates their differentiation *in vitro*. The extent of germ cell differentiation under various culture conditions can thus be evaluated. Using this technique, the differentiation of resting spermatocytes up to the late pachytene stage of the meiotic division has been shown to take place in organ culture of testes from sexually mature rats (see Figs. 33-3a,b,c), mouse, guinea pig, rabbit, monkey, and man (A. Steinberger and E. Steinberger 1965, 1967; E. Steinberger, 1967).

Phagocytic Activity of Sertoli Cells

When all germinal cell types are present initially in the tissue, a considerable degree of degeneration takes place soon after the initiation of the cultures. The necrotic cells, however, are very rapidly cleared from the lumen of the seminiferous tubules. *In vivo* the Sertoli cells are known to engulf necrotic cells and the residual bodies liberated by the spermatids. The removal of the necrotic cells *in vitro* is also apparently accomplished by the Sertoli cells. Electron microscopic studies of cultured rat testis revealed numerous necrotic cells and cell debris within the cytoplasm of the Sertoli cells (Vilar *et al.,* 1967).

The extent of degeneration is diminished considerably when testes from less mature animals are used; this is because they contain fewer germ cells and those present are in less advanced stages. Also, because the diameter of the seminiferous tubules in immature testes is smaller, and there are more

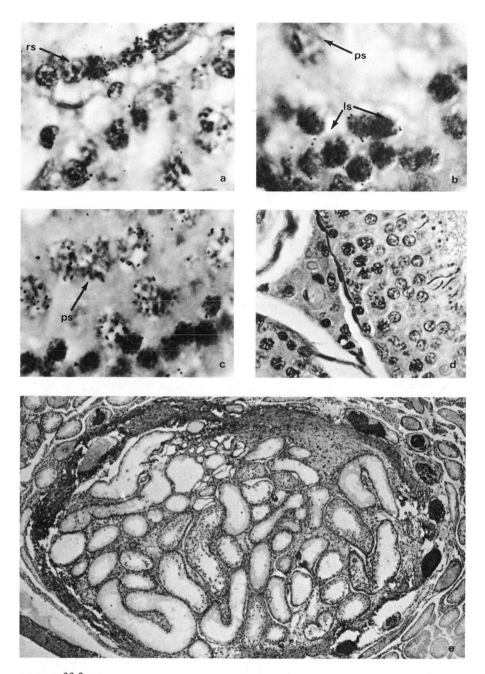

FIGURE 33-3

(a) *Radioautogram of adult rat testes three hours following pulse of tritiated thymidine. The only germinal cells labelled are spermatogonia and resting spermatocytes (rs). ×600. (b) Similar tissue following 3 to 7 days of cultivation. Labelled cells progressed to leptotene or zygotene stage (ls). ×600. (c) Similar tissue following 2 to 3 weeks of cultivation. Progression of labelled cells to stage of pachytene spermatocytes (ps). ×600. (d) Complete spermatogenesis in a testicular fragment transplanted into adult homologous testes following two weeks of cultivation. ×360. (e) General view of a transplant within the host's tissue. Note the variation in the appearance and cellular contents of the seminiferous tubules in the transplant. ×20. (a) through (c) from* J. Reprod. Fertility, *reproduced with the permission of Blackwell Scientific Publications.*

Sertoli cells per unit length of the basement membrane, the tubules have a lesser tendency to "collapse" in culture. For these reasons, testes from immature animals give the appearance of better over-all survival in culture when compared with cultures of testes from adult animals.

Under no culture conditions tried so far could the entire spermatogenic process be reproduced *in vitro*. It has been shown, however, that rat testes grown as organ cultures for several weeks retain their capacity for normal differentiation. This capacity was tested by transplanting the cultured fragments into testes of adult homologous hosts (Steinberger and Steinberger, 1967). The following procedure was used.

Intratesticular Transplants

The testes of an anesthetized host animal were exposed through an abdominal incision and the transplants were inserted under the tunica albuginea with a biopsy needle. The host testes were then returned to the scrotum and the abdominal incision was sutured. After specific time intervals, the animals were killed and the tissues bearing transplants were processed histologically for microscopic examination. Progressive differentiation with accumulation of mature spermatozoa takes place in a proportion of the seminiferous tubules in the transplants after approximately eight weeks (see Figs. 33-3d,e). These results indicate that when testicular tissue is maintained *in vitro* under proper conditions it retains a potential capacity for normal spermatogenesis that resumes and reaches completion once the necessary conditions are provided.

The transplantation method can be used as an "assay" method for checking the spermatogenic capacity of testicular tissue grown under various culture conditions. Several points, however, have to be emphasized in this regard: (1) Variability exists in the differentiation among similar transplants. (2) Occasionally, transplants become heavily infiltrated by leucocytes that seem to interfere with normal spermatogenesis; this can probably be avoided by using a genetically pure strain of animals as hosts and donors. (3) In well maintained transplants, completion of spermatogenesis takes place in approximately 30% of the tubules (see Fig. 33-3e).

CELL CULTURE

Testicular cell cultures have been initiated from a variety of laboratory and domestic animals. The original cell suspensions were obtained from minced testicular tissue by either mechanical agitation (Yao and Pace, 1964; Kodani and Kodani, 1966), or enzymatic digestion with trypsin (Dulbecco and Vogt, 1954; Ferris and Plowright, 1958; Hess *et al.*, 1963; Steinberger and Steinberger, 1966b).

It has been observed that when the cell suspensions are grown in stationary flasks, the germinal elements remain free-floating in the medium, and cells of nongerminal origin attach themselves to the surface of the culture vessel, multiply, and eventually form a monolayer. Mature spermatids, if present in the initial inoculum, are also often found in the monolayer because of their tendency to adhere to surfaces (Steinberger and Steinberger, 1966b).

An outgrowth of cells from small tissue fragments was reported also to be composed of nongerminal elements (Champy and Morita, 1928; Michailow, 1937; Dux, 1940). In some cases, the investigators were primarily concerned with cultures of rapidly dividing cells of nongerminal origin (Dulbecco and Vogt, 1954; Ferris and Plowright, 1958; Hess *et al.*, 1963; Kodani and Kodani, 1966), while in others the survival and differentiation of germinal cells was also investigated (Jordan *et al.*, 1961; Yao and Pace, 1964; Kodani, 1962; Steinberger and Steinberger, 1966b).

Cultures of the two cell types will be considered under separate headings.

CULTURE OF NONGERMINAL CELLS

Morphology and origin

Although it has been established that nongerminal cells can grow in a monolayer, there is disagreement about their morphology and origin. The monolayers have been described as epithelial-like (Kodani and Kodani 1966); fibroblast-like (Hess *et al.*, 1963; Steinberger and Steinberger, 1966b), or a mixture of both (Ferris and Plowright, 1958; Champy and Morita, 1928; Dux, 1940; Jordan *et al.*, 1961). The epithelial-like growth has been attributed usually to Sertoli cells, and the fibroblast-like growth to the connective tissue or Leydig cells. The growth of Sertoli cells in culture that has been reported by several investigators (Champy and Morita, 1928; Jordan *et al.*, 1961; Kodani and Kodani, 1966) is somewhat questionable since in these studies testes from sexually mature or young adult animals were used, and it is known that *in vivo* these cells divide only in very young animals (Clermont and Perey, 1957; Steinberger and Steinberger, in press). The identification of cells in the monolayer was based by most investigators on morphologic appearance viewed with a bright-light microscope. Although most cell types can be recognized easily by light microscopy, in histologic sections of testes—due to their characteristic location and appearance—they lack enough specific morphologic markers which could be used for their identification in cell suspension or in monolayer. Also, morphologic appearance is not a reliable criterion for identification of cultured cells, because it changes easily due to variation in temperature, pH, cell density, composition of media and many other factors.

Recent studies using electron microscopy (Vilar *et al.*, 1966) have shown that the ultrastructure of various cell types present in testes does not change significantly during *in vitro* growth. Since many of these characteristics are unique for specific cell types, they are extremely useful in the identification of these cells in culture. Thus, monolayers originating from testes of 20-day-old rats were found to be composed exclusively of fibroblasts and Leydig cells.

The monolayer cultures were studied also by cytochemical methods (Steinberger *et al.*, 1966). Since in testicular sections only Leydig cells give a positive reaction for 3-beta-hydroxysteroid dehydrogenase (Wattenberg

1958; Baillie and Griffith, 1964), this reaction was utilized to determine the presence of Leydig cells in monolayer cultures.

Gas atmosphere

In some instances, cultures were incubated in a gas atmosphere of 5% CO_2 in air (Ferris and Plowright, 1958; Steinberger and Steinberger, 1966b). Atmosphere containing pure oxygen instead of air was found to be unfavorable (Steinberger and Steinberger, 1966b). Monolayer cultures have also been grown successfully in closed containers without gassing. Even suspension cultures maintained in spinner flasks apparently did not require gassing (Hess *et al.*, 1963).

Incubation temperature

Monolayer cultures have been grown successfully at temperatures ranging from 37°C (Ferris and Plowright, 1958; Hess *et al.*, 1963) to 31°C (Steinberger and Steinberger 1966b). Growth of nongerminal cells of the testes at temperatures higher than the scrotal temperature is not unusual; it is known from *in vivo* studies that nongerminal elements survive under cryptorchid conditions (Steinberger and Nelson, 1955).

Culture medium

A number of various media were used for testicular cell culture. Good results have been achieved using various chemically defined media supplemented with animal sera. Since growth was obtained with complex media only, the nutritional requirements for these cultures have not been determined. Attempt to maintain growth of rat testicular cells in completely chemically defined medium was not successful (Steinberger and Steinberger, 1966b).

Subcultures

Growth of nongerminal cells has been maintained in culture through numerous passages for prolonged periods of time. An alteration of chromosomal complement to heteroploidy was found in one cell line (Kodani and Kodani, 1966), although in others the cells remained euploid after many subcultures (Hess *et al.*, 1963; Steinberger and Steinberger, 1966b).

The monolayer growth becomes more uniform in appearance with each subsequent subculture due to elimination of various nondividing elements and pieces of tubules which are usually present in the primary cultures. Mature spermatids found in the monolayers when cultures are initiated with testes from adult animals may remain morphologically intact even after several passages.

CULTURE OF GERMINAL CELLS

Differentiation of germinal cells to mature spermatozoa *in vitro* has been reported by several investigators (Jordan *et al.*, 1961; Kodani,

1962; Yao and Pace, 1964). It is not clear from these studies, however, whether germinal cells observed in culture were formed from less mature elements *in vitro* or survived from the initial inoculum, since testes from sexually mature animals were used to initiate the cultures. Yao and Pace (1964) reported the transformation to mature spermatozoa of cells growing on a glass surface. This finding is unique in view of the numerous studies showing that only nongerminal elements grow upon glass surfaces (Jordan *et al.,* 1961; Kodani and Kodani, 1966; Steinberger and Steinberger, 1966b). In addition, mature spermatids, when present in the initial inoculum, have a tendency to adhere to glass surfaces and thus may lead to erroneous interpretation of results.

Occurrence of complete spermatogenesis in culture of mature guinea pig testes reported by Jordan *et al.* (1961) also seems unlikely, since the cultures were grown at a temperature (37°C) that is very damaging to germ cells.

Steinberger and Steinberger (1966b) grew cultures of testes from rats of various ages. The ages were selected to represent specific stages of the spermatogenic development and were precisely determined by microscopic examination of the tissues prior to cultivation. Minced testicular tissue was treated with a 0.25% trypsin solution of *p*H 7.2 for ten minutes at 31°C. Repeated manual agitation of the cell suspension increased the dissociation of the tissue. The remaining cell clusters and fibrous strands were allowed to settle by gravity and the supernatant fluid, containing predominantly single cells, was spun at 900 rpm for ten minutes. The cells were resuspended in culture medium and counted in a hematocytometer. Culture vessels were seeded with one to two million cells/ml and were incubated in a stationary position at 31°C. The *p*H was maintained between 6.8 and 7.0 in a gas atmosphere of 5% CO_2 and 95% air. Every week one-half of the culture medium was replenished. The free-floating cells were spun gently in the centrifuge, resuspended in fresh culture medium and returned to the culture vessels. At frequent intervals the cultures were examined by phase-contrast microscopy and bright-light microscopy of histologic sections of fixed cell pellets prepared as follows:

The free-floating cells were sedimented by centrifugation. The supernatant was decanted and the cells were resuspended in Bouin's fixative. After one hour, the cells were spun to form a firm pellet, which was embedded in paraffin and cut at 4 microns. The histologic sections were stained with periodic acid Schiff-hematoxylin and examined microscopically.

Various culture media, including a completely chemically defined medium, were tested for their ability to support spermatogenesis *in vitro*. The chemically defined medium of Eagle (1955), supplemented with 1.0 mM of sodium pyruvate and 0.1 mM of alanine, asparagine, aspartic acid, glycine, proline and serine, did not support the survival of the germinal cells. Survival for a limited time was observed when the above medium was enriched with 10% calf serum.

In cultures of sexually mature testes, many young spermatids showed degenerative changes after three days of cultivation. Mature spermatids, spermatocytes, and possibly spermatogonia remained in good condition for several weeks. In testes from younger animals, more advanced germ cells than those initially present were not observed in culture. These authors, therefore, found no evidence of progressive differentiation of germinal elements in cultures of dissociated cells in contrast to organ cultures. Perhaps proper histoorganization of the testicular tissue is essential for the differentiation of germ cells.

CONCLUDING REMARKS

The following points should be considered when evaluations of gonadal cultured tissues are made: (1) Spermatogenesis in mammals is a slow process—48 days in the rat (Clermont *et al.,* 1959), 36 days in the mouse (Oakberg, 1956), and 64 days in the human (Heller and Clermont, 1963). (Since there is no evidence that spermatogenesis is accelerated *in vitro,* cultures used for study of this process must be maintained in viable condition for a long time.) (2) When culture of gonadal tissue is described, one has to be very precise in defining the cell types which are being maintained, grown, or differentiated. For example, it is generally not difficult to maintain gonadal organ cultures for many months if one refers to the maintenance of tubular histo-architecture, supporting cells (Sertoli cells) and primitive type-A spermatogonia. On the other hand, spermatids have been maintained in culture only for several days, and spermatocytes for three to four weeks. (3) the presence of germ cells in cultures following a certain period of growth does not always constitute evidence for their *in vitro* formation. As mentioned earlier, different germ cell types survive in culture for varying periods of time. One must, therefore, distinguish between cell survival from the initial inoculum and their *in vitro* formation.

Much remains to be learned about the factors necessary for successful maintenance of mammalian spermatogenesis in culture. However, with continuous improvements of tissue culture methods and inexhaustable sources of human imagination, this should be accomplished successfully. Thus, an invaluable tool would become available for the study of numerous problems of male fertility and infertility.

BIBLIOGRAPHY

Baillie, A. H., and S. K. Griffith (1964) 3β-Hydroxysteroid dehydrogenase activity in the mouse Leydig cells. J. Endocrinol., 29:9–17.

Champy, Ch. (1920) De la méthode de culture des tissus. VI. Le testicule. Arch. Zool. Exptl. Gen., 60:461–500.

Champy, Ch., and J. Morita (1928) Recherches sur les culture de tissus. Arch. Exptl. Zellforsch., 5:308–340.

Clermont, Y., and B. Perey (1957) Quantitative study of the cell population of the seminiferous tubules in immature rats. Am. J. Anat., 100:241–267.

Clermont, Y., C. P. Leblond, and B. Messier (1959) Durée du cycle de l'épithélium séminal du rat. Arch. Anat. Microscop. Morphol. Exptl., 48:37–56.

Dulbecco, R., and M. Vogt (1954) Plaque formation and isolation of pure lines with poliomyelitis viruses. J. Exptl. Med., 99:167–182.

Dux, C. (1940) Recherches sur les cultures du testicule hors de l'organisme action de l'hormone. Arch. Anat. Microscop. Morphol. Exptl., 35:391–413.

Eagle, H. (1959) Amino acid metabolism in mammalian cell cultures. Science, 130:432.

Esaki, S. (1928) Über Kulturen des Hodengewebes der Säugetiere und über die Natur des interstitiellen Hodengewebes und Zwischenzellen. Z. Mikroscop. Anat. Forsch., 15:368–404.

Ferris, R. D. and W. Plowright (1958) Simplified method for the production of monolayers of testis cells from domestic animals and for the serial examination of monolayer cultures. J. Pathol. Bacteriol., 75:313–318.

Gaillard, P. J., and W. W. Varossieau (1938) The structure of explants from different stages of development of the testis on cultivation in media obtained from individuals of different ages. Acta Neurol. Morphol. Norm. Pathol., 1:313–327.

Gaillard, P. J., and W. W. Varossieau (1940) Der Einflus des Hypophysen-vorderlappens auf die morphologishe Entwicklung des Hodens (studiert mit Hilfe der kombinierten Gewebzüchtung *in vitro*.) Arch. Exptl. Zellforsch., 24:141–168.

Heller, C. G., and Y. Clermont (1963) Spermatogenesis in man: an estimate of its duration. Science, 140:184–186.

Hess, W. R., H. J. May, and R. E. Patty (1963) Serial cultures of lamb testicular cells and their use in virus studies. Am. J. Vet. Res., 24:59–64.

Jordan, R. T., S. Katsh, and N. Stackelburg (1961) Spermatocytogenesis with possible spermiogenesis of guinea pig testicular cells grown *in vitro*. Nature, 192:1053–1055.

Kodani, M. (1962) Long term *in vitro* maintenance of mammalian spermatogenesis. Genetics, 47:965.

——— and Kodani, K. (1966) The *in vitro* cultivation of mammalian Sertoli cells. Proc. Natl. Acad. Sci., U.S.A., 56:1200–1206.

Lostroh, A. J. (1960) *In vitro* response of mouse testis to human chorionic gonadotropin. Proc. Soc. Exptl. Biol. Med., 103:25–27.

Martinovitch, P. N. (1937) Development *in vitro* of the mammalian gonad. Nature, 139:413.

Michailow, W. (1937) Experimentell-histologische Untersuchungen über die Elemente der Hodenkanälchen. Z. Zellforsch. Microskop. Anat., 26:174–201.

Oakberg, E. F. (1956) Duration of spermatogenesis in the mouse and timing of stages of the cycle of the seminiferous epithelium. Am. J. Anat., 99:507–516.

Ortavant, R. (1954) Contribution à l'étude de la durée du processus spermatogenetique du Belier à l'aide du ^{32}P. C.R. Soc. Biol. (Paris), 148:804–806.

Steinberger, A., E. Steinberger, and W. H. Perloff (1964c) Mammalian testes in organ culture. Exp. Cell Res., 36:19–27.

——— and E. Steinberger (1965) Differentiation of rat seminiferous epithelium in organ culture. J. Reprod. Fertility, 9:243–248.

——— and E. Steinberger (1966a) Stimulatory effect of vitamins and glutamine on the differentiation of germ cells in rat testes organ culture grown in chemically defined media. Exp. Cell Res., 44:429–435.

——— and E. Steinberger (1966b) *In vitro* culture of rat testicular cells. Exp. Cell Res., 44:443–452.

———— and E. Steinberger (1967) Factors affecting spermatogenesis in organ culture of mammalian testes. J. Reprod. Fertility, (Suppl.) 2:117–124.

———— and E. Steinberger (in press) Growth kinetics of Sertuli cells in maturing rat testes. Exptl. Cell Res.

Steinberger, E. (1967) Maintenance of adult human testicular tissue in culture. Anat. Record, 157:327.

———— and W. O. Nelson (1955) The effect of hypophysectomy, cryptorchidism, estrogen and androgen upon the level of hyaluronidase in the rat testis. Endocrinology, 56:429–444.

———— and W. J. Dixon (1959) Some observations of the effect of heat on the testicular germinal epithelium. Fertility Sterility, 10:578–595.

————, A. Steinberger, and O. Vilar (1966) Cytochemical study of Δ^5–3β hydroxysteroid dehydrogenase in testicular cells grown *in vitro*. Endocrinology, 79:406–410.

————, A. Steinberger, and W. H. Perloff (1964a) Initiation of spermatogenesis *in vitro*. Endocrinology, 74:788–792.

————, A. Steinberger, and W. H. Perloff (1964b) Studies on growth in organ culture of testicular tissue from rats of various ages. Anat. Record, 148:581–589.

Trowell, O. A. (1959) The culture of mature organs in a synthetic medium. Exp. Cell Res., 16:118–147.

Vilar, O., A. Steinberger, and E. Steinberger (1966) Electron microscopy of isolated rat testicular cells grown *in vitro*. Zeitschr. Zellforsch., 74:529–538.

———— (1967) An electron microscopic study of cultured rat testicular fragments. Zeitschr. Zellforsch., 78:221–233.

Yao, K. T. S., and D. M. Pace (1964) Mammalian germinal cells grown *in vitro*. Excerpta Medica 18, no. 1, sec. 1, 17.

Wattenberg, L. W. (1958) Microscopic histochemical demonstration of steroid 3-beta-ol-dehydrogenase in tissue sections. J. Histochem. Cytochem., 6:225–232.

34

Preparation of Primary Cell Cultures from Preimplantation Embryos

R. J. COLE
School of Biological Sciences
University of Sussex
Brighton, England

In many experimental situations, especially in the study of cytodifferentiation, the ability to use cell cultures has several advantages over the use of whole embryos. Cultures are easily accessible for *in vivo* observation, for still photography or cinematography, and for histological, histochemical, and autoradiographic analysis. Culture techniques may permit the accumulation of large numbers of an experimentally valuable cell type, and facilitate experiments on cellular interactions which could not be achieved in the intact organism.

This chapter describes the application of some routine cell culture techniques to preimplantation embryos of rabbit and mouse. To ensure success it is essential that throughout the manipulations described the embryos or cells should be subjected to a minimum of stress, particular attention being paid to control of temperature, illumination, carbon dioxide tension, pH, and sterility. During manipulation or culture the temperature should not be allowed to rise above 37.5°C. Satisfactory growth occurs at 36.5°C and this allows a useful margin of safety. Exposure to light, both from room illumination and from microscope lamps, should be kept to a minimum. Detailed observations on light sensitivity of rabbit ova have been reported by Daniel (1964). Generally, illumination by red or orange light appears to be the least damaging. All stages of preimplantation mammalian embryos appear to be rapidly and irreversibly damaged by exposure to alkaline

The methods described here were developed in the Department of Biochemistry, University of Glasgow, and supported by Grant CA 05855 from the United States Public Health Service to Dr. John Paul.

conditions. If bicarbonate-buffered media are being used for manipulations, the initial pH and the speed of the manipulations should be such that loss of CO_2 due to exposure to air does not allow pH 7.4 to be exceeded. Alternatively, media buffered with Tris/sodium citrate (Paul, 1965) may be used for manipulating embryos. However, while not apparently harmful on short exposure, these media will not support continued growth of embryos *in vitro*. Routine aseptic precautions used in cell culture techniques should be used here also. Adequate protection against airborne contamination during isolation and manipulation of embryos is given by surrounding the dissecting microscope with a hood which can be presterilized by a germicidal lamp or by washing with alcohol. The general techniques employed in support of the methods described here are those recommended by Paul (1965).

PREPARATION OF CELL CULTURES
FROM RABBIT BLASTOCYSTS

ISOLATION OF BLASTOCYSTS

Blastocysts can be obtained from does mated naturally with fertile males, or the mating stimulus reinforced by intravenous injection of 50 I.U. of human chorionic gonadotrophin immediately following 2 or 3 forced matings. Alternatively, superovulation with anterior pituitary extract can be employed (Pincus, 1940; Kennelly and Foote, 1965). However, if advanced expanded blastocysts are to be used, the large numbers of embryos obtained from superovulation may hinder dissection from the uterus.

At the required time after mating the does are killed by cervical dislocation or preferably by an intravenous injection of Nembutal or a similar narcotic. The uteri are removed aseptically, placed cleanly into Hanks balanced salt solution (Paul, 1965), and then washed several times to remove blood and debris. Blastocysts degenerate rapidly in protein-free solution so it is best to isolate them in culture medium. Early blastocysts may be flushed carefully from the uteri; later, expanded blastocysts, when in contact with or extending the walls of the uterus, must be dissected free by immersing the whole implantation site in medium, opening it with fine instruments, and teasing the blastocyst directly into medium. Blood cells and debris are removed by carefully transferring the blastocysts through several changes of the medium.

Intact blastocysts or fragments of blastocysts are transferred best in Pasteur pipettes adjusted to a suitable size; this avoids tearing the embryos and transfers a minimum of medium. The pipettes are controlled easily via a mouthpiece, a rubber tube approximately 30 inches long, and a sterile interchangeable aseptic filter that is plugged with cotton wool.

Dissections, enzyme treatments, and other manipulations may be carried out in embryonic (solid) watch glasses, disposable polystyrene Petri dishes (e.g., Falcon Plastics) or in droplets of medium under sterile equilibrated

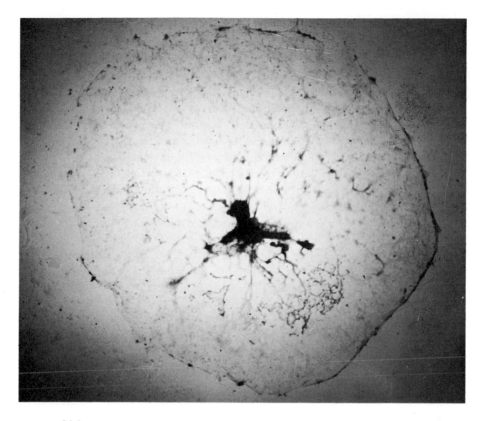

FIGURE 34-1

A culture grown from a whole five-day-old rabbit blastocyst after attachment of the trophoblast cells. Fixed after 18 days in vitro. ×13.

liquid paraffin (Wildy and Stoker, 1958). This method limits evaporation and loss of CO_2, and is therefore particularly useful for small fragments. To avoid the formation of a skin at the medium/paraffin interface the sterilized paraffin should be shaken vigorously with 5 to 10% of its volume of culture medium and the resulting mixture kept at 37°C until the paraffin clears; it should also be equilibrated with the gas phase. The paraffin is sterilized by steaming or autoclaving to 120°C. Higher temperatures may cause toxicity to cultured cells to develop.

MEDIA AND CULTURE CONDITIONS

In my experience, media based on Waymouth's MB 752/1 (Waymouth, 1959) have been most satisfactory for mouse and rabbit embryos, and embryonic cells: e.g., mouse cleavage stages, blastocysts and postimplantation stages (Cole and Paul, 1965; Cole, 1967); rabbit cleavage stages, blastocysts, and isolated cells (Cole, Edwards, and Paul, 1966); and mouse, rat, and rabbit fetal erythropoietic cells (Cole and Paul, 1966; Cole, Hunter and Paul, 1968).

Satisfactory development of rabbit blastocysts has also been reported in medium 199 by Glenister (1965) and in more complex media by Daniel (1965).

Media must be supplemented by the addition of serum. Satisfactory growth of rabbit blastocysts and cells has been obtained with 5% neonatal calf serum and 2% unfiltered human serum. No requirement for homologous serum could be demonstrated and fetal calf serum had no advantages over neonatal serum. The presence of low concentrations of unfiltered serum enhances early cell adhesion and growth in small explants and in dissociated cells from rabbit blastocysts. Heat inactivation of human serum did not improve its growth-promoting capacity. Low concentrations (e.g., 50 units/ml) of penicillin and/or streptomycin may be routinely added to media without damaging blastocysts or cells. All media should be throughly equilibrated with the required gas phase, usually $5\%\,CO_2$ in air, and adjusted to a pH of 7.2 to 7.4 before use. Cultures should be maintained in an incubator flushed with $5\%\,CO_2$ in air; any other means of maintaining an atmosphere of that composition is also acceptable.

PREPARATION OF THE SUBSTRATUM

Except for the trophoblast cells, rabbit and mouse early embryonic cells do not adhere to glass or polystyrene surfaces. It is essential to provide an organic substratum for the earliest stages of growth *in vitro,* and reconstituted collagen membranes prepared by a modification of the method of Ehrmann and Gey (1956) provide a satisfactory and easily prepared surface.

Rat tail tendons are dissected free of muscle and, if necessary, are washed in Hanks BSS. The tendons are then extracted in 0.5% acetic acid at 4°C until the collagen is soluble (the preparation becomes viscous at this time). This is usually achieved after 48 hours of extraction. Debris is removed by hand and by centrifugation. The collagen solution is then dialysed against distilled water at 4°C until the desired viscosity is reached. The solution should remain as a discrete drop when placed on the surface to be coated. The collagen is then reconstituted on the coated surface by exposure to ammonia vapor (in a dessicator or similar container) until a strong gel is formed. The ammonia is then removed by a thorough washing with balanced salt solution, taking care not to tear the gel. For some applications the gel may be used without drying, but for the uses detailed here the gel is dried completely to the surface, and the resulting thin membrane is sterilized by exposure to a UV germicidal lamp. Alternatively, the original tendon may be sterilized by immersion in ethyl alcohol prior to extraction; all subsequent manipulations are carried out aseptically with sterilised solutions. This procedure has the advantage that the dialysed collagen solution can be stored under refrigeration for a long time but the technique is cumbersome and has a high probability of contamination. Collagen solutions may be used to coat both glass and plastic surfaces but for initiating the cultures coated cover slips are best. Plastic or glass culture bottles may also be coated; they can be sterilized by immersion in 70%

rectified ethyl alcohol, followed by drying and washing with sterile BSS in aseptic conditions.

REMOVAL OF ZONA PELLUCIDA

The blastocysts can not be dissected until the zona pellucida and mucin layer are removed. In early blastocysts with a thick mucin layer, removal is best achieved by use of 1% pronase (Calbiochem) solution (Mintz, 1962) in Hanks BSS, with intermittent observation under a dissection microscope. The degraded mucin can be removed from the vicinity of the blastocyst by gently swirling. Pronase does not appear to be inhibited rapidly by serum; thus prolonged activity can be prevented only by washing through several changes of medium. To prevent damage during this process it is advantageous to remove the blastocysts from the pronase solution before the zona has completely disappeared, and finally to free the blastocysts by carefully tearing the thinned zona with glass needless or by gentle pipetting. The addition of serum to the pronase solution may help to prevent cell damage during the relatively longer exposure to enzyme necessary to remove the mucin layer in early blastocysts. Pronase concentrations of less than 1% may be used with later blastocysts when the zona is degraded more rapidly.

PREPARATION OF EXPLANTS

Whole blastocysts

If, even prior to the appearance of the embryonic disc, zona-free blastocysts are cultured in contact with a suitable substratum the trophoblastic cells adhere to the substratum, migrate outwards, and pull the blastocyst into a flattened sheet of cells. Primary cultures of this type survived and may continue to develop for 40 to 50 days and may therefore be of use in some studies (Cole, Edwards, and Paul, 1965).

PREPARATION OF DISSOCIATED CELL CULTURES

Following removal of the zona pellucida, dissociated cell preparations are obtained most easily by treatment with 0.25% Difco 1/250 trypsin in isotonic sodium chloride (0.6%) sodium citrate (0.296%) adjusted to pH 7.7. Addition of 0.3% carboxy methyl cellulose helps prevent damage to cell membranes during exposure to trypsin. The blastocysts are washed in one or two changes of this solution to remove serum before being incubated at 37°C to allow cellular adhesions to be disrupted. The trypsin exposure necessary to achieve disaggregation has been found to vary widely. Ideally the blastocysts should be removed from the trypsin and placed in medium while they are still intact and just as soon as disaggregation can be completed by mild pipetting. The time should be determined by preliminary tests on a few of the blastocysts to be disaggregated. Trophoblastic cells from later blastocysts rapidly become established in culture and they displace the more slowly growing cells of the embryonic disc. To establish cell lines from

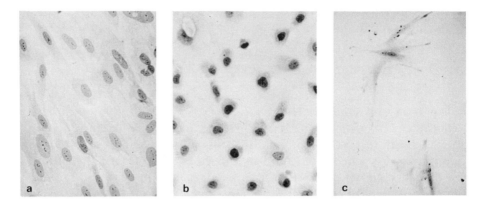

FIGURE 34-2

Three cell lines grown from rabbit blastocysts. (a) *RB1: Established from an*
explanted fragment of an embryonic disc from a 5½ day blastocyst that had been
maintained in vitro *for 24 hours before dissection.* ×208. (b) *RB3/3: Established*
from a dissociated cell suspension that was produced by trypsin treatment of a six-day
embryonic disc. ×208. (c) *RB2: Established from an explanted fragment of the*
embryonic disc of a six-day blastocyst that had been maintained in vitro *for 24 hours*
before dissection. ×52.

dissociated cells the embryonic disc must be separated from the trophoblastic portion of the blastocyst. After removal of the zona pellucida the
blastocyst is opened with glass needles and the trophoblast is spread on the
bottom of the culture vessel thus displaying the embryonic disc. The edges
of the trophoblastic region will adhere to polystyrene if gentle pressure is
applied with the needles. The embryonic region can then be excised with
glass needles and be transferred to a drop of medium under liquid paraffin
for further treatment. Dissociation by trypsinization is performed most
easily in this kind of culture drop. When it is wished to isolate known
portions of the embryonic area of later blastocysts this can be carried out by
a process similar to that used for the whole discs.

For culturing, the dissociated cells may be transferred by micropipette to
droplets of medium or collagen-coated coverslips under equilibrated liquid
paraffin and cultured in 5% CO_2 in air. These microcultures can be fed
through the paraffin by first withdrawing and then replacing a portion of
the medium, or by merely increasing the size of the drop until the whole
coverslip is covered by medium. Alternatively, especially when the dissociated cells from an embryonic disc are to be cultured together, the cell
suspension can be transferred to a Petri dish containing a collagen-coated
coverslip flooded with medium. The dissociated embryonic discs of 5½ to
6½ day blastocysts provide enough cells for an initial culture of about 3 ml
of medium. With smaller cell numbers the initial volume should be reduced
accordingly. Successful cultures may be subcultured either by trypsinising
the cell layer *in situ* or by transferring the collagen-coated coverslip with
cells attached to another container. This is most easily done by increasing
the size of the drop of medium and removing the paraffin. As the paraffin is
reduced it will leave the medium in the centre of the dish with a clear

surface through which the coverslip may be removed with sterile forceps, without contamination with paraffin. A similar technique may be used for histological procedures.

PREPARATION OF CULTURES FROM EXPLANTS

When is it desirable to avoid the trauma of dissociation, primary cell cultures may be obtained from explants of embryonic discs, but one should avoid the inclusion of trophoblastic cells. The fragments of the embryonic discs are prepared and cultured by a technique similar to that used for dissociated cells. The explant is placed in contact with the collagen surface of the cover slip and spread by gentle pressure. Contact between explant and collagen can be enhanced by withdrawing medium from the culture drop until the medium/paraffin interface holds the explant in place. When cellular attachment of the explant to the collagen is established, usually after 10 to 15 hours, more medium may be added, and the drop is expanded. Where this technique is successful cells migrate from the explant and form a monolayer on the collagen. As this increases more medium may be added, and the preparation may be subcultured as necessary by trypsinization.

It appears that only the earliest stages of growth *in vitro* require the collagen substratum, and cell strains originated by either of the techniques described above can, once established, be subcultured routinely onto plastic or glass.

Embryonic discs can be prepared for culture immediately after isolation of the blastocyst or the blastocysts may be cultured intact, with or without experimental treatment, for some time. Five-day-old blastocysts continue to expand in culture to Day 7 or Day 8, but if the zona pellucida is left intact both the zona and the blastocyst burst and the blastocyst degenerates. Therefore, if the blastocysts are to be cultured beyond this stage before dissection, the zona must first be removed. Trophoblastic cells tend to attach to the surface of glass or plastic culture vessels and thus deform the blastocyst, but this may be prevented by a thin layer of agar. The embryonic area will continue to develop if it is near the medium surface, but it otherwise degenerates.

If it is desired to establish large-scale cell cultures by these methods the characteristics of the cells should be checked as soon as possible after explantation. The karyotypes of the cells are easily determined and provide a check against contamination from other cell lines, especially in laboratories where cell cultures from several species are present. This may also be checked routinely by immunological methods. In the case of cell lines of XY chromosome constitution, karyotype analysis also indicates that the cells did in fact arise from an embryonic source, and were not carried over from the mother.

Some indication of the histiotypic capacity of the cell lines may be determined by reimplanting cells into an immunologically tolerant site, e.g., the anterior chamber of the eye, or by *in vitro* techniques. Small aggregates using approximately 1×10^6 cells may be formed by gentle centrifugation at

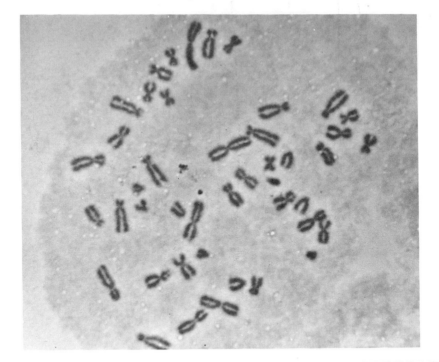

FIGURE 34-3

Chromosome preparations from strain RB1. The cells have a male karyotype. ×1200.

4°C in a Pasteur pipette plugged with gelatin. These aggregates may be cultured on organ culture grids, or on the surface of nutrient agar, and subsequently analysed by histological or histochemical techniques.

PREPARATION OF CELL CULTURES FROM MOUSE BLASTOCYSTS

Because of their small size, mouse blastocysts are less easily manipulated than rabbit blastocysts, and specific areas of the embryonic region cannot be separated from the trophoblast before cultures are initiated.

However, as in the case of explanted whole rabbit blastocysts, the presence of trophoblast cells appears to inhibit the development of embryonic cells even on a reconstituted collagen stubstratum.

Embryos may be obtained from naturally mated females, timed by observation of the mating plug, or from females superovulated with gonadotrophins (Runner and Palm, 1953; Fowler and Edwards, 1957). The blastocysts are isolated, using the same general techniques and precautions as those listed previously, by flushing the uteri with culture medium, or they may be derived from cleaving ova cultivated *in vitro* (Brinster, 1965; Cole and Paul, 1965; see also Biggers, Chap. 6 above).

Early mouse blastocysts continue to develop satisfactorily in Waymouth's medium MB 752/1 supplemented with 5 to 10% neonatal calf serum. Cultures should be grown at 36.5°C in 5% CO_2 in air. These cultures are maintained easily as drops of medium under liquid paraffin, or if adequate protection is provided against drying, a small culture chamber may be formed by a glass ring with a ground surface that is sealed against a coverslip with sterile silicone grease. The zona pellucida may be removed by treatment with pronase or the blastocysts be allowed to escape naturally (Cole and Paul, 1965; Cole, 1967). Eight to twenty hours after escape from the zona the blastocysts attached to the substratum, and the trophoblast migrates rapidly outward forming a flat sheet of cells. Usually the cells of the embryonic area remain as a discrete mass in the center, which subsequently tends to spread over the upper surface of the trophoblast but not beyond it. Cultures of this type may survive for 15 or 16 days and may provide a very convenient approach to the study of the cellular properties of the implanting embryo, especially of the trophoblastic portions. One or two days after the spreading of the trophoblast the mass of embryonic cells may be dissected easily, but attempts to establish cell cultures from this have not been successful.

Some of the other requirements for attachment and outgrowth of the mouse blastocyst *in vitro* have been examined by Gwatkin (1966). He has demonstrated an absolute requirement for arginine, cystine, histidine, leucine, and methionine; lack of one or more of these amino acids in the medium can cause a reversible failure of the trophoblast cells to attach and spread.

Primary cultures of mouse blastocysts of the type described can be adapted easily for autoradiographic, histochemical, or karyotypic studies, and for photography, but attempts to derive cultures from enzymatically disaggregated blastocysts have not been successful.

Cultures of mouse trophoblastic cells may also be initiated by trypsin treatment of trophoblastic tumors, originating from blastocysts artificially implanted beneath the kidney capsule (Schlesinger and Koren, 1967).

Primary cell cultures maintained by established procedures may therefore be derived from intact preimplantation embryos, from small explants, or from individually isolated cells. In the latter case a homogeneous clone of cells may be isolated at a specific stage of development and removed from inductive and other stimuli to which it would normally be subject in the course of differentiation. Investigations of cells derived by the techniques

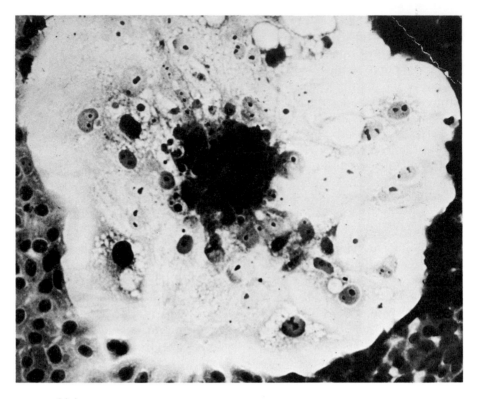

FIGURE 34-4

A culture derived from a mouse blastocyst that was fixed five days after trophoblast attachment. This blastocyst was grown in vitro *from a four-cell ovum in the presence of a feeder layer of irradiated cells that are visible at the edge of the trophoblastic area. Cells derived from the inner cell mass appear darkly staining in the center of the trophoblast cells. ×158.*

described here indicate that they are able to maintain specific morphological, biochemical, and karyotypic properties for considerable times in culture, and it is hoped that they may provide particularly favorable material for studies on the mechanism of cytodifferentiation.

BIBLIOGRAPHY

Brinster, R. L. (1963) A method for *in vitro* cultivation of mouse ova from 2 cell to blastocyst. Exp. Cell Res., 32:205.

Cole, R. J. (1967) Cinemicrographic observations on the trophoblast and zona pellucida of the mouse blastocyst. J. Embryol. Exp. Morphol., 17:481.

————, R. G. Edwards, and J. Paul (1966) Cytodifferentiation and embryogenesis in cell colonies and tissue cultures derived from ova and blastocysts of the rabbit. Develop. Biol., 13:385.

―――― and J. Paul (1965) Properties of cultured pre-implantation mouse and rabbit embryos and cell strains derived from them. In: Ciba Foundation Symposium on Preimplantation Stages of Pregnancy. Ed. by G. E. Wolstenholme and M. O'Connor. Churchill, London.

―――― and J. Paul (1966) The effects of erythropoietin on haem synthesis in mouse yolk sac and cultured foetal liver cells. J. Embryol. Exp. Morphol., 15:245.

――――, J. Hunter, and J. Paul (1968) Hormonal regulation of prenatal haemoglobin synthesis by erythropoietin. Brit. J. Haematol., 14:477.

Daniel, J. C. (1964) Cleavage of mammalian ova inhibited by visible light. Nature, 201:316.

―――― (1965) Studies on the growth of five-day rabbit blastocysts *in vitro*. J. Embryol. Exp. Morphol., 13:83.

Ehrmann, R. L., and G. O. Gey (1956) The growth of cells on a transparent gel of reconstituted rat tail collagen. J. Nat. Cancer Inst., 16:1375.

Fowler, R. E., and R. G. E. Edwards (1957) Induction of superovulation and pregnancy in mature mice by gonadotrophins. J. Endocrinol., 15:374.

Glenister, T. W. (1963) Observations on mammalian blastocysts implanting in organ cultures. In: Delayed Implantation. Ed. by A. C. Enders. Rice University Press, Houston.

Gwatkin, R. B. L. (1966) Amino acid requirements for attachment and outgrowths of the mouse blastocyst *in vitro*. J. Cell Physiol., 68:335.

Kennelly, J. J., and R. H. Foote (1965) Supervulatory response of pre- and post-pubertal rabbits to commercially available gonadotrophins. J. Reprod. Fertility, 9:177.

Mintz, B. (1962) Experimental study of the mammalian egg, removal of the zona pellucida. Science, 138:594.

Paul, J. (1965) Cell and Tissue Culture. 3rd Ed. Livingstone, Edinburgh.

Pincus, G. (1940) Superovulation in rabbits. Anat. Record, 77:1.

Schlesinger, M., and Z. Koren (1967) Mouse trophoblastic cells in tissue culture. Fertility Sterility, 18:95.

Waymouth, C. (1959) Rapid proliferation of sublines of NCTC clone 929 (strain L) mouse cells in a simple chemically defined medium. J. Natl. Cancer Inst., 22:1003.

Wildy, P., and M. Stoker (1958) Multiplication of solitary Hela cells. Nature, 181:1407.

Index